Generalized Integral Transforms in Mathematical Finance

Generalized Integral Transforms in Mathematical Finance

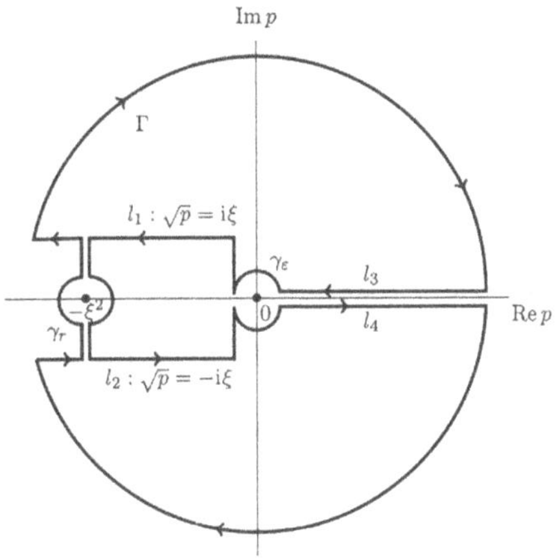

Andrey Itkin
New York University, USA

Alexander Lipton
Hebrew University of Jerusalem, Israel

Dmitry Muravey
Moscow State University, Russia

World Scientific

NEW JERSEY • LONDON • SINGAPORE • BEIJING • SHANGHAI • HONG KONG • TAIPEI • CHENNAI • TOKYO

Published by

World Scientific Publishing Co. Pte. Ltd.
5 Toh Tuck Link, Singapore 596224
USA office: 27 Warren Street, Suite 401-402, Hackensack, NJ 07601
UK office: 57 Shelton Street, Covent Garden, London WC2H 9HE

Library of Congress Control Number: 2021943643

British Library Cataloguing-in-Publication Data
A catalogue record for this book is available from the British Library.

GENERALIZED INTEGRAL TRANSFORMS IN MATHEMATICAL FINANCE

ISBN 978-981-123-173-5 (hardcover)
ISBN 978-981-123-174-2 (ebook for institutions)
ISBN 978-981-123-175-9 (ebook for individuals)

For any available supplementary material, please visit
https://www.worldscientific.com/worldscibooks/10.1142/12147#t=suppl

Desk Editors: Balasubramanian Shanmugam/Lai Ann

Typeset by Stallion Press
Email: enquiries@stallionpress.com

Foreword

As its name suggests, mathematical finance merges the lofty world of pure and applied mathematics with the down-in-the-trenches, belt and suspenders world of practical finance. The result of this shotgun marriage can range anywhere from prestigious journal publication extolling category theory (abstractions of abstractions) for mathematical finance to OCaml code implementing high-frequency trading strategies in hedge funds. The content of the present book firmly occupies the middle ground between these two extremes. A revolution occurred in mathematical finance in the late 1980s and early 1990s when it was realized that integral transforms can be used to get an analytical handle on models that capture market realities such as stochastic volatility and jumps in asset prices. While the inversion step is still done numerically, the speedup over pure numerical methods such as Monte Carlo or finite difference meant that these realistic models could be calibrated to market data in real time. The semi-analytical nature of these models also lead to greater error control. Finally, as practitioners and academics internalized this brave new world of integral transforms, their simpler algebraic structure permitted even further modifications such as stochastic time changes that would have been impossible to fathom working with so-called closed-form formulas such as Black Scholes. I personally found that the insights afforded by the switch in perspective (and basis elements) were so great that at one point I resolved to never again work with probability densities per se, even when they were explicit and simple.

The current book goes well beyond the first applications of integral transforms to derivatives pricing, which are now standard fare in math-finance curricula. Since integral transforms associated with the names of Fourier and Laplace are based on translation invariance, they were ill-suited to solve problems that violate homogeneity in either time or space. A practitioner faced with the arduous and difficult task of quickly calibrating to term and strike structures of multiple financial instruments will eschew homogeneity in a heartbeat. The new generation of integral transforms presented in this book are designed to handle the arbitrary time-dependent and/or spot-dependent geometric structures that permeate the working lives of the modern quant. As such, these methods will likely generate more light than heat, despite the monikers with which they are endowed. As the technological arms race moves from countries to companies, quants must always be prepared to embrace new methodologies. Just as back-slapping cigar-chomping high-fiving bankers and traders have now been exiled to the dustbins of history, quants who fail to keep pace with new developments are destined to join those dearly departed brethren. Fortunately, the authors of this book are supremely qualified to present these latest advances. All of them combine a highly prestigious academic pedigree with the nitty-gritty know-how that can only be procured while toiling as the braintrust of financial institutions, past and present. It has been my privilege to co-author academic papers with all of the authors of this book. I have also previously asked two of the three authors to teach their craft to Masters' students at NYU. The experience they gained in explaining their advances to these fledgling quants permeates the pages of this exciting tome. Just as quants again turn the page on tried and true methods, I ask you gentle reader to also turn the page and begin a new chapter of mathematical discovery.

Prof. Peter Carr
Department Chair,
Tandon School of Engineering
Department of Finance and Risk Engineering
New York University

Preface

Classical problems of financial mathematics recently attracted a lot of attention due to several factors. Among them, one could mention:

- Very small or even negative interest rates observed at the market caused by the Federal Reserve's attempts to achieve its macro-economic goals, see, e.g. Itkin *et al.* (2020) and reference therein. Therefore, financial models that allow negative rates and reflect fiscal moves by central banks became real necessities.
- Negative oil prices caused by the COVID-19 pandemic and the subsequent economic recession, see Bouchouev (2020), Farrington and Cesa (2020).
- Dramatic shift to electronic trading since major options exchanges temporarily closed their floors due to COVID-19 restrictions, and brokers and market makers were forced to work from home. That raised the need for real-time tools for fast calculating the option prices and Greeks (Brogan, 2020).

These developments and some other factors forced the financial mathematics community to critically reassess the simplest classical one-factor mathematical finance models. For example, the Bachelier and Hull-White models, traditionally viewed as inadequate or ill-posed, got the new lease of life. In Doff (2020) it is advocated that risk managers could even use Black-Scholes to help drive strategy. Therefore, the fast pricing of barrier options for both the most straightforward and more advanced models became increasingly important.

For fast pricing, analytical and semi-analytical solutions of various pricing and hedging problems are exceedingly helpful. However, a minimal number of such solutions are known in mathematical finance, especially if compared with physics where similar problems (e.g., initial-boundary value problem for parabolic partial differential equations (PDEs)) are also of fundamental interest.

This book aims to present new methods and approaches developed in a series of papers written by the authors and their associates to solve various initial-boundary value problems for PDEs with moving boundaries. Such problems often appear in mathematical finance when pricing barriers and American options, finding the hitting time probability distribution for some stochastic process, etc. As mentioned, aside from the financial interpretation, mathematically similar problems have been studied in physics for a long time. Similar problems arise in nuclear power engineering and safety of nuclear reactors; combustion in solid-propellant rocket engines; laser action on solids; phase transitions (the Stefan and Verigin problems in fluid dynamics); in the processes of sublimation in freezing and melting; in the kinetic theory of crystal growth; etc., see Kartashov (1999) and references therein. Analytical solutions to these problems require non-traditional and sometimes sophisticated methods.

The methods we discuss in this book belong to a class of integral transforms. However, they are not classical transforms widely used in science. Instead, they are broad generalizations of their classical counterparts. Strictly speaking, they are further extensions of heat potentials (HP) and generalized integral transforms (GIT) methods.

The HP method is uniquely well-suited to solving rather challenging problems occurring routinely in applied mathematics in general and in financial engineering in particular. In a nutshell, this method allows one to reduce a complicated partial differential equation of the parabolic type with a time-dependent boundary to a much simpler Volterra integral equation. A detailed introduction to heat potentials can be found in Watson (2012) and Tikhonov and Samarskii (1963). The first use of the method of heat potentials in mathematical finance is due to one of the authors Lipton (2001), who used it for pricing path-dependent options with curvilinear barriers.

The GIT method is capable of solving the same problem by using a slightly different approach. To the best of our knowledge, initially, the GIT method was developed by A.V. Luikov, B.Ya. Lyubov, E.M. Kartashov, and is described in detail in Kartashov (1999). However, this development covers only the domain $[0, y(t)]$, where $y(t) < \infty$ is the time-dependent moving boundary. As applied to finance, for the first time, this method was used in Carr and Itkin (2021) for pricing American and barrier options. Later in a series of papers by two of the authors of this book and Peter

Carr, similar problems for other time-dependent domains, including those with two moving boundaries, have been solved. It is worth noting that the GIT approach for the Bessel PDE was not known in the literature in full (i.e., up to the final formula) until (Carr *et al.*, 2020), despite some comments on possible ways of achieving this could be found in Kartashov (1999) (but only for the domain $x \in [0, y(t)]$).

We emphasize that due to the generality of the proposed methods, they could be used not only in finance but in physics, biology, and applied mathematics to solve various initial-boundary value problems of interest. We present some of the solutions to such representative non-financial problems in this book for completeness.

In this book, our primary focus is on one-dimensional initial-boundary value problems, which correspond to various one-factor models in financial mathematics. One can develop a similar approach for solving multi-dimensional problems, which is a subject of our ongoing work.

Since this book is about finance, integral transforms, and integral equations, we wanted to make it self-contained and provide readers with various backgrounds all necessary information right in the book. Therefore, to give the reader a smooth introduction to the subject, Part 1 provides a short overview of the basics of financial mathematics. Chapter 1 describes the mathematical apparatus needed for our investigation of continuous-time markets with a particular emphasis on the stochastic engines used as sources of randomness in continuous time. Chapter 2 provides a short overview of the financial models used in the book. These models cover various asset classes: Equity, Fixed Income, Commodities, and Credit. We do not claim to have given an extensive introduction to this vast subject. Instead, we assume that the readers are already familiar with some basic concepts and gently point them in the right direction.

Part 2 makes a short introduction to integral equations. We introduce basic information about Fredholm and Volterra equations of the first and second kind, give main theorems on their solutions' uniqueness and existence, and describe some analytical and numerical methods to solve them. This material is helpful to understand better various numerical examples provided later in the book.

Since the proposed methods are generalizations of classical transforms, in Part 3 we shortly describe several classical integral transforms well-known in mathematics. Then, we explain the GIT and HP methods in detail. We represent various new results and provide the basis for solving multiple financial and physics problems in the book's remainder. We are not familiar with such a general and detailed description of these approaches in the literature.

Parts 4–6 cover various financial problems for one-factor models with moving boundaries. In Part 4 we consider problems specific to Equities, FX, and Commodities. We present semi-analytical solutions for barrier and American options in Bachelier and Black-Scholes models. We derive a semi-analytical representation for the first hitting time density for reducible diffusion processes, including time-dependent OU processes. Also, we closely follow (Lipton and Lopez de Prado, 2020b) to describe how to use the method of heat potentials to construct a closed-form solution for optimal mean-reverting trading strategies. All these results are new and obtained in the papers of the authors (with co-authors).

In Part 5 our approach is further extended to Fixed Income problems. We consider the Constant elasticity of variance (CEV) model, the Cox-Ingersoll-Ross (CIR) model, the Black-Karasinski (BK) model, and the Verhulst model. The first two models require a generalization of the HP method since the PDE we need to solve cannot be transformed to the heat equation but rather to the Bessel PDE. This approach is new and has not been developed yet in the literature until (Carr *et al.*, 2020).

In Part 6 the same ideas are used to solve various problems related to Credit and various physics problems. We show how to find the default boundary corresponding to a constant default intensity, an old problem, which was open for several decades. Besides, we present an analysis of a representative bank's stability in the structural default framework (Kaushansky *et al.*, 2018). We show how to use our approach to solve several physics problems: the supercooled Stefan problem, heat-exchange problems with a time-dependent coefficient of heat exchange, and the integrate-and-fire neuron excitation model for the steady and unsteady cases.

Part 7 is devoted to the generalization of our approach to problems with several moving boundaries. First, we solve the pricing problem for double barrier options for one-factor models reducible to the heat equation. Second, we develop a Multilayer (ML) method for solving one-factor parabolic equations using our results. Our approach provides a powerful alternative to the well-known finite difference and Monte Carlo methods. We discuss various advantages of this approach, which judiciously combines semi-analytical and numerical techniques and provides a fast and accurate way of finding solutions to the corresponding equations. To introduce the core of the method, we consider multilayer heat equations, known in physics for a relatively long time but never used in financial problems. Thus, we expand the analytic machinery of quantitative finance by augmenting it with the ML method. We demonstrate how one can solve various problems of mathematical finance by using our approach. Specifically, we develop efficient algorithms for pricing barrier options for time-dependent one-factor short-rate models, such as Black-Karasinski and Verhulst. Besides, we show how to solve the well-known Dupire equation quickly and accurately. Numerical examples confirm that our approach is considerably more efficient for solving the corresponding partial differential equations than the conventional finite difference method by being much faster and more accurate than the known alternatives.

Overall, the book will help readers who want to learn modern analytical tools for solving various applied problems of mathematical finance and physics. From the mathematical perspective, the level of details is closer to the applied rather than to the abstract or pure theoretical mathematics.

The book could also be a subject of a semester course on integral transforms and their financial applications for Master and Ph.D. programs in computational finance or financial engineering or, possibly, applied mathematics. It can also be used to train practitioners who want to extend their knowledge of modern tools of mathematical and computational finance.

Andrey Itkin
Tandon School of Engineering
New York University, USA

Alexander Lipton
Hebrew University of Jerusalem, Israel
Massachusetts Institute of Technology, USA

Dmitry Muravey
Moscow State University, Russia

About the Authors

Dr. Andrey Itkin is an Adjunct Professor at NYU, Department of Risk and Financial Engineering and Director, Senior Research Associate at Bank of America. He received his PhD in physics of liquids, gases and plasma, and a degree of Doctor of Science in computational physics. During his academic career, he published four books and more than 100 papers, first on chemical and theoretical physics and astrophysics, and later on computational and mathematical finance. Andrey has occupied various research and managerial positions in the financial industry and also is a member of multiple professional associations in finance and physics. He is also an associate editor of several journals on computer mathematics and mathematical finance and on the board of the Center for Financial Professionals (CeFPro). He has been a guest editor of several special issues on computational finance, and serves as a reviewer of multiple journals. He has also been an organizer of two international workshops on computational and algorithmic finance. He is a permanent speaker at various international financial conferences.

Alexander Lipton is Co-Founder and Chief Information Officer at Sila, Partner at Numeraire Financial, Visiting Professor and Dean's Fellow at the Hebrew University of Jerusalem, and Connection Science Fellow at MIT. He is a Board Member at Sila (Oregon) and an Advisory Board Member at numerous Fintech companies worldwide.

In 2016, he left Bank of America Merrill Lynch, where he had served for 10 years in various senior managerial roles, including Quantitative Solutions Executive and Co-Head of the Global Quantitative Group. Earlier, he held senior administrative positions at Citadel Investment Group, Credit Suisse, Deutsche Bank, and Bankers Trust. In parallel, Alex had several prestigious professorial appointments at École Polytechnique Fédérale de Lausanne, NYU, Oxford University, Imperial College, and the University of Illinois. Before switching to finance, he was a Full Professor of Mathematics at the University of Illinois and a Consultant at Los Alamos National Laboratory.

His current professional interests include FinTech, particularly applications of distributed ledger technology to payments and banking, digital currencies, including stablecoins and asset-backed cryptocurrencies, robust asset allocation, automated investing, balance sheet optimization, and industrial-strength risk management systems for large systemically important financial institutions.

His scientific interests are centered on cryptocurrencies, distributed ledgers, decentralized finance, quantitative development of asset allocation theory, modern monetary circuit theory, central clearing, and related topics.

In 2000, Alex was awarded the first-ever Quant of the Year Award by *Risk Magazine*; in 2021, he was awarded the Buy-Side Quant of the Year Award (jointly with Marcos Lopez de Prado).

Alex has published eight books covering thermonuclear fusion, foreign exchange trading, equity and credit derivatives, and related topics. His forthcoming books *Blockchain and Distributed Ledgers: Mathematics, Technology, and Economics* (with A. Treccani), *Building the New Economy* (with A. Pentland and T. Hardjono), and *Generalized Integral Transforms in Mathematical Finance* (with A. Itkin and D. Muravey) will be published in August, September and October of 2021, respectively.

In addition, Alex has published more than a hundred papers on financial engineering and the future of money and hydrodynamics, magnetohydrodynamics, astrophysics, and chemical physics. He has published numerous opinion pieces on the most vital topics in quantitative finance, including the cover article "Breaking the Bank" for *Scientific American* (jointly with Sandy Pentland).

Alex is an Associate Editor of several journals, including *Finance and Stochastics, Journal of FinTech, International Journal of Theoretical and Applied Finance, and Quantitative Finance.* In addition, he is a frequent keynote speaker at Quantitative Finance and FinTech conferences and forums worldwide.

Dmitry Muravey is a Research Fellow at Lomonosov State University, Moscow, Russia. He holds a PhD in Operations Research from Lomonosov State University. He has worked as a quantitative researcher in market making and statistical arbitrage and as an algorithms researcher in high-dimensional optimization problems and inverse problems in geophysics. He has authored numerous research papers in finance, stochastic analysis and partial differential equations. He has also won awards from Machine Learning and Data Science competitions.

Acknowledgments

Various people contributed to forming our views on the topics presented in this book via multiple discussions and comments. Although we cannot mention every person, it is our pleasure to gratefully acknowledge Peter Carr, Vadim Kaushansky, Marsha Lipton, Marcos Lopez de Prado, Christoph Reisinger, Igor Halperin, and Alexander Veygman. We also thank Ben Hambly, Andreas Sojmark, and Sam Howison for discussions on their model's theoretical properties and links to the Stefan problem.

We presented the contents of this book at multiple industrial and academic conferences. We are grateful to all organizers for providing us with numerous opportunities to discuss our results. We also acknowledge participants of these conferences and our conversations with them.

We are thankful to the editors of this book, Ms. Yubing Zhai and Ann Lai, whose tremendous efforts made this book a reality.

Finally, writing this book would have been impossible without our families, particularly our wives' great help. We greatly appreciate the support of Katya Itkin, Marsha Lipton, and Anna Muravey. Also, we are thankful to Andrey's daughter Anastasia for advice and revision of the book cover design.

Contents

PART 5. Fixed Income 229

List of Figures

PART 1

One-Factor Financial Models and Problems

Chapter 1

Stochastic Engines and Partial Differential Equations

1.1 Introduction

In this chapter, we describe the mathematical apparatus which is needed for our investigation of continuous-time markets with a particular emphasis on the stochastic engines which are used as sources of randomness in continuous time.[1] We define random processes on infinite probability spaces as families of random variables depending on a continuous parameter which we interpret as time. We introduce finite-dimensional distributions associated with a random process and formulate Kolmogorov's fundamental theorem which shows that without loss of generality we can consider random processes on the space of functions on an interval of the real axis. We define the so-called Markov random processes which have the property that the future states are independent of the past states provided that the present state is known. We discuss the transition probabilities (t.p.'s) and transition probability density functions (t.p.d.f.'s) associated with Markov processes and formulate the Chapman-Kolmogorov equation which restricts possible choices of t.p.'s. Next we introduce diffusion processes which are Markov processes with sufficiently regular trajectories and discuss Kolmogorov's backward and the Fokker-Planck forward equations which govern the t.p.d.f.'s for diffusion processes. We emphasize their duality which is particularly important for such tasks of financial engineering as model calibration. As examples of diffusion processes we discuss the so-called Wiener processes which are continuous-time limits of Bernoulli processes with fixed probabilities of success. Even though a rigorous description of such processes is somewhat involved, it is easier to deal with them than with Bernoulli processes. We interpret Wiener processes in terms

[1]The chapter is an abridged and streamlined version of Chapter 4 of Lipton (2001).

of a Brownian (random) walk of a particle on the real axis. Since knowing properties of Wiener processes is necessary in order to obtain an adequate description of markets with continuously evolving observables, we expose the key features of such processes in detail. In particular, we show how to construct the corresponding functional probability space and endow it with an appropriate filtration which is used in order to describe the flow of information revealed during the observation of the movements of a Brownian particle. We calculate the t.p.d.f.'s for Wiener processes as well as some other useful distributions associated with such processes. We also discuss the martingale property of the Brownian motion. We use the stochastic engines in order to define stochastic integrals and stochastic differential equations (SDEs) which are needed in order to construct financially relevant stochastic processes such as arithmetic and geometric Brownian motions and Bessel processes out of Wiener processes. We discuss in detail Ito's lemma. Both are used on numerous occasions below. We briefly discuss some general properties of SDEs such as existence and uniqueness of solutions and then introduce some special classes of SDEs which are particularly useful for the purposes of financial engineering including linear SDEs in the narrow and broad sense. We show that solutions of SDEs can be considered as diffusions and derive the corresponding backward and forward equations for their t.p.d.f.'s.

The chapter is organized as follows. In Section 1.2 we define stochastic processes which are continuous time counterparts of time series. In Section 1.3 we introduce and study Markov processes which form a special class of stochastic processes which are particularly important for our purposes. In Section 1.4 we study diffusions which are Markov processes whose trajectories are regular. Section 1.5 covers Wiener processes which are used later as "stochastic engines" for various financial quantities of interest. In Section 1.6 we study stochastic differential equations (SDEs) which are used in order to construct new stochastic processes from the standard Wiener processes as well as partial differential equations (PDEs) which describe the corresponding t.p.d.f.'s. In Section 1.7 we show how to solve linear SDEs which form a very important special class of SDEs.

1.2 Stochastic processes

We define a *scalar stochastic (or random) process* as a family of scalar random variables $\xi(t)$ or ξ_t depending on a continuous time parameter t varying on an interval $I = [t_0, t_1]$ which can be semi-infinite. We define a sample path (trajectory) of a random process $\xi_t(\omega)$ in the same way as before, such a path is a function from I to $\mathbb{R}$, i.e., a scalar function of a continuous scalar argument.

Recall that with a random variable we associate a distribution function. Similarly, with a random process we associate finite-dimensional distributions which are defined as follows

$$F_{t_1}(x^1) = P\{\xi_{t_1} \le x^1\}, \quad F_{t_1,t_2}(x^1, x^2) = P\{\xi_{t_1} \le x^1, \xi_{t_2} \le x^2\},$$

etc., where times $t_1, \ldots, t_n, \ldots$ are chosen in the increasing order. Finite-dimensional distributions have to satisfy the *compatibility condition*:

$$F_{t_1,\ldots,t_m,\ldots,t_n}(x^1, \ldots, x^m, \infty, \ldots, \infty) = F_{t_1,\ldots,t_m}(x^1, \ldots, x^m).$$

A random process is called *stationary* if

$$F_{t_1+t,\ldots,t_n+t}(x^1, \ldots, x^n) = F_{t_1,\ldots,t_n}(x^1, \ldots, x^n), \tag{1.1}$$

i.e., if its distribution functions are invariant with respect to time shifts. The vector case is studied by analogy.

Kolmogorov's fundamental theorem (Karatzas and Shreve, 1991), states that for every family of finite-dimensional distributions which satisfies the compatibility condition there is a probability space and a stochastic process having this family of distributions. Moreover, we can choose as a space of events Ω the set of all functions $\omega(t)$ from I to $\mathbb{R}$, assume that the corresponding sigma algebra $\mathfrak{U}$ is generated by the so-called cylinder sets $\mathfrak{g}$ of the form

$$\mathfrak{g} = \{\omega : \omega(t_1) \in B_1, \ldots, \omega(t_n) \in B_n\},$$

where B_n are Borel sets in $\mathbb{R}$ and assign probabilities $P(\mathfrak{g})$ to these sets in accordance with the given family of distributions,

$$P\{\omega : \omega(t_1) < x^1, \ldots, \omega(t_n) < x^n\} = F_{t_1,\ldots,t_n}(x^1, \ldots, x^n).$$

For a function $\omega(t)$ the value of the corresponding stochastic processes $\xi_t(\omega)$ at time t is the value of this function,

$$\xi_t(\omega) = \omega(t).$$

This construction shows that in dealing with stochastic processes we can restrict ourselves to functional spaces.

In order to describe the flow of information associated with a given random process we first define the set $\mathfrak{G}_t$ of cylinder sets of the form

$$\mathfrak{g} = \{\omega : \omega(t_1) \in B_1, \ldots, \omega(t_n) \in B_n\},$$

where $0 \le t_1 < t_2 \cdots < t_n \le t$, and then introduce the sigma algebra $\mathfrak{F}_t$ generated by $\mathfrak{G}_t$. The corresponding sigma algebras are ordered as follows

$$\mathfrak{F}_t \subseteq \mathfrak{F}_{t'}, \quad t \le t'.$$

We say that these sigma algebras form a filtration generated by the random process $\xi(t)$.

1.3 Markov processes

Deterministic dynamical systems are said to have the Markov property when their future states are independent of their past states provided that their present state is known. It is possible to extend this property to stochastic dynamical systems whose behavior is described by stochastic processes on an interval $I = [0, T]$, where T can be infinite. We already know that in order to describe the flow of information associated with a particular stochastic process we need to construct the filtration $\mathfrak{F}_t$ associated with this process and condition all probabilities on this filtration. We say that a scalar stochastic process ξ_t has the Markov property if

$$P\{\xi_{t'} \in B' | \mathfrak{F}_t\} = P\{\xi_{t'} \in B' | \xi_t\},$$

where $0 \leq t \leq t' \leq T$ and B' is a Borel set in $\mathbb{R}$. Such processes are called Markov processes. Broadly speaking, this is the only class of processes used in this book.

For Markov processes we can define the so-called t.p.'s $P(t, x, t', B')$ as follows:

$$P(t, x, t', B') = P\{\xi_{t'} \in B' | \xi_t = x\}.$$

T.p.'s have several useful properties. For instance, for fixed t, x, t' they define a probability measure on $\mathbb{R}$. Another property, which is very important for us, is the fact that for all times $t \leq t' \leq t'' \leq T$, Borel sets B'' and almost all x they satisfy the Chapman-Kolmogorov equation which states that

$$P(t, x, t'', B'') = \int_{\mathbb{R}^N} P(t, x, t', dx') P(t', x', t'', B''). \tag{1.2}$$

We can always modify the t.p.d.f. in such a way that the above equation is satisfied for all x. The Eq. (1.2) is a mathematical expression of the fact that trajectories starting at a point x at time t and finishing in a set B'' at time t'' do so by visiting intermediate points x' at time t'. The initial condition for this equation is

$$P(t, x, t, B) = \chi_B(x),$$

where, as usual, $\chi_B(.)$ is the characteristic function of a set.

In view of definition Eq. (1.1) we define homogeneous Markov processes as processes with stationary t.p.'s such that

$$P(t, x, t', B') = P(t' - t, x, B'),$$

so that t.p.'s depend only on the interval of time between observations $\tau = t' - t$, rather than the moments t, t' when these observations are made. For homogeneous Markov processes the Chapman-Kolmogorov equation can be written as follows (Van Kampen, 2007)

$$P(t'' - t, x, B'') = \int_{\mathbb{R}^N} P(t' - t, x, dx') P(t'' - t', x', B'').$$

We say that the t.p. P has t.p.d.f. p if

$$P(t, x, t', B') = \int_{B'} p(t, x, t', x') dx'.$$

In terms of t.p.d.f.'s the Chapman-Kolmogorov equation can be written as

$$p(t, x, t'', x'') = \int_{\mathbb{R}^N} p(t, x, t', x') p(t', x', t'', x'') dx'. \tag{1.3}$$

The corresponding homogeneous equation has the form

$$p(t'' - t, x, x'') = \int_{\mathbb{R}^N} p(t' - t, x, x') p(t'' - t', x', x'') dx'.$$

Example 1.1. The principal example of a Markov process with continuous trajectories is the ubiquitous scalar Wiener process W_t. This process is characterized by its non-random initial value $W_0 = 0$ and the stationary t.p.d.f.

$$p(\tau, x, x') = \frac{e^{-(x'-x)^2/2\tau}}{\sqrt{2\pi\tau}} \equiv H(\tau, x' - x). \tag{1.4}$$

It is easy to verify that this t.p.d.f. satisfies the Chapman-Kolmogorov equation. Moreover, by using the so-called Kolmogorov's condition, one can show that trajectories of W_t are continuous. As we will see shortly, these trajectories are nowhere differentiable. For fixed t the scalar random variable W_t is distributed normally, $W_t \sim N(0, t)$, where $N(x)$ is the normal cumulative distribution function (CDF). For historic reasons the function $H(\tau, x)$ is called the one-dimensional heat kernel.

1.4 Diffusions

In principle, trajectories of Markov processes can have rather peculiar properties. Diffusion processes form a special class of Markov processes such that all the trajectories are continuous and have other useful properties. (Markov processes with continuous trajectories which are not diffusions do exist.) They model a great many physically and financially important phenomena from Brownian motion of small particles in water, to propagation of heat in solids, to evolution of quantities of financial interest such as share prices and foreign exchange rates. For simplicity, we consider Markov processes with t.p.d.f.'s.

A scalar Markov process is called a diffusion provided that the following conditions are satisfied:

$$\lim_{t'\to t}\left[\frac{1}{t'-t}\int_{|x'-x|<\upsilon}\left\{\begin{matrix}(x'-x)\\(x'-x)^2\\(x'-x)^3\end{matrix}\right\}p\left(t,x,t',x'\right)dx'\right]=\left\{\begin{matrix}b\left(t,x\right)\\a\left(t,x\right)\\0\end{matrix}\right\}. \quad (1.5)$$

If a particle moves from (t,x) to (t',x'), then the drift coefficient $b\left(t,x\right)$ and the diffusion coefficient $a\left(t,x\right)$ measure the regular velocity and the degree of stochasticity of its motion, respectively.

It turns out that for diffusion processes t.p.d.f.'s are completely determined by the drift $b\left(t,x\right)$ and diffusion $a\left(t,x\right)$. Moreover, one can derive the so-called Kolmogorov's backward and Fokker-Planck forward partial differential equations for $p\left(t,x,t',x'\right)$ (Van Kampen, 2007).

In order to derive the Fokker-Planck forward PDE we multiply the Chapman-Kolmogorov equation by an arbitrary smooth function $\Psi\left(x''\right)$ vanishing outside of a certain domain and integrate the product with respect to x'' to get

$$\begin{aligned}&\int p\left(t,x,t'',x''\right)\Psi\left(x''\right)dx'' \\ &\quad-\int p\left(t,x,t',x'\right)dx'\int p\left(t',x',t'',x''\right)\Psi\left(x''\right)dx''=0.\end{aligned} \quad (1.6)$$

Next, we expand $\Psi\left(x''\right)$ in the second integral in a Taylor series centered at x',

$$\Psi\left(x''\right)=\Psi\left(x'\right)+\Psi_{x'}\left(x'\right)\left(x''-x'\right)+\frac{1}{2}\Psi_{x'x'}\left(x'\right)\left(x''-x'\right)^2+\ldots,$$

keep the first three terms and rewrite Eq. (1.6) as

$$\begin{aligned}&\int p\left(t,x,t'',x''\right)\Psi\left(x''\right)dx''-\int p\left(t,x,t',x'\right)\Psi\left(x'\right)dx' \\ &\quad-\int p\left(t,x,t',x'\right)\left(\int p\left(t',x',t'',x''\right)\left(x''-x'\right)dx''\right)\Psi_{x'}\left(x'\right)dx' \\ &\quad-\frac{1}{2}\int p\left(t,x,t',x'\right)\left(\int p\left(t',x',t'',x''\right)\left(x''-x'\right)^2dx''\right)\Psi_{x'x'}\left(x'\right)dx'=0.\end{aligned}$$

Thus,

$$\int (p(t,x,t'',y) - p(t,x,t',y)) \Psi(y)\, dy - \int p(t,x,t',x')$$
$$\times \left\{ \int p(t',x',t'',x'')(x''-x')\, dx'' \Psi_{x'}(x') \right.$$
$$\left. + \frac{1}{2} \int p(t',x',t'',x'')(x''-x')^2\, dx'' \Psi_{x'x'}(x') \right\} dx' = 0,$$

where y is a dummy variable. We divide the above equation by $t''-t'$, let $t'' \to t'$, and use conditions (Eq. (1.5)) to obtain

$$\int p_{t'}(t,x,t',x') \Psi(x')\, dx' - \int p(t,x,t',x')$$
$$\times \left\{ b(t',x') \Psi_{x'}(x') + \frac{1}{2} a(t',x') \Psi_{x'x'}(x') \right\} dx' = 0.$$

Finally, integration by parts yields

$$\int \left\{ p_{t'}(t,x,t',x') + (b(t',x') p(t,x,t',x'))_{x'} \right.$$
$$\left. - \frac{1}{2} (a(t',x') p(t,x,t',x'))_{x'x'} \right\} \Psi(x')\, dx' = 0.$$

Since $\Psi(x)$ is arbitrary, this equation implies that $p(t,x,t',x')$ as a function of t',x' satisfies the forward Fokker-Plank PDE (Risken and Haken., 1989)

$$p_{t'}(t,x,t',x') - \frac{1}{2} (a(t',x') p(t,x,t',x'))_{x'x'} + (b(t',x') p(t,x,t',x'))_{x'} = 0. \tag{1.7}$$

This is equation should be solved subject to the initial condition

$$p(t,x,t' \to t,x') = \delta(x'-x),$$

where, as usual, $\delta(\cdot)$ is the Dirac delta function.

Straightforward integration by parts shows that as a function of t,x the t.p.d.f. $p(t,x,t',x')$ satisfies the backward Kolmogorov equation (Van Kampen, 2007)

$$p_t(t,x,t',x') + \frac{1}{2} a(t,x) p(t,x,t',x')_{xx} + b(t,x) p(t,x,t',x')_x = 0. \tag{1.8}$$

The terminal condition reads

$$p(t \to t',x,t',x') = \delta(x-x').$$

Since the backward and forward equations govern the behavior of the same t.p.d.f. they are called dual.

For financial applications we need to consider more general dual backward and forward equations, namely

$$\begin{aligned} p_t\left(t,x,t',x'\right) + \frac{1}{2}a\left(t,x\right)p\left(t,x,t',x'\right)_{xx} & \\ + b\left(t,x\right)p\left(t,x,t',x'\right)_x - c\left(t,x\right)p\left(t,x,t',x'\right) &= 0, \\ p\left(t\to t',x,t',x'\right) = \delta\left(x-x'\right), & \end{aligned} \tag{1.9}$$

and

$$\begin{aligned} p_{t'}\left(t,x,t',x'\right) - \frac{1}{2}\left(a\left(t',x'\right)p\left(t,x,t',x'\right)\right)_{x'x'} & \\ + \left(b\left(t',x'\right)p\left(t,x,t',x'\right)\right)_{x'} + c\left(t',x'\right)p\left(t,x,t',x'\right) &= 0, \\ p\left(t,x,t'\to t,x'\right) = \delta\left(x'-x\right), & \end{aligned} \tag{1.10}$$

where $c>0$. These equations can be interpreted in probabilistic terms. To do so, we start with a process ξ_t with the diffusion coefficient a and drift coefficient b and construct a new process $\tilde{\xi}_t$ by killing ξ_t at a certain random stopping (killing) time ζ,

$$\tilde{\xi}_t = \begin{cases} \xi_t, & t<\zeta, \\ \maltese, & t>\zeta. \end{cases}$$

Here $\maltese \notin \mathbb{R}$ is an artificial "coffin" state. If we assume that killing occurs at the rate c, i.e.,

$$\lim_{t'\to t}\left[\frac{1}{t'-t}\left(1-P\left(t,x,t',\mathbb{R}\right)\right)\right] = c\left(t,x\right),$$

then the corresponding backward and forward equations for the t.p.d.f. have the form Eqs. (1.9) and (1.10), respectively.

Due to the linearity of the general terminal value problem

$$\begin{aligned} U_t\left(t,x\right) + \frac{1}{2}a\left(t,x\right)U_{xx}\left(t,x\right) + b\left(t,x\right)U_x\left(t,x\right) - c\left(t,x\right)U\left(t,x\right) &= 0, \\ U\left(T,x\right) = u\left(x\right), & \end{aligned} \tag{1.11}$$

we can represent its solution in the form

$$U\left(t,x\right) = \int_{\mathbb{R}} p\left(t,x,T,\xi\right)u\left(\xi\right)d\xi. \tag{1.12}$$

Similarly, the solution of the initial value problem

$$\begin{aligned} V_t\left(t,x\right) - \frac{1}{2}\left(a\left(t,x\right)V\left(t,x\right)\right)_{xx} + \left(b\left(t,x\right)V\left(t,x\right)\right)_x + c\left(t,x\right)U\left(t,x\right) &= 0, \\ V\left(0,x\right) = v\left(x\right), & \end{aligned}$$

can be written as

$$V(t,x) = \int_{\mathbb{R}^N} p(0,\xi,t,x)\, v(\xi)\, dx.$$

In the PDE literature p is called Green's function for the above problems (Kythe, 2011)

On several occasions below we need to solve inhomogeneous backward problems of the form

$$U_t(t,x) + \frac{1}{2} a(t,x) U_{xx}(t,x) + b(t,x) U_x(t,x) - c(t,x) U(t,x) = W(t,x),$$
$$U(T,x) = u(x),$$

and similar forward problems. This task can be accomplished via Duhamel's principle. This principle (which can be proved by direct substitution) states that

$$U(t,x) = \int_{\mathbb{R}} p(t,x,T,\xi)\, u(\xi)\, d\xi - \int_t^T \int_{\mathbb{R}} p(t,x,t',\xi)\, W(t',\xi)\, dt' d\xi.$$

For homogeneous killed diffusions with $p(t,x,t',x') = p(\tau,x,x')$, where $\tau = t' - t$, $a(t,x) = a(x)$, $b(t,x) = b(x)$, $c(t,x) = c(t)$, the backward and forward equations assume the form

$$\begin{aligned} p_\tau(\tau,x,x') - \frac{1}{2} a(x)\, p_{xx}(\tau,x,x') \\ - b(x)\, p_x(\tau,x,x') + c(x)\, p(\tau,x,x') = 0, \end{aligned} \tag{1.13}$$

and

$$\begin{aligned} p_\tau(\tau,x,x') - \frac{1}{2} \left(a(x')\, p(\tau,x,x')\right)_{x'x'} \\ + \left(b(x')\, p(\tau,x,x')\right)_{x'} + c(x')\, p(\tau,x,x') = 0, \end{aligned} \tag{1.14}$$

respectively. Both of them are forward parabolic equations augmented with the initial condition

$$p(0,x,x') = \delta(x - x'). \tag{1.15}$$

When the coefficients a, b, c are constant, we can write $p(\tau,x,x')$ as $p(\tau,y)$, where $y = x' - x$. The problem for $p(\tau,y)$ reads

$$\begin{aligned} p_\tau(\tau,y) - \frac{1}{2} a p_{yy}(\tau,y) + b p_y(\tau,y) + c p(\tau,y) = 0, \\ p(0,y) = \delta(y). \end{aligned} \tag{1.16}$$

Example 1.2. In this example we treat the scalar Wiener process as a diffusion process. Since this process is homogeneous, we know that

$p = p(\tau, x, x'), a = a(x), b = b(x)$. The killing rate is zero, $c(x) = 0$. Straightforward calculation which is left to the reader as an exercise shows that $a(x) = 1, b(x) = 0$. Thus, both the backward and forward problems reduce to the standard heat equation

$$p_\tau - \frac{1}{2} p_{xx} = 0, \tag{1.17}$$

supplied with the initial condition

$$p(0, x, x') = \delta(x - x').$$

So far, we considered processes which take values on the entire real line. Often, it is necessary to consider processes taking values on the interval $\mathcal{I} = (e_1, e_2)$, where $-\infty \le e_1 < e_2 \le \infty$. The boundaries can appear due to singularities in coefficients a, b, c or be introduced for other reasons. When the process $\tilde{\xi}$ reaches the boundaries of the interval $\mathcal{I}$ it is necessary to augment backward and forward PDEs with appropriate boundary conditions. A very detailed classification of the boundary points is due to Feller. To be concrete, we consider the left boundary e_1. In general, a diffusion process can exhibit different types of behavior at e_1. The simplest possibility is that the process never reaches the value e_1. In this case, no boundary conditions at $x = e_1$ are needed and the corresponding point is called natural. Alternatively, the diffusion process can reach the value e_1 with positive probability. In this case two possibilities have to be distinguished: (A) the boundary point e_1 absorbs the process, so that no boundary conditions can be imposed there; (B) the process behaves like the standard diffusion in the vicinity of e_1, so that different boundary conditions can be imposed there. The corresponding boundary is called an exit in case (A) and regular in case (B). In order to classify the boundaries, introduce the so-called *natural scale density* $\mathsf{s}(x)$, *speed measure* $\mathsf{m}(x)$ and *killing measure* $\mathsf{k}(x)$ as follows

$$\mathsf{s}(x) = \exp\left\{-2 \int^x \frac{b(y)}{a(y)} dy\right\},$$

$$\mathsf{m}(x) = \frac{1}{a(x)\,\mathsf{s}(x)},$$

$$\mathsf{k}(x) = \frac{c(x)}{a(x)\,\mathsf{s}(x)},$$

and choose an arbitrary point x inside the interval $\mathcal{I}$. The left boundary point e_1 is called entrance if

$$J_{11} = \int_{e_1}^{x} \left(\int_{y}^{x} \mathsf{s}(z)\, dz \right) (\mathsf{m}(y) + \mathsf{k}(y))\, dy < \infty,$$

and exit if

$$J_{12} = \int_{e_1}^{x} \left(\int_{y}^{x} (\mathsf{m}(z) + \mathsf{k}(z))\, dz \right) \mathsf{s}(y)\, dy < \infty.$$

For the right boundary point the definition is analogous. The process can reach an exit boundary and can be started at an entrance boundary. The boundary point is called regular if it is both exit and entrance, and natural if it is neither exit nor entrance.

Example 1.3. We consider the standard Wiener process on the semi-infinite interval $(0, \infty)$. We already know that the Kolmogorov and Fokker-Plank equations have the same form Eq. (1.17). Since we consider these equations on the half-axis, we have to augment them with appropriate boundary conditions. By definition, $\mathsf{s}(x) = 1$, $\mathsf{m}(x) = 1$, $\mathsf{k}(x) = 0$. For the left end point $e_1 = 0$ both integrals J_{11}, J_{12} are clearly finite, so that this point is regular. For the right end point $e_2 = \infty$, both integrals J_{21}, J_{22} are infinite, so that this point is natural. Accordingly, we have to impose only one boundary condition at $e_1 = 0$. There are several possibilities. The Wiener process can be

(a) reflected at zero,

$$p_{x'}(\tau, x, 0) = 0; \tag{1.18}$$

(b) killed at zero,

$$p(\tau, x, 0) = 0; \tag{1.19}$$

(c) killed elastically at zero,

$$p_{x'}(\tau, x, 0) = \alpha p(\tau, x, 0). \tag{1.20}$$

In the first case the process is immediately returned back to the interval $(0, \infty)$, in the second case it is terminated as soon as it reaches zero, finally, in the third case a combination of the above occurs. It is easy to find the corresponding t.p.d.f.'s in the first and second case by the method of images. In the case of reflecting boundary we represent $p(\tau, x, x')$ as a sum

$$p(\tau, x, x') = H(\tau, x - x') + H(\tau, x + x'),$$

which guarantees that p solves Eq. (1.17) and satisfies boundary conditions Eq. (1.18). For the killing boundary we write $p(\tau, x, x')$ as a difference

$$p(\tau, x, x') = H(\tau, x - x') - H(\tau, x + x'),$$

which guarantees that p solves Eq. (1.17) and satisfies boundary conditions Eq. (1.19). The case of elastic killing Eq. (1.20) is more complex and will be studied later in the specific financial context.

Example 1.4. We consider the so-called Bessel process (Revuz and Yor, 1999), with

$$a(x) = 1, \quad b(x) = \frac{(1-2\nu)}{2x}, \qquad c(x) = 0.$$

It is natural to consider such a process on the semi-infinite interval $(0, \infty)$. The corresponding Kolmogorov equation is

$$p_{\mathcal{T}} - \frac{1}{2} p_{xx} - \frac{(1-2\nu)}{2x} p_x = 0.$$

The expression in parenthesis is the radial part of the Laplace operator in $2(1-\nu)$ dimensional space. The natural scale density and the speed measure are

$$\mathsf{s}(x) = x^{-(1-2\nu)}, \qquad \mathsf{m}(x) = x^{(1-2\nu)}.$$

For the left boundary we have (for simplicity we assume that $\nu \neq 1$)

$$\begin{aligned} J_{11} &= \int_0^x \left(\int_y^x z^{2\nu-1} dz \right) y^{1-2\nu} dy \\ &= \frac{1}{2\nu} \int_0^x \left(x^{2\nu} - y^{2\nu} \right) y^{1-2\nu} dy \\ &= \left(\frac{x^{2\nu} y^{2-2\nu}}{2\nu (2-2\nu)} - \frac{y^2}{4\nu} \right) \Big|_0^x, \end{aligned}$$

so that

$$\begin{cases} J_{11} < \infty, & \nu < 1, \\ J_{11} = \infty, & \nu \geq 1. \end{cases}$$

Similarly,

$$\begin{cases} J_{12} < \infty, & \nu > 0, \\ J_{12} = \infty, & \nu \leq 0. \end{cases}$$

Thus, 0 is an entrance-not-exit boundary when $\nu \leq 0$, exit-not-entrance boundary when $\nu \geq 1$, and a regular boundary when $0 < \nu < 1$. The right boundary is always natural.

1.5 Wiener processes

Scalar Wiener processes which are the most important stochastic engines for continuous-time financial models have several important properties. They were introduced in Section 1.3 as examples of Markov processes. The t.p.d.f. for such processes is given by Eq. (1.4). Trajectories of Wiener processes are continuous. For the standard Wiener process starting at zero, $W_0 = 0$ we have

$$P\Big\{W_{t_1} \in B_1, \ldots, W_{t_N} \in B_N\Big\} \tag{1.21}$$
$$= \int_{B_1} \cdots \int_{B_N} H(t_1, x_1) H(t_2 - t_1, x_2 - x_1)$$
$$\times \ldots \times H(t_N - t_{N-1}, x_N - x_{N-1})\, dx_1 \ldots dx_N,$$

where $t_0 = 0 < t_1 < t_2 \ldots < t_N = T$, and $H(t, x)$ is the one-dimensional heat kernel. This formula defines a probability measure P on the space Ω of continuous functions on the interval $[0, T]$.

The Eq. (1.21) shows that the Wiener process has independent normally distributed increments, i.e., that random variables W_{t_1}, $W_{t_2} - W_{t_1}, \ldots,$ are independent normal variables $N(0, t_1), N(0, t_2 - t_1), \ldots$. In fact, this property defines the Wiener process. Symbolically,

$$W_{t'} - W_t \sim \sqrt{t' - t}. \tag{1.22}$$

More formally,

$$\mathbb{E}\{W_{t'} - W_t\} = 0, \tag{1.23}$$
$$\mathbb{E}\left\{(W_{t'} - W_t)^2\right\} = t' - t.$$

One more useful formula which is used below is

$$\mathbb{E}\{|W_{t'} - W_t|\} = \sqrt{\frac{2(t' - t)}{\pi}}. \tag{1.24}$$

The characteristic function of the Wiener process is given by

$$\Phi(t, \varsigma) = e^{-t\varsigma^2/2},$$

while its covariation and correlation functions have the form

$$\Psi(t_1, t_2) = \mathbb{E}\{W_{t_1} W_{t_2}\} = \min\{t_1, t_2\},$$
$$\tilde{\Psi}(t_1, t_2) = \frac{\mathbb{E}\{W_{t_1} W_{t_2}\}}{\sqrt{\mathbb{E}\{W_{t_1}^2\}\mathbb{E}\{W_{t_2}^2\}}} = \min\left\{\sqrt{\frac{t_1}{t_2}}, \sqrt{\frac{t_2}{t_1}}\right\}.$$

The Wiener process W_t is a martingale, and so is the process $W_t^2 - t$. Levy's theorem states that every continuous scalar process ξ_t, such that ξ_t and $\xi_t^2 - t$ are martingales, is a Wiener process.

The Wiener process has important symmetry properties which show that the processes

$$W_t^{sym} = -W_t, \quad \text{(symmetry)}, \tag{1.25}$$

$$W_t^{sca} = \sqrt{c}W_{t/c}, \quad \text{(scaling)},$$

$$W_t^{inv} = \begin{cases} 0, & t = 0, \\ tW_{1/t}, & t > 0 \end{cases} \quad \text{(inversion)},$$

$$W_t^{ref} = \begin{cases} W_t, & t \leq T, \\ 2W_T - W_t, & t \geq T, \end{cases} \quad \text{(reflection)},$$

are Wiener processes as well. In Eq. (1.25) T is a stopping time.

Due to the fact that the increments of the Wiener process are independent, we can use the strong law of large numbers and show that

$$\text{a.c. } \lim \frac{W_t}{t} = 0.$$

The behavior of the Wiener process can be described more precisely by the law of the iterated logarithm which states that asymptotically W_t oscillates between two parabolas,

$$\begin{Bmatrix} \limsup_{t\to\infty} \\ \liminf_{t\to\infty} \end{Bmatrix} \frac{W_t}{\sqrt{2t\ln(\ln t)}} = \begin{Bmatrix} 1 \\ -1 \end{Bmatrix}.$$

For small t we can use the scaling properties of the Wiener process and the law of iterated logarithm to get

$$\begin{Bmatrix} \limsup_{t\to 0} \\ \liminf_{t\to 0} \end{Bmatrix} \frac{W_t}{\sqrt{2t\ln(\ln 1/t)}} = \begin{Bmatrix} 1 \\ -1 \end{Bmatrix}.$$

Since a similar relation describes the behavior of the increment $W_{t'} - W_t$, we can conclude that trajectories of W_t are nowhere differentiable and have unbounded variation on any interval $[t, t']$, i.e.,

$$\lim_{N\to\infty} \sum_{n=1}^{N} |W_{t_n} - W_{t_{n-1}}| \to \infty,$$

where $t_0 = t < t_1 < t_2 \ldots < t_N = t'$. However, their quadratic variation is finite

$$\lim_{N\to\infty} \sum_{n=1}^{N} |W_{t_n} - W_{t_{n-1}}|^2 = t' - t.$$

Scalar Wiener processes with drift are deterministic functions of the standard Wiener processes,

$$W_t^{(\mu)} = \mu t + W_t.$$

For fixed t we have $W_t^{(\mu)} \sim N(\mu t, t)$. Such processes are supermartingales when $\mu < 0$ and submartingales when $\mu > 0$. Their t.p.d.f.'s have the form

$$p(\tau, x, x') = \frac{e^{-(x'-x-\mu\tau)^2/2\tau}}{\sqrt{2\pi\tau}} = H(\tau, x' - x - \mu\tau),$$

so that

$$\begin{aligned} &P\left\{W_{t_1}^{(\mu)} \in B_1, \dots, W_{t_N}^{(\mu)} \in B_N\right\} \\ &= \int_{B_1} \dots \int_{B_N} H(t_1, x_1 - \mu t_1) H(t_2 - t_1, x_2 - x_1 - \mu(t_2 - t_1)) \\ &\quad \times \dots \times H(t_N - t_{N-1}, x_N - x_{N-1} - \mu(t_N - t_{N-1})) dx_1 \dots dx_N. \end{aligned} \tag{1.26}$$

This formula defines a probability measure $P^{(\mu)}$ on the space Ω of continuous functions on the interval $[t, t']$.

The properties of $W_t^{(\mu)}$ are broadly similar to the properties of W_t. Moreover, by changing the measure on the space Ω of continuous real-valued functions on the interval $[0, T]$, we can transform $W_t^{(\mu)}$ into the standard Wiener process. Of course, in view of our treatment of random walks in Section 1.3, such a possibility does not come as a real surprise. The exact adjustment is described by the following theorem.

1.6 SDE and mappings

For the purposes of financial engineering we need to map the Wiener process which is the principal "stochastic engine" for continuous-time markets onto financially observable quantities. This can be done via stochastic differential equations.

Recall that a typical ordinary differential equation (ODE) can be written is several equivalent forms given x_0, namely

$$\begin{aligned} \frac{dx_t}{dt} &= b(t, x_t), \\ dx_t &= b(t, x_t) dt, \\ x_t &= x_0 + \int_0^t b(s, x_s) ds. \end{aligned} \tag{1.27}$$

According to Picard's existence theorem (Coddington and Levinson, 1955), in a sufficiently small neighborhood of the point $(0, x_0)$ the evolution of x_t is deterministic and is completely specified by the initial condition x_0, provided that b is a continuous function that satisfies the Lipschitz condition

$$|b(t, x) - b(t, y)| \leq L|x - y|.$$

We can introduce stochasticity into the picture by generalizing equations Eq. (1.27) as follows. Given x_0 we get

$$\begin{aligned}\frac{dx_t}{dt} &= b(t,x_t) + \sigma(t,x_t)\varpi_t, \\ dx_t &= b(t,x_t)\,dt + \sigma(t,x_t)\,dW_t, \\ x_t &= x_0 + \int_0^t b(s,x_s)\,ds + \int_0^t \sigma(s,x_s)\,dW_s,\end{aligned} \tag{1.28}$$

where ϖ_t is the so-called white noise formally defined as the derivative of the standard Wiener process with respect to time,

$$\varpi_t = dW_t/dt.$$

Solutions of the SDEs in Eq. (1.28) exhibit sufficient randomness to describe the continuous evolution of financially relevant quantities. In situations when jumps are important, we need to incorporate Poisson process into our framework. The coefficients b and σ are called the drift and volatility of the process x_t. An SDE is called autonomous if b and σ are time-independent.

The first generalization, which is due to Langevin is much less straightforward than the other two because of the highly irregular properties of the white noise and we do not consider it here. An obvious difficulty with formalizing the concept of SDEs written either in the differential form (the second line in Eq. (1.28)), or the integral form (the last line in Eq. (1.28)) is that W_s is so irregular that dW_s does not exist in the usual sense. Equation (Eq. (1.22)) suggests that, broadly speaking, $dW_s \sim \sqrt{ds}$. To illustrate the implications of this fact, we consider two integrals

$$I_{0,T} = \int_0^T f(s)\,dW_s,$$

where $f(t)$ is a non-random function, and

$$\tilde{I}_{0,T} = \int_0^T W_s dW_s.$$

Let $t_0 = 0, t_1, \ldots, t_{k-1}, t_K = T$ be a uniform partition of the interval $[0,T]$ into K subintervals, $t_k = kT/K$, and τ_k be some intermediate points, $t_{k-1} \leq \tau_k \leq t_k$. The first integral can be understood in the usual Riemann-Stieltjes sense

$$I_{0,T} = \lim_{K\to\infty} I_{0,T}^K,$$

where

$$I_{0,T}^{K} = \sum_{k=1}^{K} f(\tau_k)\left(W_{t_k} - W_{t_{k-1}}\right),$$

as expected. This integral is independent of the choice of intermediate points τ_i. However, this is not the case for the second integral. We have

$$\tilde{I}_{0,T} = \lim_{K\to\infty} \tilde{I}_{0,T}^{K},$$

where

$$\begin{aligned}\tilde{I}_{0,T}^{K} &= \sum_{k=1}^{K} W_{\tau_k}\left(W_{t_k} - W_{t_{k-1}}\right)\\ &= \frac{1}{2}W_T^2 - \frac{1}{2}\sum_{k=1}^{K}\left(W_{t_k} - W_{t_{k-1}}\right)^2 + \sum_{k=1}^{K}\left(W_{\tau_k} - W_{t_{k-1}}\right)^2\\ &\quad + \sum_{k=1}^{K}\left(W_{t_k} - W_{\tau_k}\right)\left(W_{\tau_k} - W_{t_{k-1}}\right).\end{aligned}$$

By using Eq. (1.23) and the fact the process W_t has independent increments, we obtain

$$\begin{aligned}\underset{K\to\infty}{\text{ms-lim}}\,\tilde{I}_{0,T}^{K} &= \frac{1}{2}W_T^2 - \frac{1}{2}\lim_{K\to\infty}\sum_{k=1}^{K}(t_k - t_{k-1}) + \lim_{K\to\infty}\sum_{k=1}^{K}(\tau_k - t_{k-1})\\ &= \frac{1}{2}W_T^2 - \frac{1}{2}T + \lim_{K\to\infty}\sum_{k=1}^{K}(\tau_k - t_{k-1}).\end{aligned}$$

It is obvious that the corresponding limit depends of the choice of τ_k. For instance, if we follow Ito and choose τ_k in a non-anticipating manner, $\tau_k = t_{k-1}$, we get

$$\tilde{I}_{0,T} \overset{\text{Ito}}{=} \frac{1}{2}W_T^2 - \frac{1}{2}T. \tag{1.29}$$

On the other hand, if we choose τ_k according to Stratonovich (Van Kampen, 2007), $\tau_k = (t_k + t_{k-1})/2$, we get

$$\tilde{I}_{0,T} \overset{\text{Stratonovich}}{=} \frac{1}{2}W_T^2.$$

Stratonovich's stochastic integrals have many nice properties. However, below we always use Ito's integrals because they are based on choosing the integrand in the nonanticipating way and are better suited for the purposes of financial engineering. There are relatively simple rules for transforming Ito's integrals into Stratonovich's integrals and *vice versa*.

Inspired by the above example, we define the general Ito's integral as a mean-square limit of non-anticipating integrals over step functions

$$\begin{aligned} I_{0,T} &= \int_0^T f(s,\omega)\, dW_s(\omega) \\ &= \operatorname*{ms-lim}_{K\to\infty} \sum_{k=1}^{K} f(t_{k-1},\omega)\left(W_{t_k}(\omega) - W_{t_{k-1}}(\omega)\right), \end{aligned}$$

where ω is a particular trajectory (realization) of the Wiener process and $W_s(\omega)$ is its value at time s. It is clear that I is a random variable on Ω. Due to the nonanticipating choice of intermediate points τ_k, $I_{0,t}$ is a martingale as function of t, so that

$$\mathbb{E}\left\{I_{0,t'}|\mathfrak{F}_t\right\} = I_{0,t},$$

where $\mathfrak{F}_t$ is the filtration generated by the Wiener process W_t. The martingale property is arguably the single most important property of Ito's integrals.

Below we understand an SDE as an integral equation (the last line in Eq. (1.28)) with integration performed according to Ito. Conditions guarantying existence and uniqueness of the solution are the Lipschitz condition

$$|b(t,x) - b(t,y)| + |\sigma(t,x) - \sigma(t,y)| \le C|x-y|,$$

which guarantees that the coefficients of the SDE are sufficiently smooth, and the growth condition

$$|b(t,x)|^2 + |\sigma(t,x)|^2 \le C\left(1 + |x|^2\right),$$

which guarantees that the coefficients grow at most linearly when $x \to \infty$. When these conditions are violated, solutions can have undesirable properties, for instance, they can blow up in finite time or even instantaneously.[2]

Since the rules of Ito's integration deviate from the familiar rules of classical analysis, so do the rules of Ito's differentiation. These are expressed by the celebrated Ito's lemma (Karatzas and Shreve, 1991)

Lemma 1.1 (Scalar Ito's lemma). *Let x_t be a scalar stochastic process with stochastic differential of the form*

$$dx_t = b(t,x_t)\,dt + \sigma(t,x_t)\,dW_t,$$

[2]This is not just a theoretical possibility, the blow up does occur when one solves the so-called Heath-Jarrow-Morton equations describing the dynamics of forward interest rates in fixed income (Andersen and Piterbarg, 2010).

where the coefficients are sufficiently regular. If $f(t,x)$ is a continuous functions with continuous derivatives f_t, f_x, f_{xx}, then the process $f(t,x_t)$ also has stochastic differential and

$$df(t,x_t) = \left[f_t(t,x_t) + b(t,x_t)f_x(t,x_t) + \frac{1}{2}a(t,x_t)f_{xx}(t,x_t)\right]dt$$
$$+\sigma(t,x_t)f_x(t,x_t)dW_t,$$

where $a(t,x) = \sigma^2(t,x)$. Below we call $a(t,x)$ the diffusion coefficient.

Heuristically, Ito's lemma can be understood as a second-order Taylor expansion formula based on the following multiplication rule

$$dt\cdot dt = 0, \quad dt\cdot dW_t = 0, \quad dW_t\cdot dW_t = dt.$$

For two processes x^1, x^2 governed by SDEs Eq. (1.28) the product rule consistent with Ito's calculus reads

$$d\left(x^1x^2\right) = x^1dx^2 + x^2dx^1 + \sigma^1\sigma^2dt.$$

Example 1.5. If we choose $x_t = W_t$ and $f(t,x) = x^2/2$, we get

$$d\left[\frac{1}{2}W_t^2\right] = \frac{1}{2}dt + W_tdW_t,$$

so that

$$I_{0,t} = \int_0^t W_sdW_s = \frac{1}{2}W_t^2 - \frac{1}{2}t,$$

in agreement with Eq. (1.29).

Ito's lemma for scalar processes can be used provided that $f(t,x)$ can be differentiated ones with respect to t and twice with respect to x.

Solutions of SDEs represent new stochastic processes generated by the standard Wiener process. These processes are martingales provided that $b = 0$. It can be shown that these processes are diffusions, so that they can be described in terms of the t.p.d.f.'s. The corresponding Kolmogorov's backward and Fokker-Plank forward equations $p(t,x,t',x')$ have the form Eqs. (1.8) and (1.7) with a, b being the diffusion and drift, respectively.

In a typical financial engineering context we have to evaluate the conditional expectation

$$U(t,x) = \mathbb{E}\{f(x_T)|x_t = x\} \tag{1.30}$$

for a given function $f(x)$. It can be shown that $U(t,x)$ solves the equation

$$U_t(t,x) + \frac{1}{2}a(t,x)U_{xx}(t,x) + b(t,x)U_x(t,x) = 0, \tag{1.31}$$

supplied with the terminal condition

$$U(T,x) = f(x). \tag{1.32}$$

Thus, we can either compute the corresponding expectation Eq. (1.30) directly or solve the problem Eqs. (1.31) and (1.32) (if necessary, numerically).

If killing (or discounting) is important, the expectation assumes the form

$$U(t,x) = \mathbb{E}\left\{ f(x_T) \exp\left(-\int_t^T c(s,x_s)\,ds \right) |x_t = x \right\}, \tag{1.33}$$

where $c(t,x) \geq 0$. The corresponding backward problem reads

$$U_t(t,x) + \frac{1}{2}a(t,x)U_{xx}(t,x) + b(t,x)U_x(t,x) - c(t,x)U(t,x) = 0,$$
$$U(T,x) = f(x). \tag{1.34}$$

Formula Eq. (1.33) is known as the Feynman-Kac formula (Karatzas and Shreve, 1991). It should be compared with Eq. (1.12).

1.7 Linear SDEs

In this section we consider linear SDEs which very frequently appear in financial engineering problems. Scalar SDEs linear in the broad sense have the form

$$dx_t = (\beta_0 + \beta_1 x_t)\,dt + (\sigma_0 + \sigma_1 x_t)\,dW_t,$$

where $\beta_0, \beta_1, \sigma_0, \sigma_1$ are scalar functions of time. Scalar SDEs linear in the narrow sense have the form

$$dx_t = (\beta_0 + \beta_1 x_t)\,dt + \sigma_0 dW_t.$$

Surprisingly enough, even in the scalar case general linear SDEs cannot be solved explicitly. At the same time, an explicit solution for scalar SDEs linear in the narrow sense can easily be found.

1.7.1 *Arithmetic Brownian motion*

We start with the following SDE linear in the narrow sense

$$dx_t = \beta_0 dt + \sigma_0 dW_t,$$

where β_0, β, σ_0 are constant parameters. This equation describes the arithmetic Brownian motion. Its solution is a Wiener process with drift starting at x_0,

$$x_t = x_0 + \beta_0 t + \sigma_0 W_t.$$

For time-dependent β_0, σ_0 the corresponding solution is

$$x_t = x_0 + \int_0^t \beta_0 (s) \, ds + \int_0^t \sigma_0 (s) \, dW_s.$$

The arithmetic Brownian motion plays an important role in financial engineering where it is used to model the dynamics of returns on risky investments. For fixed t the random variable x_t is distributed normally,

$$x_t \sim N \left(x_0 + \beta_0 t, \sigma_0^2 t \right).$$

Typical trajectories of an arithmetic Brownian motion without and with drift are shown in Figure 1.1.

1.7.2 *Geometric Brownian motion*

The geometric Brownian motion is described by the following scalar SDE which is linear in the broad sense

$$dx_t = \beta_1 x_t dt + \sigma_1 x_t dW_t, \quad \beta_1, \sigma_1 - const. \tag{1.35}$$

We introduce

$$y_t = \ln \left(\frac{x_t}{x_0} \right),$$

and use Ito's lemma to get the following equation for y_t

$$dy_t = \left(\beta_1 - \frac{1}{2} \sigma_1^2 \right) dt + \sigma_1 dW_t.$$

Accordingly,

$$y_t = \left(\beta_1 - \frac{1}{2} \sigma_1^2 \right) t + \sigma_1 W_t, \tag{1.36}$$

$$x_t = x_0 \exp \left(\left(\beta_1 - \frac{1}{2} \sigma_1^2 \right) t + \sigma_1 W_t \right).$$

The Eq. (1.35) is extremely important in the financial context since it describes the evolution of prices of risky assets. The fact that it has an explicit solution explains why so many problems of financial engineering can be solved in the closed form. The corresponding model is called the Black-Scholes-Merton model and is discussed in Chapter 2.

Since y_t is distributed normally, x_t is distributed lognormally,

$$\ln x_t \sim N \left[\ln x_0 + \left(\beta_1 - \frac{1}{2} \sigma_1^2 \right) t, \sigma_1^2 t \right],$$

Typical trajectories of a geometric Brownian motion are shown in Figure 1.2.

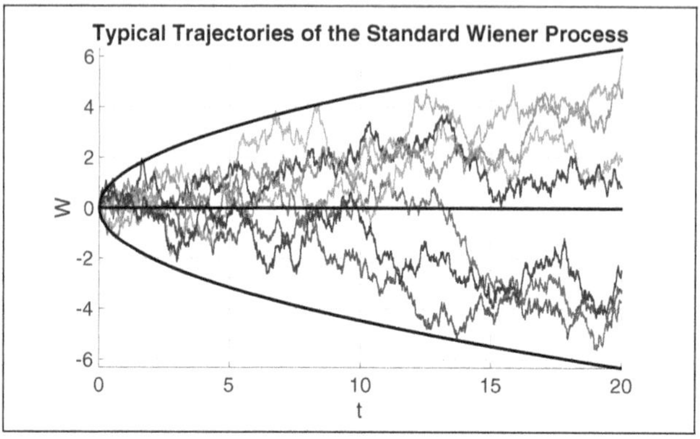

(a)

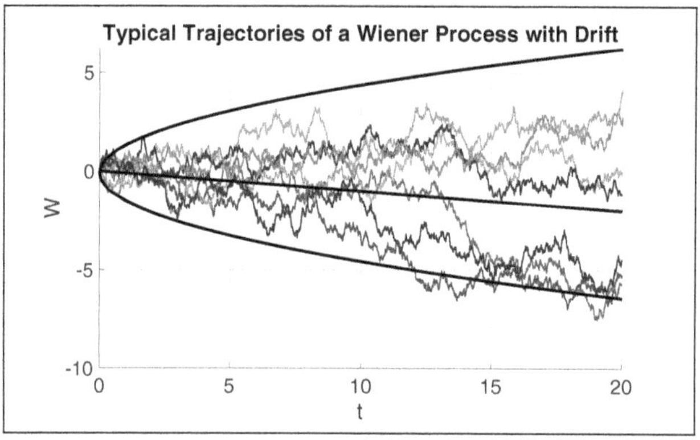

(b)

Figure 1.1: Typical trajectories of the standard Brownian motion (or Wiener process) as a function of time t (in years) with $\sigma_0 = \sqrt{2}$. (a) without drift, the limiting curves $W = \pm\sqrt{2t}$ and the mean $W = 0$ are shown in boldface; (b) with drift $\mu = -0.1$, the limiting curves $W = \mu \pm \sqrt{2t}$ and the mean $W = \mu t$ are shown in boldface.

1.7.3 *Mixed Brownian motion*

The mixed Brownian motion is described by the following scalar SDE which is linear in the broad sense

$$dx_t = \beta_1 x_t dt + \sigma_0 dW_t. \tag{1.37}$$

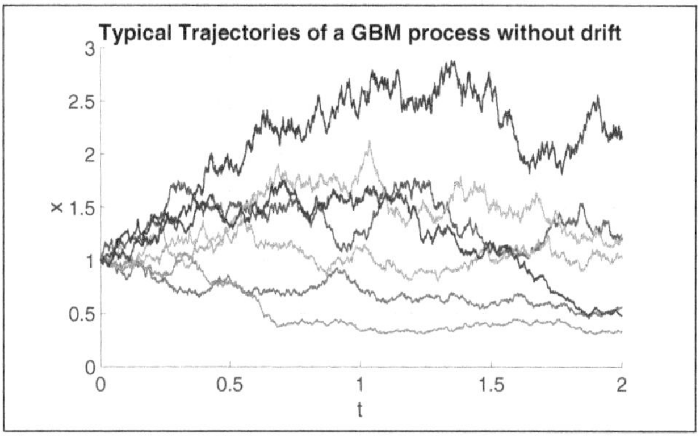

(a)

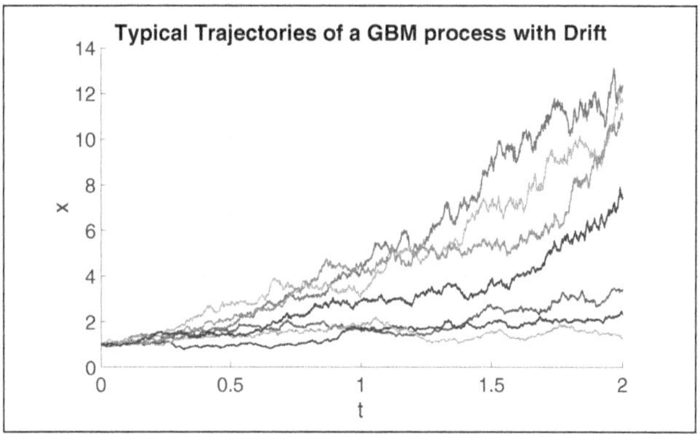

(b)

Figure 1.2: Typical trajectories of a geometric Brownian motion process as a function of time t (in years) with $x_0 = 1,\ \sigma_1 = 0.4$; (a) $\beta_1 = 0$, (b) $\beta_1 = 0.05$.

We introduce

$$y_t = e^{-\beta_1 t} x_t,$$

and get the following equation for y_t

$$dy_t = e^{-\beta_1 t} \sigma_0 dW_t.$$

Thus, y_t is driven by an arithmetic Brownian motion without the drift, so that

$$y_t \sim N\left(x_0, \frac{\sigma_0^2\left(1 - e^{-2\beta_1 t}\right)}{2\beta_1}\right),$$

$$x_t \sim N\left(e^{\beta_1 t}x_0, \frac{\sigma_0^2\left(e^{2\beta_1 t}-1\right)}{2\beta_1}\right).$$

The Eq. (1.37) is not as important in the financial context, compared to Eq. (1.35), it is used to describe the evolution of prices of risky assets, which can have negative values. The corresponding model is called the Bachelier model and is discussed in Chapter 2.

When $\beta_1 > 0$, Eq. (1.37) describes mean-repulsion away from zero, when $\beta_1 < 0$ it describes mean-reversion to zero, and, as such is a special case of the Ornstein-Uhlenbeck process considered in the next section. Typical trajectories of a lognormal process are shown in Figure 1.3.

1.7.4 *The Ornstein-Uhlenbeck process*

Another important scalar linear SDE describes the so-called Ornstein-Uhlenbeck (OU) process (Van Kampen, 2007). This SDE has the form

$$dx_t = (\beta_0 - \beta_1 x_t)\, dt + \sigma_0 dW_t, \tag{1.38}$$

where $\beta_1 > 0, \sigma_0 > 0$ are constants. Its deterministic counterpart has the form

$$dx_t = (\beta_0 - \beta_1 x_t)\, dt.$$

The solution of Eq. (1.38) reads

$$x_t = \frac{\beta_0}{\beta_1} + e^{-\beta_1 t}\left(x_0 - \frac{\beta_0}{\beta_1}\right),$$

and describes the transition of x_t from its initial value x_0 to its asymptotic value β_0/β_1. The corresponding time scale is $1/\beta_1$. We solve the Ornstein-Uhlenbeck equation via the method of variation of constants. Namely, we represent x_t in the form $x_t = e^{-\beta_1 t}y_t$ and verify that y_t follows the arithmetic Brownian motion,

$$dy = \beta_0 e^{\beta_1 t}dt + \sigma_0 e^{\beta_1 t}dW_t.$$

Accordingly,

$$y_t = x_0 + \frac{\beta_0\left(e^{\beta_1 t}-1\right)}{\beta_1} + \int_0^t \sigma_0 e^{\beta_1 s}dW_s,$$

$$x_t = \frac{\beta_0}{\beta_1} + e^{-\beta_1 t}\left(x_0 - \frac{\beta_0}{\beta_1}\right) + \int_0^t \sigma_0 e^{-\beta_1 (t-s)}dW_s.$$

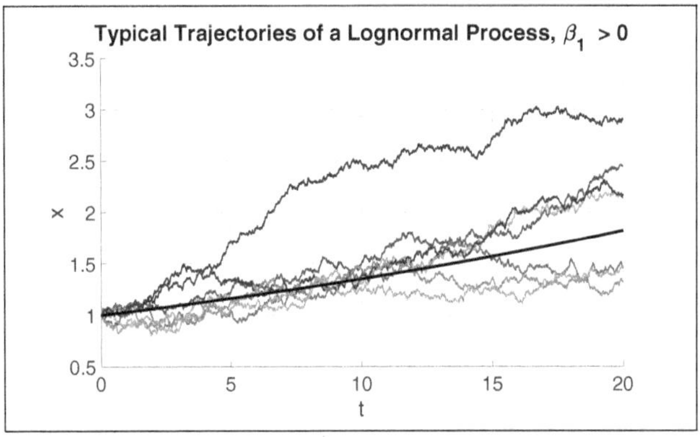

(a)

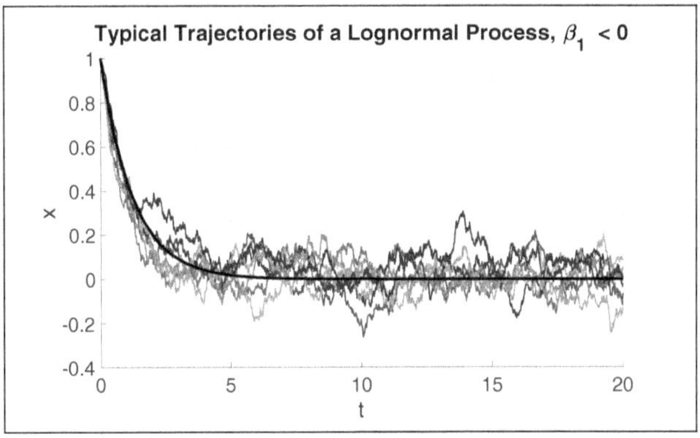

(b)

Figure 1.3: Typical trajectories of a lognormal process as a function of time t (in years) with $x_0 = 1$, $\sigma_0 = 0.1$; (a) $\beta_1 = 0.03$, (b) $\beta_1 = -0.8$. The exponential trajectory $x = \exp(\beta_1 t)$ is shown in boldface.

Thus, for fixed t the corresponding x_t is normal,

$$x_t \sim N\left(\frac{\beta_0}{\beta_1} + e^{-\beta_1 t}\left(x_0 - \frac{\beta_0}{\beta_1}\right), \frac{\sigma_0^2\left(1 - e^{-2\beta_1 t}\right)}{2\beta_1}\right).$$

The corresponding stochastic process is called mean-reverting since its behavior is determined by two competing trends, namely, the deterministic attraction to the asymptotic level β_0/β_1 and the stochastic repulsion from this level. We emphasize that the variance of x_t does not become infinite

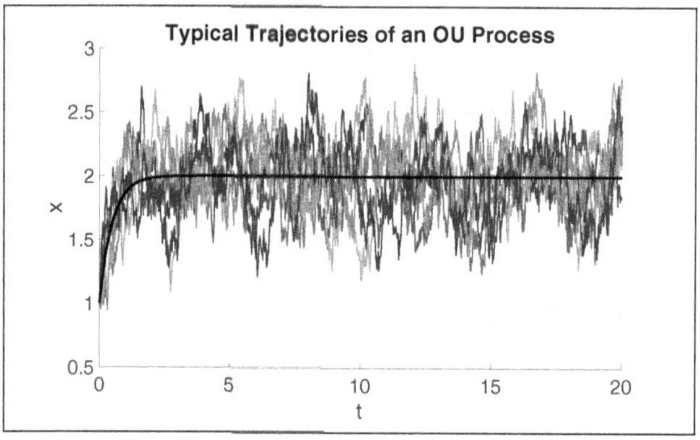

Figure 1.4: Typical trajectories of an Ornstein-Uhlenbeck process as a function of time t (in years) with $x_0 = 1$, $\beta_0 = 4$, $\beta_1 = 2$, $\sigma_0 = 0.6$. The deterministic trajectory corresponding to $\sigma = 0$ is shown in boldface.

when $t \to \infty$ (in sharp contrast with the standard Wiener process). The limiting random variable x_∞ has the form

$$x_\infty \sim N\left(\frac{\beta_0}{\beta_1}, \frac{\sigma_0^2}{2\beta_1}\right).$$

Trajectories of a typical Ornstein-Uhlenbeck process are shown in Figure 1.4.

Chapter 2

Popular One-Factor Models by Asset Classes

2.1 Introduction

This book is devoted to developing new analytical methods to solve various unsteady finance (and physics) problems efficiently. These problems' common attribute is that they are defined at domains with time-dependent boundaries and have time-dependent boundary conditions on these boundaries. Also, all parameters of the models considered in this book could be time-dependent. These problems naturally occur in mathematical finance as time-dependence of the model parameters and domains of definition is necessary for calibrating this model to market data. Various examples of such models for various asset classes could be found in the financial literature, see e.g., Brigo and Mercurio (2006); Andersen and Piterbarg (2010); Hull (2011); Lipton (2001); Geman (2009) and Bossu (2014) among others.

In this book, our primary focus is on one-factor (or one-dimensional) financial models. These models have been used for modeling derivatives on equities, foreign exchange, fixed income, and commodities for a long time. However, later, these models have been supplanted by more sophisticated multifactor models, which are better for modeling the market. Nevertheless, one-factor models are still in use in many financial firms. Therefore, their fast and accurate pricing and hedging, especially for exotic options, are still important. More generally, the development of analytical and semi-analytical methods for one-factor problems (not necessarily financial) is very desirable. The necessity for such methods is made clear by the rise of electronic and high-frequency trading, and this trend is still upward over the last few years.

With these preliminaries in mind, below, we briefly discuss one-factor models used throughout the book. This introduction is made for the reader's convenience and doesn't aim at being comprehensive. Still, it could help make the book self-contained. For additional information, we refer the reader to the cited literature.

2.2 Equities, foreign exchange, and commodities

2.2.1 *The Black-Scholes-Merton model*

The celebrated Black-Scholes-Merton (BSM) model is perhaps the most popular and best-known mathematical finance model. In 1968, based on research previously developed by academicians and practitioners, such as L. Bachelier, S. Kassouf and E. Thorp among others, Fischer Black and Myron Scholes first showed the existence of dynamic hedging leading to risk neutrality of the portfolio consisting of a stock and a certain amount of options written on this stock an underlying. In 1970 the first attempt to apply their formula to the markets resulted in financial losses due to lack of risk management in their trades. And only in three years the Black-Scholes formula, named after them, was published in Black and Scholes (1973). The mathematical understanding of this option pricing model was provided by Robert Merton and coined the term "Black–Scholes options pricing model" (Merton, 1974). Merton and Scholes received the 1997 Nobel Memorial Prize in Economic Sciences for their work, the committee citing their discovery of the risk-neutral dynamic revision as a breakthrough that separates the option from the risk of the underlying security. Although ineligible for the prize because of his death in 1995, Black was mentioned by the Swedish Academy as a critical contributor to the model.

The key idea behind the model is to hedge the option by buying and selling the underlying asset in just the right way and, as a consequence, to eliminate risk. This type of hedging is called "continuously revised delta hedging" and is the basis of more complicated hedging strategies such as those engaged in by investment banks and hedge funds.

Since this model got such great attention from both researchers and practitioners over the years, we only shortly describe its main features by following (Hull, 2011). The main concepts underlying the BSM model are as follows:

(1) The option price and the stock price depend on the same underlying source of uncertainty.

(2) The stock price follows the geometric Brownian motion process

$$dS_t = \mu S_t dt + \sigma S_t dW_t, \quad S\Big|_{t=0} = S. \tag{2.1}$$

where S_t is the stock price, t is the time. μ, σ are the drift and volatility, W_t is the Brownian motion. Thus, S_t is lognormally distributed.

(3) .We can form a portfolio consisting of the stock and the option, eliminating the source of uncertainty.

(4) The portfolio is instantaneously riskless and must instantaneously earn the risk-free rate r.

Using the hedging argument that the portfolio's return must be risk-free rate, it is possible to derive a PDE (the Black-Scholes PDE). Any security (contingent claim) whose price depends on the stock price satisfies this PDE. The Black-Scholes (BS) PDE reads

$$\left[\frac{\partial}{\partial t} + \mu S\frac{\partial}{\partial S} + \frac{1}{2}\sigma^2 S^2 \frac{\partial^2}{\partial S^2}\right] f(t,S) = rf(t,S). \tag{2.2}$$

In what follows, we will call this type of PDE as the *pricing equation*.

The corresponding drift is risk-neutralized, so $\mu = r - q$, where q is the dividend rate. When dealing with foreign exchange, we interpret r, q as domestic and foreign risk-neutral rates, respectively. When modeling commodities, we typically put $\mu = 0$ and interpret S as a forward price, in the spirit of the celebrated (Black and Cox, 1976a) model.

The particular security is determined by the initial and boundary conditions of this PDE. For instance, for a European Call option with the strike K and maturity T, the payoff at maturity is (the terminal condition) $C(T,S) = (S-K)^+$, where $C(t,S)$ is the Call option price. And the boundary conditions are

$$C(t,0) = 0, \quad C(t,S)\Big|_{S\uparrow\infty} = S - K. \tag{2.3}$$

We proceed by assuming that the expected return from the stock price is the risk-free rate (which determines the drift function μ), discount the option payoff at the risk-free rate, and solve Eq. (2.2) in closed-form. The result is known as the BS formula for the European Call option price. For the sake of convenience, let us introduce new variables $y = \log K/F_T$, where $F_T = Se^{(r-q)T}$ is the forward stock price, q is the continuous dividend, and $w = \sigma^2 T$ which is the total implied variance. In terms of these variables, the Black-Scholes formula for the future value of the European Call option price becomes (Lipton, 2002; Gatheral, 2006).

$$
\begin{aligned}
C_{BS}(F_T, y, w) &= D(t,T)F_T[N(d_1) - e^y N(d_2)] \qquad (2.4)\\
&= D(t,T)F_T\left[N\left(-\frac{y}{\sqrt{w}} + \frac{\sqrt{w}}{2}\right) - e^y N\left(-\frac{y}{\sqrt{w}} - \frac{\sqrt{w}}{2}\right)\right],
\end{aligned}
$$

where $D(t,T) = e^{-r(T-t)}$ and $N(x)$ is the normal CDF, and

$$
d_1 = \frac{\ln\frac{F_T}{K} + \frac{1}{2}\sigma_{BS}^2 T}{\sigma_{BS}\sqrt{T}}, \qquad d_2 = d_1 - \frac{1}{2}\sigma_{BS}\sqrt{T}. \qquad (2.5)
$$

In above it is assumed that parameters of the model - r, q, σ are constant. However, the BS formula remains valid in case of time-dependent parameters $r(t), q(t), \sigma(t)$ if instead of r, q, σ one uses the averages

$$
\bar{r} = \frac{1}{T}\int_0^T r(s)ds, \quad \bar{q} = \frac{1}{T}\int_0^T q(s)ds, \quad \bar{\sigma}^2 = \frac{1}{T}\int_0^T \sigma^2(s)ds. \qquad (2.6)
$$

The prices of barrier options with constant barriers can also be obtained in closed form in the BSM model, see Lipton (2001) and Howison (1995). However, no closed-form solutions are known for prices of barrier options with time-dependent barriers. We will consider this problem in Chapter 9.

2.2.2 *The Bachelier model*

The Bachelier model uses mixed Brownian motion as a stochastic driver and assumes that the stock price is governed by the SDE of the form

$$
dS_t = \mu S_t dt + \sigma dW_t, \quad S\Big|_{t=0} = S. \qquad (2.7)
$$

The corresponding pricing equation has the form

$$
\left[\frac{\partial}{\partial t} + \mu S\frac{\partial}{\partial S} + \frac{1}{2}\sigma^2\frac{\partial^2}{\partial S^2}\right] f(t,S) = rf(t,S), \qquad (2.8)
$$

where $\mu = r - q$.

As discussed in the previous chapter, the mixed Brownian motion can be reduced to the Wiener process. By doing so, we can derive the Bachelier pricing formula for the call option:

$$
C_B(F_T, y, w) = D(t,T)\left((F_T - K)N(z) + \Sigma n(z)\right), \qquad (2.9)
$$

where $N(x)$ is the normal CDF, $n(x)$ is the normal PDF, and

$$
z = \frac{F_{t,T} - K}{\Sigma}, \quad \Sigma = \sigma\sqrt{\frac{e^{2(r-q)T} - 1}{2(r-q)}}. \qquad (2.10)
$$

The Bachelier model recently became very popular in conjunction with pricing of commodities due to the fact that it is capable of producing negative prices. To the great surprise of almost everybody, such prices were recently observed in the oil market, albeit briefly.

2.2.3 The CEV model

The time-dependent constant elasticity of variance (CEV) model is a one-dimensional diffusion process that solves a stochastic differential equation

$$dS_t = \mu(t) S_t dt + \sigma(t) S_t^{\beta+1} dW_t, \qquad S_{t=0} = S_0. \tag{2.11}$$

Here $t \geq 0$ is the time, S_t is the stochastic stock price, $\mu(t)$ is the drift, $\sigma(t)$ is the volatility and β is the elasticity parameter such that $\beta < 1$, $\beta \neq \{0, -1\}$[1], W_t is the standard Brownian motion. It is known, that under risk-neutral measure $\mu(t) = r(t) - q(t)$ where $r(t)$ is the deterministic short interest rate, and $q(t)$ is the continuous dividend. We assume that all parameters of the model are known either as a continuous functions of time $t \in [0, \infty)$, or as a discrete set of N values for some moments t_i, $i = 1, \ldots, N$.

The CEV model with constant coefficients has been introduced in Cox (1975) as an alternative to the geometric Brownian motion for modeling asset prices. Despite some sophistication as compared with the Black-Scholes model, the model is still analytically tractable, and prices of the European options can be obtained in closed form. That is because the CEV process with constant coefficients is related to the Bessel process (Linetsky and Mendoza, 2010). Also, as mentioned in that reference, the elasticity parameter β controls the steepness of the skew (the larger the $|\beta|$ — the steeper the skew), while the volatility (scale) parameter σ fixes the at-the-money volatility level. This ability to capture the skew has made the CEV model popular in equity options markets.

For the standard CEV process with constant parameters it is known that change of variable $z_t = 1/(\sigma|\beta|) S_t^{-\beta}$ reduces the CEV process without drift ($\mu = 0$) to the standard Bessel process of order $1/2\beta$ (see Revuz and Yor, 1999; Davydov and Linetsky, 2001). Then the continuous part of the risk-neutral density of S_t, conditional on $S_0 = S$, is obtained from the well-known expression for transition density of the Bessel process. If $\mu \neq 0$, using the result of Goldenberg (1991) this CEV process could be obtained from the process without drift via a scale and time change

$$S_t^{\mu} = e^{\mu t} S^0_{\tau(t)}, \qquad \tau(t) = \frac{1}{2\mu\beta}\left(e^{2\mu\beta t} - 1\right).$$

Also, as shown in Davydov and Linetsky (2001), at $\beta > 0$ according to Feller's classification the origin $S_t = 0$ is a natural boundary, and infinity is an entrance boundary.

[1] In case $\beta = 0$ this model is the Black-Scholes model, while for $\beta = -1$ this is the Bachelier, or time-dependent Ornstein-Uhlenbeck (OU) model.

From the volatility smile prospective, when $\beta = 0$ the CEV model translates into the celebrating Black-Scholes model. When $\beta < 0$ the implied volatility falls as stock price raises which generates a fatter left tail. This tail can be observed at market, and so the CEV model behaves realistically. When $\beta > 0$ the implied volatility rises as stock price raises — the property inherent for options on futures. Thus, the CEV model is able to better catch market behavior.

European option prices in the CEV model with constant coefficients are also known, so the model is tractable. In Lipton (2001) the following theorem is proved:

Theorem 2.1. *The call option price $C(t, S)$ for the CEV process $S_t = S$ is given by*

$$C(t,S) = e^{-r(T-t)} S \int_{\breve{K}}^{\infty} \left(\frac{\hat{S}}{y}\right)^{\frac{1}{2(2-\theta)}} e^{-(\hat{S}+y)} I_{\frac{1}{2-\theta}}(2\sqrt{\hat{S}y})dy \tag{2.12}$$
$$+ e^{-r(T-t)} K \int_{\breve{K}}^{\infty} \left(\frac{y}{\hat{S}}\right)^{\frac{1}{2(2-\theta)}} e^{-(\hat{S}+y)} I_{\frac{1}{2-\theta}}(2\sqrt{\hat{S}y})dy.$$

Here

$$\hat{S} = \frac{2Se^{r(2-\theta)(T-t)}}{(2-\theta)^2\chi}, \quad \chi = \frac{\sigma^2}{(2-\theta)r}\left(e^{r(2-\theta)T} - e^{r(2-\theta)t}\right) \tag{2.13}$$
$$\breve{K} = \frac{2K^{2-\theta}}{(2-\theta)^2\chi}, \quad \theta = 2(\beta+1),$$

K, T are the option strike and maturity, r is the deterministic interest rate, and $I_\alpha(x)$ is the modified Bessel function of the first kind (Abramowitz and Stegun, 1964).

2.2.4 The local volatility model

All the models considered above, use SDEs with parametric volatilities as stochastic drivers. Unfortunately, such a simple choice contradicts market reality. In order to match given market prices of European options, we need to assume that the stock price is governed by the generic SDE of the form

$$dS_t = \mu S_t dt + \sigma(t, S_t)dW_t, \quad S\Big|_{t=0} = S, \quad . \tag{2.14}$$

where $\sigma(t, S_t)$ is a given generic function. This model was invented independently by B. Dupire and E. Derman/I. Kani around 1994 as a relatively simple extension of the Black-Scholes model with the aim to explain the volatility smile and skew. They can be observed by comparing the implied volatilities of options written on the same underlying asset and having the same expiration date but different strikes. As mentioned in Itkin (2020) with a reference to, there exists an interesting opinion (Gatheral, 2002), that Dupire, Derman and Kani unlikely ever thought of local volatility as representing a model of how volatilities actually evolve. Rather, perhaps they introduced a notion of local volatilities as representing some kind of average over all possible instantaneous volatilities in a stochastic volatility world (an "effective theory"). As such, this model allows practitioners to price exotic options consistently with the known prices of vanilla options. This means that using the so-called Dupire equation one can find a unique local volatility function $\sigma(K,T)$ from the market prices of standard options, and then construct, e.g., an implied tree that incorporates these local volatilities to value exotic options and to hedge standard options (Derman *et al.*, 2016). Later this approach was further extended by introducing the local stochastic volatility models, see, e.g., Lipton (2002), Bergomi (2016) and references therein.

The corresponding pricing PDE in variables (t, S) for the local volatility model reads

$$\left[\frac{\partial}{\partial t} + \mu S\frac{\partial}{\partial S} + \frac{1}{2}\sigma^2(t, S_t)\frac{\partial^2}{\partial S^2}\right] f(t, S) = rf(t, S), \tag{2.15}$$

where $\mu = r - q$. In general, it is not possible to solve the corresponding pricing problem explicitly. An extended discussion of the model is provided in Chapter 19.

2.3 Fixed Income

2.3.1 *The Hull-White model*

The Hull-White model, invented in Hull and White (1990), is a one-factor model for the stochastic short interest rate r_t of the form

$$dr_t = \kappa(t)[\theta(t) - r_t]dt + \sigma(t)dW_t. \tag{2.16}$$

Thus, the model uses an Ornstein-Uhlenbeck process as a stochastic driver. This model is actually a time-dependent version of the earlier Vasicek model; see Andersen and Piterbarg (2010). Since invented, it became to be very popular among practitioners for modeling interest rates and credit. That is because it is relatively simple and allows for negativity prices (while for a long time this behaviour was treated as a deficiency of the model, nowadays this became its advantage). The model could be calibrated to the given term-structure of interest rates and to the prices or implied volatilities of caps, floors or European swaptions since the mean-reversion level and speed $\theta(t), \kappa(t)$ and volatility $\sigma(t)$ are functions of time. Under the Hull-White model the prices of Zero-coupon bonds (ZCB) and European Vanilla options are known in closed form, if $\kappa(t) = \kappa = const$ (Andersen and Piterbarg, 2010). Indeed, suppose also that $\sigma(t) = \sigma = const$. With allowance for the affine term structure of the solution, the latter can be represented as

$$P(S,T) = A(S,T)e^{-B(S,T)r(S)}, \tag{2.17}$$

where $P(S,T)$ is the price of the ZCB with the time-S value and the maturity T, and

$$\begin{aligned} B(S,T) &= \frac{1-\exp(-\kappa(T-S))}{\kappa}, \\ A(S,T) &= \frac{P(0,T)}{P(0,S)} \exp\Bigg[- B(S,T)\frac{\partial \log(P(0,S))}{\partial S} \\ &\quad - \frac{\sigma^2(\exp(-\kappa T) - \exp(-\kappa S))^2(\exp(2\kappa S)-1)}{4\kappa^3}\Bigg]. \end{aligned} \tag{2.18}$$

Note that the terminal distribution of $P(S,T)$ is lognormal.

However, for exotic options, e.g., barrier options, these prices are not known yet in closed form. Therefore, various numerical methods are used to obtain such prices, see survey in Kuan and Webber (2003).

Since the process in Eq. (2.16) is Gaussian, the model is tractable for pricing European plain vanilla options. However, for exotic options, e.g., liquid barrier options, or for American options these prices are not known yet in closed form. Therefore, various numerical methods are used to obtain them, that sometimes could be computationally expensive. Note, that it is known that simple one factor models of the type considered in this book are not well-suited to replicate the implied volatility surface of the exotic options, and instead more sophisticated models which treats volatility as a stochastic variable, should be used in this case. Still, construction of a semi-analytical solution even for our simple model is useful as this is discussed in Part 4.

2.3.2 *The Cox-Ingersoll-Ross model*

The Cox-Ingersoll-Ross (CIR) model was invented in Cox *et al.* (1985) and called after the inventors' names. It assumes that the short-rate, r_t, satisfies

$$dr_t = \kappa(\theta - r_t)\,dt + \sigma\sqrt{r_t}dW_t, \quad r_0 > 0. \tag{2.19}$$

where κ and θ are positive constants. The principal advantage of the CIR model over the Vasicek model is that the short rate can be made to remain non-negative, if the so-called Feller condition is satisfied. The latter can be expressed in terms of the model parameters and reads $2\kappa\theta > \sigma^2$ (Karatzas and Shreve, 1991).

Unlike the Vasicek model, however, the CIR model is not Gaussian and is therefore considerably more difficult to analyze. Perhaps surprisingly, zero-coupon bond prices can still be computed analytically[2], and are given by

$$P(t,T) = \mathbb{E}_{\mathbb{Q}}\left[e^{-\int_t^T r_s ds}\right] = e^{A(T-t)-B(T-t)r_t}, \tag{2.20}$$

where

$$\begin{aligned} A(\tau) &= \frac{2\kappa\theta}{\sigma^2}\log\left(\frac{2\gamma e^{(\gamma+\kappa)\tau/2}}{(\gamma+\kappa)(e^{\gamma\tau}-1)+2\gamma}\right), \\ B(\tau) &= \frac{2(e^{\gamma\tau}-1)}{(\gamma+\kappa)(e^{\gamma\tau}-1)+2\gamma}, \\ \gamma &= \sqrt{\kappa^2+2\sigma^2}. \end{aligned} \tag{2.21}$$

Closed form solutions for option prices on zero-coupon bonds can also be found in this model. In general, derivatives prices can be estimated by either numerically solving the corresponding PDE with the terminal condition

$$P(T,T) = 1, \tag{2.22}$$

and the boundary conditions in Eq. (2.39), or by using Monte-Carlo methods.

Also in this model the mean and variance of the chi-square distribution can be computed in closed form and read (Andersen and Piterbarg, 2010)

$$\begin{aligned} \mathbb{E}[r_t \mid r_s] &= r_s e^{-\kappa(t-s)} + \theta\left(1 - e^{-\kappa(t-s)}\right), \quad s < t, \\ \mathrm{V}[r_t \mid r_s] &= r_s\frac{\sigma^2}{\kappa}\left(e^{-\kappa(t-s)} - e^{-2\kappa(t-s)}\right) + \frac{\theta\sigma^2}{2\kappa}\left(1 - e^{-\kappa(t-s)}\right)^2. \end{aligned} \tag{2.23}$$

The transition density function of r_t at time t given r_s at time s is

$$p(r_s, s, r_t, t) = ce^{-(u+v)}\left(\frac{v}{u}\right)^{\frac{q}{2}} I_q(2\sqrt{uv}), \tag{2.24}$$

[2]The volatility function is chosen to be level-dependent and proportional to $\sqrt{r}$ since then the model remains affine.

where

$$c = \frac{2\kappa}{\sigma^2\left(1 - e^{-\kappa\Delta t}\right)}, \quad u = cr_s e^{-\kappa\Delta t}, \tag{2.25}$$
$$v = cr_t, \quad q = \frac{2\kappa\theta}{\sigma^2} - 1, \quad \Delta t = t - s.$$

Here $I_q(\cdot)$ is the modified Bessel function of the first kind of order q which is defined as (Abramowitz and Stegun, 1964)

$$I_q(x) = \sum_{j=0}^{\infty}\left(\frac{x}{2}\right)^{2j+q}\frac{1}{k!\Gamma(j+q+1)}, \tag{2.26}$$

and $\Gamma(x)$ is the Gamma function.

Let $z_t = 2cr_t$. Then the conditional distribution of z_t at time t given z_s at time s is non-central $\chi^2_d(2u)$ distributed with $d = 4\alpha\mu/\sigma^2$ degrees of freedom and non-central parameter $\lambda = 2u$, which is proportional to the current spot price (Cox *et al.*, 1985). This means that z_t conditioned on z_s has the same distribution as

$$\chi^2_d(\lambda) = (Z + \sqrt{\lambda})^2 + \chi^2_{d-1}, \tag{2.27}$$

where χ^2_{d-1} is a χ^2 random variable with $d-1$ degrees of freedom, and Z is an independent standard $\mathcal{N}(0,1)$ random variable. We can write this as

$$z_t \mid z_s \sim \chi^2_d(\lambda), \tag{2.28}$$

where

$$\lambda = 2u = 2cr_s e^{-\alpha\Delta t} = \frac{4\alpha}{\mu^2\left(1 - e^{-\alpha\Delta t}\right)}e^{-\alpha\Delta t}r_s, \tag{2.29}$$

and $\Delta t = t - s$. Since $z_t = 2cr_t, r_t$ conditioned on r_s has the same distribution as $z_t/2c$ conditioned on $z_s/2c$. In other words

$$r_t \Big| r_s \sim \frac{z_t}{2c}\Big|\frac{z_s}{2c} \sim \frac{1}{2c}\chi^2_d(\lambda). \tag{2.30}$$

2.3.3 *The Black-Karasinski model*

The Black-Karasinski (BK) model is also widely used by practitioners for modeling interest rates and credit. A similar model, known as the Schwartz one-factor model, is often used to model commodities. The enduring popularity of the model is because, despite some lack of tractability, it is relatively simple and guarantees non-negativity of rates. Besides, one can calibrate it to a given term-structure of interest rates and prices or implied volatilities of caps, floors, or European swaptions, provided that the mean-reversion level and volatility are functions of time.

The Black-Karasinski (BK) model was introduced in Black and Karasinski (1991), see also Brigo and Mercurio (2006) for a more detailed discussion. The BK is a one-factor short-rate model of the form

$$\begin{aligned} dz_t &= \kappa(t)[\theta(t) - z_t]dt + \sigma(t)dW_t, \qquad r \in \mathbb{R},\ t \geq 0, \\ r_t &= s(t) + Re^{z_t}, \qquad r(t=0) = r_0. \end{aligned} \tag{2.31}$$

Here t is the time, r_t is the short interest rate, $\kappa(t) > 0$ is the speed of mean-reversion, $\theta(t)$ is the mean-reversion level, $\sigma(t)$ is the volatility, R is some constant with the same dimensionality as r_t, e.g., it can be $1/(1 \text{ year})$, W_t is the standard Brownian motion. This model is similar to the Hull-White model but preserves the positivity of r_t by exponentiating the OU random variable z_t. Frequently, practitioners add a deterministic function (shift) $s(t)$ to the definition of r_t to address possible negative rates and be more flexible when calibrating the term-structure of the interest rates.

By the Itô's lemma the short rate $\bar{r}_t = (r_t - s(t))/R$ in the BK model solves the following stochastic differential equation (SDE)

$$d\bar{r}_t = [\kappa(t)\theta(t) + \frac{1}{2}\sigma(t)^2 - \kappa(t)\log \bar{r}_t]\bar{r}_t dt + \sigma(t)\bar{r}_t dW_t. \tag{2.32}$$

This SDE can be explicitly integrated. Let $0 \leq s \leq t \leq T$, with T being the maturity of a ZCB, and suppose $\kappa(t) = k = const$. Then r_t can be represented as (Brigo and Mercurio, 2006)

$$\bar{r}_t = \exp\left[e^{-k(t-s)}\log \bar{r}_s + k\int_s^t e^{-k(t-u)}\theta(u)du + \int_s^t \sigma(u)e^{-k(t-u)}dW_u\right], \tag{2.33}$$

and thus, conditionally on filtration $\mathcal{F}_s$ is lognormally distributed and always stays positive. The expectation $\mathbb{E}_{\mathbb{Q}}[r_t|\mathcal{F}_s]$ and variance $\mathrm{V}[r_t|\mathcal{F}_s]$ can be found analytically (Brigo and Mercurio, 2006)

$$\begin{aligned} \mathbb{E}_{\mathbb{Q}}[\bar{r}_t|\mathcal{F}_s] &= \exp\left[e^{-k(t-s)}\log \bar{r}_s + k\int_s^t e^{-k(t-u)}\theta(u)du \right. \\ &\quad \left. + \frac{1}{2}\int_s^t \sigma^2(u)e^{-2k(t-u)}du\right], \\ \mathrm{V}[\bar{r}_t|\mathcal{F}_s] &= \exp\left[2e^{-k(t-s)}\log \bar{r}_s + 2k\int_s^t e^{-k(t-u)}\theta(u)du\right] \\ &\quad \times \left[e^{2\mathcal{I}(s,t)} - e^{\mathcal{I}(s,t)}\right], \\ \mathcal{I}(s,t) &= \int_s^t \sigma^2(u)e^{-2k(t-u)}du. \end{aligned} \tag{2.34}$$

Despite as follows from Eq. (2.31), the short-interest rate r_t in this model is positive, if necessary, it can be made negative by using a deterministic shift $s(t)$. Initially, this positivity was one of the significant advantages of the BK model. However, in the current environment, this feature seems to be less useful. Another problem with the model is in the lognormality of r_t. Indeed, the lognormality means the CDF of the distribution is right-skewed. Therefore, the time a typical path stays in the lower rate area is short, because the short-rate quickly moves to the mean-reversion level. Accordingly, we need to choose a low mean-reversion speed to rectify this behavior, but the qualitative behavior of r_t remains the same. In contrast, for the current market conditions we would prefer an interest rate model with fat tails at the lower end.

In addition to the structural drawbacks, the BK model is not sufficiently tractable, especially when its coefficients are time-dependent. For instance, prices of zero-coupon bonds (ZCB) are not known in closed form. It is known that, written in terms of variable $z = \log(\bar{r})$, the corresponding PDE for the ZCB price $F(t,z)$ reads (Andersen and Piterbarg, 2010)

$$0 = \frac{\partial F}{\partial t} + \frac{1}{2}\sigma^2(t)\frac{\partial^2 F}{\partial z^2} + \kappa(t)[\theta(t) - z]\frac{\partial F}{\partial z} - [s(t) + Re^z]F, \tag{2.35}$$

This equation should be solved subject to the terminal and boundary conditions (Andersen and Piterbarg, 2010) (see also discussion in Carr and Itkin (2021))

$$F(T,z) = 1, \qquad F(t,z)\Big|_{z\to\infty} = 0. \tag{2.36}$$

This problem is not affine, and, therefore, the analytical solution is not know yet. A similar problem should be solved when pricing liquid barrier options written on the underlying which follows the BK model. Multiple good approximations have been developed in the literature using asymptotic expansions of various flavors; see, e.g., Antonov and Spector (2011); Capriotti and Stehlikova (2014); Horvath *et al.* (2017), and also survey in Turfus (2020). Alternatively, we have to find these prices numerically by solving the corresponding partial differential equations (PDEs), see Eq. (14.2), either via finite differences or asymptotically.

2.3.4 *The modified Black-Karasinski (Verhulst) model*

The Global Economic Crisis (GEC) of 2008-2010 caused unprecedented changes in the way central banks in general, and the mighty Fed in particular, conduct their business. The Quantitative Easing (QE) resulted in central banks embracing the fractional reserve *modus operandi.* At the same time, commercial banks switched to the narrow bank model, partly by choice and partly by necessity. The Federal Reserve has used short-term interest rates as the policy tool for achieving its macroeconomic goals. As a result, short rates were close to zero for much of the past decade, reflecting the effects of QE, low inflation caused by an aging population, and low productivity growth; see Rudebusch (2018). The current economic recession due to the COVID-19 pandemic forces the Fed to push the short interest rates into extremely low or outright negative territory. Given the unprecedented level of unemployment, the economic recession is likely to pave the way for further use of the Fed's unconventional monetary policy, resulting in meager short rates.

Typically, short-rate interest models use stochastic drivers governed by an Ornstein-Uhlenbeck (OU) process and transform these drivers into the actual rates via suitable mappings. For instance, the Vasicek-Hull-White model uses a linear mapping, while the Black-Karasinski (BK) uses an exponential mapping. The Cox-Ingersoll-Ross model is an exception, which uses a driver governed by a Feller process.

For short-rate models driven by OU processes, the rate spends an approximately equal amount of time below and above its equilibrium level. This assumption was valid for decades. However, as was mentioned earlier, it is no longer adequate due to the nontraditional interventions of central banks. Once the rate becomes low, it tends to stay low for a very long time. Under these circumstances, we have to revisit short-rate interest models and modify them to reflect prevailing current market conditions better.

In Itkin *et al.* (2020) we proposed a modification of the BK model, which organically resolves the lower end fat tail issue, and improves the model tractability. This model is shortly discussed below and also in more detail — in Chapter 14. There we also present an integral equation for the ZCB price for our model and provide a closed-form solution of the model PDE in a particular case. We demonstrate that this solution accelerates the computation of the ZCB prices and provides a basis for efficient calibration.

Since the BK model doesn't support fat tails at the lower end, besides not being analytically tractable, we introduce its modified version of the form

$$
\begin{aligned}
dz_t &= \kappa(t)[\bar{\theta}(t) - e^{z_t}]dt + \sigma(t)dW_t, \qquad (2.37)\\
r_t &= s(t) + Re^{z_t}, \quad R = r_0 - s(0), \quad (t, z_t) \in [0, T] \times (-\infty, \infty).
\end{aligned}
$$

In other words, we modify the dynamics of the stochastic variable z_t in Eq. (2.31) in the mean-reversion term by replacing z_t with e^{z_t}.

In Eq. (2.37) r_t is the short interest rate, t is the time, W_t is the standard Brownian motion, $\kappa(t) > 0$ is the speed of mean-reversion, $\bar{\theta}(t)$ is the mean-reversion level, $\sigma(t)$ is the volatility, R is some constant with the same dimensionality as r_t, e.g., it can be $R = r(0) - s(0)$, T is the maturity. This model is similar to the Hull-White model, but preserves positivity of r_t by exponentiating the OU random variable z_t. Because of that, usually practitioners add a deterministic function (shift) $s(t)$ to the definition of r_t to address possible negative rates and be more flexible when calibrating the term-structure of the interest rates.

It can be seen, that at small t $|z_t| \ll 1$, and so choosing $\bar{\theta}(t) = 1 + \theta(t)$ replicates the BK model in the linear approximation on z_t. Similarly, the choice $\bar{\theta}(t) = e^{\theta(t)}$ replicates the BK model at z_t close the mean-reversion level $\theta(t)$. Thus, the modified BK model acquires the properties of the BK model while is a bit more tractable as this will be seen below.

By Itô's lemma and the Feynman–Kac formula any contingent claim written on the r_t as the underlying (for instance, the price $F(\bar{r}, t, T)$ of a Zero-coupon bond (ZCB) with maturity T) solves the following partial differential equation

$$
\begin{aligned}
0 &= \frac{\partial F}{\partial t} + \frac{1}{2}\sigma^2(t)\bar{r}^2 \frac{\partial^2 F}{\partial \bar{r}^2} + \theta(t)\bar{r}[\tilde{\theta}(t) - \bar{r}]\frac{\partial F}{\partial \bar{r}} - (s(t) + R\bar{r})F, \qquad (2.38)\\
\bar{r}_t &= \frac{r_t - s(t)}{r_0 - s(0)} = e^{z_t}, \quad \tilde{\theta}(t) = \bar{\theta}(t) + \frac{\sigma^2(t)}{2\kappa(t)}.
\end{aligned}
$$

This equation should be solved subject to the same terminal and boundary conditions as in Eq. (2.36)

$$
F(T, \bar{r}) = 1, \qquad F(t, \bar{r})\Big|_{\bar{r}\to\infty} = 0. \qquad (2.39)
$$

Note that since $\bar{r} \in (0, \infty)$, i.e., the boundary $\bar{r} = 0$ is not attainable, Eq. (2.38) doesn't need the boundary condition at the left boundary $\bar{r} \to 0$, as this is discussed in Itkin and Muravey (2020b) with a reference to Fichera theory (in other words, the PDE itself with substituted $\bar{r} = 0$ serves as the boundary condition). If, however, the boundary $\bar{r} = 0$ is attainable, the boundary condition at this point is necessary, see, e.g., an example in Eq. (14.35). This is also applicable to all the below PDEs obtained from Eq. (2.38) by transformations.

It is worth noting that Eq. (2.38) is the stochastic Verhulst or stochastic logistic model, which are well-known in the population dynamics and epidemiology; see, e.g., Verhulst (1838), Bacaer (2011), Giet *et al.* (2015) and references therein. In the past, several authors attempted to use this model in finance; see, e.g., Chen (2010); Londono and Sandoval (2015); Halperin and Feldshteyn (2018). In our case, the stochastic Verhulst equation has the form

$$\begin{aligned} d\bar{r}_t &= k(t)\bar{r}_t[\tilde{\theta}(t) - \bar{r}_t]dt + \sigma(t)\bar{r}_t dW_t, \\ \bar{r}_t &= [r_t - s(t)]/R, \qquad r(t=0) = r_0. \end{aligned} \tag{2.40}$$

The Eq. (2.40) can be explicitly solved (for the time-homogeneous coefficients this is done, e.g., in Giet *et al.* (2015), Proposition 3.3). The following Proposition holds.

Proposition 2.1. *The Eq. (2.40) admits a unique positive solution* $\bar{r}_t$

$$\bar{r}_t = \frac{1}{\kappa(t)} \frac{X_t}{\frac{T_0}{\kappa(0)\bar{r}_0} + \int_0^t X_q dq}, \qquad t \geq 0, \tag{2.41}$$

where X_t *solves the lognormal SDE*

$$dX_t = \theta(t)X_t dt + \sigma(t)X_t dW_t, \qquad \theta(t) = \frac{\kappa'(t)}{\kappa(t)} + \kappa(t)\tilde{\theta}(t). \tag{2.42}$$

Also,

(1) *The diffusion* $\bar{r}_t$ *is recurrent if and only if* $q(t) \leq 0$, *where* $q(t) = \frac{1}{2} - \frac{\kappa(t)\tilde{\theta}(t)}{\sigma^2(t)}$, $\forall t \in [0, T]$.
(2) *If* $q(t) < 0$, $\forall t \in [0, T]$, *assuming that the limits of* $q(t)$ *and* $\sigma^2(t)/(2\kappa(t)$ *exist at* $t \to \infty$, *the diffusion* $\bar{r}_t$ *converges in law towards the unique stationary Gamma probability distribution* $\Gamma\left(-2q(t), \sigma^2(t)/(2\kappa(t))\right)_{t\to\infty}$.
(3) *If* $\exists t_* \geq 0 : q(t_*) > 0$, *the diffusion goes a.s. to zero when time goes to infinity.*

Proof. The proof can be obtained by applying Itô's lemma to Eq. (2.41) and using Eq. (2.42). The second part follows from Proposition 3.3 in Giet *et al.* (2015). It is interesting to note, that the condition $q < 0$ is precisely the Feller condition for the famous CIR model (Andersen and Piterbarg, 2010). □

Thus, the stationary distribution for the Verhulst model is the Gamma distribution. It is easy to check that as compared with the mean-reversion lognormal model (the BK model) with the same parameters, the former has much fatter tails at the lower end, while the latter has the fatter tails when $\bar{r}_t \to \infty$. However, since, under the current market conditions, we are

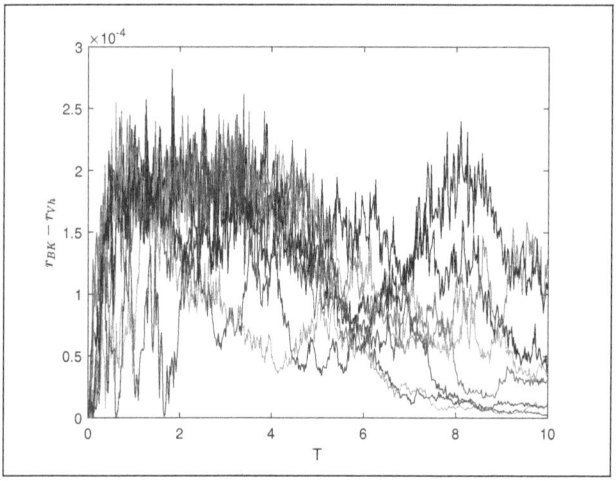

Figure 2.1: Sample paths of $r_{BK} - r_{Vh}$ as a function of time, r_{BK} is the interest rate for the Black-Karasinski model and r_{Vh} — for the Verhulst model.

interested in modeling the lower end in the first place, the Verhulst model has a distinct advantage compared with the BK model. In other words, the probability of having lower rates for the Verhulst model is much higher than for the BK model, and comes naturally.

To illustrate this in a slightly different way, we produce a set of Monte-Carlo paths for both models which have the same volatility and mean-reversion rate, while the mean-reversion level $\bar{\kappa}(t)$ in Eq. (2.37) is chosen as $\bar{\kappa}(t) = 1 + \kappa(t)$, so the dynamics Eq. (2.37) corresponds to the BK dynamics in Eq. (2.31) for small z_t. The results obtained by using parameters given in Section 14.4 are presented in Fig. 2.1, which shows that r_{BK} is always higher than r_{Vh}, which confirms our theoretical observation in above.

2.4 Credit

Merton developed the original version of the so-called structural default model in his celebrated paper (Merton, 1974). Merton assumed that the firm's value V is driven by a log-normal diffusion and that the firm, which borrowed a zero-coupon bond with face value N and maturity T, defaults at time T when the value of the firm V is less than the bond's face value N. Following this pioneering insight (Black and Cox, 1976b) and many others proposed various extensions of the basic model; see Bielecki and Rutkowski (2013); Lipton and Sepp (2011a) and references therein. The main idea is to consider more complicated forms of debt and assume that the default event may be triggered continuously up to the debt maturity, rather than at the moment when debt matures.

Structural default framework is widely used for assessing credit risk of a corporate debt. In contrast to reduced-form models (see, e.g., Bielecki and Rutkowski, 2013) structural default models suffer from the curse of dimensionality when the number of counterparties grows; however, these models provide a more detailed and financially meaningful description of the default event for a typical firm.

Simple structural credit models assume that the value of a firm's assets V are driven by the geometric Brownian motion

$$dV_t = \mu V_t dt + \sigma V_t dW_t, \quad V\Big|_{t=0} = v. \tag{2.43}$$

The firm defaults prior to some given time T provided that V_τ touches the default boundary $b(\tau)$ at $\tau < T$. The choice of the default boundary is a topic of debate. For instance, Merton assumes that

$$b(\tau) = 0, \quad 0 < \tau < T, \quad b(T) = N, \tag{2.44}$$

where N is the notional of the firm's debt. Thus, in Merton's framework, default is only possible at the debt's maturity. On the other hand, Black-Cox assume that

$$b(\tau) = RN, \quad 0 < \tau \le T, \tag{2.45}$$

where R is the recovery rate. Hence, in the Black-Cox framework, default is possible prior to the debt's maturity. The truth is probably in between. We return to the determination of the boundary in Chapter 15.

More general and realistic SDEs, often incorporating jumps, are used on occasion (Lipton and Sepp, 2011a; Itkin and Lipton, 2015). For instance, in Itkin and Lipton (2015) the structural default model of Lipton and Sepp (2011a) is generalized for a set of banks with mutual interbank liabilities whose assets are driven by correlated Lévy processes with idiosyncratic and common components. This is done since it is well-known that pure diffusion asset dynamics is manifestly inadequate for relatively short time-scales, and we need to introduce jumps into the model, see, e.g., Zhou (2001) and Lipton (2002). Therefore, for the asset dynamics we choose a Lévy driver and which contains also a diffusion component.

However, despite credit models with jumps are of a great interest, for the purpose of this book the model in Eq. (2.43) is sufficient.

Below, we discuss the calibration of the default boundary in a structural default model to a constant default intensity; see Chapter 15. Also, we discuss the structural default model, which is generalized for a set of banks with interbank mutual liabilities whose assets are driven by diffusions; see Chapter 16. Here the point is that no bank is an island — they operate as a group. Tangible links between banks manifest themselves via interbank loans; intangible links are manifold — overall sentiment, ease of doing business, and others. Hence, to build a meaningful structural default model for a bank, one needs to consider this bank's interactions with all the banks it lends to or borrows from. Eisenberg and Noe developed a Merton-like model of the bank default (default can happen only at maturity) in the seminal paper (Eisenberg and Noe, 2001). The Eisenberg-Noe model can be extended to the Black-Cox setting (default can happen at any time before maturity provided that debt covenants are violated); see Lipton (2016). Lipton's work was subsequently generalized in Itkin and Lipton (2015, 2017). Recently, several authors considered the interconnected banking system in the mean-field framework and studied a representative bank; see Hambly *et al.* (2018); Ichiba *et al.* (2018); Kaushansky and Reisinger (2019); Nadtochiy and Shkolnikov (2017, 2018) among many others.

2.5 Analytical solution of financial PDEs

2.5.1 *General considerations*

Analytical solution of the PDEs of financial engineering derived in the previous sections is a challenging task. Problems which can be solved analytically are not only interesting in their own right but also serve as useful benchmarks for other problems which can only be solved numerically. In this section we discuss the reduction method which is repeatedly used below.

Consider a process which is governed by the SDE

$$dS_t = b\left(t, S_t\right) dt + \sigma\left(t, S_t\right) dW_t, \quad S_0 = s, \tag{2.46}$$

and is killed at the rate $c\left(t, S_t\right)$. The corresponding backward Kolmogorov equation with a killing term reads

$$V_t\left(t, S\right) + \frac{1}{2} a\left(t, S\right) V_{SS}\left(t, S\right) + b\left(t, S\right) V_S\left(t, S\right) - c\left(t, S\right) V\left(t, S\right) = 0, \tag{2.47}$$

where $a\left(t, S\right) = \sigma^2(t, S)$. The main idea of the reduction method is to transform this equation into the simplest possible form, preferably, into the standard backward heat equation, or, more generally, to the Bessel equation, by using changes of independent and dependent variables. In probabilistic terms it means that we transform a given killed stochastic process into a known stochastic process. We emphasize that for pricing vanilla options, this method often does the trick. For example, it can be used to derive the Black-Sholes-Merton and Bachelier formulas. However, for exotic options, such as barrier options, the method results in the backward heat or Bessel equations in a time-dependent strip. To solve the corresponding problems, we have to use either the method of heat potentials, or the method of generalized integral transforms described below.

The reduction method is due to Carr *et al.* (2018). We also describe the so-called Cherkasov's method, which guarantees that, under certain conditions, a stochastic process can be transformed into the standard Wiener process. The Cherkasov's method is a particular case of the more general one, since it assumes that there is no killing term, $c(t, S_t) = 0$.

2.5.2 *Reducibility to the Wiener process*

First, we derive conditions on the coefficients of the backward parabolic equation which guarantee that it is reducible to the standard backward heat equation

$$U_\vartheta(\vartheta, X) + \frac{1}{2} U_{XX}(\vartheta, X) = 0,$$

by virtue of the following transformations

$$U(\vartheta, X) = e^{-\mathtt{A}(t,S)} V(t, S), \quad \vartheta = \mathtt{B}(t), \quad X = \mathtt{C}(t, S). \tag{2.48}$$

The change of dependent variable yields the equation for $U(t, S)$,

$$\begin{aligned}
&U_t + \frac{1}{2} a U_{SS} + \tilde{b} U_S - \tilde{c} U = 0, \\
&\tilde{b} = a\mathtt{A}_S + b, \\
&\tilde{c} = -\left(\mathtt{A}_t + \frac{1}{2} a\left(\mathtt{A}_{SS} + \mathtt{A}_S^2\right) + b\mathtt{A}_S\right) + c.
\end{aligned}$$

Here we use the relations

$$\begin{aligned}
\partial_S e^{\mathtt{A}} &= e^{\mathtt{A}}\left(\partial_S + \mathtt{A}_S\right), \\
\partial_S^2 e^{\mathtt{A}} &= e^{\mathtt{A}}\left(\partial_S^2 + 2\mathtt{A}_S\partial_S + \mathtt{A}_{SS} + \mathtt{A}_S^2\right), \\
\partial_t e^{\mathtt{A}} &= e^{\mathtt{A}}\left(\partial_t + \mathtt{A}_t\right).
\end{aligned}$$

Also, using the identities

$$\partial_S = \mathtt{C}_S\partial_X, \quad \partial_S^2 = \mathtt{C}_S^2\partial_X^2 + \mathtt{C}_{SS}\partial_X, \quad \partial_t = \mathtt{B}_t\partial_\vartheta + \mathtt{C}_t\partial_X,$$

the change of independent variables yields the equation for $U(\vartheta, X)$

$$U_\vartheta + \frac{1}{2}\frac{a\mathtt{C}_S^2}{\mathtt{B}_t}U_{XX} + \frac{\mathtt{C}_t + a\mathtt{C}_{SS}/2 + \tilde{b}\mathtt{C}_S}{\mathtt{B}_t}U_X - \frac{\tilde{c}}{\mathtt{B}_t}U = 0.$$

It is important to emphasize that the derivatives of $\mathtt{B}, \mathtt{C}$ are taken with respect to the original variables t, S rather than the modified variables ϑ, X. In order to obtain the heat equation we have to impose the following reducibility conditions

$$\begin{aligned} &\frac{a\mathtt{C}_S^2}{\mathtt{B}_t} = 1, \\ &\mathtt{C}_t + \frac{1}{2}a\mathtt{C}_{SS} + \left(a\mathtt{A}_S + b\right)\mathtt{C}_S = 0, \\ &\mathtt{A}_t + \frac{1}{2}a\left(\mathtt{A}_{SS} + \mathtt{A}_S^2\right) + b\mathtt{A}_S - c = 0. \end{aligned} \tag{2.49}$$

By using the first condition we can rewrite the other two as follows

$$\begin{aligned} &\mathtt{C}_t + \frac{1}{2}a\mathtt{C}_{SS} + b\mathtt{C}_S + P\sigma\mathtt{A}_S = 0, \\ &\mathtt{A}_t + \frac{1}{2}a\mathtt{A}_{SS} + b\mathtt{A}_S - c + \frac{1}{2}a\mathtt{A}_S^2 = 0, \end{aligned}$$

where $\mathtt{P} = \sqrt{\mathtt{B}_t}$. The first reducibility condition yields

$$\begin{aligned} &\mathtt{C}\left(t, S\right) = \mathtt{P}\left(t\right) I\left(t, S\right) + \mathtt{Q}\left(t\right), \\ &I\left(t, S\right) = \int \frac{d\zeta}{\sigma\left(t, \zeta\right)}. \end{aligned} \tag{2.50}$$

We introduce $\Omega = -\mathtt{A}_S$, differentiate the last line in Eq. (2.49) with respect to S and obtain

$$\Omega_t + \left(\frac{1}{2}a\left(\Omega_S - \Omega^2\right) + b\Omega + c\right)_S = 0. \tag{2.51}$$

By using Eqs. (2.49) and (2.50) we get

$$\Omega = \frac{\left(\mathtt{P}I + \mathtt{Q}\right)_t + \mathtt{P}\left(\frac{1}{2}aI_{SS} + bI_S\right)}{\mathtt{P}\sigma} = \frac{\mathtt{P}_t I}{\mathtt{P}\sigma} + \frac{I_t}{\sigma} + \frac{\mathtt{Q}_t}{\mathtt{P}\sigma} - \frac{\sigma_S}{2\sigma} + \frac{b}{\sigma^2},$$

thus obtaining the reducibility condition written in terms of the coefficients $a = \sigma^2, b, c$ and two arbitrary functions of time $\mathtt{P}, \mathtt{Q}$. We emphasize that the condition Eq. (2.51) is constructive since it not only guarantees that the reduction is possible, but also shows how it can be achieved.

To illustrate the efficiency of the condition Eq. (2.51), we show how to use it to prove the well-known Bluman theorem (Bluman and Cole, 1974). This theorem deals with the reducibility of the killed Wiener process to the standard Wiener process, or, to put it differently, the reducibility of the equation

$$V_t + \frac{1}{2}V_{SS} - cV = 0,$$

to the standard backward heat equation. In the case in question we have $a = 1, b = 0$, so that

$$\vartheta = \mathtt{B}(t) = \int \mathtt{P}^2(t)\,dt,$$

$$X = \mathtt{C}(t,S) = \mathtt{P}(t)\,S + \mathtt{Q}(t),$$

$$\Omega = \frac{\mathtt{C}_t + \frac{1}{2}\mathtt{C}_{SS}}{\mathtt{P}} = \frac{\mathtt{P}_t}{\mathtt{P}}S + \frac{\mathtt{Q}_t}{\mathtt{P}}.$$

Accordingly, the reducibility condition can be written in the form

$$\left(\left(\frac{\mathtt{P}_t}{\mathtt{P}}\right)_t - \frac{\mathtt{P}_t^2}{\mathtt{P}^2}\right)S + \left(\left(\frac{\mathtt{Q}_t}{\mathtt{P}}\right)_t - \frac{\mathtt{P}_t\mathtt{Q}_t}{\mathtt{P}^2}\right) + c_S = 0. \tag{2.52}$$

For solvability of this equation, it is necessary and sufficient that

$$\begin{aligned} c &= -\frac{1}{2}\left(\left(\frac{\mathtt{P}_t}{\mathtt{P}}\right)_t - \frac{\mathtt{P}_t^2}{\mathtt{P}^2}\right)S^2 - \left(\left(\frac{\mathtt{Q}_t}{\mathtt{P}}\right)_t - \frac{\mathtt{P}_t\mathtt{Q}_t}{\mathtt{P}^2}\right)S - R \\ &= \frac{1}{2}\mathtt{P}\left(\frac{1}{\mathtt{P}}\right)_{tt}S^2 - \mathtt{P}\left(\frac{\mathtt{Q}_t}{\mathtt{P}^2}\right)_t S + R, \end{aligned}$$

where R is an arbitrary function of time (on a par with $\mathtt{P}, \mathtt{Q}$). Thus, a Wiener process killed at the rate

$$c = \frac{1}{2}\alpha(t)\,S^2 + \beta(t)\,S + \gamma(t),$$

is equivalent to the standard Wiener process. To carry out the corresponding reduction for a given c we have to find $\mathtt{P}, \mathtt{Q}$. For a given quadratic c, Eq. (2.52) yields the following linear equations for $1/\mathtt{P}$ and $\mathtt{Q}_t$

$$\left(\frac{1}{\mathtt{P}}\right)_{tt} - \frac{\alpha}{\mathtt{P}} = 0, \quad \left(\frac{\mathtt{Q}_t}{\mathtt{P}^2}\right)_t + \frac{\beta}{\mathtt{P}} = 0.$$

Bluman's theorem has interesting applications in the fixed income context which will be described elsewhere.

2.5.3 *Reducibility to the Bessel process*

We now study conditions under which a diffusion can be reduced to a Bessel process of order ν, or, equivalently, equation Eq. (2.47) can be reduced to the backward parabolic equation of the form

$$U_\vartheta(X,\vartheta) + \frac{1}{2}U_{XX}(X,\vartheta) + \frac{1-2\nu}{2X}U_X(X,\vartheta) = 0.$$

We note in passing that this equation describes the propagation of heat from a spherically symmetric source in the $(2-2\nu)$-dimensional space. Conditions guaranteeing that a diffusion is reducible to the Bessel process via transformations Eq. (2.48) have the form

$$\frac{a\mathtt{C}_S^2}{\mathtt{B}_t} = 1,$$

$$\frac{\mathtt{C}_t + \frac{1}{2}a\mathtt{C}_{SS} + (a\mathtt{A}_S + b)\,\mathtt{C}_S}{\mathtt{B}_t} = \frac{1-2\nu}{2\mathtt{C}},$$

$$\mathtt{A}_t + \frac{1}{2}a\left(\mathtt{A}_{SS} + \mathtt{A}_S^2\right) + b\mathtt{A}_S - c = 0.$$

By using the first condition, we rewrite the second and the third conditions as

$$\mathtt{C}_t + \frac{1}{2}a\mathtt{C}_{SS} + b\mathtt{C}_S + \mathtt{P}\sigma\mathtt{A}_S = \frac{(1-2\nu)\,\mathtt{P}^2}{2\mathtt{C}},$$

$$\mathtt{A}_t + \frac{1}{2}a\mathtt{A}_{SS} + b\mathtt{A}_S - c + \frac{1}{2}a\mathtt{A}_S^2 = 0.$$

As before $\mathtt{C}$ has the form Eq. (2.50), while the reducibility condition has the form Eq. (2.51). However, the expression for Ω is different,

$$\begin{aligned}\Omega &= \frac{(\mathtt{P}I + \mathtt{Q})_t + \mathtt{P}\left(\frac{1}{2}aI_{SS} + bI_S\right)}{\mathtt{P}\sigma} - \frac{(1-2\nu)\,\mathtt{P}}{2\,(\mathtt{P}I + \mathtt{Q})\,\sigma} \\ &= \frac{\mathtt{P}_t I}{\mathtt{P}\sigma} + \frac{I_t}{\sigma} + \frac{\mathtt{Q}_t}{\mathtt{P}\sigma} - \frac{\sigma_S}{2\sigma} + \frac{b}{\sigma^2} - \frac{(1-2\nu)\,\mathtt{P}}{2\,(\mathtt{P}I + \mathtt{Q})\,\sigma}.\end{aligned}$$

2.5.4 *Cherkasov's transformation*

Let us consider a general diffusion process X_t governed by Eq. (2.46). In this section we assume that killing rate is zero. We want to establish certain conditions such that this process can be transformed to the standard Wiener process. These conditions have been studied in Cherkasov (1957); Ricciardi (1976); Bluman (1980) and are called Cherkasov's conditions. Applications of these conditions in financial mathematics were studied in Lipton (2001) and Lipton (2018) (see Chapter 8).

Let us introduce new functions

$$\beta(t,S) = \sigma(t,S)\int^x \frac{1}{\sigma(t,\zeta)}d\zeta,$$

$$\gamma(t,S) = 2b(t,S) - \sigma(t,S)\,\sigma_S(t,S) - 2\sigma(t,S)\int^S \frac{\sigma_t(t,\zeta)}{\sigma^2(t,\zeta)}d\zeta,$$

where the lower limit of the integral is chosen as convenient. Let

$$P(t,S) = \begin{vmatrix} \beta(t,S) & \gamma(t,S) \\ \beta_S(t,S) & \gamma_S(t,S) \end{vmatrix}, \quad Q(t,S) = \begin{vmatrix} \sigma(t,S) & \gamma(t,S) \\ \sigma_S(t,S) & \gamma_S(t,S) \end{vmatrix},$$

$$R(t,S) = \begin{vmatrix} \sigma(t,S) & \beta(t,S) & \gamma(t,S) \\ \sigma_S(t,S) & \beta_S(t,S) & \gamma_S(t,S) \\ \sigma_{SS}(t,S) & \beta_{SS}(t,S) & \gamma_{SS}(t,S) \end{vmatrix}.$$

The Cherkasov condition stays that if

$$R(t,S) \equiv 0,$$

then we can transform X_t into the standard Wiener process via the following mapping

$$\vartheta = \vartheta(t,S) = \int_0^t \Phi^2(u,S)du, \tag{2.53}$$

$$X = X(t,S) = \Phi(t,S)\frac{\beta(t,S)}{\sigma(t,S)} + \frac{1}{2}\int_0^t \Phi(u,S)\frac{P(u,S)}{\sigma(u,S)}du,$$

where

$$\Phi(t,S) = \exp\left[-\frac{1}{2}\int_0^t \frac{Q(u,S)}{\sigma(u,S)}du\right]. \tag{2.54}$$

The corresponding transition probability density transforms as follows

$$p(t,S;s) = \left|\frac{\partial X(t,S)}{\partial S}\right| \tilde{p}(\vartheta, X; x), \tag{2.55}$$

$$x = \frac{\beta(0,s)}{\sigma(0,s)}. \tag{2.56}$$

Moreover, the boundary $b(t)$ transforms to

$$b(t) \to B(\vartheta),$$

where

$$B(\vartheta) = \Phi(t,b(t))\frac{\beta(t,b(t))}{\sigma(t,b(t))} + \frac{1}{2}\int^t \Phi(u,b(t))\frac{P(u,b(t))}{\sigma(u,b(t))}du. \tag{2.57}$$

Since the method of HP is specifically designed for dealing with curvilinear boundaries, we get a solvable problem. A powerful application of the above approach is demonstrated in Chapter 10 where the hitting time probability distribution for an Ornstein-Uhlenbeck process is studied.

PART 2

Integral Equations

Chapter 3

Fredholm Integral Equations

Throughout this book for solving various problems of mathematical finance and physics with moving boundaries we use a powerful technique that briefly can be described as follows. We start by assuming that by using the Feynman-Kac theorem (Karatzas and Shreve, 1991), the original problem can be formulated in terms of a parabolic PDE with some initial/terminal condition and boundary conditions at the moving boundaries. Using an appropriate integral transform we solve this PDE and obtain the solution in a semi-analytical form. Usually, this is an integral with its kernel being proportional to some function $\Psi(t)$ which is the gradient of the solution at the moving boundary (in the one-dimensional case). This function can be found by solving some integral equation which also can be explicitly obtained for each particular problem. This technique could also be generalized for a multi-dimensional case, e.g., see Part 7.

To make the content of this book self-consistent, in this part we provide some basic facts about integral equations (Chapters 3 and 4), and also about various numerical method to solve them in Chapter 5. Readers who are willing to get additional knowledge of integral equations could look, e.g., at Kanwal (1997), Polyanin and Manzhirov (2008), Wazwaz (2011) and Brunner (2017) among others.

Below in this chapter we consider the Fredholm equations.

3.1 Basic definitions

The most general type of linear integral equation is of the form (Kanwal, 1997)

$$h(s)g(s) = f(s) + \lambda \int_a K(s,t)g(t)dt, \tag{3.1}$$

where the upper limit may be either variable or fixed. The functions f, h and K are known functions, while g is to be determined; λ, a are real or complex parameters, and $\lambda \neq 0$. The function $K(s,t)$ is called the kernel.

The Eq. (3.1) is called the Fredholm integral equation if the upper limit of integration b, is fixed. Then a more granular classification can be provided which stays:

(1) If $b = const$ and $h(s) = 0$, Eq. (3.1) is called the Fredholm integral equation of the first kind, i.e.

$$0 = f(s) + \lambda \int_a^b K(s,t)g(t)dt. \tag{3.2}$$

(2) If $b = const$ and $h(s) = 1$, Eq. (3.1) is called the Fredholm integral equation of the second kind, i.e.

$$g(s) = f(s) + \lambda \int_a^b K(s,t)g(t)dt. \tag{3.3}$$

(3) The homogeneous Fredholm integral equation of the second kind is a special case of Eq. (3.3). In this case $f(s) = 0$ and

$$g(s) = \lambda \int_a^b K(s,t)g(t)dt. \tag{3.4}$$

Throughout this chapter the kernel $K(s,t)$ is assumed measurable and square integrable on the domain $a \leq s,t \leq b$

$$\int_a^b \int_a^b |K(s,t)|^2 ds\, dt = k < \infty. \tag{3.5}$$

3.2 Fredholm equations with degenerate kernels

In this section, we study Eq. (3.3) where $\lambda = 1$ and the kernel $K(s,t)$ has the special form

$$K(s,t) = \sum_{i=1}^{n} P_i(s)Q_i(t), \quad n \geq 1. \tag{3.6}$$

Here P_i and Q_i are some functions that belong to the space of square integrable functions L^2. The kernel defined by Eq. (3.6) is called a degenerate or separable kernel. Thus, such a kernel is a function that can be expressed as the sum or product of two functions each depends only on one variable.

Without loss of generality, we assume linear independency of the functions P_i. Under standard regularity assumption, substitution of Eq. (3.6) into Eq. (3.3) yields

$$g(s) = \sum_{i=1}^{n} q_i P_i(s) + f(s), \quad q_i = \int_a^b Q_i(t) g(t) dt. \tag{3.7}$$

Plugging Eq. (3.7) into Eq. (3.3), we obtain the identity

$$\sum_{i=1}^{n} q_i P_i(s) + f(s) = \sum_{i=1}^{n} P_i(s) \int_a^b Q_i(t) \left[\sum_{j=1}^{n} q_j P_j(t) + f(t) \right] dt + f(s),$$

or, in an equivalent form

$$\sum_{i=1}^{n} q_i P_i(s) = \sum_{i=1}^{n} P_i(s) \left[\sum_{j=1}^{n} a_{ij} q_j + b_i \right],$$

where the coefficients a_{ij} and b_j are defined as

$$a_{ij} = \int_a^b Q_i(t) P_j(t) dt, \qquad b_i = \int_a^b Q_i(t) f(t) dt. \tag{3.8}$$

Since the functions P_i are linearly independent, Eq. (3.8) is equivalent to the system of linear equations

$$q_i = \sum_{j=1}^{n} a_{ij} q_j + b_i, \qquad i = 1, 2, \ldots, n. \tag{3.9}$$

Once it is solved, the explicit representation for $g(s)$ comes from Eq. (3.7). Therefore, the Fredholm integral equation of the second kind Eq. (3.3) with the degenerate kernel Eq. (3.6) is equivalent to the system of linear equations. The importance of this result is twofold: first, it provides a simple method of solving the integral equation in this case; also it gives the exact conditions of uniqueness and existence of the solution. The latter is due to the fact that for the systems of linear equations these conditions are well known, e.g. Strang (2006).

Uniqueness and existence conditions

(1) The system of linear equations

$$Tx = y, \quad T = \|a_{ki}\|, \quad x = (x_1, \ldots, x_n), \ y = (y_1, \ldots, y_n)$$

is solvable if and only if the vector y is orthogonal to the complete set of the solutions z of the adjoint system

$$T^* z = 0, \qquad T^* = \|\bar{a_{ki}}\|.$$

(2) If $\det T \neq 0$, the equation $Tx = y$ has a unique solution for any right-hand-side vector y. If $\det T = 0$ the homogeneous equation $Tx = 0$ has a non-zero solutions.
(3) Since the matrix T and its adjoint T^* have a common rank, the homogeneous systems $Tx = 0$ and $T^*z = 0$ have the same number of linearly independent solutions.

3.3 Fredholm equations with non-degenerate kernels

The existence and uniqueness conditions discussed in the previous section, can be generalized to the case of Fredholm integral equations of the second kind with non-degenerate kernels. Let us denote by A any compact operator in a Hilbert space $\mathcal{H}$, and by T — the operator $I - A$, where I is a identity operator. The uniqueness and existence conditions for the equation $Tg = f$ are called the Fredholm theorems (Khvedelidze, 2001).

Theorem 3.1 (Fredholm theorem). *Inhomogeneous equation $Tg = f$ is solvable if and only if the function f is orthogonal to the complete set of the solutions to the adjoint homogeneous equation $T^*h = 0$.*

Theorem 3.2 (Fredholm alternative). *Either the equation $Tg = f$ has a unique solution for any $f \in \mathcal{H}$, or the homogeneous equation $Th = 0$ has a non-zero solution.*

Theorem 3.3. *The homogeneous equations $Tg = 0$ and $T^*h = 0$ have the same (finite) number of linearly independent solutions.*

Proof. Proofs of all these theorems can be found in Khvedelidze (2001). □

Under the assumption Eq. (3.5) the kernel $K(s,t)$ becomes a compact operator, so the Fredholm theorems are valid for the equation Eq. (3.4).

3.4 The Fredholm determinant: Resolvent

Consider a Fredholm equation of the second kind

$$(I - \lambda A)\,\phi = f, \tag{3.10}$$

with the integral operator A defined as

$$A\phi = \int_a^b K(s,t)\phi(t)dt. \tag{3.11}$$

The solution of Eq. (3.10) can be found by a formal inversion

$$\phi = (I - \lambda A)^{-1} f. \tag{3.12}$$

It is known, that under the assumption $\|\lambda A\| < 1$ the inverse operator $(I - \lambda A)^{-1}$ is bounded and can be formally expanded into series

$$(I - \lambda A)^{-1} = I + \lambda A + \lambda^2 A^2 + \ldots + \lambda^n A^n + \ldots \tag{3.13}$$

The series Eq. (3.13) converges if $|\lambda| < 1/\|A\|$, where $\|A\|$ is some norm of the operator A. Then there exists the following representation of the solution ϕ

$$\phi = \sum_{k=0}^{\infty} \lambda^k \phi_k, \tag{3.14}$$

$$\phi_0 = f, \quad \phi_1 = Af, \ldots, \phi_n = A\phi_{n-1} = A^n f, \ldots, \tag{3.15}$$

where the operator A^n

$$A^n \phi = \int_a^b K_n(s,t)\phi(s)ds \tag{3.16}$$

is associated with the so called iterated kernels K_n

$$K_1 = K(s,t), \quad K_n = \int_a^b K_{n-1}(s,u)K(u,t)du. \tag{3.17}$$

Since the kernel $K(s,t)$ satisfies the condition Eq. (3.5), the following inequality holds

$$\int_a^b \int_a^b |K_n(s,t)|^2 dsdt \leq k^{2n}, \tag{3.18}$$

where k is defined in Eq. (3.5). Therefore, the series

$$K(s,t) + \lambda K_2(s,t) + \ldots + \lambda^{n-1} K_n(s,t) + \ldots \tag{3.19}$$

converges. Let us denote its limit by $\Gamma(s,t;\lambda)$, so

$$\lambda\Gamma(s,t;\lambda) = \sum_{n=1}^{\infty} \lambda^n K_n(s,t). \tag{3.20}$$

The integral operator $\lambda\Gamma(\lambda)$ corresponding to the kernel $\lambda\Gamma(s,t;\lambda)$ obviously obeys the equation

$$I + \lambda\Gamma(\lambda) = (I - \lambda A)^{-1}. \tag{3.21}$$

The condition $|\lambda| < 1/k$ is sufficient for convergence of Eq. (3.20), but is not necessary. For instance, the Volterra integral operators with the kernel $|K(s,t)| \leq M$ generate the series Eq. (3.20) which converges for any values of λ, see Chapter 4. It is known that Eq. (3.10) can also be solved for any λ such that $1/\lambda$ is not an eigenvalue of the operator A.

It was shown by Fredholm that the solution of Eq. (3.10) can be explicitly found if the corresponding kernel $K(s,t)$ is bounded and continuous. Similar to matrix operators, the inverse integral operator can be found in terms of the determinant and minors. This generalizations of the determinant and minors for integral operators are called the Fredholm determinant and the Fredholm minor, respectively, which are denoted by $D(\lambda)$ and $D(s,t;\lambda)$. Thus, the kernel $\Gamma(s,t;\lambda)$ allows the following representation

$$\Gamma(s,t;\lambda) = \lambda \frac{D(s,t;\lambda)}{D(\lambda)}, \tag{3.22}$$

where

$$\begin{aligned} D(\lambda) = 1 &- \lambda \int_a^b K\begin{pmatrix}\xi_1\\ \xi_1\end{pmatrix} d\xi_1 + \frac{\lambda^2}{2!}\int_a^b\int_a^b K\begin{pmatrix}\xi_1\ \xi_2\\ \xi_1\ \xi_2\end{pmatrix} d\xi_1 d\xi_2 + \dots \\ &+ (-1)^n \frac{\lambda^n}{n!}\int_a^b \cdots \int_a^b K\begin{pmatrix}\xi_1 \dots \xi_n\\ \xi_1 \dots \xi_n\end{pmatrix} d\xi_1 \dots d\xi_n + \dots, \end{aligned} \tag{3.23}$$

$$\begin{aligned} D(s,t;\lambda) = K\begin{pmatrix}s\\ t\end{pmatrix} &- \lambda \int_a^b K\begin{pmatrix}s\ \xi_1\\ t\ \xi_1\end{pmatrix} d\xi_1 \\ &+ \frac{\lambda^2}{2!}\int_a^b\int_a^b K\begin{pmatrix}s\ \xi_1\ \xi_2\\ t\ \xi_1\ \xi_2\end{pmatrix} d\xi_1 d\xi_2 + \dots \\ &+ (-1)^n \frac{\lambda^n}{n!}\int_a^b \cdots \int_a^b K\begin{pmatrix}s\ \xi_1 \dots \xi_n\\ t\ \xi_1 \dots \xi_n\end{pmatrix} d\xi_1 \dots d\xi_n + \dots, \end{aligned}$$

and the function K is defined as

$$K\begin{pmatrix}s_1 \dots s_n\\ t_1 \dots t_n\end{pmatrix} = \begin{vmatrix} K(s_1,t_1) & \dots & K(s_1,t_n)\\ \dots & \dots & \dots \\ K(s_n,t_1) & \dots & K(s_n,t_n)\end{vmatrix}. \tag{3.24}$$

3.5 Fredholm equations of the first kind

It is known, that solving the Fredholm equation of the first kind Eq. (3.2) is a much more complicated problem than that for the Fredholm equation of the second kind. There are several reasons for why this happens. First, for any bounded kernel satisfying Eq. (3.5), the solution of Eq. (3.2) doesn't exist for an arbitrary function $f \in \mathcal{H}$, see Theorem 3.4 below and also discussion after Eq. (3.45). Second, the inverse to any compact operator is an unbounded operator (Fomin and Kolmogorov, 1999), hence the inverse procedure described in Section 5.5 is unstable. In other words, small perturbations in f might lead to large variations of the solution g, so this problem is ill-posed unless constraints on the function spaces exist or be imposed. Therefore, solving the both linear and nonlinear Fredholm integral equations requires special techniques, which are discussed in Chapter 5.

To understand this in more detail, let us rewrite Eq. (3.2) in the form

$$f = Kg, \quad (K\circ)(s) = -\lambda \int_a^b K(s,t) \circ dt, \quad s \in [a,b], \tag{3.25}$$

so K is a linear integral operator. If the kernel $K(s,t)$ is continuous, smooth, and quadratic bounded, then K is a compact operator. Following Verges-Liahi (2005), let us also introduce a transposed operator K^*

$$(K^*\circ)(t) = -\lambda \int_a^b K(s,t) \circ ds, \quad t \in [a,b]. \tag{3.26}$$

It is similar to K, but the variable x has been interchanged with the variable y.

Applying Eq. (3.26) to Eq. (3.25) yields the equation $h = K^*f = K^*Kg$, where

$$(K^*K\circ)(z) = \int_a^b \left[\int_a^b K(s,z)K(s,t)dt \right] \circ ds, \quad z \in [a,b]. \tag{3.27}$$

This operator is symmetric, and hence, its eigenvalues are real and non-negative. An analogous operator KK^* also has same eigenvalues despite $KK^* \neq K^*K$ in general. Nevertheless, the eigenfunctions of the two operators differ in general.

Let us denote as λ_i the positive eigenvalues of K^*K, as v_i - the corresponding eigenfunctions, and as $s_i = \sqrt{\lambda_i}$ - the corresponding set of singular values. This implies

$$K^*Kv_i = s_i^2 v_i. \tag{3.28}$$

We can also define the function

$$u_i = \frac{1}{s_i} K v_i. \tag{3.29}$$

Applying the transposed operator K^* to u_i we obtain

$$K^* u_i = \frac{1}{s_i} K^* K v_i = s_i v_i. \tag{3.30}$$

The Eqs. (3.29) and (3.30) are dual. They also yield

$$KK^* u_i = s_i K v_i = s_i^2 u_i. \tag{3.31}$$

Thus, functions u_i defined in Eq. (3.29) are eigenfunctions of the operator KK^*. It can be shown that Eq. (3.31) generates all such eigenfunctions. On the other hand, one can do in the opposite direction, i.e. start with the eigenfunctions u_i of KK^*, reverse all the steps and show that the functions v_i defined by Eq. (3.31) are the eigenfunctions of K^*K.

To recall, two sets of functions, v_i and u_i have been associated with the operator K, which are eigenfunctions of K^*K and KK^*, respectively. The following expressions encompass their properties (Verges-Liahi, 2005)

$$
\begin{aligned}
K^*u_i &= s_i v_i, \quad KK^*u_i = s_i^2 u_i, \\
Kv_i &= s_i u_i, \quad K^*Kv_i = s_i^2 v_i.
\end{aligned}
\tag{3.32}
$$

Functions u_i and v_i are called *singular functions* of the kernel K, and values s_i are referred to as *singular values*. These functions are orthogonal and we can assume them to be also normalized. A special case of functions v such that $K^*Kv = 0$, is included into the singular functions with the corresponding singular value being zero. The same is true zero. The same is true for the functions u such that $KK^*u = 0$.

This also allows the following expansion for the kernel $K(s,t)$

$$
K(s,t) = \sum_{i=1}^{\infty} s_i u_i(s) v_i(t). \tag{3.33}
$$

Having that, we can represent the solution of Eq. (3.25) in the form of series over the function f. it is clear that if there exists a solution g such that $Kg = f$, then f must be expressed in terms of a series as follows:

$$
f = \sum_{i=1}^{\infty} s_i \left(g, v_i\right) u_i. \tag{3.34}
$$

We also need to expand the solution g into series on the singular functions v_i to get

$$
g = \sum_{i=1}^{\infty} g_i v_i \Rightarrow Kg = \sum_{i=1}^{\infty} g_i K v_i = f. \tag{3.35}
$$

This, in turn can be represented as

$$
Kg = \sum_{i=1}^{\infty} g_i s_i u_i = \sum_{i=1}^{\infty} \left(f, u_i\right) u_i = f. \tag{3.36}
$$

Identifying the coefficients of these two series, we can determine f_i

$$
f_i = (f, u_i) = g_i s_i \Rightarrow g_i = \frac{(f, u_i)}{s_i}. \tag{3.37}
$$

Using the above, finally we can find a formal solution of Eq. (3.25)

$$g = \sum_{i=1}^{\infty} \frac{(f, u_i)}{s_i} v_i. \tag{3.38}$$

Two important conclusion follows from the analysis of this solution.

(1) The space to which the solution g belongs, is not arbitrary, rather it is spanned by v_i functions. Therefore, in order for the solution to exist, both functions g and f can not be taken arbitrary. Moreover, this is the kernel $K(s,t)$ which determines the right spaces for g and f.

(2) It also clarifies why solving the Fredholm equation of the first kind is an ill-posed problem. The terms $g_i = (f, u_i)/s_i$ in Eq. (3.38) could diverge if there are infinitely many singular values s_i that tend to zero. To make this series convergent, the Picard theorem requires the kernel to be $K(s,t) \in L^2$.

Thus, we managed to demonstrate main problems that arise when solving the Fredholm integral equation of the first kind. However, in some limited cases it is possible to solve Eq. (3.2) analytically. An example of equation of the type Eq. (3.2) solvable in closed form is that with $a = 0$ and $b = \infty$ which is generated by any classical integral transform, see Chapter 6. Also, under some mild assumptions about the function f, it is also possible to explicitly solve Eq. (3.2) with finite a and b, see Chapter 7.

Moreover, in some cases the ill-posed equation Eq. (3.2) can be reduced to the well-posed one. To illustrate this, let us consider the following Fredholm equation of the first kind with regard to the unknown function $\Psi(\tau)$

$$\int_0^{\infty} e^{-p\tau - \sqrt{p} y(\tau)} \Psi(\tau) d\tau = -e^{-\sqrt{p} z_0} \tag{3.39}$$

Here the function $y(\tau) \in C^1$, $y_0 < z(0)$ is given as well as the constant z_0.

Suppose that $y(\tau) = \alpha\tau + \beta$, with α, β being constants. Then the integral operator in Eq. (3.39) becomes the Laplace transform Eq. (6.18) with the spectral parameter $\lambda = p + \alpha\sqrt{p}$. As shown in Chapter 6, this transform can be inverted analytically. For an arbitrary dependence of $y(\tau)$ the analysis of Eq. (3.39) can be found in Chapter 12.

Also, in some cases the Fredholm equations of the first kind can be transformed to the Volterra integral equations of the second kind which are discussed in Chapter 4. Solving the latter is a well-posed problem as described in Chapter 5. Other examples of the reduction to the Volterra equations of the second kind are presented in Chapters 9 and 13.

3.6 Nonlinear Fredholm equations

The nonlinear Fredholm integral equations of the second kind are characterized by fixed limits of integration of the form

$$u(t) = f(t) + \lambda \int_a^b K(t,s)F(u(s))ds, \tag{3.40}$$

where $F(u(s))$ is some nonlinear real-valued function, the unknown real-valued function $u(t)$ occurs inside and outside the integral sign, $\lambda \in \mathbb{R}$ is a parameter, and $a, b \in \mathbb{R}$ are constants. Here we always assume that the kernel $K(x,t)$ and the function $f(x)$ are given real-valued functions.

Several analytic and numerical methods have been used to handle the nonlinear Fredholm integral equations. A short exposition of the corresponding numerical methods is given in Chapter 5, while here we are concentrated on analytical and semi-analytical approaches.

For nonlinear Fredholm equation there exist an existence theorem similar to that for the linear counterparts.

Theorem 3.4. *Let us rewrite Eq. (3.40) in the form*

$$u(t) = f(t) + \lambda \int_a^b G(t,s,u(s))ds. \tag{3.41}$$

Solution of Eq. (3.41) exists if the following conditions are satisfied

- *The function $f(t)$ is bounded, $|f(t)| < R$ in $a \leqslant t \leqslant b$.*
- *The function $G(t,s,u(s))$ is integrable and bounded, i.e. $|G(t,s,u(s))| < K$ $\forall a \leqslant t, s \leqslant b$.*
- *The function $G(t,s,u(s))$ satisfies the Lipschitz condition*

$$|G(t,s,z) - G(t,s,z')| < M|z - z'|. \tag{3.42}$$

Proof. The proof can be found in Davis (1962), see also Wazwaz (2011) for a more detailed description. □

Equations of the second kind. Here we consider only one approach to solving Eq. (3.40) — the successive approximation method. Other methods can be found in Wazwaz (2011).

The successive approximation method is also known as the Picard iteration method. This method uses an iterative procedure to find the solution of Eq. (3.40) starting from some initial guess $u_0(t)$. Obviously, the method can be applied to solving the Fredholm integral equation of the second kind, while the equations of the first kind need special treatment.

Given the initial guess $u_0(t)$ the successive iterations method introduces the recurrent relation for the solution of Eq. (3.40)

$$u_{n+1}(x) = f(x) + \lambda \int_a^b K(x,t)u_n(t)dt, \quad n \geqslant 0. \tag{3.43}$$

It was proven in Davis (1962) that this series converges if

$$\lambda < \frac{1}{k(b-1)},$$

where k is the larger of the two numbers $K\left(1 + \frac{R}{|\lambda|K(b-a)}\right)$ and M. The final solution of is determined as

$$u(x) = \lim_{n\to\infty} u_{n+1}(x).$$

Equations of the first kind. The nonlinear Fredholm equations of the first kind is usually written in the form

$$f(t) = \int_a^b K(t,s)F(u(s))ds, \tag{3.44}$$

where $F(u(s))$ is some nonlinear real-valued function, the unknown real-valued function $u(t)$ occurs inside and outside the integral sign, and $a, b \in \mathbb{R}$ are constants. We also assume that the kernel $K(x,t)$ and the function $f(x)$ are given real-valued functions.

To solve Eq. (3.44) we convert it to a linear Fredholm integral equation of the first kind

$$f(t) = \int_a^b K(t,s)v(t)dt, \quad x \in D, \tag{3.45}$$

where D is the domain of definition of the variable t, by using the transformation $v(t) = F(u(t))$. Assuming that $F(u(t))$ is invertible, we can represent $u(t)$ as $u(t) = F^{-1}(v(t))$.

Equations of type Eq. (3.45) have been already considered in Section 3.5. It is important to emphasize that the function $f(t)$ must lie in the range of the kernel $K(t,s)$, see Wazwaz (2011) and references therein. For instance, suppose we define the kernel by $K(t,s) = e^t \sin(s)$, Then, if we substitute any integrable function $F(u(t))$ into Eq. (3.44) and evaluate the integral, the resulting function $f(t)$ must clearly be a multiple of e^t. This implies that if $f(t)$ is not a multiple of the t component of the kernel, then the solution of Eq. (3.44) does not exist. This is the necessary condition on $f(t)$, which could be generalized for other kernels.

It is known that the nonlinear Fredholm equation of the first kind is an ill-posed problem since it doesn't satisfy all three Hadamard necessary and sufficient conditions formulated in above, i.e. (a) the existence of the solution, (b) the uniqueness of the solution, (c) the continuous dependence of the solution on $f(t)$. The latter means that the small errors in $f(t)$ cause small errors in the solution $f(t)$, while for ill-posed problem this is not true. To resolve this issue, various regularization methods have been proposed in the literature. We discuss some of them in Chapter 5.

Chapter 4

Volterra Integral Equations

As mentioned at the beginning of the previous chapter, using the HP and GIT methods various problems with moving boundaries occurring in mathematical finance and physics with moving boundaries can be reduced to suitable integral equation. Chapter 3 discussed Fredholm equations, while here our primary focus is on Volterra equations.

As their name suggests, these integral equations were introduced and investigated by Vito Volterra who presented his theory in four papers back to 1896. By definition, the linear Volterra integral equations for the unknown function $u(t)$ can be represented as Eq. (3.1) except in this case the upper limit of integration $b = t$. Accordingly, we obtain the Volterra integral equation of the first kind

$$f(t) = \int_0^t K(t,s)u(s)ds, \tag{4.1}$$

and the Volterra integral equation of the second kind

$$u(t) = f(t) + \int_0^t K(t,s)u(s)ds, \tag{4.2}$$

where for our purposes in this book we assume $t \in [0,T]$ and $T = const > 0$.

The Volterra integral equations can be solved analytically only in a very limited number of cases. Therefore, in this Chapter our main goal is to demonstrate that under some mild assumptions on functions $K(t,s), f(t)$ the solution $u(t)$ of the problem Eq. (4.2) exists and is unique. Various numerical approaches to solving Eq. (4.2) are presented in Chapter 5.

4.1 Picard iterations and the resolvent formalism

Let $f(t) \in \mathcal{C}^0, K(t,s) \in \mathcal{C}^0$ on $D = \{(t,s) : 0 \le s \le t \le T\}$. Since the unknown solution $u(t)$ appears in both parts of Eq. (4.2), one can try to solve this equation by using the fix-point Picard iteration algorithm (which relies on the Banach fixed-point theorem (Granas and Dugundji, 2003)).

We begin iterations by setting $u_0(t) = f(t)$, and define the iterative sequence $\{u_n(t)\}$ by

$$u_n(t) = f(t) + \int_0^t K(t,s)u_{n-1}(s)\,ds, \quad n \geq 1. \tag{4.3}$$

Using some algebra to exclude $u_{n-1}(s)$ from the RHS of Eq. (4.3), we arrive at the explicit representation of the solution at the n-th step

$$u_n(t) = f(t) + \int_0^t \left(\sum_{\nu=1}^{n} K_\nu(t,s)\right) f(s)ds, \tag{4.4}$$

where $K_1(t,s) = K(t,s)$ and

$$K_n(t,s) = \int_s^t K_1(t,v)K_{n-1}(v,s)dv, \quad n \geq 2. \tag{4.5}$$

Alternatively, it is easy to show that the following identity holds for the iterated kernels

$$K_n(t,s) = \int_s^t K_r(t,v)K_{n-r}(v,s)dv, \quad 1 \leq r < n, \quad n \geq 2. \tag{4.6}$$

Since $K(t,s) \in \mathcal{C}^0$ and D is a compact set, the following estimate for the iterated kernels $K_n(t,s)$ holds $\forall n \geq 1$

$$|K_n(t,s)| \leq \bar{K}^n \frac{(t-s)^{n-1}}{(n-1)!} \leq \bar{K}^n \frac{T^{n-1}}{(n-1)!}, \quad (t,s) \in D, \tag{4.7}$$

where $\bar{K} = \max\{|K(t,s)| : (t,s) \in D\}$. Therefore, if $n \to \infty$, the series in Eq. (4.4) becomes the Neumann series, and

$$\sum_{n=1}^{\infty} K_n(t,s) = \lim_{v \to \infty} \sum_{n=1}^{v} K_n(t,s) = R(t,s), \tag{4.8}$$

converges absolutely and uniformly in D. This implies that $R(t,s)$, the so-called resolvent kernel associated with the given kernel $K(t,s)$, is continuous in D. Hence, we obtain

$$u(t) = \lim_{n \to \infty} u_n(t) = f(t) + \int_0^\infty R(t,s)ds. \tag{4.9}$$

The uniform convergence of the Neumann series also implies that the resolvent kernel $R(t,s)$ satisfies a resolvent equation

$$R(t,s) = K(t,s) + \int_s^t K(t,v)R(v,s)dv. \tag{4.10}$$

This result can also be represented in an alternative form known as a complimentary resolvent equation

$$R(t,s) = K(t,s) + \int_s^t R(t,v)K(v,s)dv. \tag{4.11}$$

4.2 Convolution kernels

Often in various applied problems the kernel $K(t,s)$ becomes a function of just one variable which is the difference $t-s$: $K(t,s)=k(t-s)$. In this case Eq. (4.2) takes a form of convolution

$$u(t)=f(t)+\int_0^t k(t-s)u(s)ds=f(t)+(k*u)(t). \tag{4.12}$$

Here $*$ denotes the convolution operator, see e.g., Eq. (6.21). By letting $r(\cdot)$ be the corresponding resolvent, and again using Picard iterations, we obtain the following integral equation for $r(\cdot)$

$$r(z)=k(z)+(k*r)(z), \quad z\in[0,T]. \tag{4.13}$$

Under the assumptions $k\in\mathcal{C}^0, f\in\mathcal{C}^0$, Eq. (4.12) possesses a unique solution $u\in\mathcal{C}^0$ given by

$$u(t)=f(t)+\int_0^t r(t-s)f(s)ds, \quad t\in[0,T]. \tag{4.14}$$

A nice feature of Eq. (4.12) is that $u(t)$ can be explicitly represented in terms of the solution $w(t)$ to the problem Eq. (4.12) with $f(t)\equiv 1$. Indeed, let us consider the following pair of the Volterra equations

$$u(t)=f(t)+\int_0^t k(t-s)u(s)ds, \tag{4.15}$$
$$w(t)=1+\int_0^t k(t-s)w(s)ds.$$

Suppose that $f\in\mathcal{C}^1$ and $k\in L^1$. Then the (unique) solutions u,w are connected by

$$u(t)=f(0)w(t)+\int_0^t w(t-s)f'(s)ds. \tag{4.16}$$

Moreover, if $w'\in L^1$, we also have

$$u(t)=f(t)w(0)+\int_0^t w'(t-s)f(s)ds. \tag{4.17}$$

4.2.1 *Solution by using the Laplace transform*

Another approach to solving Eq. (4.12) is based on using the Laplace transform $\mathfrak{L}$. Since the Laplace transform maps the convolution operator in the original space to multiplication in the image space, the image of Eq. (4.12) becomes an algebraic equation. Indeed, introducing the notation

$$\bar{u}(\lambda)=\mathfrak{L}u, \qquad \bar{f}(\lambda)=\mathfrak{L}f, \qquad \bar{k}(\lambda)=\mathfrak{L}k, \tag{4.18}$$

and using the convolution operator properties discussed in Section 6.2, yields

$$\bar{u}(\lambda) = \frac{\bar{f}}{1 - \bar{k}}. \tag{4.19}$$

Therefore,

$$u(t) = \mathfrak{L}^{-1}\left\{\frac{\bar{f}}{1 - \bar{k}}\right\}(t). \tag{4.20}$$

For some particular specifications of the functions f and k the inversion in Eq. (4.20) can be found explicitly. In contrast, for the general case this can be done numerically, for instance by using the Gaver-Stehfest algorithm, see Chapter 5 in more detail. Another approach could rely on expanding $1/(1 - \bar{k})$ into series and again using the convolution property. As can be checked, this gives rise to the representation of the resolvent in Eq. (4.8).

4.3 Equations of the second kind with weakly singular kernels

Many integral equations arising in applications have a kernel with singularities. A typical example can be found in Chapter 8 where the density of a double layer heat potential $\Psi(t)$ associated with the function $y(t) \in \mathcal{C}^1$ solves the equation

$$2\phi(t) = \Psi(t) + \frac{1}{2\sqrt{\pi}} \int_0^t \Psi(s) \frac{y(t) - y(s)}{\sqrt{(t-s)^3}} e^{-\frac{y(t)-y(s)^2}{4(t-s)}} ds. \tag{4.21}$$

More generally, the equation of the form

$$u(t) = f(t) + \int_0^t k_\alpha(t-s) K(t,s) u(s) ds \tag{4.22}$$

with $(t,s) \in D$, $K(t,s) \in \mathcal{C}(D)$, and the weakly singular function $k_\alpha(t-s)$

$$k_\alpha(t-s) = \begin{cases} (t-s)^{\alpha-1}, & 0 < \alpha < 1, \\ \log(t-s), & \alpha = 0, \end{cases} \tag{4.23}$$

is called the Volterra integral equation of the second kind with a weakly singular kernel.

It is easy to see that Eq. (4.21) obeys the definition in Eq. (4.22) with

$$k_{1/2}(t-s) = \frac{1}{\sqrt{t-s}}, \quad K(t,s) = -\frac{1}{2\sqrt{\pi}} \frac{y(t) - y(s)}{t-s} e^{-\frac{y(t)-y(s)^2}{4(t-s)}}, \tag{4.24}$$
$$f(t) = 2\phi(t).$$

To illustrate regularity of $K(t,s)$ in Eq. (4.24), observe that $K(t,s)$ is regular if $s \neq t$. Since $y(t) \in \mathcal{C}^1$, the limit $s \to t$ can be computed as follows:

$$\lim_{s\to t} \frac{1}{2\sqrt{\pi}} \frac{y(t)-y(s)}{t-s} e^{-\frac{y(t)-y(s)^2}{4(t-s)}} = \frac{y'(t)}{2\sqrt{\pi}} < \infty.$$

At the same time the existence and uniqueness of the solution of Eq. (4.22) still hold. Since $K(t,s) \in \mathcal{C}^0$, $f \in \mathcal{C}^0$ and $0 < \alpha < 1$, Eq. (4.22) possesses a unique solution $u \in \mathcal{C}^0$) given by

$$u(t) = f(t) + \int_0^t R_\alpha(t,s) f(s) ds. \tag{4.25}$$

Here $R_\alpha(t,s)$ is the resolvent kernel corresponding to

$$K_\alpha(t,s) = K(t,s) k_\alpha(t-s).$$

The resolvent $R_\alpha(t,s)$ inherits the weak singularity $(t-s)^{\alpha-1}$ and can be represented via the Neumann series

$$R_\alpha(t,s) = (t-s)^{\alpha-1} \sum_{n=1}^{\infty} Q_{\alpha,n}(t,s), \quad 0 \leq s < t \leq T, \quad (t,s) \in D, \tag{4.26}$$

where $Q_{\alpha,n}(t,s)$ is defined as

$$\begin{aligned} Q_{\alpha,1}(t,s) &= K(t,s), \quad Q_{\alpha,n}(t,s) = (t-s)^{(n-1)\alpha} \Phi_{\alpha,n}(t,s), \quad n \geq 2, \\ \Phi_{\alpha,n}(t,s) &= \int_0^1 (1-z)^{\alpha-1} z^{(n-1)\alpha-1} \\ &\quad \times Q_{\alpha,n-1}(t, s+(t-s)z) K(s+(t-s)z, s) dz. \end{aligned}$$

Moreover, the functions $Q_{\alpha,n}$ are uniformly bounded on D

$$|Q_{\alpha,n}(t,s)| \leq \bar{K}^n T^{(n-1)\alpha} \frac{\Gamma^n(\alpha)}{\Gamma(n\alpha}, \quad (t,s) \in D, \quad n \geq 1, \tag{4.27}$$

with $\bar{K} = \max\{|K(t,s)| : (t,s) \in D\}$. The resulting uniform and absolute convergence of the series

$$\sum_{n=1}^{\infty} Q_{\alpha,n}(t,s) = Q_\alpha(t,s), \quad (t,s) \in D \tag{4.28}$$

implies that $Q_\alpha \in C^0, \forall \alpha \in (0,1)$.

4.4 Systems of linear Volterra integral equations

When the problem under consideration has several moving boundaries (for instance, two as in Chapter 18, or many as in Chapter 19) instead of a single Volterra integral equation the corresponding PDEs can be reduced to a system of the Volterra equations of a similar structure. In this case the resolvent formalism discussed in the previous sections can be naturally extended.

Let us consider a system of the Volterra integral equations of the second kind

$$\mathbf{u}(t) = \mathbf{f}(t) + \int_0^t \mathbf{K}(t,s)\mathbf{u}(s)ds, \tag{4.29}$$

where $\mathbf{u}(t) = (u_1(t), \dots, u_M(t))^\top \in \mathbb{R}^M$, $\mathbf{f}(t) = (f_1(t), \dots, f_M(t))^\top \in \mathbb{R}^M$ is continuous in $[0,T]$, and $\mathbf{K}(t,s) = (K_{ij}(t,s)) \in \mathbb{R}^{M\times M}$ is an $M \times M$ matrix whose elements K_{ij} are either in $\mathcal{C}^0$ or contain weak singularities. By analogy to the scalar case, we define the resolvent $\mathbf{R}(t,s)$ either by

$$\mathbf{R}(t,s) = \mathbf{K}(t,s) + \int_s^t \mathbf{K}(t,v)\mathbf{R}(v,s)ds, \quad (t,s) \in D, \tag{4.30}$$

or by

$$\mathbf{R}(t,s) = \mathbf{K}(t,s) + \int_s^t \mathbf{R}(t,v)\mathbf{K}(v,s)ds, \quad (t,s) \in D. \tag{4.31}$$

If $K_{ij} \in \mathcal{C}^0$ the resolvent $\mathbf{R}(t,s)$ can also be represented via the Neumann series

$$\mathbf{R}(t,s) = \sum_{n=1}^{\infty} \mathbf{K}_n(t,s), \tag{4.32}$$

that also converges absolutely and uniformly. Then the solution $\mathbf{u}(t)$ can be represented as

$$\mathbf{u}(t) = \mathbf{f}(t) + \int_0^t \mathbf{R}(t,s)\mathbf{f}(s)ds. \tag{4.33}$$

For the details of the proof we refer to Brunner (2017).

4.5 Nonlinear Volterra equations

Here we consider the nonlinear Volterra integral equations of the first and the second kind. The difference with the linear equations already studied in this chapter consists in the main fact that the kernel now is multiplied not by the solution $u(s)$ but rather by some nonlinear function of the solution $F(u(t))$. Thus, the nonlinear Volterra integral equation of the second kind is represented by the form

$$u(t) = f(t) + \int_0^t K(t,s)F(u(s))ds. \tag{4.34}$$

Similarly, the nonlinear Volterra integral equations of the first kind reads

$$f(t) = \int_0^t K(t,s)F(u(s))ds. \tag{4.35}$$

For these two kinds of equations, the kernel $K(t,s)$ and the function $f(t)$ are given real or complex-valued functions.

Let us denote $G(t,s,u(s)) = K(t,s)F(u(s))$. The specific conditions under which the solution of Eq. (4.34) exists are (Wazwaz, 2011)

(1) The function $f(t)$ is integrable and bounded in $a \leqslant t \leqslant b$.
(2) The function $f(t)$ must satisfy the Lipschitz condition in the interval (a,b). This means that

$$|f(x) - f(y)| < k|x-y|. \tag{4.36}$$

(3) The function $G(t,s,u(s))$ is integrable and bounded $|G(t,s,u(s))| < K$ in $a \leqslant t, s \leqslant b$.
(4) The function $G(t,s,u(s))$ must satisfy the Lipschitz condition

$$|G(t,s,z) - G(t,s,z')| < M|z - z'|. \tag{4.37}$$

Let us consider Eq. (4.34). Usually, to solve this nonlinear Volterra integral equation of the second kind, three basic semi-analytical methods are used (while various numerical methods can be equally efficient, see e.g., Brunner and van der Houwen, 1986; Brunner, 1985, 2004 and references therein). These three methods are the successive approximations method, the series solution method, and the Adomian decomposition method (ADM).

The successive approximations method is also known as the method of Picard iterations and was already explained in the previous sections as applied to the linear equations. For the nonlinear equation it can be constructed along the same lines. Same is true for the series method which was also explained for the linear equations. In more detail, see Wazwaz (2011). Therefore, below we focus on a more detail description of the ADM method.

The idea of the ADM method is to represent the solution $u(x)$ as an infinite sum of components that will be determined recursively by iterations.

This is easy to do for the term $u(x)$ in the LHS of Eq. (4.34), however, for the nonlinear term $F(u(s))$ this is not straightforward. Therefore, a special representation, called the Adomian polynomials A_n, $n \geq 0$ is utilized for this purpose. A formal algorithm to construct a reliable representation for all forms of nonlinear terms was established in 1980 by George Adomian (1923–1996) (Adomian, 1994), so the method is called after his name.

Following this idea, we assume

$$u(t) = \sum_{u=0}^{\infty} u_n(t), \quad n \geq 0. \tag{4.38}$$

In turn, the Adomian polynomials used to approximate the nonlineal term $F(u(s))$ are defined as

$$A_n = \frac{1}{n!} \frac{d^n}{d\lambda^n} \left[F\left(\sum_{i=0}^{n} \lambda^i u_i \right) \right]_{\lambda=0}, \quad n = 0, 1, ... \tag{4.39}$$

It can be seen that the Adomian polynomials A_n can be evaluated for any form of the nonlinear function $F(u(s))$.

We proceed as follows. Given the nonlinear function $F(u(s))$, we can decompose Eq. (4.39) into a sequence

$$\begin{aligned}
A_0 &= F(u_0), \\
A_1 &= u_1 F'(u_0), \\
A_2 &= u_2 F'(u_0) + \frac{1}{2!} u_1^2 F''(u_0), \\
A_3 &= u_3 F'(u_0) + u_1 u_2 F''(u_0) + \frac{1}{3!} u_1^3 F'''(u_0), \\
A_4 &= u_4 F'(u_0) + \left(\frac{1}{2!} u_2^2 + u_1 u_3 \right) F''(u_0) + \frac{1}{2!} u_1^2 u_2 F'''(u_0) \\
&\quad + \frac{1}{4!} u_1^4 F^{(\text{iv})}(u_0),
\end{aligned} \tag{4.40}$$

etc. One can observe that A_0 depends only on u_0, A_1 depends only on (u_0, u_1), A_2 depends only on (u_0, u_1, u_2), and so on. Next, we can substitute Eq. (4.40) into Eq. (4.39) to obtain

$$\begin{aligned}
F(u) &= A_0 + A_1 + A_2 + A_3 + \cdots \\
&= F(u_0) + (u_1 + u_2 + u_3 + \cdots) F'(u_0) \\
&\quad + \frac{1}{2!} \left(u_1^2 + 2u_1 u_2 + 2u_1 u_3 + u_2^2 + \cdots \right) F''(u_0) \\
&\quad + \frac{1}{3!} \left(u_1^3 + 3u_1^2 u_2 + 3u_1^2 u_3 + 6u_1 u_2 u_3 + \cdots \right) F'''(u_0) + \cdots \\
&= F(u_0) + (u - u_0) F'(u_0) + \frac{1}{2!} (u - u_0)^2 F''(u_0) + \cdots
\end{aligned} \tag{4.41}$$

The last line in Eq. (4.41) shows that the representation of F via the Adomian polynomials A_n is a Taylor series expansion of the functions $F(u(s))$ about the point s_0 (so about the function $u_0 = u(s_0)$), see e.g., Ernzerhof (1994). For instance, the first four Adomian polynomials for $F(u) = sin(u)$ read (Wazwaz, 2011)

$$A_0 = \sin u_0, \qquad A_1 = u_1 \cos u_0,$$

$$A_2 = u_2 \cos u_0 - \frac{1}{2!}u_1^2 \sin u_0, \qquad A_3 = u_3 \cos u_0 - u_1 u_2 \sin u_0 - \frac{1}{3!}u_1^3 \cos u_0.$$

With these definitions in mind, we can use the ADM for solving Eq. (4.34). Using the representations in Eqs. (4.38) and (4.41), Eq. (4.34) can be re-written in the form

$$\sum_{n=0}^{\infty} u_n(t) = f(t) + \int_0^t K(t,s) \sum_{n=0}^{\infty} A_n(s) ds, \tag{4.42}$$

where $A_n(s)$ are the Adomian polynomials for $F(u(s))$ as shown above. Using the ADM we set

$$u_0(t) = f(t), \quad u_{k+1}(t) = \int_0^x K(t,s) A_k(s) ds, \quad k \geqslant 0, \tag{4.43}$$

so the solution in a series form is given by Eq. (4.38) with $u_n(t)$ determined by Eq. (4.43) and computed iteratively.

As mentioned in Wazwaz (2011), the Adomian decomposition method provides the solution in a form of infinite series, where the series components u_j, $j \geqslant 0$ can be easily computed if the inhomogeneous term $f(t)$ in Eq. (4.34) is a single polynomial. However, if $f(x)$ is a combination of two or more of polynomials, trigonometric functions, hyperbolic functions, and others, the evaluation of u_j, $j \geqslant 0$ could be a tedious work. In this case a modified decomposition method was proposed in Wazwaz (2011) which facilitates the computational process and accelerates the convergence of the series solution.

The main idea behind this modified method is as follows. In many cases, the function $f(x)$ can be represented as a sum of two functions: $f(t) = f_1(t) + f_2(t)$. Then, the modified decomposition method uses the modified recurrence relation

$$\begin{aligned} u_0(t) &= f_1(t), \quad u_1(t) = f_2(t) + \int_0^t K(t,s)A_0(s)ds, \\ u_{k+1}(t) &= \int_0^t K(t,s)A_k(s)ds, \quad k \geqslant 1. \end{aligned} \tag{4.44}$$

The difference between the ADM and modified ADM rests only in the formation of the first two components $u_0(t)$ and $u_1(t)$ only. It turns out, that despite this variation is slight, it plays a significant role in accelerating the convergence of the solution and in minimizing the size of computational work. Moreover, reducing the number of terms in $f_1(x)$ affects not only the component $u_1(x)$, but also the other components as well. This result was confirmed by several research works, again in more detail see Wazwaz (2011).

Chapter 5

Solving Integral Equations Numerically

As follows from the description of the HP and GIT methods in Chapters 7 and 8 and various examples considered later in Parts 4–7, the HP and GIT methods are highly robust and versatile approaches. The HP method was frequently exploited in mathematical physics; see, e.g., Tikhonov and Samarskii (1963), Rubinstein (1971), Kartashov (2001) and Watson (2012) among others. It is essential in numerous vital fields, such as thermal engineering, nuclear engineering, and material science. However, it is not particularly well-known in mathematical finance, even though the first meaningful use in this context was described in Lipton (2001) almost twenty years ago. As far as the GIT method is concerned, aside of some physics application related to the domain $x \in [0, y(t)]$ described in Kartashov (2001), no solutions have been known for other domains. Only recently these problems were solved in various papers of the authors (also with co-authors), in the context of mathematical finance. These solutions can be found in Parts 4–7 of this book.

Common property of both HP and GIT methods is that the solution of the initial-boundary value problem with moving boundaries and time-dependent boundary conditions is reduced to either solution of the Volterra (or Fredholm) integral equation (in case of one moving boundary), or a system of Volterra (Fredholm) equations. These equations have to be solved numerically. Therefore, in this chapter we discuss various numerical approaches to solving this type of equations.

Generally, numerical methods used for solving the integral equations could be classified into four basic categories:

- quadrature methods;
- collocation-based methods;

- successive iteration methods;
- non-standard methods.

Below we shortly describe all these approaches in more detail.

5.1 Approximation of the integral by quadratures

Obviously, there exist numerous approaches to solving Volterra equations, see, Linz (1985), among many others. We start with the most straightforward approach and show how to solve the following archetypal Volterra equation with weak singularity numerically

$$\nu(t) + \int_0^t \frac{K(t,t')}{\sqrt{t-t'}}\nu(t')\,dt' = f(t), \tag{5.1}$$

where t is the time, $K(t,t')$ is a non-singular kernel, and $\nu(t)$ is the function in question, i.e., the solution of Eq. (5.1) . The integral in Eq. (5.1) can be re-written in the form

$$\int_0^t \frac{K(t,t')\nu(t')}{\sqrt{t-t'}}dt' = -2\int_0^t K(t,t')\nu(t')\,d\sqrt{t-t'}. \tag{5.2}$$

To solve this equation we discretize the time t at a temporal grid $0 = t_0 < t_1 < \ldots < t_N = T$, where $t \in [0,T]$, and $T = const$ (the maturity). Accordingly, we introduce the following notation

$$f_k = f(t_k), \quad \nu_k = \nu(t_k), \quad K_{k,l} = K(t_k, t_l), \quad \Delta_{k,l} = t_k - t_l. \tag{5.3}$$

Then, Eq. (5.1) can be approximated by, e.g., the trapezoidal rule as

$$f_k = \nu_k + \sum_{l=1}^{k}(K_{k,l}\nu_l + K_{k,l-1}\nu_{l-1})\,\Pi_{k,l} = 0, \tag{5.4}$$

$$\Pi_{k,l} = \frac{\Delta_{l,l-1}}{\left(\sqrt{\Delta_{k,l-1}} + \sqrt{\Delta_{k,l}}\right)},$$

so that

$$\nu_k = \frac{\left(f_k - K_{k,k-1}\nu_{k-1} - \sum_{l=1}^{k-1}(K_{k,l}\nu_l + K_{k,l-1}\nu_{l-1})\,\Pi_{k,l}\right)}{\left(1 + K_{k,k}\sqrt{\Delta_{k,k-1}}\right)}. \tag{5.5}$$

Thus, ν_k can be found, e.g., by Picard iterations starting with $\nu_0 = f_0$.

The approximation error of the integrals is of order $O(\Delta^2)$, where $\Delta = \max_{k,i}\sqrt{\Delta_{k,i-1}} - \sqrt{\Delta_{k,i+1}}$ is the step size. Hence, on the uniform grid $t_i = i\Delta$, the convergence rate is of order $O(\Delta)$.

5.1.1 *Block-by-block method based on quadratic interpolation*

Now we consider a block-by-block method based on quadratic interpolation from Linz (1985). Using piece-wise quadratic interpolation (Linz, 1985) derived

$$\begin{aligned} F_{2m+1} = g(t_{2m+1}) &+ (1-\delta_{m0})\sum_{i=0}^{2m} w_{2m+1,i}K(t_{2m+1},t_i)F_i \\ &+ \alpha\left(t_{2m+1},t_{2m},\frac{h}{2}\right)K(t_{2m+1},t_{2m})F_{2m} + \beta\left(t_{2m+1},t_{2m},\frac{h}{2}\right) \\ &\times K(t_{2m+1},t_{2m+1/2})\left(\frac{3}{8}F_{2m}+\frac{3}{4}F_{2m+1}-\frac{1}{8}F_{2m+2}\right) \\ &+ \gamma\left(t_{2m+1},t_{2m},\frac{h}{2}\right)K(t_{2m+1},t_{2m+1})F_{2m+1}, \end{aligned} \tag{5.6}$$

and

$$\begin{aligned} F_{2m+2} = g(t_{2m+2}) &+ (1-\delta_{m0})\sum_{i=0}^{2m} w_{2m+2,i}K(t_{2m+2},t_i)F_i \\ &+ \alpha\left(t_{2m+2},t_{2m},h\right)K(t_{2m+2},t_{2m})F_{2m} \\ &+ \beta\left(t_{2m+2},t_{2m},h\right)K(t_{2m+2},t_{2m+1})F_{2m+1} \\ &+ \gamma\left(t_{2m+2},t_{2m},h\right)K(t_{2m+2},t_{2m+2})F_{2m+2}. \end{aligned} \tag{5.7}$$

Here

$$w_{n,i} = \begin{cases} (1-\delta_{i,n-1})\alpha(t_n,t_i,h) + (1-\delta_{i,0})\gamma(t_n,t_i-2h,h), & i \text{ is even}, \\ \beta(t_n,t_i-h,h), & i \text{ is odd}, \end{cases}$$

and δ_{ij} is a Kronecker delta.

The functions α, β, and γ are defined as

$$\begin{aligned} \alpha(x,y,z) &= \frac{z}{2}\int_0^2 \frac{(1-s)(2-s)}{\sqrt{x-y-sz}}ds, \\ \beta(x,y,z) &= z\int_0^2 \frac{s(2-s)}{\sqrt{x-y-sz}}ds, \\ \gamma(x,y,z) &= \frac{z}{2}\int_0^2 \frac{s(s-1)}{\sqrt{x-y-sz}}ds. \end{aligned}$$

Note, that α, β, and γ depend only on $(x-y,z)$, and can be written as a function of two variables. Moreover, these integrals can be computed analytically.

Algorithm 1 Block-by-block method based on quadratic interpolation

Require: N — number of time steps: $0 = t_0 < t_1 < t_2 < \ldots < t_N = T$

$F_0 = f(0)$
for $m = 0 : N/2$ **do**
  Compute $w_{2m+1,i}$ for $i = 0, \ldots, 2m$
  Compute $w_{2m+2,i}$ for $i = 0, \ldots, 2m$
  Get $[F_{2m+1}, F_{2m+2}]$ by solving Eq. (5.6) and Eq. (5.7)
end for

Based on this representation, one can find $[F_{2m+1}, F_{2m+2}]$ by solving the system of two linear equations (5.6) and (5.7). We present the numerical algorithm in Algorithm 1.

Under some assumptions on regularity, the convergence rate of this method is $O(\Delta^{3.5})$. If these assumptions are not satisfied (Lipton and Kaushansky, 2018) empirically confirm the convergence of order $O(\Delta^{1.5})$.

5.1.2 *Higher order quadratures*

To increase the accuracy of the solution instead of the trapezoidal rule one can use higher order quadratures, e.g., Simpson quadratures, etc. This approach is discussed throughout the whole book when we present examples of solving integral equations for each particular problem. By using these quadratures, e.g., Eq. (5.1) can be transformed to a system of linear equations on a grid in time with matrix of the system being lower triangular. The complexity of solving this system is $O(M^2)$ where M is the number of quadrature points in the interval $[0, T]$. This can be compared with, for instance, a finite-difference method which provides linear complexity in time.

However, the accuracy of the solution of the Volterra integral equation is regulated by a quadrature rule used to discretize the kernel. Therefore, the accuracy of the method can be easily increased by using higher order quadratures. For instance, using the Simpson instead of the trapezoid rule provides the accuracy $O(h^4)$ instead of $O(h^2)$ for the trapezoidal rule, where $h = T/M$ is the step. Accordingly, same accuracy can be achieved by using $\sqrt{M}$ points instead of M, which also reduces complexity of the method to $O(M)$. Doing the same trick for the finite-difference method is not trivial, see, e.g., Itkin (2017), and discussion in Parts 4–7.

5.2 Approximation by an Abel integral equation

Consider, e.g., the integral equations, which is discussed in more detail in Chapter 10

$$\nu(\theta) + \frac{2b}{\sqrt{\pi}} \int_0^\theta \frac{\exp\left(-b^2 \frac{(\theta-\theta')}{(2+\theta+\theta')}\right)(1+\theta')\,\nu(\theta')}{\sqrt{(2+\theta+\theta')^3(\theta-\theta')}}\, d\theta' \tag{5.8}$$

$$+ \frac{\exp\left(-\frac{((1+\theta)b-z)^2}{((1+\theta)^2-1)}\right)}{\sqrt{\pi\left((1+\theta)^2-1\right)}} = 0.$$

Here the independent variable is denoted as $\theta \in [0, T]$, so $\nu = \nu(\theta)$, $b = b(\theta)$, and $z > b(0)$ is parameter of the model. For small values of θ, Eq. (5.8) can be approximated as

$$\nu(\theta) + \frac{b}{\sqrt{2\pi}} \int_0^\theta \frac{1}{\sqrt{\theta-\theta'}} \nu(\theta')\, d\theta' + \frac{\exp\left(-\frac{(b-z)^2}{2\theta}\right)}{\sqrt{2\pi\theta}} = 0. \tag{5.9}$$

This equation is an Abel integral equation of the second kind (Polyanin and Manzhirov, 2008), and can be solved analytically using direct and inverse Laplace transforms. Application of the direct Laplace transform yields

$$\bar{\nu}(\Lambda) + b\frac{\bar{\nu}(\Lambda)}{\sqrt{2\Lambda}} + \frac{e^{-\sqrt{2\Lambda}(z-b)}}{\sqrt{2\Lambda}} = 0.$$

Then, $\bar{\nu}(\Lambda)$ can be expressed as

$$\bar{\nu}(\Lambda) = -\frac{e^{-\sqrt{2\Lambda}(z-b)}}{\sqrt{2\Lambda}+b}.$$

Taking the inverse Laplace transform, we get the final expression for $\nu(\theta)$

$$\nu(\theta) = b e^{\frac{b^2}{2}\theta + b(z-b)} N\left(-\frac{b\theta + z - b}{\sqrt{\theta}}\right) - \frac{\exp\left(-\frac{(b-z)^2}{2\theta}\right)}{\sqrt{2\pi\theta}}, \tag{5.10}$$

where $N(x)$ is the normal cumulative distribution function (CDF).

Now consider another equation which is also discussed in Chapter 10

$$\nu(\vartheta) - \frac{2b}{\sqrt{\pi}} \int_0^\vartheta \frac{\exp\left(-b^2 \frac{(\vartheta-\vartheta')}{(2-\vartheta-\vartheta')}\right)(1-\vartheta')\,\nu(\vartheta')}{\sqrt{(2-\vartheta-\vartheta')^3}\sqrt{\vartheta-\vartheta'}} d\vartheta' - 1 = 0. \tag{5.11}$$

For small values of ϑ, Eq. (5.11) can be approximated by

$$\nu(\vartheta) - \frac{b}{\sqrt{2\pi}} \int_0^{\vartheta} \frac{1}{\sqrt{\vartheta - \vartheta'}} \nu(\vartheta')\, d\vartheta' - 1 = 0. \tag{5.12}$$

Similar to Eq. (5.8) we solve it by taking direct and inverse Laplace transforms. The direct Laplace transform yields

$$\bar{\nu}(\Lambda) - b\frac{\bar{\nu}(\Lambda)}{\sqrt{2\Lambda}} - \frac{1}{\Lambda} = 0.$$

Hence,

$$\bar{\nu}(\Lambda) = \frac{1}{\sqrt{\Lambda}(\sqrt{\Lambda} - b/\sqrt{2})}.$$

Taking the inverse Laplace transform, we obtain

$$\nu(\vartheta) = 2e^{\frac{b^2\vartheta}{2}} N(b\sqrt{\vartheta}). \tag{5.13}$$

Alternatively, one can find an analytical solution using the following results in Polyanin and Manzhirov (2008). As shown there, the solution of the Abel equation of the second kind

$$y(t) + \xi \int_0^t \frac{y(s)ds}{\sqrt{t-s}} = f(t)$$

has the form

$$y(t) = F(t) + \pi\xi^2 \int_0^t \exp[\pi\xi^2(t-s)]F(s)\, ds,$$

where

$$F(t) = f(t) - \xi \int_0^t \frac{f(s)\, ds}{\sqrt{t-s}}.$$

In some cases where the inverse Laplace transform cannot be found analytically, one can invert it numerically. An efficient algorithm of such an inversion is the Gaver-Stehfest algorithm, see Kuznetsov (2013); Abate *et al.* (2000) and references therein. Using this method, the inverse transform $g(t)$ can be expressed via the Laplace image $u(p)$ as

$$g(t) \approx \frac{\ln 2}{t} \sum_{k=1}^{2m} \omega_k u\left(\frac{k \ln 2}{t}\right),$$

where

$$\omega_k = (-1)^{m+k} \sum_{j=\lfloor (k+1)/2 \rfloor}^{\min(k,m)} \frac{j^{m+1}}{m} C_m^j C_{2j}^j C_{k-j}^j,$$

and C^j_{k-j} are the binomial coefficients (Abramowitz and Stegun, 1964). As one can see, the method only requires the values of u on the positive real semi-axis.

This algorithm was studied in many papers (see, e.g., Kuznetsov, 2013 and references therein), and, provided that the resulting function is non-oscillatory, converges very quickly. For instance, choosing $N = 12$ is usually sufficient. Obviously, the coefficients ω_k can be precomputed. It is also known that in many cases this algorithm requires high-precision arithmetic for its implementation. This effect is especially pronounced for small t, so the inversion can become numerically unstable unless a sufficient number of significant digits is used.

5.3 The differential transform method

In Chapter 14 we discuss pricing of Zero-coupon bonds (ZCB) under the Black-Karasinski and Verhulst models that are described in Chapter 2. It turns out that the ZCB price solves the following two-dimensional Volterra integral equation of the second kind

$$v(\tau, x) = 1 - \frac{1}{2\sqrt{\pi}} \int_{-\infty}^{\infty} \int_{\tau_0}^{\tau} \frac{\beta(t(k)) e^{\bar{a}(t(k))\zeta}}{\sqrt{\tau - k}} e^{-\frac{(x-\zeta)^2}{4(\tau-k)}} v(k, \zeta) d\zeta dk. \quad (5.14)$$

The Eq. (5.14) is a two dimensional integral Volterra equation of the second kind. Various authors have proposed efficient numerical methods for solving this type of equations. These methods include the block-by-block method, collocation and iterated collocation methods, the differential transform method (DTM), Galerkin and spectral Galerkin methods, multi-step collocation methods, and several other, see Torabi *et al.* (2019) and references therein. However, the complexity of the numerical methods (excluding the DTM) is at least $O(N^3)$, where N is the number of computational nodes. On the other hand, N could be taken relatively small compared, e.g., with the corresponding finite-difference (FD) method, if the high order quadrature rules are used when approximating the integrals.

Also, when applying all the methods mentioned above, the infinite interval should be replaced with a finite one. Another change of variables can do this, e.g., $\zeta \mapsto \tanh(\zeta)$. Then another class of methods can be used where the unknown function $v(\tau, x)$ is expanded into series on some basis. This basis could be a set of orthogonal functions, or Taylor series, etc.

However, a quick estimation of the solution of Eq. (5.14) can be obtained along the lines of the reduced differential transform method (RDTM)

(Abazari and Kilicman, 2013). The RDTM can be considered as an asymptotic solution of Eq. (5.14) around some time $t = t_0$. It can be constructed with arbitrary precision. It is worth mentioning that the RDTM can not be directly applied to Eq. (5.14) as the kernel in Eq. (5.14) depends on τ itself. Therefore, we propose a modification of the RDTM suitable to handle this situation as well.

Next, we briefly present basic definitions of the RDTM and some theorems from Abazari and Kilicman (2013) necessary to use this method for solving Eq. (5.14).

Consider a function of two variables $w(x,t)$, and suppose that it can be represented as a product of two single-variable functions $w(x,t) = f(x)g(t)$. The function $w(x,t)$ can be represented as

$$w(x,t) = \sum_{i=0}^{\infty} F(i)x^i \sum_{j=0}^{\infty} G(j)t^j = \sum_{i=0}^{\infty}\sum_{j=0}^{\infty} W_j(i)(i,j)x^i t^j, \tag{5.15}$$

where $W(i,j) = F(i)G(j)$ is called the spectrum of $w(x,t)$.

To start with, we briefly present basic definitions of the RDTM and some theorems from Abazari and Kilicman (2013) necessary to use this method for solving Eq. (5.14).

Consider a function of two variables $w(x,t)$, and suppose that it can be represented as a product of two single-variable functions $w(x,t) = f(x)g(t)$. The function $w(x,t)$ can be represented as.

If the double sum in Eq. (5.15) is truncated to the N terms in each variable, this expressions is the Poisson series of the input expression $w(x,t)$ with respect to the variables (x,t) to order N using the variable weights $W(i,j)$.

If $w(x,t)$ is an analytic function in the domain of interest, then the spectrum function

$$W_k(x) = \frac{1}{k!}\left[\frac{\partial^k}{\partial t^k} w(x,t)\right]_{t=t_0} \tag{5.16}$$

is called the reduced transformed function of $w(x,t)$. We use the notation where the lowercase $w(x,t)$ denotes the original function while the uppercase $W_k(x)$ stands for the reduced transformed function. The differential inverse transform of $W_k(x)$ is defined as

$$w(x,t) = \sum_{k=0}^{\infty} W_k(x)(t-t_0)^k. \tag{5.17}$$

Combining Eqs. (5.16) and (5.17) one can get

$$w(x,t) = \sum_{k=0}^{\infty} \frac{1}{k!}\left[\frac{\partial^k}{\partial t^k} w(x,t)\right]_{t=t_0} (t-t_0)^k. \tag{5.18}$$

To proceed, we need the following fragment of Theorem 7 in Abazari and Kilicman (2013).

Theorem 5.1. *Assume that $U_k(x), H_k(x)$ and $W_k(x)$ are the reduced differential transforms of the functions $u(x,t), h(x,t)$ and $w(x,t)$, respectively. If*

$$w(x,t) = \int_{t_0}^{t} \int_{x_0}^{x} h(y,z) u(y,z) dy dz, \tag{5.19}$$

then

$$W_k(x) = \frac{1}{k} \int_{x_0}^{x} \left(\sum_{\nu=0}^{k-1} H_\nu(y) U_{k-\nu-1}(y) \right) dy, \qquad k = 1, 2, \ldots. \tag{5.20}$$

Proof. See Abazari and Kilicman (2013). □

Now one can observe that Eq. (5.14) actually has the form of Eq. (5.19) with $x = \infty, x_0 = -\infty, t_0 = 0$, and thus

$$u(k,\zeta) = -\beta(t(k)) e^{\zeta \bar{a}(t(k))} \frac{1}{2\sqrt{\pi(\tau - k)}} e^{-\frac{(x-\zeta)^2}{4(\tau-k)}}. \tag{5.21}$$

Our modification of the RDTM consists in eliminating the definition in Eq. (5.16) for the function $u(k,\zeta)$. Then, based on the definition of the reduced differential transform in Eq. (5.16), we get

$$\begin{aligned} U_0(s,\zeta) &= -\beta(t(s)) e^{\zeta \bar{a}(t(s))} \frac{1}{2\sqrt{\pi(\tau - s)}} e^{-\frac{(x-\zeta)^2}{4(\tau-s)}}, \\ U_1(s,\zeta) &= -\int_0^s \beta(t(k)) e^{\zeta \bar{a}(t(k))} \frac{1}{2\sqrt{\pi(s - k)}} e^{-\frac{(x-\zeta)^2}{4(s-k)}} dk, \end{aligned} \tag{5.22}$$

and so on.

Let us denote the reduced differential transform of $v(\tau, x)$ as $W_k(x)$. From Eq. (5.14) it follows that $W_0(x) = 1$, as the double integral vanishes at $\tau = 0$, and the following properties of the RDT hold

$$\begin{aligned} w_k(x) = u_k(x) \pm v_k(x) \quad &\longmapsto \quad W_k(x) = U_k(x) \pm V_k(x), \\ w(x,t) = A = const, \quad &\longmapsto \quad W_0(x) = 1, \quad W_k(x) = 0, \ k > 0. \end{aligned}$$

Then from Eqs. (5.20) and (5.22) we have[1]

$$W_1(x) = \int_0^\tau \int_{-\infty}^{\infty} W_0(s,\zeta) U_0(s,\zeta) d\zeta ds = \int_0^\tau e^{\bar{a}(s)(\bar{a}(s)(\tau-s)+x)} ds. \tag{5.23}$$

[1]This expression now contains an integral in time since we eliminated the Taylor series expansion in Eq. (5.16).

The next iteration reads

$$W_2(x) = \int_0^\tau \int_{-\infty}^{\infty} [W_1(\zeta)U_0(\zeta) + W_0(\zeta)U_1(\zeta)]\, d\zeta\, ds = I_1 + I_2, \tag{5.24}$$

$$I_1 = \int_0^\tau \int_0^s \beta(k)\beta(s)e^{a(k)(2a(s)(\tau-s)+x)+a(k)^2(\tau-k)+a(s)(a(s)(\tau-s)+x)} dk\, ds,$$

$$I_2 = -\int_0^\tau \int_0^s \beta(t(k))e^{\zeta\bar{a}(t(k))} \frac{1}{2\sqrt{\pi(s-k)}} e^{-\frac{(x-\zeta)^2}{4(s-k)}} dk\, ds,$$

etc.

Once all the terms $W_k(x),\ k = 1, \dots, N$ are found, the final representation of the solution follows from the inverse formula Eq. (5.17) changed according to our modification of the RDTM

$$v(\tau, x) = \sum_{k=0}^{N} W_k(x). \tag{5.25}$$

The time-integrals in Eqs. (5.23) and (5.24) can be computed either numerically, or analytically if functions $\bar{a}(\tau), \beta(\tau)$ could be expanded into series on τ around some τ_0. In the latter case the method becomes almost identical to the original RDTM. When doing so, one has to remember that derivatives of $a(\tau), \beta(\tau)$ are the derivatives on τ while the definitions of these functions in Eq. (14.4) are given in terms of $t = t(\tau)$. The latter map is also given in Eq. (14.4).

5.4 Other methods

In Itkin *et al.* (2021) (see also Chapter 19) an efficient approach is proposed to solve the integral equations of the type

$$2\phi(\tau) = \Psi(\tau) + \frac{1}{2\sqrt{\pi}} \int_0^\tau \Psi(k) \frac{y(\tau) - y(k)}{\sqrt{(\tau-k)^3}} e^{-\frac{(y(\tau)-y(k))^2}{4(\tau-k)}} dk, \tag{5.26}$$

where $\phi(\tau)$ is the known function of τ, $\Psi(\tau)$ is the unknown function of τ to be determined, and $y(\tau)$ is the known moving boundary.

Extending the idea discussed in Section 5.2, we can construct approximate the moving boundary $y(\tau)$ in such a way, that the resulting integral becomes a convolution. For instance, we can use polynomial functions, such as $y_i(s) = a_i s^2 + b_i s + c_i$, where a_i, b_i, c_i are some constants. Then $y_i(t) - y_i(s)$ can also be represented as a certain function $g(t - s)$. Indeed

$$y_i(t) - y_i(s) = -a_i(t-s)^2 + (b_i + 2a_i t)(t - s). \tag{5.27}$$

This approach is not restricted just by parabolas. Given $y(\tau)$ we can find a polynomial of the necessary degree to guarantee that our boundaries don't cross each other unless the external curves have a very peculiar shape, which does not happen in the context of financial applications. In more detail, see Chapter 19.

With this approximation Eq. (5.2) becomes an Abel equation of the second kind which can be solved by taking the Laplace transform of both parts, solving for the image of $\Psi(\tau)$ and then taking the inverse Laplace transform. The order of approximation depends on the number of points taken along the boundary (the degree of the approximating polynomial).

Another popular numerical approach for solving Volterra integral equations consists in using collocation based methods. As mentioned in Nedaiasl *et al.* (2019), classically, collocation discretization which is based on interpolatory projection of $C(X)$ — the space of continuous functions on — onto a finite-dimensional subspace is widely used in the numerical solution of integral and differential equations. In the literature an approximation by various kind of exponential functions (multi-piece exponential functions) is popular. Some researchers replace piecewise exponential functions with linear splines to get more accuracy and speed-up gains. Recently, a polynomial spectral collocation method has also been considered. In more detail, see (Brunner, 2004).

To solve nonlinear Volterra equations in Nedaiasl *et al.* (2019) a product integration method is proposed which belongs to the category of numerical schemes based on rational barycentric interpolation. The method is applied to the Volterra equation with respect to the American option Put price. The main problem with this equation is treatment of the nonlinear part which is weakly singular, thus producing a singular behavior of the exercise boundary. Below we provide a short exposition of the method as per (Nedaiasl *et al.*, 2019).

As an example, consider an integral equation

$$u(t) = g(t, u(t)) + \int_0^t \frac{1}{(t-s)^\alpha} K_1(t, s, u(t), u(s)) ds + \int_0^t K_2(t, s, u(t), u(s)) ds, \tag{5.28}$$

where $0 < \alpha < 1$, and the function $g(t)$ as well as kernels K_1, K_2 are given.

Product integration method is a kind of quadrature methods where the smooth part of the kernel is interpolated in order to manage the weak singularity of the kernel. As usual, first $N+1$ distinct discretization points in the interval $[0,T]$ are chosen in the interval $[0,T]$ and then collocate (2.15) at these nodes to obtain

$$u(t_i) = g(t_i, u(t_i)) + \int_0^{t_i} \frac{1}{(t_i - s)^\alpha} K_1(t_i, s, u(t_i), u(s))ds \qquad (5.29)$$
$$+ \int_0^{t_i} K_2(t_i, s, u(t_i), u(s))ds, \quad i = 0, 1, \ldots, N.$$

The second integral has a smooth kernel and can be approximated by some direct quadrature rule. For the first integral the interpolation of the smooth part of the first kernel, K_1 can be used to cope with the weakly singular term (Baratella and Orsi, 2004). For doing that (Nedaiasl *et al.*, 2019) project the functions

$$\kappa_i(s, u(s)) = K_1(t_i, s, u(t_i), u(s)), \quad i = 0, 1, \ldots, N,$$

into the space $V_n = \mathrm{Span}\left\{\mathcal{L}_j(s)\right\}_{j=0}^N$ for appropriate basis functions $\mathcal{L}_j(s),\ j = 0, \ldots, N$ which yields

$$(\mathcal{P}_n \kappa_i)(s) = \sum_{j=0}^{N} \kappa_i(t_j, u(t_j))\mathcal{L}_j(s).$$

Next, substitution of this approximation of the kernel into Eq. (5.29) yields a system of equations

$$u_i = g(t_i, u_i) + \sum_{j=0}^{i} \omega_{i,j} K_1(t_i, t_j, u_i, u_j) + \sum_{j=0}^{i} \omega_j K_2(t_i, t_j, u_i, u_j), \quad (5.30)$$

$$\omega_{i,j} = \int_0^{t_i} \frac{\mathcal{L}_j(s)}{\sqrt{t_i - s}} ds, \qquad i = 0, 1, \ldots, N,$$

and ω_j are the corresponding quadrature weights. Both sets of weights $\omega_{i,j}, \omega_j$ should be computed numerically by using some quadrature rule, e.g., the Gauss-Legendre or Clenshaw-Curtis method. Once this is done, the solution $u_i \approx u(t_i)$ is found as the solution of the nonlinear system of equations in Eq. (5.30).

As mentioned in Nedaiasl *et al.* (2019), a natural choice is to interpolate the kernel with the Lagrange polynomials and use interpolatory quadrature rules. A barycentric form of interpolation is a viable variant of Lagrange's classic polynomial interpolation which has desirable features such as stability and computational speed (Trefethen, 2013). Therefore, they use this method and analyze the convergence of the approximation scheme. Some numerical experiments based on the introduced method are presented and compared to some exiting approaches demonstrating the advantage of the proposed approach.

5.5 Solving Fredholm equations of the first kind

In Chapter 9, we discuss pricing of barrier and American options where the underlying follows a time-dependent OU process. It is shown there that using the GIT method one can obtain the solution in semi-analytical form (represented via an integral), where the integrand also depends on the gradient of the solution at the boundary of the problem domain of definition. This gradient solves an integral equation, which is a linear Fredholm integral equation of the first kind for the case of a barrier option, and a non-linear Fredholm integral equation of the first kind in case of the American option. Also, these Fredholm equations can be transformed to the Volterra equations of the second kind. This is important, since solving the Fredholm equation of the first kind is an ill-posed problem, while solving the Volterra equation of the second kind is a regular procedure. However, here for completeness we discuss how the Fredholm equation for the first kind can be solved numerically.

As shown in Chapter 9, the gradient of the solution at the moving boundary $y(\tau)$ solves the equation

$$\int_0^\infty \Psi(\tau)e^{-p\tau}\sinh(y(\tau)\sqrt{p})d\tau = F(p), \tag{5.31}$$

where $F(p)$ is known. This is a linear Fredholm integral equation of the first kind (Polyanin, 2002). The solution $\Psi(\tau)$ can be found numerically on a grid by solving a system of linear equations. However, it is known that this problem is ill-posed (a discrete ill-posed problem). This means that the solution to this discrete ill-posed problems exhibits the severe sensitivity to errors. All discrete ill-posed problems obey the following criteria: a) the solution is sensitive to perturbations in the data (initial and boundary conditions and their numerical approximations); b) the singular values of the matrix gradually decay to zero.

As mentioned, e.g., in Fuhry (2001), the numerical solution of linear discrete ill-posed problems typically requires regularization, i.e., replacement of the available ill-conditioned problem by a nearby better conditioned one. The most popular regularization methods for problems of small to moderate size, which allow evaluation of the singular value decomposition of the matrix defining the problem, are the truncated singular value decomposition and Tikhonov regularization.

In more detail, consider a general linear Fredholm integral equation of the first kind in Eq. (3.2), and rewrite it in the form

$$f(x) = \int_a^b K(x, s)p(s)ds, \tag{5.32}$$

where $p(s)$ is a real-valued unknown solution, $a, b \in \mathbb{R}$ are constants, and the kernel $K(x, t)$ and the function $f(x)$ are given real-valued functions. This equation can be solved, e.g., by discretizing the integrand in the RHS of Eq. (5.32) on a gird in the x space. For doing so, the integral is approximated by a Riemann summation

$$f(x_i) = \sum_j \Delta_s K(x_i, s_j)\, p(s_j), \tag{5.33}$$

where Δ_s is the grid size along the dimension s and x_i, s are the grid points with i and j indicating their indices. It is known, that at $\Delta_s \to 0$, this sum converges to the true integral.

The Eq. (5.33) can be rewritten in the matrix form to yield

$$\boldsymbol{\Delta}_s \mathbf{K}\mathbf{p} = \mathbf{f}, \tag{5.34}$$

where

$$\mathbf{f} = (f(x_1), f(x_2), \cdots, f(x_n))^{\mathrm{T}}, \tag{5.35}$$

$$\mathbf{K} = \begin{pmatrix} K(x_1, s_1) & K(x_1, s_2) & \cdots & K(x_1, s_m) \\ K(x_2, s_1) & K(x_2, s_2) & \cdots & K(x_2, s_m) \\ \vdots & \vdots & \ddots & \vdots \\ K(x_n, s_1) & K(x_n, s_2) & \cdots & K(x_n, s_m) \end{pmatrix},$$

$$\mathbf{p} = (p(s_1), p(s_2), \cdots, p(s_m))^{\mathrm{T}},$$

and $\boldsymbol{\Delta}_s = \Delta_s \mathbf{I}$ with $\mathbf{I}$ being the identity matrix of dimension $n \times n$. Therefore, solving Eq. (5.32) is equivalent to solving a system of linear equations in Eq. (5.34). However, this problem may not have a unique solution, and, hence, it is ill-posed. In particular, if the problem is solved by using some optimization methods by minimizing the norm $\|\boldsymbol{\Delta}_s \mathbf{K}\mathbf{p} - \mathbf{f}\|$, the numerical solution turns out to be very sensitive to small errors in $\mathbf{f}$ and often diverges. The problem becomes even worse if some additional constrains should be imposed for the solution of its derivatives, so the unconstrained optimization should be replaced by the constrained one.

Various regularization method have been proposed in the literature to improve this including Tikhonov's regularization, the truncated singular value decomposition (TSVD) among others (Groetsch, 1984; Hansen, 1998). In Fuhry (2001) a novel choice of regularization matrix for Tikhonov regularization is proposed that bridges the gap between Tikhonov regularization and truncated singular value decomposition. Computed examples illustrate the benefit of the proposed method.

Another approach is considered in Maleknejad *et al.* (2016) where an advanced numerical model is proposed for solving the Fredholm integral equations of the first kind, by using Sinc basis functions. Some of the important advantages of this method are rate of convergence of an approximate solution and simplicity for performing even in the presence of singularities. The convergence analysis of the proposed method is proved by preparing the theorems which show the errors decay exponentially and guarantee the applicability of that. Finally, several numerical examples are considered to show the capabilities, validity, and accuracy of the numerical scheme.

A similar approach can be utilized for solving nonlinear Fredholm equations of the first kind. In this case instead of a system of linear equations one has to solve a system of nonlinear equations. Still regularization is required to get the method to converge.

PART 3
Integral Transforms

Chapter 6

Classical Integral Transforms

As follows from the title of this book, it describes generalized integral transforms as applied to mathematical finance. However, before we explain what is a generalized integral transform in Chapter 7, it would be useful to give a brief overview of classical integral transforms. A nice introduction into integral transforms can be found in Debnath and Bhatta (2014).

As mentioned in Debnath and Bhatta (2014), historically, the concept of an integral transform originated from the celebrated Fourier integral formula. The importance of integral transforms is that they provide powerful operational methods for solving initial value problems and initial-boundary value problems for linear differential and integral equations. In fact, one of the main impulses for the development of the operational calculus of integral transforms was the study of differential and integral equations arising in applied mathematics, mathematical physics, and engineering science; it was in this setting that integral transforms arose and achieved their early successes. With ever greater demand for mathematical methods to provide both theory and applications for science and engineering, the utility and interest of integral transforms seems more clearly established than ever. In spite of the fact that integral transforms have many mathematical and physical applications, their use is still predominant in advanced study and research.

In a nutshell, any integral transform is a map of a given function $f(x)$ into another function $\bar{f}(\xi)$

$$\bar{f}(\xi) = \int_{-\infty}^{\infty} K(x,\xi) f(x) dx, \tag{6.1}$$

such that some properties of the function $f(x)$ might be more easily characterized and manipulated. The function $K(x, \xi)$ is called the kernel of the transform.

6.1 The Fourier transform

Historically, J. Fourier first introduced the remarkable idea of expansion of a function in terms of trigonometric series without giving any attention to rigorous mathematical analysis. It is interesting, that the integral formulas for the coefficients of the Fourier expansion were already known to L. Euler and others. In fact, Fourier developed his new idea for finding the solution of heat (or Fourier) equation in terms of Fourier series so that the Fourier series can be used as a practical tool for determining the Fourier series solution of partial differential equation under prescribed boundary conditions. Again, in more detail, see Debnath and Bhatta (2014).

The main idea of the Fourier transform can be easily illustrated by the Fourier series. Let us consider a function $f(x) : [-\pi/\omega, \pi/\omega] \mapsto \mathbb{R}$. Under some mild assumptions, the function $f(x)$ can be represented by the following infinite trigonometric series (Fourier series)

$$f(x) = \sum_{n=0}^{\infty} a_n \cos(n\omega x) + b_n \sin(n\omega x) = \sum_{n=-\infty}^{\infty} c_n e^{\mathrm{i}n\omega x}, \tag{6.2}$$

where the coefficients c_n are defined as

$$c_n = \frac{\omega}{2\pi} \int_{-\pi/\omega}^{\pi/\omega} e^{-\mathrm{i}n\omega\xi} f(\xi) d\xi. \tag{6.3}$$

Combining Eqs. (6.2) and (6.3) together and letting $\omega \to 0$ yields the following identity

$$f(x) = \frac{1}{2\pi} \lim_{\omega\to 0} \sum_{n=-\infty}^{\infty} \omega \int_{-\pi/\omega}^{\pi/\omega} e^{\mathrm{i}n\omega x} e^{-\mathrm{i}n\omega\xi} f(\xi) d\xi,$$

or

$$f(x) = \frac{1}{2\pi} \int_{-\infty}^{\infty} d\omega e^{\mathrm{i}\omega x} \int_{-\infty}^{\infty} e^{-\mathrm{i}\omega\xi} f(\xi) d\xi. \tag{6.4}$$

The inner integral can be considered as a direct Fourier transform and the outer integral as an inversion.

Definition 6.1. The operator $\mathfrak{F}$ acting on function $f(x) : \mathbb{R} \mapsto \mathbb{R}$ and defined by the following expression

$$\bar{f}(\omega) = \{\mathfrak{F} f\}(\omega) = \int_{-\infty}^{+\infty} e^{-\mathrm{i}\omega x} f(x) dx, \tag{6.5}$$

is called the Fourier transform. The inverse transform is defined as

$$f(x) = \{\mathfrak{F}^{-1} \bar{f}\}(x) = \frac{1}{2\pi} \int_{-\infty}^{+\infty} e^{\mathrm{i}\omega x} \bar{f}(\omega) d\omega, \tag{6.6}$$

respectively.

Below we list some properties of the Fourier transform $\mathfrak{F}$

$$\begin{aligned}\mathfrak{F}(\alpha f) &= \alpha\mathfrak{F}(f), \\ \mathfrak{F}(f+g) &= \mathfrak{F}f + \mathfrak{F}g, \\ \mathfrak{F}(f * g) &= \mathfrak{F}f \cdot \mathfrak{F}g, \\ \mathfrak{F}(f') &= -\mathrm{i}\omega\mathfrak{F}(f), \\ \mathfrak{F}(f^{(n)}) &= (-\mathrm{i}\omega)^n \mathfrak{F}(f).\end{aligned} \tag{6.7}$$

Here the convolution $(*)$ is defined for two functions $f(x) : \mathbb{R} \mapsto \mathbb{R}$ and $g(x) : \mathbb{R} \mapsto \mathbb{R}$ as follows

$$(f * g)(x) = \int_{-\infty}^{\infty} f(x-y)g(y)dy. \tag{6.8}$$

The above properties can be easily proved (Debnath and Bhatta, 2014).

In context of problems described by ordinary differential equations (ODE), and partial differential equations (PDE), the properties in Eq. (6.7) are essential. That is because the Fourier transform $\mathfrak{F}$ reduces the dimensionality of the equation by one, and so maps an ODE to an algebraic equation, and a PDE — to another PDE with the reduced spatial dimensionality.

To illustrate the method of the Fourier transform, consider the following initial-boundary value problem for the heat equation

$$\begin{aligned}\left(\frac{\partial^2}{\partial x^2} - \frac{\partial}{\partial t}\right) G(t_0, x_0 \,|\, t, x) &= 0, \quad (t,x) \in \mathbb{R}_+ \times \mathbb{R}, \\ G(t_0, x_0 \,|\, s, x) &= \delta(x - x_0), \quad x \in \mathbb{R}, \\ \lim_{x \to \pm\infty} G(t_0, x_0 \,|\, t, x) &= 0, \quad \forall t > 0.\end{aligned} \tag{6.9}$$

Here the initial condition is given by the Dirac delta function $\delta(x)$. The function $G(t_0, x_0 \,|\, t, x)$ is known as the Green function or the fundamental solution of the heat equation. Assuming $g(t,\omega) = \mathfrak{F}G$ and applying the Fourier transform to Eq. (6.9), we obtain the following ODE

$$g_t + \omega^2 g = 0, \tag{6.10}$$

that should be solved subject to the initial condition

$$g(s,\omega) = e^{-\mathrm{i}\omega x_0}, \tag{6.11}$$

and vanishing boundary conditions at $x \to \pm\infty$. The explicit solution of the problem Eqs. (6.10) and (6.11) reads

$$g(t,\omega) = e^{-\omega^2(t-t_0) - \mathrm{i}\omega x_0}.$$

Using the inversion formula Eq. (6.5) we get

$$
\begin{aligned}
G(t_0, x_0 \,|\, t, x) &= \frac{1}{2\pi} \int_{-\infty}^{\infty} e^{-\omega^2 (t-t_0) + \mathrm{i}\omega(x-x_0)} d\omega \\
&= \frac{1}{2\pi} \int_{-\infty}^{\infty} e^{-\omega^2 (t-t_0)} \left[\cos\left(\omega(x - x_0 s)\right) + \mathrm{i}\sin\left(\omega(x_0 - x)\right)\right] d\omega \\
&= \frac{1}{2\pi} \int_{-\infty}^{\infty} e^{-\omega^2 (t-t_0)} \cos\left(\omega(x - x_0)\right) d\omega.
\end{aligned}
$$

The last integral can be computed in the explicit form, see Gradshtein and Ryzhik (2007), and the final formula for $G(t_0, x_0 \,|\, t, x)$ reads

$$
G(t_0, x_0 \,|\, t, x) = \frac{1}{2\sqrt{\pi(t - t_0)}} e^{-\frac{(x-x_0)^2}{4(t-t_0)}}. \tag{6.12}
$$

This expression is known in the literature as the heat kernel. Suppose also that W_t is the standard Brownian motion with zero mean and unit variance. It is known that $W_t \sim \mathcal{N}(0, t)$ for each $t > 0$ (Karatzas and Shreve, 1991). We can also define Brownian motion starting at $x \in \mathbb{R}$ by setting $B_t = W_t + x$ for each $t \geq 0$.

Suppose that $\{B_t, t \geq 0\}$ is a Brownian motion starting at x. Since B_t is a normal variable with the mean x and the variance t, i.e., $B(t) \sim \mathcal{N}(x, t)$, we define the transition density by

$$
p(t; x, y) = \frac{1}{\sqrt{2\pi t}} \exp\left\{-\frac{1}{2t}(y - x)^2\right\}, \quad -\infty < y < \infty. \tag{6.13}
$$

Thus, $p(t; x, y)$ represents the probability that Brownian motion starting at x will be at position y at time t. Of course, $p(t; x, y)$ is a density function for a continuous random variable and so we must interpret it to mean

$$
P\left\{B_t \in V \mid B_0 = x\right\} = \int_V p(t; x, y) \mathrm{d}y, \tag{6.14}
$$

for any set $V \subset \mathbb{R}$. Hence, the transition density for the process B_t is also given by the heat kernel.

6.2 The Laplace transform

Another popular integral transform with an exponential kernel is the Laplace transform. We show that by a series of transformations, the Fourier transform Eq. (6.5) can be reduced to the bilateral (two sided) Laplace transform. Making the change of variables in Eq. (6.5)

$$
\omega = \omega' + \mathrm{i}\gamma, \quad (\Im\omega' = -\gamma), \quad f_1(x) = f(x)e^{\gamma x}, \quad \bar{f}_1 = \bar{f}(\omega + \mathrm{i}\gamma)
$$

yields the following form of the Fourier transform:

$$\bar{f}_1(\omega') = \int_{-\infty}^{\infty} e^{-\mathrm{i}\omega' x} f_1(x)dx, \quad \Im\omega' = \gamma,$$

$$f_1(x) = \frac{1}{2\pi}\int_{-\infty-\mathrm{i}\gamma}^{+\infty-\mathrm{i}\gamma} e^{\mathrm{i}\omega' x}\bar{f}_1(\omega')d\omega', \qquad \Im x = 0.$$

Doing another change of variables

$$\lambda = \mathrm{i}\omega' = \mathrm{i}\omega + \gamma, \quad f_1(-\mathrm{i}\lambda) = \frac{\bar{h}(\lambda)}{2\pi},$$

after some algebra we obtain the following integral transform

$$\bar{h}(\lambda) = \int_{-\infty}^{\infty} e^{\lambda x} h(x)dx, \quad \Im\lambda = \gamma,$$

$$h(x) = \frac{1}{2\pi\mathrm{i}}\int_{\gamma-\mathrm{i}\infty}^{\gamma+\mathrm{i}\infty} e^{\lambda x}\bar{h}(\lambda)d\lambda, \qquad \Im x = 0.$$

The condition $\Im\lambda = \gamma$ for the spectral parameter λ can be relaxed, in more detail see van der Pol and Bremmer (1950)

$$\bar{h}(\lambda) = \int_{-\infty}^{\infty} e^{\lambda x} h(x)dx, \quad \alpha < \Re\lambda < \beta, \tag{6.15}$$

$$h(x) = \frac{1}{2\pi\mathrm{i}}\int_{\gamma-\mathrm{i}\infty}^{\gamma+\mathrm{i}\infty} e^{\lambda x}\bar{h}(\lambda)d\lambda, \qquad \alpha < \gamma < \beta.$$

Here α and β are some constants. The transform in Eq. (6.15) is called the bilateral (or two sided) Laplace transform. The bilateral transform also covers the classical (one-sided) Laplace transform as a special case. Indeed, if the function $h(x)$ has the following structure

$$h(x) = \mathbf{1}_x f(x), \qquad f(x) : \mathbb{R}_+ \mapsto \mathbb{R}, \tag{6.16}$$

then the parameters α, β take the values $\alpha = 0, \beta \to \infty$, and consequently $\gamma > 0$, so the bilateral Laplace transform is reduced to the one-sided Laplace transform. Here $\mathbf{1}_{x>0}$ denotes the Heaviside step function

$$\mathbf{1}_x = \begin{cases} 1, & x \geq 0, \\ 0, & x < 0. \end{cases} \tag{6.17}$$

Definition 6.2. The operator $\mathfrak{L}$ acting on functions $f(x) : \mathbb{R}_+ \to \mathbb{R}$ and defined as

$$\bar{f}(\lambda) = \{\mathfrak{L}f\}(\lambda) = \int_0^{\infty} e^{-\lambda x} f(x)dx, \tag{6.18}$$

is called the Laplace transform. The inverse transform is given by

$$f(x) = \{\mathfrak{L}^{-1}\bar{f}\}(x) = \frac{1}{2\pi\mathrm{i}}\int_{\gamma-\mathrm{i}\infty}^{\gamma+\mathrm{i}\infty} e^{\lambda x}\bar{f}(\lambda)d\lambda, \quad \gamma > 0. \tag{6.19}$$

In this book by the Laplace transform we always mean the one-sided Laplace transform. The so-called Laplace–Carson transform can be beneficial when solving various problems, see e.g., Lipton and Sepp (2011b) and Itkin and Lipton (2018)

$$\bar{f}(\lambda) = \lambda \int_0^\infty e^{-\lambda x} f(x) dx, \tag{6.20}$$

$$f(x) = \left\{\mathfrak{L}^{-1}\bar{f}\right\}(x) = \frac{1}{2\pi \mathrm{i}} \int_{\gamma - \mathrm{i}\infty}^{\gamma + \mathrm{i}\infty} e^{\lambda x} \frac{\bar{f}(\lambda)}{\lambda} d\lambda.$$

Here γ is a real-valued constant chosen in such a way that all poles of the function under the last integral reside in a complex plane to the left of a straight line parallel to the imaginary axis $x \in (\gamma, -\mathrm{i}\infty) - (\gamma, \mathrm{i}\infty)$.

Let us introduce the convolution Eq. (6.8) of two functions defined on positive half-axis $f(x) : \mathbb{R}_+ \mapsto \mathbb{R}$ and $g(x) : \mathbb{R}_+ \mapsto \mathbb{R}$ as

$$(f * g)(x) = \int_0^x f(x-y) g(y) dy, \tag{6.21}$$

The following properties of the Laplace transform are inherited from the Fourier transform

$$\begin{aligned} \mathfrak{L}(\alpha f) &= \alpha \mathfrak{L}(f), \\ \mathfrak{L}(f+g) &= \mathfrak{L}f + \mathfrak{L}g, \\ \mathfrak{L}(f * g) &= \mathfrak{L}f \, \mathfrak{L}g. \end{aligned} \tag{6.22}$$

However, the properties corresponding to the Laplace transform of the functions' derivatives differ

$$\begin{aligned} \mathfrak{L}(f') &= \lambda \mathfrak{L}(f) - f(0), \\ \mathfrak{L}(f^{(n)}) &= \lambda^n \mathfrak{L}(f) - \sum_{k=1}^n \lambda^{n-k} f^{(k-1)}(0). \end{aligned} \tag{6.23}$$

Let us apply the method of the Laplace transform to solving an initial-boundary value problem for the heat equation

$$\begin{aligned} \left(\frac{\partial^2}{\partial x^2} - \frac{\partial}{\partial t}\right) U(t,x) = 0, & \quad (t,x) \in \mathbb{R}_+ \times \mathbb{R}_+, \\ U(0,x) = \delta(x - x_0), & \quad x \in \mathbb{R}_+, \quad x_0 > 0, \\ \lim_{x \to \infty} U(t,x) = 0, & \quad \forall t > 0, \\ U(t,0) = 0, & \quad \forall t > 0. \end{aligned} \tag{6.24}$$

It is worth mentioning, that we can not solve this problem by directly using the Fourier transform with respect to the spatial variable x. However, it can be solved by using the Sine Fourier transform, see Section 6.3 in more detail.

Here we proceed by applying the Laplace transform $\bar{u}(\lambda, x) = \mathfrak{L}U$ to Eq. (6.24)

$$\mathfrak{L}\frac{\partial^2 U}{\partial x^2} = \frac{\partial^2 \bar{u}}{\partial x^2},$$
$$\mathfrak{L}\frac{\partial U}{\partial t} = \lambda\bar{u} - U(0,x) = \lambda\bar{u} - \delta(x - x_0).$$

The function $\bar{u}(\lambda, x)$ solves the following inhomogeneous second order ODE

$$\frac{d^2\bar{u}}{dx^2} - \lambda\bar{u} = -\delta(x - x_0). \tag{6.25}$$

This ODE should be solved subject to the boundary conditions

$$\bar{u}(\lambda, 0) = 0, \qquad \lim_{x\to\infty} \bar{u}(\lambda, x) = 0. \tag{6.26}$$

The solution of Eqs. (6.25) and (6.26) can be found explicitly. We seek for it in the form

$$C_-(x)u_-(x) + C_+(x)u_+(x).$$

Here $u_\pm$ denote two linearly independent solutions of the homogeneous problem in Eq. (6.25)

$$u_+ = e^{\sqrt{\lambda}x}, \qquad u_- = e^{-\sqrt{\lambda}x},$$

and the functions $C_\pm(x)$ can be found from the following system of differential equations

$$\begin{bmatrix} u_-(x) & u_+(x) \\ u'_-(x) & u'_+(x) \end{bmatrix} \begin{bmatrix} C'_-(x) \\ C'_+(x) \end{bmatrix} = \begin{bmatrix} 0 \\ -\delta(x - x_0) \end{bmatrix}$$

Denoting by $\mathcal{W}(u_-, u_+)$ the Wronskian of the functions u_- and u_+, i.e.,

$$\mathcal{W}(u_-, u_+) = \det \begin{bmatrix} u_-(x) & u_+(x) \\ u'_-(x) & u'_+(x) \end{bmatrix} = \det \begin{bmatrix} e^{-\sqrt{\lambda}x} & e^{\sqrt{\lambda}x} \\ -\sqrt{\lambda}e^{-\sqrt{\lambda}x} & \sqrt{\lambda}e^{\sqrt{\lambda}x} \end{bmatrix} = 2\sqrt{\lambda}$$

and representing the inverse matrix in terms of the adjugate matrix, yields

$$\begin{bmatrix} C'_-(x) \\ C'_+(x) \end{bmatrix} = \frac{1}{\mathcal{W}(u_-, u_+)} \begin{bmatrix} u'_+(x) & -u_+(x) \\ -u'_-(x) & u_-(x) \end{bmatrix} \begin{bmatrix} 0 \\ -\delta(x - x_0) \end{bmatrix}$$
$$= \frac{1}{\mathcal{W}(u_-, u_+)} \begin{bmatrix} u_+(x_0)\delta(x - x_0) \\ -u_-(x_0)\delta(x - x_0) \end{bmatrix}$$

$$= \frac{1}{2\sqrt{\lambda}} \begin{bmatrix} e^{\sqrt{\lambda}x_0}\delta(x - x_0) \\ -e^{-\sqrt{\lambda}x_0}\delta(x - x_0) \end{bmatrix}.$$

Therefore,

$$\begin{bmatrix} C_-(x) \\ C_+(x) \end{bmatrix} = \frac{1}{2\sqrt{\lambda}} \begin{bmatrix} e^{\sqrt{\lambda}x_0} \int_{-\infty}^{x} \delta(\xi - x_0)d\xi \\ e^{-\sqrt{\lambda}x_0} \int_{x}^{+\infty} \delta(\xi - x_0)d\xi. \end{bmatrix}$$

$$= \frac{1}{2\sqrt{\lambda}} \begin{bmatrix} e^{\sqrt{\lambda}x_0} \mathbf{1}_{x>x_0} \\ e^{-\sqrt{\lambda}x_0} \mathbf{1}_{x<x_0} \end{bmatrix}$$

and the function $\bar{u}(\lambda, x)$ can be represented as

$$\bar{u}(\lambda, x) = \frac{e^{-\sqrt{\lambda}|x-x_0|}}{2\sqrt{\lambda}} + C_1 e^{-\sqrt{\lambda}x} + C_2 e^{\sqrt{\lambda}x}. \tag{6.27}$$

By taking into account the boundary conditions Eq. (6.26) we find that

$$C_1 = -\frac{e^{-\sqrt{\lambda}(x+x_0)}}{2\sqrt{\lambda}}, \qquad C_2 = 0,$$

and, hence, $\bar{u}(\lambda, x)$ takes the form

$$\bar{u}(\lambda, x) = \frac{e^{-\sqrt{\lambda}|x-x_0|}}{2\sqrt{\lambda}} - \frac{e^{-\sqrt{\lambda}|x+x_0|}}{2\sqrt{\lambda}}.$$

Using the well-known formula

$$\mathfrak{L}^{-1}\left\{\frac{e^{-\sqrt{\lambda}|z|}}{2\sqrt{\lambda}}\right\}(t) = \frac{e^{-\frac{z^2}{4t}}}{2\pi\sqrt{t}}, \tag{6.28}$$

we arrive at the final formula for $U(t, x)$ (since x and x_0 are non-negative then $x - x_0 = |x - x_0|$)

$$U(t, x) = \frac{e^{-\frac{(x-x_0)^2}{4t}}}{2\pi\sqrt{t}} - \frac{e^{-\frac{(x+x_0)^2}{4t}}}{2\pi\sqrt{t}}. \tag{6.29}$$

6.3 The Hankel transform

Another popular integral transform that has a non-exponential kernel is the Hankel transform. This transform arises when solving a two-dimensional Laplace equation in cylindrical coordinates.

Definition 6.3. The operator $\mathfrak{H}_\nu$ acting on functions $f(x) : \mathbb{R}_+ \to \mathbb{R}$ and defined by

$$\bar{f}(\xi) = \{\mathfrak{H}_\nu f\}(\xi) = \int_0^\infty x f(x) J_\nu(x\xi)dx, \tag{6.30}$$

is called the Hankel trasnform. The inverse transform is given by

$$\left\{\mathfrak{H}_\nu^{-1}\bar{f}\right\}(x) = \int_0^\infty \xi\bar{f}(\xi)J_\nu(x\xi)d\xi. \tag{6.31}$$

Here $J_\nu(x)$ denotes the Bessel function of the first kind (Abramowitz and Stegun, 1964), which can be defined as follows

$$J_\nu(z) = \frac{z^\nu}{2^\nu}\sum_{k=0}^\infty(-1)^k\frac{z^{2k}}{2^{2k}k!\Gamma(\nu+k+1)}, \qquad |\arg z| < \pi, \tag{6.32}$$

and $\Gamma(\cdot)$ denotes the Euler Gamma function (Abramowitz and Stegun, 1964).

It is well-known that the Bessel function $J_\nu(z)$ solves the ODE

$$\frac{d^2Z}{dz^2} + \frac{1}{z}\frac{dZ}{dz} + \left(1 - \frac{\nu^2}{z^2}\right)Z = 0. \tag{6.33}$$

Also, for the special values of indexes $\nu = \pm 1/2$ the Bessel function $J_\nu(z)$ is reduced to the sine and cosine functions

$$J_{1/2}(z) = \sqrt{\frac{2}{\pi z}}\sin z, \qquad J_{-1/2}(z) = \sqrt{\frac{2}{\pi z}}\cos z. \tag{6.34}$$

In this case the Hankel transform Eq. (6.30) is also reduced to the Sine and Cosine Fourier transforms, respectively.

Let us use the Hankel transform to solve the following problem

$$\begin{gathered}\left(\frac{\partial^2}{\partial x^2} + \frac{1-2\nu}{x}\frac{\partial}{\partial x} - \frac{\partial}{\partial t}\right)U = 0, \quad (x,t) \in \mathbb{R}_+ \times \mathbb{R}_+, \\ U(t,0) = 0, \quad \lim_{x\to\infty} U(t,x) = 0, \quad t > 0, \\ U(0,x) = \delta(x - x_0), \quad x \in \mathbb{R}_+.\end{gathered} \tag{6.35}$$

We proceed with the following change of variables

$$\begin{gathered}U = x^\nu V(t,x), \qquad \frac{\partial U}{\partial x} = x^\nu\left[\frac{\nu}{x}V + \frac{\partial V}{\partial x}\right], \\ \frac{\partial^2 U}{\partial x^2} = x^\nu\left[\frac{\nu(\nu-1)}{x^2}V + \frac{2\nu}{x}\frac{\partial V}{\partial x} + \frac{\partial^2 V}{\partial x^2}\right], \qquad \frac{\partial U}{\partial t} = x^\nu\frac{\partial V}{\partial t},\end{gathered}$$

to get

$$\begin{gathered}\frac{\partial V}{\partial t} = \left(\frac{\partial^2}{\partial x^2} + \frac{1}{x}\frac{\partial}{\partial x} - \frac{\nu^2}{x^2}\right)V, \\ \lim_{x\to 0} x^{-\nu}V(t,x) = 0, \quad \lim_{x\to\infty} x^{-\nu}V(t,x) = 0, \quad t > 0, \\ V(0,x) = x_0^{-\nu}\delta(x - x_0).\end{gathered} \tag{6.36}$$

Now we introduce the Hankel transform $\mathfrak{H}_\nu$ of the function $V(t,x)$

$$\bar{v}(t,\xi) = \int_0^\infty xV(t,x)J_\nu(x\xi)d\xi.$$

By multiplying both parts of Eq. (6.36) by $xJ_\nu(x\xi)$ and integrating on x from 0 to ∞, we obtain

$$\begin{aligned}
&\int_0^\infty \left[\frac{\partial^2 V}{\partial x^2} + \frac{1}{x}\frac{\partial V}{\partial x} - \frac{\nu^2}{x^2}V\right] xJ_\nu(x\xi)dx \\
&= \int_0^\infty xV(t,x)\left[\frac{\partial^2 (xJ_\nu(x\xi))}{\partial x^2} - \frac{\partial J_\nu(x\xi)}{\partial x} - \frac{\nu^2}{x^2}J_\nu(x\xi)\right] dx \\
&= \xi^2 \int_0^\infty xV(t,x)\left[J_\nu''(x\xi) + \frac{1}{x\xi}J_\nu'(x\xi) - \frac{\nu^2}{(x\xi)^2}J_\nu(x\xi)\right] dx \\
&= -\xi^2 \int_0^\infty xV(t,x)J_\nu(x\xi)dx \\
&= -\xi^2\bar{v} \int_0^\infty x\frac{\partial V(t,x)}{\partial t}J_\nu(x\xi)dx = \frac{d\bar{v}}{dt}.
\end{aligned}$$

By taking into account the initial condition in Eq. (6.36), we obtain the following Cauchy problem for the function $\bar{v}(t,\xi)$

$$\begin{aligned}
\frac{d\bar{v}}{dt} &= -\xi^2\bar{v}, \\
\bar{v}(0,\xi) &= x_0^{1-\nu}J_\nu(x_0\xi),
\end{aligned} \tag{6.37}$$

which can be solved via standard techniques. The solution reads

$$\bar{v}(t,\xi) = x_0^{1-\nu}J_\nu(x_0\xi)e^{-\xi^2 t}. \tag{6.38}$$

Applying the inversion formula Eq. (6.31) to this solution we obtain

$$V(t,x) = x_0^{1-\nu}\int_0^\infty \xi e^{-\xi^2 t}J_\nu(x_0\xi)J_\nu(x\xi)d\xi.$$

Now, using well-known identity from (Gradshtein and Ryzhik, 2007)

$$\int_0^\infty e^{-q^2x^2}J_p(\alpha x)J_p(\beta x)xdx = \frac{1}{2q^2}e^{-\frac{\alpha^2+\beta^2}{4q^2}}I_p\left(\frac{\alpha\beta}{2q^2}\right),$$

$$\Re p > -1, \quad |\arg q| < \frac{\pi}{4}, \quad \alpha > 0, \quad \beta > 0,$$

and returning back to the original variables, the final representation for the function $U(t,x)$ is obtained

$$U(t,x) = \frac{\sqrt{xx_0}}{2t}\left(\frac{x}{x_0}\right)^{\nu-1/2} e^{-\frac{x_0^2+x^2}{4t}}I_\nu\left(\frac{xx_0}{2t}\right). \tag{6.39}$$

Here $I_\nu(z)$ denotes the modified Bessel function of the first kind of the index ν, which can be defined as

$$I_\nu(z) = \sum_{k=0}^{\infty} \frac{1}{k!\Gamma(\nu+k+1)} \left(\frac{z}{2}\right)^{\nu+2k}. \tag{6.40}$$

The Eq. (6.39) with $\nu = 1/2$ coincides with Eq. (6.29). Indeed, the function $I_{1/2}(z)$ can be represented in terms of the hyperbolic sine

$$\begin{aligned} I_{1/2}(z) &= \sum_{k=0}^{\infty} \frac{1}{k!\Gamma(1/2+k+1)} \left(\frac{z}{2}\right)^{2k+1/2} \\ &= \frac{1}{\sqrt{\pi z}} \sum_{k=0}^{\infty} \frac{z^{2k+1}}{k!(2k+1)!!2^{k-1/2}} \\ &= \frac{2}{\sqrt{2\pi z}} \sum_{k=0}^{\infty} \frac{z^{2k+1}}{(2k+1)!} = \frac{e^z - e^{-z}}{\sqrt{2\pi z}}, \end{aligned}$$

where the following identity for the Euler Gamma function is used (Olver *et al.*, 2020)

$$\Gamma\left(n + \frac{1}{2}\right) = \frac{\sqrt{\pi}}{2^n}(2n-1)!!, \qquad n \in \mathbb{N}.$$

Therefore,

$$\begin{aligned} \frac{\sqrt{xx_0}}{2t} e^{-\frac{x_0^2+x^2}{4t}} I_{1/2}\left(\frac{xx_0}{2t}\right) &= \frac{e^{-\frac{x_0^2+x^2}{4t}}}{2\sqrt{\pi t}} \left(e^{\frac{xx_0}{2t}} - e^{-\frac{xx_0}{2t}}\right) \\ &= \frac{e^{-\frac{(x-x_0)^2}{4t}} - e^{-\frac{(x+x_0)^2}{4t}}}{2\sqrt{\pi t}}. \end{aligned}$$

6.4 The Index transform

It was demonstrated in the previous sections that the kernels of the integral transforms solve a linear ODE of the second order (different ODEs correspond to different transforms). For a theoretical study of how such transforms arise from the second order ODEs we refer the reader to Mercer (1962) and Titchmarsh (1962).

Here we consider integral transforms related to cylinder and confluent hypergeometric functions, see Yakubovich (1996) and references therein. The name "index" emphasizes the fact that the transformation is made with respect to the index of a cylinder or hypergeometric function.

The discussed below integral transform was originally proposed by the Soviet physicists Kontorovich and Lebedev (1939).

Definition 6.4. The operator $\mathfrak{K}$ acting on functions $f(x) : \mathbb{R}_+ \to \mathbb{R}$ and defined as

$$\bar{f}(\omega) = \{\mathfrak{K}f\}(\omega) = \int_0^\infty f(x) K_{\mathrm{i}\omega}(x) dx, \tag{6.41}$$

is called the Kontorovich–Lebedev transform. The inverse transform is given by

$$f(x) = \left\{\mathfrak{K}^{-1}\bar{f}\right\}(x) = \frac{2}{\pi x^2} \int_0^\infty \omega \sinh(\pi\omega) K_{\mathrm{i}\omega}(x) \bar{f}(\omega) d\omega. \tag{6.42}$$

Here $K_{\mathrm{i}\omega}(z)$ is the Macdonald function (the modified Bessel function) of an imaginary index, which can be represented by the following integral

$$K_{\mathrm{i}\omega}(z) = \int_0^\infty e^{-z\cosh\xi} \cos(\omega\xi) d\xi. \tag{6.43}$$

An interesting generalization of the transform in Eq. (6.41) was proposed in Wimp (1964).

Definition 6.5. The operator $\mathfrak{W}_\mu$ with $\mu < 1/2$ acting on functions $f(x)$ and defined by

$$\bar{f}(\omega) = \{\mathfrak{W}_\mu f\}(\omega) = \int_0^\infty f(x) W_{\mu,\mathrm{i}\omega}(x) dx, \tag{6.44}$$

is called the Wimp transform. The inverse transform is given by

$$f(x) = \left\{\mathfrak{W}_\mu^{-1}\bar{f}\right\}(x) = \frac{1}{(\pi x)^2} \int_0^\infty \Bigg\{\omega \sinh(2\pi\omega) \tag{6.45}$$

$$\Gamma\left(\frac{1}{2} - \mu - \mathrm{i}\omega\right) \Gamma\left(\frac{1}{2} - \mu + \mathrm{i}\omega\right) W_{\mu,\mathrm{i}\omega}(x) \bar{f}(\omega)\Bigg\} d\omega, \quad \mu < 1/2.$$

Here $W_{\lambda,\mu}(x)$ denotes the Whittaker function of the second kind. The Whittaker functions $M_{\lambda,\mu}(z)$ and $W_{\lambda,\mu}(z)$ can be defined as a linearly independent pair of the solutions to the following ODE:

$$Z'' + \left(-\frac{1}{4} + \frac{\lambda}{z} + \frac{1/4 - \mu^2}{z^2}\right) Z = 0. \tag{6.46}$$

The necessary condition for the existence of the direct and inverse Wimp transform of the function f can be found in Yakubovich (1996).

The Wimp transform $\mathfrak{W}_0$ with the parameter $\mu = 0$ coincides with the Kontorovich–Lebedev transform. It is easy to see by using the well-known relationship between the Whittaker function $W_{0,\mathrm{i}\omega}(2z)$ and the Macdonald function $K_{\mathrm{i}\omega}(z)$ together with the following property of the Euler Gamma function (Olver *et al.*, 2020)

$$W_{0,\mathrm{i}\omega}(2z) = \sqrt{\frac{2z}{\pi}} K_{\mathrm{i}\omega}(z), \quad \Gamma\left(\frac{1}{2} - \mathrm{i}\omega\right)\Gamma\left(\frac{1}{2} + \mathrm{i}\omega\right) = \frac{\pi}{\cosh(\pi\omega)}.$$

We use the Wimp transform to solve the following problem:

$$0 = \left(\frac{\partial^2}{\partial w^2} + \left[-\frac{1}{4} + \frac{C_\gamma}{w}\right] - \frac{1}{w^2}\frac{\partial}{\partial \tau}\right) u. \tag{6.47}$$

$$u(0, w) = w^{C_\alpha} e^{-w/2}, \quad u(\tau, w)\Big|_{w\to\infty} = 0.$$

The Eq. (6.47) is used to get price of a Zero coupon bond in the Verhulst model, which is discussed in Chapters 2 and 14. We show that under some mild assumptions on parameters C_α and C_γ the problem in Eq. (6.47) can be efficiently solved via the Wimp transform Eq. (6.44).

In doing so, we wish to obtain a first order ODE for the function

$$\bar{u}(\tau, \omega) = \left\{\mathfrak{W}_{C_\gamma}\frac{u}{w^2}\right\}(\omega). \tag{6.48}$$

Applying the identity (Gradshtein and Ryzhik, 2007)

$$\int_0^\infty e^{-x/2} x^{\nu-1} W_{\chi,\mu}(x)dx = \frac{\Gamma\left(\nu + \frac{1}{2} - \mu\right)\Gamma\left(\nu + \frac{1}{2} + \mu\right)}{\Gamma(\nu - \chi + 1)}, \tag{6.49}$$

$$\Re\left(\nu + \frac{1}{2} \pm \mu\right) > 0,$$

we obtain the initial value $\bar{u}(0, \omega)$

$$\bar{u}(0, \omega) = \frac{\Gamma^-_{1-C_\alpha}(\omega)\Gamma^+_{1-C_\alpha}(\omega)}{\Gamma(C_\alpha - C_\gamma)}. \tag{6.50}$$

Here

$$\Gamma^\pm_y(\omega) = \Gamma\left(\frac{1}{2} - y \pm \mathrm{i}\omega\right). \tag{6.51}$$

According to Eq. (6.45), the first index of the Whittaker function should be less than 1/2, i.e., $C_\gamma \leq 1/2$. Also, the integral in Eq. (6.49) should converge, hence $C_\alpha \geq 1/2$. Therefore, the necessary condition for the direct Eq. (6.44) and inverse Eq. (6.45) Wimp transforms to exist is

$$C_\gamma \leq \frac{1}{2} \leq C_\alpha. \tag{6.52}$$

It is worth mentioning, that the problem in Eq. (6.47) can be explicitly solved by using the Laplace transform with respect to the time variable τ (under much more relaxed conditions for C_γ and C_α), as this is discussed in Chapter 14.

Let us multiply both parts of Eq. (6.47) by $W_{C_\gamma, i\omega}(w)$ and integrate on w

$$\int_0^\infty \left[\frac{\partial^2 u(\tau, w)}{\partial w^2} + \left[-\frac{1}{4} + \frac{C_\gamma}{w}\right] u(\tau, w) - \frac{1}{w^2}\frac{\partial u(\tau, w)}{\partial \tau}\right] W_{C_\gamma, i\omega}(w) dw = 0,$$

$$\int_0^\infty \left(W''_{C_\gamma, i\omega}(w) + \left[-\frac{1}{4} + \frac{C_\gamma}{w}\right] W_{C_\gamma, i\omega}(w) \right) u(\tau, w) dw$$
$$= \int_0^\infty \frac{1}{w^2}\frac{\partial u(\tau, w)}{\partial \tau} W_{C_\gamma, i\omega}(w) dw.$$

Since the function $W_{C_\gamma, i\omega}(w)$ solves the Whittaker equation Eq. (6.46), the last identity is simplified to

$$\int_0^\infty \left[-\frac{1/4 + \omega^2}{w^2} W_{C_\gamma, i\omega}(w) u(\tau, w) - \frac{1}{w^2}\frac{\partial u(\tau, w)}{\partial \tau} W_{C_\gamma, i\omega}(w)\right] dw = 0,$$

which is the first order ODE for the function $\bar{u}$

$$\frac{d\bar{u}}{d\tau} = -\left(\frac{1}{4} + \omega^2\right)\bar{u}. \tag{6.53}$$

Solving this ODE with the initial condition in Eq. (6.50) and applying the inversion Eq. (6.45), we immediately obtain the final formula for $u(\tau, w)$

$$u(\tau, w) = \frac{1}{\pi^2 \Gamma(C_\alpha - C_\gamma)} \int_0^\infty \left\{ \omega e^{-(\omega^2 + 1/4)\tau} \sinh(2\pi\omega) \Gamma^-_{C_\gamma}(\omega) \Gamma^+_{C_\gamma}(\omega) \right.$$
$$\left. \times \Gamma^-_{1-C_\alpha}(\omega) \Gamma^+_{1-C_\alpha}(\omega) W_{C_\gamma, i\omega}(w) \right\} d\omega.$$

6.5 The Mellin transform

Another popular integral transform is the Mellin transform. It was developed by Finnish mathematician Hjalmar Mellin and is widely used in the number theory and probability and asymptotic analysis. The Mellin transform can be considered as a multiplicative version of the bilateral Laplace transform. Indeed, by replacing the function $f(x)$ in Eq. (6.15) using a map $f(-\ln x) \mapsto f(x)$, we arrive at the Mellin transform.

Definition 6.6. The operator $\mathfrak{M}$ acting on functions $f(x) : \mathbb{R}_+ \to \mathbb{R}$ and defined by

$$\bar{f}(p) = \{\mathfrak{M}f\}(p) = \int_0^\infty x^{p-1} f(x)dx, \quad \alpha < \Re(p) < \beta \tag{6.54}$$

is called the Mellin transform. The inverse transform is given by

$$\{\mathfrak{M}^{-1}\bar{f}\}(x) = \frac{1}{2\pi\mathrm{i}} \int_{c-\mathrm{i}\infty}^{c+\mathrm{i}\infty} x^{-p}\bar{f}(p)dp, \qquad \alpha < c < \beta. \tag{6.55}$$

In general, the integral exists only for complex values of $p = a + b\mathrm{i}$ such that $\alpha < a < \beta$, where α, β depend on the function $f(t)$ to transform. This introduces what is called the strip of definition of the Mellin transform that will be denoted by $S(\alpha, \beta)$. In some cases, this strip may extend to a half-plane ($\alpha = -\infty$ or $\beta = \infty$) or to the whole complex p-plane ($\alpha = -\infty$ and $\beta = \infty$).

The important property of the Mellin transform $\mathfrak{M}$ is that the differential operator $x^n \partial^n / \partial x^n$ is mapped to an algebraic operator.

We illustrate application of the transform Eq. (6.54) using the following example. Let X_t be a Geometric Brownian Motion (GBM) with the constant drift μ and volatility σ

$$dX_t = \mu X_t dt + \sigma X_t dW_t,$$

where W_t is a standard Brownian motion. Our aim is to calculate the following conditional expectation of the random variable X_T

$$g(x, t; T) = \mathbb{E}[f(X_T)|X_t = x].$$

Here $f(\cdot)$ is a continuous function of a polynomial growth. By a standard result of probability theory, the function $g(x, t; T)$ solves the terminal-boundary value problem for the backward Kolmogorov equation

$$\begin{aligned}
&\Big(\frac{\sigma^2 x^2}{2}\frac{\partial^2}{\partial x^2} + \mu x \frac{\partial}{\partial x} + \frac{\partial}{\partial t}\Big) g(x, t; T) = 0, \quad (t, x) \in \mathbb{R}_+ \times \mathbb{R}_+, \qquad (6.56)\\
&g(x, T; T) = f(x), \quad x \in \mathbb{R}_+,\\
&g(0, t; T) = 0, \quad \lim_{x\to\infty} g(x, t; T) = 0, \quad 0 \le t < T.
\end{aligned}$$

Applying the Mellin transform $\bar{g}(t, p) = \mathfrak{M}g(x, t; T)$ with respect to the spatial variable x, we obtain the following ODE

$$\frac{d\bar{g}}{dt} + \left[p(p+1)\frac{\sigma^2}{2} - p\mu\right]\bar{g} = 0, \tag{6.57}$$

which should be solved subject to terminal condition

$$\bar{g}(T,p) = \int_0^\infty y^{1-p} f(y) dy. \tag{6.58}$$

The solution of Eqs. (6.57) and (6.58) is straightforward

$$\bar{g}(t,p) = \int_0^\infty y^{p-1} e^{\frac{p^2\sigma^2}{2}(T-t) - p\left(\mu - \frac{\sigma^2}{2}\right)(T-t)} f(y) dy.$$

Using the inversion formula Eq. (6.55) with $c = 0$, making change of the integration variable $p \mapsto \mathrm{i}\zeta$ and changing the order of integrations yields

$$g(x,t;T) = \int_0^\infty \left[\int_{-\infty}^{+\infty} \frac{e^{\frac{-\zeta^2\sigma^2}{2}(T-t) + \mathrm{i}\zeta\left[\ln\left(\frac{x_0}{x}\right) - \left(\mu - \frac{\sigma^2}{2}\right)(T-t)\right]}}{2\pi} d\zeta \right] \frac{f(y)}{y} dy.$$

Since the inner integral is known, see Gradshtein and Ryzhik (2007) or Eq. (6.12), the final formula for $g(x,t;T)$ reads

$$g(x,t;T) = \frac{1}{\sigma\sqrt{2\pi(T-t)}} \int_0^\infty \frac{f(y)}{y} e^{-\frac{\left[\ln\left(\frac{x}{y}\right) + \left(\mu - \frac{\sigma^2}{2}\right)(T-t)\right]^2}{2\sigma^2(T-t)}} dy.$$

Chapter 7

Generalized Integral Transforms

In this chapter, we introduce the method of generalized integral transform (GIT). This method offers an attractive possibility of solving various initial-boundary value problems for parabolic PDE in semi-analytical form for domains with curvilinear boundaries. The main idea of the GIT consists in reduction of the corresponding PDE problem to the Volterra equation of the second kind. To the best of authors' knowledge, initially the GIT method was developed in papers of A.V. Luikov, B.Ya. Lyubov, E.M. Kartashov, and is described in detail in Kartashov (1999). However, this development covers only the domain $[0, y(t)]$, where $y(t) < \infty$ is the moving boundary to be a function of time t. As applied to finance, first this method was used in Carr and Itkin (2021) for pricing American and barrier options. Later in a series of papers of the authors of this book and also Peter Carr, similar problems for other time-dependent domains including those with two moving boundaries have been solved. These approaches and solution are presented in Chapters 4–7.

Chapter 6 clearly articulates the power of classical integral transforms in solving various initial-boundary value problems with constant coefficients and fixed boundaries. However, as mentioned in Kartashov (1999), introduction of additional factors into the formulation of a boundary-value problem even of the linear type (complication of the shape of a body, moving boundaries of the domain of definition, the time or space dependence of the thermophysical characteristics of a medium, etc.) necessitates the development of a special area of mathematics that, as a rule, yields an semi-analytical (or, perhaps, approximate) solution of the problem and turns out to be efficient for obtaining an exact solution only in a certain situation.

In physics the boundary-value problems of non-stationary transfer in a non-cylindrical region include the cases where the motion of the region boundary is known, or the cases where it is required that this motion should be determined from additional conditions of the problem (the Stefan condition that expresses the energy balance in transition of the medium from one aggregate state to another one; or a Stefan-type condition in more general problems for a heat equation with a free boundary). Solution of these and similar problems requires modern analytical methods which are able to operate with curvilinear boundaries. In this Chapter we describe basic principles of construction of such a method (the GIT method). Note, that in doing so, we significantly extend and generalize the original approach described in Kartashov (2001).

7.1 PDE with moving boundaries

Let Ω be a domain with curvilinear boundaries. In the one-dimensional case Ω can be a curvilinear strip $\Omega = [y(\tau), z(\tau)]$ or a half-plane $\Omega = [y(\tau), \infty)$ with curvilinear boundary $y(\tau)$. In what follows we assume that $y(t), z(t) \in C^1$.

Consider the following initial-boundary value problem at the domain Ω with absorbing boundary conditions:

$$\sigma^2(x)\frac{\partial^2 u(x,\tau)}{\partial x^2} + \mu(x)\frac{\partial u(x,\tau)}{\partial x} + \nu(x)u(\tau,x) = \frac{\partial u(x,\tau)}{\partial \tau}, \tag{7.1}$$
$$u(\tau,x)\Big|_{x\in\partial\Omega} = 0, \quad u(0,x) = f(x), \quad (x,\tau)\in\Omega\times\mathbb{R}_+.$$

Here x, τ are the independent spatial and temporal variables, and $u(\tau, x)$ is the dependent variable such that the first derivative of u in τ and the second derivative of u in x exist and are bounded.

The PDE in Eq. (7.1) is a parabolic equation whose solution should be found at the domain with moving boundaries. These kind of problems are known in physics for a long time. As mentioned, analytical solutions of these problems require non-traditional, and sometimes sophisticated methods. Also, it is worth mentioning that the problems Eq. (7.1) are closely related to finding distributions of the first passage time of a curvilinear boundary by a diffusion processes, see Alili *et al.* (2005), Borovkov and Novikov (2005), Lipton and Lopez de Prado (2020b) and references therein.

Let us start with a simple case of the parabolic PDE Eq. (7.1) — the heat equation in the curvilinear strip $\Omega = [y(\tau), z(\tau)]$

$$\frac{\partial^2 u}{\partial x^2} = \frac{\partial u}{\partial \tau}, \quad (x,\tau)\in[y(\tau), z(\tau)]\times\mathbb{R}_+, \tag{7.2}$$

$$u(\tau, y(\tau)) = 0, \quad u(\tau, z(\tau)) = 0, \quad u(0, x) = f(x).$$

Since the boundaries $y(\tau)$ and $z(\tau)$ are time dependent, any classical integral transform will fail to solve this problem. Technically, we can augment the function $u(\tau, x)$ at the extended domains $(x, \tau) \in (-\infty, y(\tau)) \times \mathbb{R}_+$ and $(x, \tau) \in (z(\tau), \infty) \times \mathbb{R}_+$ to be

$$U(\tau, x) = \begin{cases} u(\tau, x), & (x, \tau) \in (y(\tau), z(\tau)) \times \mathbb{R}_+, \\ 0, & (x, \tau) \in (-\infty, y(\tau)) \cup (z(\tau), \infty) \times \mathbb{R}_+, \end{cases} \tag{7.3}$$

and apply the Fourier (or bilateral Laplace) transform to the augmented function $U(\tau, x)$. This is equivalent to usage of the generalized integral transform

$$\bar{u}(\tau, \lambda) = \int_{y(\tau)}^{z(\tau)} e^{-\lambda x} u(\tau, x) dx. \tag{7.4}$$

By analogy with any classical integral transform, the image $\bar{u}$ can be explicitly found by solving the corresponding first order ODE. However, the transform in Eq. (7.4) does not have a known inversion formula, hence it is not crystal clear how to reconstruct $u(\tau, x)$ from $\bar{u}(\tau, \lambda)$. Alternatively, we can design a generalized integral transform in a way similar to Eq. (7.4), which is, however, invertible under some mild assumptions about the original function.

To be precise, let us seek for the unknown function $u(\tau, x)$ in the form

$$u(\tau, x) = \sum_{n=0}^{\infty} \alpha_n(\tau) \psi_n(x), \tag{7.5}$$

or

$$u(\tau, x) = \int_0^{\infty} \alpha(\xi, \tau) \psi(\xi, x) d\xi, \tag{7.6}$$

where functions $\psi_n(x)$ are assumed to be known, and the weights $\alpha_n(\tau)$ — to be unknown. Our main idea is first, to apply the generalized integral transform to Eq. (7.5)(or Eq. (7.6)), and then to Eq. (7.2). Once this is done, we can further equalize the corresponding images and from this condition find the unknown weights $\alpha_n(\tau)$.

To rephrase it a bit differently, our scheme consists of two steps: first, we have to construct the representation Eq. (7.5) (or Eq. (7.6)) and find the eigenfunctions $\psi_n(x)$ as Eq. (7.5) can be recognized as an eigenfunction expansion (spectral expansion) of $u(\tau, u)$. Second, we have to find the integral transform such that it maps the PDE in Eq. (7.2) to the first order

ODE, and so coefficients $\alpha_n(\tau)$ can be explicitly found if images both parts of Eq. (7.5) are known.

Since Eqs. (7.5) and (7.6) are closely related to the eigenfunction expansions associated with the second order ODE (Titchmarsh, 1962) in the next section we provide a brief introduction to this theory.

7.2 Sturm–Liouville expansion

Following Titchmarsh (1962), consider the eigenvalue problem in the strip $x \in [a, b]$

$$\frac{d^2 y}{dx^2} - q(x) y = p y. \tag{7.7}$$

Here $q(x)$ is a smooth function on $[a, b]$. The function $\psi_n(x)$ associated with some value p_n is called eigenfunction if: $\psi_n(x)$ satisfies Eq. (7.7) with $p = p_n$, and $\psi_n(a) = \psi_n(b) = 0$. In other words, $\psi_n(x)$ solves the problem

$$\begin{aligned} &\frac{d^2 \psi_n}{dx^2} - q(x)\psi_n = p_n \psi_n, \\ &\psi_n(a) = 0, \quad \psi_n(b) = 0. \end{aligned} \tag{7.8}$$

A set of functions $\{\psi_n\}_{n=0}^{\infty}$ forms an orthogonal basis on $[a, b]$, indeed

$$\begin{aligned} (p_n - p_m)\psi_n\psi_m &= \psi_n''\psi_m - \psi_n\psi_m'', \\ \int_a^b (p_n - p_m)\psi_n\psi_m dx &= \int_a^b (\psi_n''\psi_m - \psi_n\psi_m'')\, dx, \\ (p_n - p_m)\int_a^b \psi_n\psi_m dx &= (\psi_n'\psi_m - \psi_n\psi_m')\Big|_{x=a}^{x=b}. \end{aligned}$$

Taking into account the boundary conditions Eq. (7.8) we obtain

$$\int_a^b \psi_n\psi_m dx = 0. \tag{7.9}$$

Therefore, any function $f : [a, b] \mapsto \mathbb{R}$ such that $f(a) = f(b) = 0$ can be represented via the following expansion

$$f(x) = \sum_{n=0}^{\infty} c_n \psi_n(x), \quad c_n = \frac{\int_a^b f(x)\psi_n(x)dx}{\int_a^b \psi_n^2(x)dx}. \tag{7.10}$$

A natural example of the expansion in Eq. (7.10) is the trigonometric Fourier series. The corresponding ODE defining this series is $y'' = py$. The well-known solution of this ODE under vanishing boundary condition at $x = a$ reads

$$\sin\left(\sqrt{p}(x-a)\right). \tag{7.11}$$

To proceed, let us set p such that the function Eq. (7.11) also vanishes at $x = b$, i.e.,

$$p = p_n = \frac{\pi^2 n^2}{(b-a)^2}, \qquad n \in \mathbb{Z}_+.$$

Then, the set of eigenfunctions ψ_n for this problem reads

$$\psi_n(x) = \sin\left(\pi n \frac{(x-a)}{b-a}\right). \tag{7.12}$$

Accordingly, the expansion Eq. (7.10) takes the following form

$$f(x) = \sum_{n=0}^{\infty} c_n \sin\left(\pi n \frac{(x-a)}{b-a}\right), \tag{7.13}$$

$$c_n = \frac{2}{b-a}\int_a^b f(x)\sin\left(\pi n \frac{(x-a)}{b-a}\right) dx.$$

In case $b = \infty$ the expansion Eq. (7.13) transforms to

$$f(x) = \int_{-\infty}^{+\infty} c_\xi \psi_\xi(x) d\xi. \tag{7.14}$$

7.3 Construction of the GIT

Let us again consider the problem in Eq. (7.2). Since any arbitrary function $f(x) : [a, b] \mapsto \mathbb{R}_+$ can be represented in the form Eq. (7.13), the function $u(\tau, x)$ can be sought in terms of the following series

$$u(\tau, x) = \sum_{n=0}^{\infty} c_n(\tau) \sin\left(\pi n \frac{x - y(\tau)}{z(\tau) - y(\tau)}\right). \tag{7.15}$$

Once the coefficients c_n are found, the problem is solved. However, it can be checked that direct application of the formula for c_n

$$c_n(\tau) = \frac{2}{z(\tau) - y(\tau)} \int_{y(\tau)}^{z(\tau)} u(\tau, x) \sin\left(\pi n \frac{(x - y(\tau))}{z(\tau) - y(\tau)}\right) dx \tag{7.16}$$

to Eq. (7.2) does not give rise to the explicit expression for c_n.

Therefore, instead we wish to construct the integral transform similar to Eq. (7.16) or Eq. (7.4), but such that the following two properties are held: (a) applying this transform to Eq. (7.2) yields the first order ODE, and (b) applying the transform to Eq. (7.15) results in a solvable system of equations for the coefficients c_n. One can see, that the transform Eq. (7.4) does not meet the second requirement, while the transform Eq. (7.16) does not meet the first one.

We underline, that it is not necessary to know the explicit inversion formula for $\bar{u}(p,x)$ because we rely on the predefined representation Eq. (7.15) of $u(\tau,x)$. The kernel of the transform depends on geometry of the domain and the explicit form of the corresponding differential operator. In particular, for the problem in Eq. (7.2) the following transform makes the deal

$$\bar{u}(\tau,p) = \int_{y(\tau)}^{z(\tau)} u(\tau,x)\sin\left(p\left[x - y(\tau)\right]\right) dx. \tag{7.17}$$

Indeed, application of Eq. (7.17) to Eq. (7.15) gives rise to the identity

$$\bar{u}(\tau,p) = \int_{y(\tau)}^{z(\tau)} \sum_{n=0}^{\infty} c_n(\tau)\sin\left(\pi n \frac{x-y(\tau)}{z(\tau)-y(\tau)}\right)\sin\left(p\left[x-y(\tau)\right]\right) dx.$$

By changing the order of integration and summation, and also making a change of variable $\zeta = (x - y(\tau))/l(\tau)$, $l(\tau) = z(\tau) - y(\tau)$, we obtain

$$\bar{u}(\tau,p) = l(\tau)\sum_{n=0}^{\infty} c_n(\tau)\int_0^1 \sin\left(\pi n\zeta\right)\sin\left(pl(\tau)\zeta\right) d\zeta. \tag{7.18}$$

Using the well-known identity

$$\int_0^1 \sin\left(\pi n\zeta\right)\sin\left(\pi m\zeta\right) d\zeta = \begin{cases} 0, & n \neq m, \\ \frac{1}{2}, & n = m, \end{cases}$$

and substituting $p = \pi n/l(\tau)$ into Eq. (7.18) we arrive at the explicit expression for the coefficients $c_n(\tau)$

$$c_n(\tau) = \frac{2}{l(\tau}\bar{u}\left(\tau, \frac{\pi n}{l(\tau)}\right).$$

As mentioned in above, the image $\bar{u}(\tau,p)$ can be found as a solution to the first order ODE.

A slightly different approach which relies on contour integration in a complex plane and application of the Cauchy's residue theorem is presented in Chapter 18. To utilize this approach one should use the following integral transform

$$\bar{u}(\tau,p) = \int_{y(\tau)}^{z(\tau)} u(\tau,x)\sinh\left(p\left[x - y(\tau)\right]\right) dx.$$

7.3.1 Another illustration of the GIT method

Let us consider the following problem

$$\frac{\partial u}{\partial \tau} = \left(\frac{\partial^2 u}{\partial x^2} + \frac{1-2m}{x}\frac{\partial u}{\partial x} \right) + f(\tau, x), \quad (x, \tau) \in [0, y(\tau)] \times \mathbb{R}_+,$$
$$u(\tau, 0) = 0, \quad u(\tau, y(\tau)) = 0, \tag{7.19}$$
$$u(0, x) = \phi(x), \quad m \geq -1/2.$$

For some special cases ($m = -1/2, 0, 1/2$) this problem was considered in Kartashov (1999). Here we attack it for an arbitrary m by applying the generalized Hankel transform

$$\bar{u}(\tau, p) = \int_0^{y(\tau)} x^{1-m} J_m(x\sqrt{p}) u(\tau, x) dx, \tag{7.20}$$

where p is a complex number, $\Re(p) \geq 0$ and $-\pi/4 \geq \arg p \geq \pi/4$, $J_m(x)$ is the modified Bessel function defined in Section 6.3. Introducing the image $\bar{f}(\tau, p)$ by

$$\bar{f}(\tau, p) = \int_0^{y(\tau)} x^{1-m} J_m(x) f(\tau, x) dx,$$

and using Eq. (6.35), we obtain the first order ODE and the initial condition for $\bar{u}(\tau, p)$

$$\frac{d\bar{u}}{d\tau} = -p\bar{u} + \bar{f}(\tau, p) + y^{1-m}(\tau) J_m(y(\tau)\sqrt{p}) \Psi(\tau), \tag{7.21}$$
$$\bar{u}(0, p) = \int_0^{y(0)} x^{1-m} J_m(y(\tau)\sqrt{p}) \phi(x) dx.$$

In Eq. (7.21) $\Psi(\tau)$ denotes the gradient of the solution $u(\tau, x)$ at the boundary $y(\tau)$

$$\Psi(\tau) = \left. \frac{\partial u}{\partial x} \right|_{x=y(\tau)}.$$

The solution of Eq. (7.21) is straightforward and reads

$$\bar{u}(\tau, p) = \int_0^{\tau} e^{-p(\tau - s)} \left[y^{1-m}(s) J_m(y(s)\sqrt{p}) \Psi(s) + \bar{f}(s, p) \right] ds \tag{7.22}$$
$$+ e^{-p\tau} \int_0^{y(0)} x^{1-m} J_m(y(\tau)\sqrt{p}) \phi(x) dx.$$

Following the main concept of the GIT method, we wish to construct an orthonormal basis on segment $[0, a]$ which consists of the Bessel functions

$J_\nu(\cdot)$. According to Section 7.2, let us consider the following ODE in a strip $x \in [0,1]$

$$\frac{d^2y}{dx^2} + \left(s^2 - \frac{\nu^2 - 1/4}{x^2}\right) y = 0. \tag{7.23}$$

It can be checked that the function $\sqrt{x}J_\nu(xs)$ solves Eq. (7.23). Moreover, this function vanishes at $x = 0$

$$\lim_{x\to 0} \sqrt{x}J_\nu(xs) = 0.$$

Therefore, functions $\psi_n(x)$ can be defined as

$$\psi_n(x) = \sqrt{x}J_\nu\left(\frac{\mu_n x}{a}\right), \tag{7.24}$$

where μ_n is an ordered sequence of positive zeros of the Bessel function $J_\nu(\mu_n) = 0$

$$J_\nu(\mu_n) = J_\nu(\mu_m), \quad \mu_n > \mu_m, \quad n > m.$$

Since the functions $\psi_n(x)$ form an orthogonal basis in the space $\mathcal{C}[0,1]$, the solution $u(\tau,x)$ of the problem in Eq. (7.19) can be found in the following form:

$$u(\tau,x) = \sum_{n=1}^{\infty} \alpha_n(\tau) x^m J_m\left(\frac{\mu_n x}{y(\tau)}\right). \tag{7.25}$$

Application of the integral transform Eq. (7.20) to Eq. (7.25) yields

$$\bar{u}(\tau,x) = \int_0^{y(\tau)} \sum_{n=1}^{\infty} \alpha_n(\tau) x J_m\left(\frac{\mu_n x}{y(\tau)}\right) J_m(x\sqrt{p})dx.$$

Further, changing the order of summation and integration, and introducing a new variable $z = x/y(\tau)$ we arrive at the following expression:

$$\frac{\bar{u}(\tau,p)}{y^2(\tau)} = \sum_{n=1}^{\infty} \alpha_n(\tau) \int_0^1 z J_m(\mu_n z) J_m(\sqrt{p}y(\tau)z)dz.$$

Now we need the orthogonality of the Bessel functions (Gradshtein and Ryzhik, 2007)

$$\int_0^1 xJ_\nu(\alpha x)J_\nu(\beta x)dx = \begin{cases} 0, & \alpha \neq \beta, \\ \frac{1}{2}J^2_{\nu+1}(\alpha), & \alpha = \beta,\ J_\nu(\alpha) = J_\nu(\beta), \end{cases}$$

that immediately gives rise to the explicit expression for the weight $\alpha_n(\tau)$

$$\alpha_n(\tau) = 2\frac{\bar{u}\left(\tau, \mu_n^2/y^2(\tau)\right)}{y^2(\tau)J^2_{m+1}(\mu_n)}. \tag{7.26}$$

Combining Eqs. (7.22) and (7.26) and substituting them into Eq. (7.25) yields a semi-analytical representation of the solution to the problem in Eq. (7.19). The unknown function $\Psi(\tau)$ can be found by solving a Fredholm equation, see Kartashov (1999). However, as we show in Chapter 13, the function $\Psi(\tau)$ also solves a Volterra integral equation of the second kind.

7.4 General case

In the previous sections we demonstrated a power of the GIT method by two particular examples associated with the heat and Bessel equations. These two cases are very important for the analysis of a general problem in Eq. (7.1). Indeed, it is well known that if Eq. (7.1) has six or four groups of Lie symmetries, it can be transformed into Eq. (7.2) or Eq. (7.19), respectively (Craddock, 2009; Craddock and Lennox, 2007). At the same time, any parabolic PDE with time homogeneous coefficients can have only zero, four or six groups of non-trivial symmetries. Therefore, the transforms Eqs. (7.17) and (7.20) can be used for a richer class of problems which have non-trivial Lie symmetries.

The problems of type Eq. (7.1) without non-trivial symmetries arise in many areas of applied science, and also are interesting in a pure mathematical context. In this section we show that the GIT method is still applicable in this general case.

Under some mild assumptions about the coefficients $\sigma(x)$, $\mu(x)$ and $\nu(x)$ the problem in Eq. (7.1) in a curvilinear strip $x \in \Omega = [y(\tau), z(\tau)]$ can be re-formulated as follows:

$$\begin{aligned} \frac{\partial u(\tau,x)}{\partial \tau} &= \frac{\partial^2 u(\tau,x)}{\partial x^2} - q(x)u(\tau,x), \quad (x,\tau) \in [y(\tau), z(\tau)] \times \mathbb{R}_+, \\ u(\tau, y(\tau)) &= 0, \quad u(\tau, z(\tau)) = 0, \\ u(0,x) &= \delta(x - x_0). \end{aligned} \tag{7.27}$$

Again, we solve this problem by using a generalized integral transform

$$\bar{u}(\tau,p) = \int_{y(\tau)}^{z(\tau)} \psi_p(x, y(\tau)) u(\tau,x) dx, \tag{7.28}$$

where the kernel $\psi_p(x, y(\tau))$ is defined as

$$\psi_p(a,b) = \varphi_p^+(a)\varphi_p^-(b) - \varphi_p^-(a)\varphi_p^+(b).$$

Here two functions $\varphi_p^-(x)$ and $\varphi_p^+(x)$ represent a pair of linearly independent solutions to the second order ODE

$$y'' - q(x)y + py = 0.$$

However, we cannot apply this transform directly to Eq. (7.27) because the kernel is time-dependent. Therefore, instead we represent $\bar{u}$ as the weighted sum of the following transforms:

$$\bar{u}(\tau,p) = \bar{u}^+(\tau,p)\varphi_p^-(y(\tau)) - \bar{u}^-(\tau,p)\varphi_p^+(y(\tau)), \tag{7.29}$$

$$\bar{u}^{\pm}(\tau,p) = \int_{y(\tau)}^{z(\tau)} \varphi_p^{\pm}(x) u(\tau,x) dx.$$

Application of Eq. (7.29) directly to Eq. (7.27) now yields a couple of first order ODEs

$$\frac{d\bar{u}^{\pm}(\tau,p)}{d\tau} = -p\bar{u}^{\pm}(\tau,p) + \left[\Psi(\tau)\varphi_p^{\pm}(y(\tau)) + \Phi(\tau)\varphi_p^{\pm}(z(\tau))\right], \tag{7.30}$$

that should be solved subject to the initial conditions

$$\bar{u}^{+}(0,p) = \varphi_p^{+}(x_0), \quad \bar{u}^{-}(0,p) = \varphi_p^{-}(x_0). \tag{7.31}$$

By $\Psi(\tau)$ and $\Phi(\tau)$, as usual, we denote the gradients of the function $u(\tau,x)$ on the boundaries $y(\tau)$ and $z(\tau)$, respectively

$$\Psi(\tau) = -\left.\frac{\partial u}{\partial x}\right|_{x=y(\tau)}, \qquad \Phi(\tau) = \left.\frac{\partial u}{\partial x}\right|_{x=z(\tau)}.$$

The solution of Eqs. (7.30) and (7.31) can be found straightforward

$$\bar{u}^{\pm}(\tau,p) = e^{-p\tau}\varphi_p^{\pm}(x_0) + \int_0^{\tau} e^{-p(\tau-s)}\left[\Psi(s)\varphi_p^{\pm}(y(s)) + \Phi(s)\varphi_p^{\pm}(z(s))\right] ds.$$

Therefore, we obtain the following representation of $\bar{u}(\tau,p)$

$$\begin{aligned}\bar{u}(\tau,p) &= e^{-p\tau}\psi_p(x_0, y(\tau)) \\ &+ \int_0^{\tau} e^{-p(\tau-s)}\left[\Psi(s)\psi_p(y(s),y(\tau)) + \Phi(s)\psi_p(z(s),y(\tau))\right] ds.\end{aligned}$$

Following Section 7.2, we construct an orthogonal basis on $[y(\tau), z(\tau)]$ which contains only the functions $\psi_p(x, y(\tau))$. Let us define $\lambda_n(\tau)$ as an n-th positive root of the following equation

$$\psi_{\lambda_n(\tau)}(z(\tau), y(\tau)) = \psi_{\lambda_m(\tau)}(z(\tau), y(\tau)) = 0, \quad \lambda_m > \lambda_n, \quad m > n.$$

It can be shown that the set $\left\{\psi_{\lambda_n(\tau)}(x, y(\tau))\right\}\big|_{n=1}^{\infty}$ forms an orthogonal basis on $[y(\tau), z(\tau)]$. By analogy with the previous examples, we are seeking the function $u(\tau,x)$ in the following form:

$$u(\tau,x) = \sum_{n=1}^{\infty} \alpha_n(\tau)\psi_{\lambda_n}(x, y(\tau)). \tag{7.32}$$

It is worth noting, that the representation Eq. (7.32) automatically respects the absorbing boundary conditions in Eq. (7.27). Taking the transform Eq. (7.28) of Eq. (7.32)

$$\bar{u}(\tau,p) = \sum_{n=1}^{\infty} \alpha_n(\tau) \int_{y(\tau)}^{z(\tau)} \psi_{\lambda_n}(x, y(\tau))\psi_p(x, y(\tau)) dx,$$

and using the orthogonality of the functions $\psi_{\lambda_n(\tau)}(x, y(\tau))$, we arrive at the explicit expression for the weight

$$\alpha_n(\tau) = \bar{u}(\tau, \lambda_n(\tau)) \left[\int_{y(\tau)}^{z(\tau)} \left[\psi_{\lambda_n(\tau)}(x, y(\tau))\right]^2 dx \right]^{-1}.$$

The functions $\Psi(\tau)$ and $\Phi(\tau)$ can be found from the following system of Volterra equations of the second kind

$$-\Psi(\tau) = \sum_{n=1}^{\infty} \alpha_n(\tau) \psi'_{\lambda_n}(y(\tau), y(\tau)),$$

$$\Phi(\tau) = \sum_{n=1}^{\infty} \alpha_n(\tau) \psi'_{\lambda_n}(z(\tau), y(\tau)).$$

Here $\psi'_p(\cdot, \cdot)$ denotes the derivative with respect to the first argument, i.e.,

$$\psi'_p(x, a) = \frac{d\psi_p(x, a)}{dx}.$$

Thus, we have managed to construct generalization of the GIT method for the heat equation with a killing term, Eq. (7.27) which is defined at the domain with two moving boundaries. A real challenge with this approach is finding the basis functions $\psi_{\lambda_n}(x, \cdot)$ and roots $\lambda(x, \cdot)$. We showed in this Chapter that this can be done analytically for the heat and Bessel PDEs. Perhaps, same approach can be utilized for some other PDEs as well where the basis functions and the roots are known, see examples in Polyanin (2002). However, for a general PDE of the form Eq. (7.27) this could be problematic.

Chapter 8

Method of Heat Potentials

In this chapter, we describe the classical method of heat potentials (HP) The HP method is uniquely well-suited to solving rather challenging problems occurring routinely in applied mathematics in general, and in financial engineering in particular. In a nutshell, this method allows one to reduce a complicated partial differential equation of the parabolic type with a time-dependent boundary to a much simpler Volterra integral equation. A detailed introduction to heat potentials can be found in Watson (2012) and Tikhonov and Samarskii (1963). The first use of the method of heat potentials in mathematical finance is found in Lipton (2001) for pricing path-dependent options with curvilinear barrier.

8.1 Statement of the problem

Let us consider a heat equation

$$\frac{\partial u}{\partial \tau} = \sigma^2 \frac{\partial^2 u}{\partial x^2}, \tag{8.1}$$

where x, τ are the spatial and temporal coordinates, $\sigma > 0$ is the constant coefficient of heat conductivity and $u(x, \tau)$ is an unknown function differentiable at least once in τ and twice in x. The Eq. (8.1) should be solved subject to the initial condition

$$u(x, 0) = f(x). \tag{8.2}$$

Suppose we are seeking for the solution of this problem at the domain $x \in \Omega : [y(\tau), \infty)$, where $y(\tau) \in \mathcal{C}^1$ is the time-dependent lower boundary of the domain, and $y'(\tau) < \infty$. Also suppose that the boundary conditions to Eq. (8.1) are homogeneous, i.e.,

$$u(x, \tau)\Big|_{x \to \infty} = 0, \qquad u(y(\tau), \tau) = 0, \tag{8.3}$$

Since this problem has an inhomogeneous initial condition, the classical method of heat potentials cannot be directly applied, see Tikhonov and Samarskii (1963). Therefore, let us represent $u(x,t)$ in the form

$$u(x,\tau) = q(x,\tau) + \frac{1}{2\sigma\sqrt{\pi\tau}} \int_{y(0)}^{\infty} u(x',0) e^{-\frac{(x-x')^2}{4\sigma^2\tau}} dx'. \tag{8.4}$$

Here the second term in the RHS is the solution of the heat equation in Eq. (8.1) at the infinite domain with the initial condition in Eq. (8.2). But since $x \in \Omega$, we moved the left boundary from $-\infty$ to $y(0)$.

The function $q(x,\tau)$ solves the problem

$$\begin{aligned}
\frac{\partial q(x,\tau)}{\partial \tau} &= \sigma^2 \frac{\partial^2 q(x,\tau)}{\partial x^2}, \\
q(x,0) &= 0, \qquad y(0) < x < \infty, \\
q(x,\tau)\Big|_{x\to\infty} &= 0, \qquad q(y(\tau),\tau) = \phi(\tau), \\
\phi(\tau) &= -\frac{1}{2\sigma\sqrt{\pi\tau}} \int_{y(0)}^{\infty} u(x',0) e^{-\frac{(y(\tau)-x')^2}{4\sigma^2\tau}} dx'.
\end{aligned} \tag{8.5}$$

This problem is similar to that in Eqs. (8.1)–(8.3), but now with a homogeneous initial condition. Therefore, the solution can be found by the method of heat potentials.

Following the general idea of this method, we are looking for the solution of Eq. (8.5) in the form of a generalized heat potential

$$q(x,\tau) = \frac{1}{4\sqrt{\pi}} \int_0^{\tau} \Psi(k) \frac{x - y(k)}{\sigma^3\sqrt{(\tau-k)^3}} e^{-\frac{(x-y(k))^2}{4\sigma^2(\tau-k)}} dk, \tag{8.6}$$

where $\Psi(k)$ is the heat potential density. It is easy to check that thus defined function $q(x,\tau)$ solves the first line of Eq. (8.5), and satisfies both the initial condition and the vanishing condition at $x \to \infty$. Indeed, at $\tau = 0$ the integral in Eq. (8.6) vanishes; at $x \to \infty$ the exponent under the integral tends to zero, while at $x = y(\tau)$ the numerator vanishes. Finally, substituting the representation of q in Eq. (8.6) into the heat equation (first line in Eq. (8.5)) we obtain

$$\begin{aligned}
\frac{\partial}{\partial \tau} q(x,\tau) &= \frac{1}{4\sqrt{\pi}} \int_0^{\tau} \Psi(k) \frac{\partial}{\partial \tau} H(x,\tau,k) dk + \Psi(\tau) H(x,\tau,\tau), \\
\frac{\partial^2}{\partial x^2} q(x,\tau) &= \frac{1}{4\sqrt{\pi}} \int_0^{\tau} \Psi(k) \frac{\partial^2}{\partial x^2} H(x,\tau,k) dk, \\
H(x,\tau,k) &= \frac{x - y(k)}{\sigma^3\sqrt{(\tau-k)^3}} e^{-\frac{(x-y(k))^2}{4\sigma^2(\tau-k)}}.
\end{aligned} \tag{8.7}$$

Obviously, $H(x, \tau, k)$ is the derivative of the heat kernel

$$H(x, \tau, k) = -2\frac{\partial}{\partial x}\left[\frac{1}{\sigma\sqrt{(\tau - k)}}e^{-\frac{(x-y(k))^2}{4\sigma^2(\tau-k)}}\right], \tag{8.8}$$

and as such — it obeys the heat equation. This implies that

$$\frac{\partial}{\partial \tau}H(x, \tau, k) = \sigma^2\frac{\partial^2}{\partial x^2}H(x, \tau, k). \tag{8.9}$$

The last thing to check is that $H(x, \tau, \tau) = 0$. This is obviously true if $x \neq y(\tau)$.

However, for $x = y(\tau)$ function $q(x, \tau)$ is discontinuous, and its limiting value at $x = y(\tau) + 0$ is equal to $\phi(\tau)$. We show this in the next section. Our aim below is to derive the value of this integral at $x \to y(\tau) \pm 0$, and the gradient $\partial q(\tau, x)/\partial x$ in the same limit, namely

$$\varphi(\tau) = \lim_{x \to y(\tau) \pm 0} q(\tau, x), \qquad \psi(\tau) = \lim_{x \to y(\tau) \pm 0} \frac{\partial q(\tau, x)}{\partial x}. \tag{8.10}$$

By doing so we closely follow Itkin and Muravey (2020a).

8.1.1 *The limiting value of* $\varphi(t)$

This result is obtained, e.g., in Tikhonov and Samarskii (1963). Consider a function $W(\tau, x) = 2\sigma^2\phi(t)$

$$W(\tau, x) = \int_0^\tau \Omega(k)\frac{x - y(k)}{2\sigma\sqrt{\pi(\tau - k)^3}}e^{-\frac{(y(\tau)-y(k))^2}{4\sigma^2(\tau-k)}}dk. \tag{8.11}$$

Also consider an auxiliary integral

$$\tilde{V}(\tau, x) = \int_0^\tau \frac{y'(k)\Omega(k)}{\sigma\sqrt{\pi(\tau - k)}}e^{-\frac{(x-y(k))^2}{4\sigma^2(\tau-k)}}dk. \tag{8.12}$$

Assume that $y(k)$ is differentiable. As shown in Tikhonov and Samarskii (1963), $\tilde{V}(\tau, x)$ is continuous along the curve $x = y(\tau)$ because it converges uniformly and $y'(k)$ is bounded, while $W(\tau, x)$ is discontinuous. To show this, first assume that $\Omega(\tau) = \Omega_0 = const$. Then the difference $W_0 - \tilde{V}_0$ (where the sub-index $_0$ means that we use Φ_0 instead of $\Phi(\tau)$ in the definitions Eqs. (8.11) and (8.12)) can be calculated directly with the change of variables $k \mapsto a = (x - y(k))/(2\sigma\sqrt{\tau - k})$

$$\begin{aligned}
W_0 - \tilde{V}_0 &= \frac{1}{2\sigma\sqrt{\pi}}\int_0^\tau \Omega_0 e^{-\frac{(x-y(k))^2}{4\sigma^2(\tau-k)}}\left[\frac{x - y(k)}{(\tau - k)^{3/2}} - \frac{2y'(k)}{(\tau - k)^{1/2}}\right]dk \\
&= \Omega_0\frac{2}{\sqrt{\pi}}\int_{\zeta^-}^{\zeta^+} e^{-a^2}da,
\end{aligned}$$

$$\zeta^- = \frac{x - y(0)}{2\sigma\sqrt{\tau}}, \qquad \zeta^+ = \begin{cases} \infty, & x > y(\tau), \\ 0, & x = y(\tau), \\ -\infty, & x < y(\tau). \end{cases} \tag{8.13}$$

Accordingly, at, say $x \to y(\tau) + 0$ we obtain

$$\begin{aligned} &\Big[W_0(\tau, y(\tau) + 0) - W_0(\tau, y(\tau))\Big] - \Big[\tilde{V}_0(\tau, y(\tau) + 0) - \tilde{V}_0(\tau, y(\tau))\Big] \\ &\quad = \Omega_0 \frac{2}{\sqrt{\pi}} \int_0^\infty e^{-a^2} da = \Omega_0. \end{aligned} \tag{8.14}$$

Since the function $\tilde{V}_0$ is continuous, the expression in the second square brackets in Eq. (8.14) vanishes, and so

$$W_0(\tau, y(\tau) + 0) - W_0(\tau, y(\tau)) = \Omega_0. \tag{8.15}$$

If $\Omega(\tau)$ is not constant, then

$$W(\tau, x) = W_0(\tau, x) - \int_0^\tau \frac{x - y(k)}{2\sigma\sqrt{\pi(\tau - k)^3}} e^{-\frac{(x - y(k))^2}{4\sigma^2(\tau - k)}} [\Omega(\tau) - \Omega(k)] dk. \tag{8.16}$$

We assume that the boundary $y(\tau)$ and the potential density $\Omega(k)$ are differentiable functions of their arguments, for instance, $y(\tau), \Omega(k) \in \mathcal{C}^1$. Then the integral in Eq. (8.16) has the same singularity as the function $\tilde{V}(\tau, x)$, converges uniformly, and thus is a continuous function on the curve $x = y(\tau)$. This implies that

$$W(\tau, x_0 + 0) - W(\tau, x_0) = W_0(\tau, x_0 + 0) - W_0(\tau, x_0) = \Omega(\tau), \tag{8.17}$$

and, in particular, this is true for $x_0 = y(\tau)$. In a similar way one can show that

$$W(\tau, x_0 - 0) = W_0(\tau, x_0) - \Omega(\tau), \tag{8.18}$$

Combining these results together, we obtain the final formula for $\varphi(t)$

$$\varphi(\tau) = \pm \frac{\Omega(\tau)}{2\sigma^2} + \int_0^\tau \Omega(k) \frac{y(\tau) - y(k)}{4\sigma^3\sqrt{\pi(\tau - k)^3}} e^{-\frac{(y(\tau) - y(k))^2}{4\sigma^2(\tau - k)}} dk. \tag{8.19}$$

8.1.2 *The limiting value of $\psi(t)$*

Using the definition of $q(\tau, x)$ in Eq. (8.6) we need an explicit formula for

$$\begin{aligned} \psi(\tau) &= \lim_{x \to y(\tau) \pm 0} \frac{\partial q(\tau, x)}{\partial x} \\ &= \lim_{x \to y(\tau) \pm 0} \frac{\partial}{\partial x} \int_0^\tau \Omega(k) \frac{x - y(k)}{4\sigma^3\sqrt{\pi(\tau - k)^3}} e^{-\frac{(x - y(k))^2}{4\sigma^2(\tau - k)}} dk. \end{aligned} \tag{8.20}$$

However, as shown in Section 8.1.1, this integral is discontinuous at $x \to y(\tau)$ (this is an improper Riemann integral of second kind). Hence, we cannot compute $\psi(\tau)$ directly by taking derivative of $q(\tau, x)$ with respect to x.

Therefore, to proceed let us represent this integral as

$$\int_0^\tau \Omega(k) \frac{x - y(k)}{4\sigma^3 \sqrt{\pi(\tau - k)^3}} e^{-\frac{(x-y(k))^2}{4\sigma^2(\tau-k)}} dk \tag{8.21}$$

$$= \Omega(\tau) \int_0^\tau \frac{x - y(k)}{4\sigma^3 \sqrt{\pi(\tau - k)^3}} e^{-\frac{(x-y(k))^2}{4\sigma^2(\tau-k)}} dk$$

$$+ \int_0^\tau [\Omega(k) - \Omega(\tau)] \frac{x - y(k)}{4\sigma^3 \sqrt{\pi(\tau - k)^3}} e^{-\frac{(x-y(k))^2}{4\sigma^2(\tau-k)}} dk = I_1 + I_2.$$

We showed in Section 8.1.1 that the second integral in Eq. (8.21) has the same singularity as the function $\tilde{V}(\tau, x)$, converges uniformly, and thus is a continuous function on the curve $x = y(\tau)$. Then, it is a continuous function for $x \in \Re$. Thus, by the standard theorem of integral calculus we can differentiate this integral by parameter x, and the result is continuous in x (Butuzov and Butuzova, 2016)

$$\lim_{x \to y(\tau) \pm 0} \frac{\partial}{\partial x} \int_0^\tau [\Omega(k) - \Omega(\tau)] \frac{x - y(k)}{4\sigma^3 \sqrt{\pi(\tau - k)^3}} e^{-\frac{(x-y(k))^2}{4\sigma^2(\tau-k)}} dk \tag{8.22}$$

$$= \lim_{x \to y(\tau) \pm 0} \int_0^\tau [\Omega(k) - \Omega(\tau)] \frac{e^{-\frac{(x-y(k))^2}{4\sigma^2(\tau-k)}}}{4\sigma^3 \sqrt{\pi(\tau - k)^3}} \left(1 - \frac{(x - y(k))^2}{2\sigma^2(\tau - k)}\right) dk$$

$$= \int_0^\tau [\Omega(k) - \Omega(\tau)] \frac{e^{-\frac{(y(\tau)-y(k))^2}{4\sigma^2(\tau-k)}}}{4\sigma^3 \sqrt{\pi(\tau - k)^3}} \left(1 - \frac{(y(\tau) - y(k))^2}{2\sigma^2(\tau - k)}\right) dk.$$

As far as the first integral I_1 in Eq. (8.21) is concerned, it was already considered in Section 8.1.1, and is denoted as $W_0(\tau, x)/2\sigma^2$ in Eq. (8.13). Since the integral in the variable a in the RHS of Eq. (8.13) can be computed explicitly, we have

$$W_0 - \tilde{V}_0 = \Omega_0 \frac{2}{\sqrt{\pi}} \int_{\zeta^-}^{\zeta^+} e^{-a^2} da = \Omega_0 \begin{cases} \mathrm{Erfc}\left(\frac{x-y(0)}{2\sigma\sqrt{\tau}}\right), & x > y(\tau), \\ -\mathrm{Erf}\left(\frac{x-y(0)}{2\sigma\sqrt{\tau}}\right), & x = y(\tau), \\ -\mathrm{Erfc}\left(-\frac{x-y(0)}{2\sigma\sqrt{\tau}}\right), & x < y(\tau). \end{cases} \tag{8.23}$$

Also, recall that the function $\tilde{V}_0(\tau, x)$ is the continuous function along the curve $x = y(\tau)$ as $y'(\tau)$ is bounded, and the integral converges uniformly. Therefore,

$$\frac{\partial W_0}{\partial x} = \frac{\partial \tilde{V}_0}{\partial x} - \Omega_0 \Lambda(\tau, x), \tag{8.24}$$

$$\Lambda(\tau, x) = \begin{cases} \frac{1}{\sigma\sqrt{\pi\tau}} e^{-\frac{(x-y(0))^2}{4\pi\sigma^2}}, & x > y(\tau), \\ \frac{1}{\sigma\sqrt{\pi\tau}} e^{-\frac{(x-y(0))^2}{4\pi\sigma^2}}, & x < y(\tau). \end{cases}$$

Thus, $\Lambda(\tau, y(\tau) - 0) = \Lambda(\tau, y(\tau) + 0)$, hence the function $\Lambda(\tau, x)$ is differentiable at this point. This implies

$$\frac{\partial W_0}{\partial x} = -\Omega_0 \int_0^\tau y'(k) \frac{x - y(k)}{2\sigma^3\sqrt{\pi(\tau - k)^3}} e^{-\frac{(x-y(k))^2}{4\sigma^2(\tau-k)}} dk - \frac{\Omega_0}{\sigma\sqrt{\pi\tau}} e^{-\frac{(x-y(0))^2}{4\sigma^2\tau}}. \tag{8.25}$$

As mentioned, the function $\tilde{V}_0(\tau, x)$ is continuous over the curve $x = y(\tau)$. However, its derivative with respect to x at $x = y(\tau)$ in Eq. (8.24) has a form of the RHS in Eq. (8.11). Therefore, according to the result of Section 8.1.1, in the limit $x \to y(\tau)$, again using Eq. (8.19), we obtain

$$\begin{aligned} \lim_{x\to y(\tau)\pm 0} \frac{\partial W_0}{\partial x} &= \mp\Omega_0 \frac{y'(\tau)}{\sigma^2} - \Omega_0 \int_0^\tau y'(k) \frac{y(\tau) - y(k)}{2\sigma^3\sqrt{\pi(\tau - k)^3}} e^{-\frac{(y(\tau)-y(k))^2}{4\sigma^2(\tau-k)}} dk \\ &\quad - \frac{\Omega_0}{\sigma\sqrt{\pi\tau}} e^{-\frac{(y(\tau)-y(0))^2}{4\sigma^2\tau}}. \end{aligned} \tag{8.26}$$

Combining Eqs. (8.22) and (8.26) together yields the final result

$$\begin{aligned} \psi(\tau) &= \int_0^\tau \Omega(k) \frac{e^{-\frac{(y(\tau)-y(k))^2}{4\sigma^2(\tau-k)}}}{4\sigma^3\sqrt{\pi(\tau - k)^3}} \left(1 - \frac{(y(\tau) - y(k))^2}{2\sigma^2(\tau - k)}\right) dk - \Omega(\tau) f(\tau), \\ f(\tau) &= \pm\frac{y'(\tau)}{2\sigma^4} + \frac{1}{2\sigma^3\sqrt{\pi\tau}} e^{-\frac{(y(\tau)-y(0))^2}{4\sigma^2\tau}} \\ &\quad + \int_0^\tau \frac{e^{-\frac{(y(\tau)-y(k))^2}{4\sigma^2(\tau-k)}}}{4\sigma^3\sqrt{\pi(\tau - k)^3}} \left\{1 + \frac{y'(k)[y(\tau) - y(k)]}{\sigma^2} - \frac{(y(\tau) - y(k))^2}{2\sigma^2(\tau - k)}\right\} dk. \end{aligned} \tag{8.27}$$

Thus, we proved that the derivative $\partial q(\tau, x)/\partial x$ is also discontinuous at $x = y(\tau)$, and obtained its explicit representation in Eq. (8.27). Note, that this derivative should not be confused with the normal (directional) derivative of $u(\tau, x)$ which is continuous at $x = y(\tau)$. Indeed, the function q, as defined in Eq. (8.6), is the double layer heat potential. The claim that this derivative is continuous is commonly referred as the Lyapunov-Tauber theorem of classic potential theory, see Lyapunov (1949) and Guinter (1967), Quaife (2011), Costabel (1990), Kristensson (2009) and references therein for the extension to the multi-dimensional case.

It is worth mentioning, that the formula for $f(\tau)$ can be further simplified. Indeed

$$
\begin{aligned}
d\left(\frac{e^{-\frac{(y(\tau)-y(k))^2}{4\sigma^2(\tau-k)}}}{\sqrt{\tau-k}}\right) &= \left[\frac{e^{-\frac{(y(\tau)-y(k))^2}{4\sigma^2(\tau-k)}}}{2\sqrt{(\tau-k)^3}} - \frac{e^{-\frac{(y(\tau)-y(k))^2}{4\sigma^2(\tau-k)}}}{\sqrt{\tau-k}}\right.\\
&\quad \left.\times\left(-\frac{y'(k)(y(\tau)-y(k))}{2\sigma^2(\tau-k)} + \frac{(y(\tau)-y(k))^2}{4\sigma^2(\tau-k)^2}\right)\right]dk\\
&= \frac{e^{-\frac{(y(\tau)-y(k))^2}{4\sigma^2(\tau-k)}}}{2\sqrt{(\tau-k)^3}}\left(1+\frac{y'(k)(y(\tau)-y(k))}{\sigma^2(\tau-k)} - \frac{(y(\tau)-y(k))^2}{2\sigma^2(\tau-k)^2}\right)dk.
\end{aligned}
$$

Therefore,

$$
\begin{aligned}
&\frac{e^{-\frac{(y(\tau)-y(k))^2}{4\sigma^2(\tau-k)}}}{2\sqrt{(\tau-k)^3}}\left(1+\frac{y'(k)(y(\tau)-y(k))}{\sigma^2(\tau-k)} - \frac{(y(\tau)-y(k))^2}{2\sigma^2(\tau-k)^2}\right)dk \qquad (8.28)\\
&\qquad = d\left(\frac{e^{-\frac{(y(\tau)-y(k))^2}{4\sigma^2(\tau-k)}}-1}{\sqrt{\tau-k}}\right) + \frac{dk}{2\sqrt{(\tau-k)^3}}.
\end{aligned}
$$

Plugging this expression into Eq. (8.27) and integrating yields an alternative representation for $f(\tau)$

$$
f(\tau) = \pm\frac{y'(\tau)}{2\sigma^4} + \frac{1}{2\sigma^3\sqrt{\pi\tau}} + \int_0^\tau \frac{dk}{4\sigma^3\sqrt{\pi(\tau-k)^3}}, \tag{8.29}
$$

and for $\psi(\tau)$, respectively

$$
\begin{aligned}
\psi(\tau) &= -\Omega(\tau)\left(\frac{1}{2\sigma^3\sqrt{\pi\tau}} \pm \frac{y'(\tau)}{2\sigma^4}\right) + \int_0^\tau \frac{\Omega(k)e^{-\frac{(y(\tau)-y(k))^2}{4\sigma^2(\tau-k)}} - \Omega(\tau)}{4\sigma^3\sqrt{\pi(\tau-k)^3}}dk\\
&\quad - \int_0^\tau \Omega(k)\frac{(y(\tau)-y(k))^2 e^{-\frac{(y(\tau)-y(k))^2}{4\sigma^2(\tau-k)}}}{8\sigma^5\sqrt{\pi(\tau-k)^5}}dk. \qquad (8.30)
\end{aligned}
$$

The last formula was first derived in Lipton *et al.* (2019) by using a different method.

8.1.3 *Other useful results*

In Section 16.4.2, when constructing a perturbation solution of the McKean-Vlasov equation based on the HP method, the following Lemma will be useful.

Lemma 8.1. *Consider a differentiable function $\Xi(t,t')$ such that $\Xi(t,t) = 1$. Then,*

$$
\frac{d}{dt}\int_0^t \frac{\Xi(t,t')\,\nu(t')}{\sqrt{2\pi(t-t')}}dt' = \frac{\nu(t)}{\sqrt{2\pi t}} \tag{8.31}
$$

$$+\frac{1}{2}\int_0^t \frac{\nu(t) - (\Xi(t,t') - 2(t-t')\Xi_t(t,t'))\nu(t')}{\sqrt{2\pi(t-t')^3}}dt'$$

$$= \int_0^t \frac{\frac{\partial}{\partial t'}((\Xi(t,t') - 2(t-t')\Xi_t(t,t'))\nu(t'))}{\sqrt{2\pi(t-t')}}dt'.$$

Proof. We start with the first term and judiciously use integration by parts several times to get

$$\frac{d}{dt}\int_0^t \frac{\Xi(t,t')\nu(t')}{\sqrt{2\pi(t-t')}}dt'$$

$$= 2\frac{d}{dt}\left(\Xi(t,0)\nu(0)\sqrt{\frac{t}{2\pi}} + \int_0^t \sqrt{\frac{(t-t')}{2\pi}}d(\Xi(t,t')\nu(t'))\right)$$

$$= \frac{(2t\Xi_t(t,0) + \Xi(t,0))\nu(0)}{\sqrt{2\pi t}} + \int_0^t \frac{1}{\sqrt{2\pi(t-t')}}d(\Xi(t,t')\nu(t') - \nu(t))$$

$$+ 2\int_0^t \sqrt{\frac{(t-t')}{2\pi}}d(\Xi_t(t,t')\nu(t'))$$

$$= \frac{\nu(t)}{\sqrt{2\pi t}} + \frac{1}{2}\int_0^t \frac{\nu(t) - (\Xi(t,t') - 2(t-t')\Xi_t(t,t'))\nu(t')}{\sqrt{2\pi(t-t')^3}}dt'$$

$$= \frac{\nu(t)}{\sqrt{2\pi t}} + \int_0^t (\nu(t) - (\Xi(t,t') - 2(t-t')\Xi_t(t,t'))\nu(t'))$$

$$\times d\left(\frac{1}{\sqrt{2\pi(t-t')}}\right)$$

$$= \int_0^t \frac{((\Xi(t,t') - 2(t-t')\Xi_t(t,t'))\nu(t'))_{t'}}{\sqrt{2\pi(t-t')}}dt'.$$

□

8.2 Solution via a Volterra integral equation

Based on the derivation in Section 8.1.1, from Eq. (8.6) at the barrier $x = y(\tau)$ we must have,

$$2\phi(\tau) = \frac{\Psi(\tau)}{\sigma^2} + \frac{1}{2\sqrt{\pi}}\int_0^\tau \Psi(k)\frac{y(\tau) - y(k)}{\sigma^3\sqrt{(\tau-k)^3}}e^{-\frac{(y(\tau)-y(k))^2}{4\sigma^2(\tau-k)}}dk, \tag{8.32}$$

The Eq. (8.32) is a linear Volterra equations of the second kind, see Polyanin and Manzhirov (2008) and also Chapter 4. Since $\phi(\tau)$ is a continuously differentiable function, Eq. (8.32) has a unique continuous solution for $\Psi(\tau)$. First, it can be efficiently solved by a standard approach, i.e., the integral in the RHS is approximated using some quadrature rule with

N nodes in k, and the solution is obtained at M nodes in τ. This discretization gives rise to a linear equation with respect to the discrete vector of values of $\Psi(\tau)$ with the matrix being lower triangular. Therefore, the solution can be obtained by a simple elimination procedure with quadratic complexity. Alternatively, since the kernel is proportional to Gaussian, the discrete sum approximating the integral can be computed with linear complexity $O(N+M)$ using the Fast Gauss Transform, see, e.g., Spivak *et al.* (2010). Another possible approach is to approximate boundary $y(\tau)$ by some polynomial function of τ. Then the Volterra equation (8.32) can be transformed to the Abel equation, see Chapter 19, and solved by using the Laplace transform.

8.2.1 *Other domains*

In the previous sections, we described the solution of the problem in Eqs. (8.1) and (8.3), Eq. (8.2) defined at the domain $x \in \Omega : [y(\tau), \infty)$ via the HP method. Same approach can be used when this problem is defined at the domain $x \in \Omega : [0, y(\tau)]$. In this case we need to solve Eq. (8.1) subject to the same initial condition in Eq. (8.2). However, the boundary conditions now read

$$u(0,\tau) = u(y(\tau),\tau) = 0. \tag{8.33}$$

Again, since this problem has an inhomogeneous initial condition, the classical method of heat potentials cannot be directly applied. Therefore, we represent $u(x,t)$ in the form

$$u(x,\tau) = q(x,\tau) + \frac{1}{2\sigma\sqrt{\pi\tau}} \int_0^{y(0)} u(x',0) e^{-\frac{(x-x')^2}{4\sigma^2\tau}} dx'. \tag{8.34}$$

The function $q(x,\tau)$ solves the problem

$$\begin{aligned} \frac{\partial q(x,\tau)}{\partial \tau} &= \sigma^2 \frac{\partial^2 q(x,\tau)}{\partial x^2}, \\ q(x,0) &= 0, \qquad 0 < x < y(0), \\ q(0,\tau) &= 0, \qquad q(y(\tau),\tau) = \phi(\tau), \\ \phi(\tau) &= -\frac{1}{2\sigma\sqrt{\pi\tau}} \int_0^{y(0)} u(x',0) e^{-\frac{(y(\tau)-x')^2}{4\sigma^2\tau}} dx'. \end{aligned} \tag{8.35}$$

Again, it can be solved by the HP method if we are looking for the solution of Eq. (8.35) in the form of a generalized heat potential

$$q(x,\tau) = \frac{1}{4\sqrt{\pi}} \int_0^\tau \Psi(k) \frac{x - y(k)}{\sigma^3\sqrt{(\tau-k)^3}} e^{-\frac{(x-y(k))^2}{4\sigma^2(\tau-k)}} dk. \tag{8.36}$$

But this problem has been already solved in Section 8.2.

We can also consider a similar problem at the domain with two moving boundaries, i.e., $x \in \Omega : [y(\tau), z(\tau)]$ where the upper boundary function $z(\tau)$ has the same properties as $y(\tau)$. Again, this problem can be solved via the HP method. This solution was obtained in Itkin and Muravey (2020b) and is presented in Chapter 18.

Finally, the homogeneous boundary conditions used in above could be replaced with their inhomogeneous counterparts. In other words, we now set

$$u(y(\tau), \tau) = f^-(\tau), \qquad u(z(\tau), \tau) = f^+(\tau). \tag{8.37}$$

Here $f^{\pm}(\tau)$ are some continuous functions of time τ such that $f^{\pm}_{\tau}(\tau) < \infty$. This case is also discussed in detail in Chapter 18.

8.3 A generalized method of heat potentials

Here we generalize the construction of the potential method originally proposed for the heat equation. For convenience, we follow the notation of Tikhonov and Samarskii (1963).

Consider a PDE

$$\frac{\partial V(t,x)}{\partial t} = \mathcal{L}(V(t,x)) \tag{8.38}$$

where the operator $\mathcal{L}$ is a linear differential operator with time-independent coefficients. An example of such the equation is the heat equation in Eq. (8.1) and the Bessel equation

$$\frac{\partial u}{\partial \tau} = \frac{1}{2}\frac{\partial^2 u}{\partial x^2} + \frac{\nu}{x}\frac{\partial u}{\partial x}, \tag{8.39}$$

where $\nu \neq 0$ is some constant (this problem is considered in detail in Chapter 13). Suppose that the fundamental solution (or the Green function) of Eq. (8.38) $G(x, t|\xi, \tau)$ is known in closed form.

Also, suppose we need to solve Eq. (8.38) subject to the homogeneous initial condition

$$V(0, x) = 0. \tag{8.40}$$

Note that this doesn't restrict our consideration, because as shown in Section 8.1, by the change of variables the problem with inhomogeneous initial condition can be transformed to the problem with $V(0, x) = 0$.

We are interesting in solving the first boundary-value problem assuming that the boundaries are some functions of time. Consider, for simplicity just a semi-bounded region with $x \geq y(t),\ x < \infty,\ t > 0$, with the following boundary conditions

$$V(t,x)\Big|_{x\to\infty} = 0, \qquad V(t,y(t)) = \phi(t). \tag{8.41}$$

As we have mentioned already, the assumption that the problem has just one time-dependent boundary can be easily relaxed.

Let us introduce the double layer potential (Tikhonov and Samarskii, 1963)

$$\Pi(x,t) = \int_0^t \Psi(\tau)\frac{\partial G}{\partial \xi}(x,t|\xi,\tau)\Big|_{\xi=y(\tau)} d\tau, \tag{8.42}$$

where $\Psi(t)$ the potential density. The double layer potential is a continuous and twice differentiable function in x, and continuous and differentiable in t up to the boundary $x = y(\tau)$ where it becomes discontinues. Then the below proposition follows,

Proposition 8.1. *The potential function in Eq. (8.42) solves Eq. (8.38). The potential density is determined by the boundary condition in Eq. (8.41) and solves the Volterra equation of second kind*

$$\phi(t) = b\Psi(t) + \int_0^t \Psi(\tau)\frac{\partial G}{\partial \xi}(y(t),t|y(\tau),\tau)\Big|_{\xi=y(\tau)} d\tau, \tag{8.43}$$

where $b \in \Re,\ b \neq 0$ is a constant.

Proof. First, it can be checked that by substituting the definition in Eq. (8.42) into Eq. (8.38) we get

$$\begin{aligned}
\frac{\partial \Pi(x,t)}{\partial t} &= \int_0^t \Psi(\tau)\frac{\partial}{\partial \xi}\frac{\partial G}{\partial t}\Big|_{\xi=y(\tau)} d\tau + \Psi(t)\frac{\partial G}{\partial y(\tau)}(x,y(t)|t,t)d\tau \\
&= \int_0^t \Psi(\tau)\frac{\partial}{\partial \xi}\frac{\partial G}{\partial t}\Big|_{\xi=y(\tau)} d\tau, \\
\mathcal{L}(\Pi(t,x)) &= \int_0^t \Psi(\tau)\frac{\partial}{\partial \xi}\mathcal{L}(G)\Big|_{\xi=y(\tau)} d\tau.
\end{aligned} \tag{8.44}$$

Combining these two expressions yields

$$\int_0^t \Psi(\tau)\frac{\partial}{\partial \xi}\left[\frac{\partial G}{\partial t} - \mathcal{L}(G)\right]\Big|_{\xi=y(\tau)} d\tau = 0. \tag{8.45}$$

The first line follows from the fact that the Green function solves Eq. (8.38), and

$$G(x,y(\tau)|t,\tau)\Big|_{\tau=t} = G(x,y(\tau)|t-\tau)\Big|_{\tau=t} = \delta(x - y(t) = 0$$

as $x \neq y(t)$. The second line is a consequence of time-independence of the operator coefficients.

To summarize what we got: the potential function satisfies the PDE for $x \geq y(t)$, is bounded at infinity, and has a zero initial value for any choice of $\Psi(t)$. Thus, it is the solution of Eq. (8.38), i.e.

$$V(t,x) = \int_0^t \Psi(\tau) \frac{\partial G}{\partial \xi}(x,t|\xi,\tau)\Big|_{\xi=y(\tau)} d\tau. \tag{8.46}$$

Now using the boundary condition at $x = y(t)$ we obtain from Eq. (8.46)

$$\phi(t) = \int_0^t \Psi(\tau) \frac{\partial G}{\partial \xi}(y(t),t|\xi,\tau)\Big|_{\xi=y(\tau)} d\tau. \tag{8.47}$$

However, as shown in Tikhonov and Samarskii (1963), for $x = y(t)$ the RHS is discontinuous, but with the finite limiting value at $x = y(t)+0$. The limiting value could be represented as

$$b\Psi(t) + \int_0^t \Psi(\tau) \frac{\partial G}{\partial \xi}(y(t),t|y(\tau),\tau)\Big|_{\xi=y(\tau)} d\tau.$$

The constant b depends on the particular form of the operator $\mathcal{L}$. In particular, if $\mathcal{L}$ is a second order parabolic operator with the diffusion coefficient a, then $b = 1/(2a)$ (Tikhonov and Samarskii, 1963). Thus, for instance, for $a = 1/2$ we have $b = 1$.

□

8.4 Connection to fractional differential equations

Consider the standard initial-value problem for the wave equation on the whole axis $-\infty < x < \infty$

$$\begin{aligned} V_{tt} - V_{xx} &= 0, \\ V(0,x) = p(x), &\quad V_t(0,x) = q(x). \end{aligned} \tag{8.48}$$

To solve it, let us split the equation as follows

$$\begin{aligned} \left(\frac{\partial^2}{\partial t^2} - \frac{\partial^2}{\partial x^2}\right) V &= \left(\frac{\partial}{\partial t} - \frac{\partial}{\partial x}\right)\left(\frac{\partial}{\partial t} + \frac{\partial}{\partial x}\right) V \\ &= \left(\frac{\partial}{\partial t} + \frac{\partial}{\partial x}\right)\left(\frac{\partial}{\partial t} - \frac{\partial}{\partial x}\right) V = 0, \end{aligned} \tag{8.49}$$

or, equivalently (Courant and Hilbert, 1962)

$$\frac{\partial^2 V(\xi,\eta)}{\partial \xi \partial \eta} = 0, \qquad \xi = x + t, \ \eta = x - t.$$

This allows representation of the solution $V(t,x)$ as a formal sum of two functions

$$V(t,x) = f(x-t) + g(x+t), \tag{8.50}$$

assuming that f, g exist and have the same properties as the solution $V(t,x)$. In other words, solutions of the one-dimensional wave equation are sums of a right traveling function f and a left traveling function g. "Traveling" means that the shape of these individual arbitrary functions with respect to x stays constant, however the functions are translated left and right with time at the unit speed. This result was first derived by Jean-Baptiste le Rond d'Alembert.

We can now satisfy the initial conditions by choosing functions f, g in such a way that

$$f(0,x) + g(0,x) = p(x), \qquad -f_t(0,x) + g_t(0,x) = q(x). \tag{8.51}$$

As the result, the solution of the problem can be expressed via d'Alembert's formula

$$V(t,x) = \frac{1}{2}\left[f(x-t) + f(x+t)\right] + \frac{1}{2}\int_{x-t}^{x+t} g(s)ds. \tag{8.52}$$

Of course, this is a textbook example, which is well-known, but it is useful as a pointer to things to come.

Consider now the standard initial-value problem for the heat equation on the whole axis $-\infty < x < \infty$

$$\begin{aligned} V_t - \sigma^2 V_{xx} &= 0, \\ V(0,x) &= p(x), \quad \sigma = const. \end{aligned} \tag{8.53}$$

We want to split the heat operator into a product, similar to Eq. (8.49). However, for doing so we need to use the definition of fractional derivatives. To this end, let us introduce the operator $(d/dt)^{1/2}$. There are several ways of doing it, and we proceed with the classical Riemann-Liouville definition (Oldham and Spanier, 1974)

$$\frac{d^{1/2}}{dt^{1/2}} f(t) = \frac{d}{dt}\left(\int_0^t \frac{f(\tau)}{\sqrt{\pi(t-\tau)}} d\tau\right). \tag{8.54}$$

With this definition in mind, similar to Eq. (8.49) we can write

$$\begin{aligned} \left(\frac{\partial}{\partial t} - \sigma^2 \frac{\partial^2}{\partial x^2}\right) V &= \left(\partial_t^{1/2} - \sigma\partial_x\right)\left(\partial_t^{1/2} + \sigma\partial_x\right) V \\ &= \left(\partial_t^{1/2} + \sigma\partial_x\right)\left(\partial_t^{1/2} - \sigma\partial_x\right) V = 0. \end{aligned} \tag{8.55}$$

Let us show that Green's function

$$H(t,x,\xi)=\frac{e^{-\frac{(x-\xi)^2}{4\sigma^2 t}}}{2\sigma\sqrt{\pi t}}, \tag{8.56}$$

satisfies one of the equations

$$\partial_t^{1/2}H\pm\sigma\partial_x H=0. \tag{8.57}$$

Here, we choose the plus sign when $x>\xi$, and the minus sign when $x<\xi$. Without loss of generality, we put $\xi=0$. This yields

$$\partial_x H(t,x,0)=-\frac{x}{2\sigma^2 t}H(t,x,0), \tag{8.58}$$

and

$$\partial_t^{1/2}H(t,x,0)=\frac{d}{dt}\int_0^t\frac{e^{-\frac{x^2}{4\sigma^2\tau}}}{2\sigma\sqrt{\pi(t-\tau)\pi\tau}}d\tau=\frac{d}{dt}\left(\frac{1}{\sqrt{\pi t}}*\frac{e^{-\frac{x^2}{4\sigma^2 t}}}{\sqrt{4\pi\sigma^2 t}}\right), \tag{8.59}$$

where $*$ denotes convolution. Taking the Laplace transform $\mathcal{L}$ of the RHS of Eq. (8.59) we obtain

$$\mathcal{L}\left(\frac{1}{\sqrt{\pi t}}\right)=\frac{1}{\sqrt{s}},\quad \mathcal{L}\left(\frac{e^{-\frac{x^2}{4\sigma^2 t}}}{\sqrt{4\pi\sigma^2 t}}\right)=\frac{e^{-\frac{|x|\sqrt{s}}{\sigma}}}{2\sigma\sqrt{s}}, \tag{8.60}$$

$$\mathcal{L}\left(\frac{1}{\sqrt{\pi t}}*\frac{e^{-\frac{x^2}{4\sigma^2 t}}}{\sqrt{4\pi\sigma^2 t}}\right)=\frac{e^{-\frac{|x|\sqrt{s}}{\sigma}}}{2\sigma s}.$$

The inverse Laplace transform $\mathcal{L}^{-1}$ yields

$$\mathcal{L}^{-1}\left(\frac{e^{-\frac{|x|\sqrt{s}}{\sigma}}}{2\sigma s}\right)=\frac{1}{2\sigma}\operatorname{erfc}\left(\frac{|x|}{2\sigma\sqrt{t}}\right). \tag{8.61}$$

Finally, using the definition of function $\operatorname{erfc}(x)$ (Abramowitz and Stegun, 1964) we obtain

$$\begin{aligned}\partial_t^{1/2}H(t,x,0)&=\frac{d}{dt}\left(\frac{1}{\sigma\sqrt{\pi}}\int_{\frac{|x|}{2\sigma\sqrt{t}}}^{\infty}e^{-y^2}dy\right)\\&=\frac{1}{\sigma\sqrt{\pi}}e^{-\frac{x^2}{4\sigma^2 t}}\frac{|x|}{4\sigma\sqrt{t^3}}=\frac{|x|}{2\sigma t}H(t,x,0).\end{aligned} \tag{8.62}$$

Combining Eqs. (8.58) and (8.62), we obtain

$$\partial_t^{1/2}H\pm\sigma\partial_x H=\frac{(|x|\mp x)}{2\sigma t}H. \tag{8.63}$$

Thus, our statement in Eqs. (8.56) and (8.57) is proved.

Next, we consider the following initial-boundary value problem on the positive semi-axis $x > 0$:

$$V_t - \sigma^2 V_{xx} = 0, \qquad (8.64)$$
$$V(0,x) = 0, \quad V(t,0) = r(t),$$

where $r(t) \in \mathcal{C}^1$ is some function of the time t. Using the HP method one can show that the corresponding solution has the form

$$V(t,x) = \int_0^t \frac{xH(t-t',x,0)}{(t-t')} r(t')\,dt'. \qquad (8.65)$$

One can observe that for $x > 0$, the singularity previously discussed in Section 8.1.1, does not appear since

$$\frac{xH(t-t',x,0)}{t-t'} \underset{t' \to t}{\to} 0. \qquad (8.66)$$

We claim that $V(t,x)$ solves the equation

$$\partial_t^{1/2} V + \sigma \partial_x V = 0. \qquad (8.67)$$

To prove this statement, let us introduce another function $K(t,x)$ as

$$K(t,x) = \frac{xH(t,x,0)}{t}. \qquad (8.68)$$

We have

$$\partial_x K(t,x) = \left(\frac{1}{t} - \frac{x^2}{2\sigma^2 t^2}\right) H(t,x,0), \qquad (8.69)$$

$$\partial_t^{1/2} K(t,x) = \frac{d}{dt}\int_0^t \frac{xe^{-\frac{x^2}{4\sigma^2\tau}}}{\sqrt{\pi(t-\tau)4\sigma^2\pi\tau^3}} d\tau = \frac{d}{dt}\left(\frac{1}{\sqrt{\pi t}} * \frac{xe^{-\frac{x^2}{4\sigma^2 t}}}{\sqrt{4\pi\sigma^2 t^3}}\right).$$

By proceeding as before, we take the Laplace transform of the RHS of Eq. (8.69)

$$\mathcal{L}\left(\frac{1}{\sqrt{\pi t}}\right) = \frac{1}{\sqrt{s}}, \quad \mathcal{L}\left(\frac{xe^{-\frac{x^2}{4\sigma^2 t}}}{\sqrt{4\pi\sigma^2 t^3}}\right) = e^{-x\frac{\sqrt{s}}{\sigma}},$$

$$\mathcal{L}\left(\frac{1}{\sqrt{\pi t}} * \frac{xe^{-\frac{x^2}{4\sigma^2 t}}}{\sqrt{4\pi\sigma^2 t^3}}\right) = \frac{e^{-x\frac{\sqrt{s}}{\sigma}}}{\sqrt{s}}.$$

Next, taking the inverse Laplace transform we obtain

$$\mathcal{L}^{-1}\left(\frac{e^{-x\frac{\sqrt{s}}{\sigma}}}{\sqrt{s}}\right) = 2\sigma H(t,x),$$

$$\partial_t^{1/2} K(t,x) = 2\sigma \frac{d}{dt} H(t,x) = \sigma \left(\frac{x^2}{2\sigma^2 t^2} - \frac{1}{t} \right) H(t,x,0).$$

Thus, using these representations and substituting them into Eq. (8.69), we find

$$\partial_t^{1/2} K(t,x) + \sigma \partial_x K(t,x) = 0. \tag{8.70}$$

Using the definition of $K(t,x)$ in Eq. (8.68) together with Eq. (8.70) we can finally prove the statement in Eq. (8.67) since

$$\begin{aligned} \partial_t^{1/2} V + \sigma \partial_x V &= \int_0^t \left(\partial_t^{1/2} K(t-t',x) + \sigma \partial_x K(t-t',x) \right) r(t')\, dt' \\ &= 0. \end{aligned} \tag{8.71}$$

The above results make connection between the HP method and fractional PDEs. This connection is known in the literature, see e.g., Oldham and Spanier (2006) and references therein. For instance, as mentioned in Zenyuk (2014), a natural extension of fractional calculus is the theory of fractional PDEs which, rigorously speaking, are integro-differential equations. A detailed discussions on this subject can be found in Samko *et al.* (1993), also see references therein. Also, for fractional ordinary differential equations an analog of the Picard–Lindelöf theorem (Coddington and Levinson, 1955), or existence and uniqueness theorem gives a set of conditions under which an initial value problem has a unique solution.

In more detail, the problem under consideration reads

$$\left(D^{\alpha}_{a+} u\right)(x) = f(x, u(x)), \quad \left(D^{\alpha-k}_{a+} u\right)(a+0) = c_k, \; k = 1, 2, ..., n, \tag{8.72}$$

where $\left(D^{\alpha}_{a+} u\right)(x)$ is a fractional Riemann-Liouville left-sided derivative defined as

$$\begin{aligned} \left(D^{\alpha}_{a+} u\right)(x) &= \frac{\partial^{n_1+n_2}}{\partial x_1^{n_1} \partial x_2^{n_2}} \left(I^{n-\alpha}_{a+} f\right)(x), \quad n = (n_1, n_2), \; n_k = [\alpha_k] + 1, \\ \left(I^{n-\alpha}_{a+} f\right)(x) &= \frac{1}{\Gamma(\alpha_1)\Gamma(\alpha_2)} \\ &\quad \times \int_{\alpha_1}^{x_1} \int_{\alpha_2}^{x_2} (x_1 - \xi_1)^{\alpha_1 - 1} (x_2 - \xi_2)^{\alpha_2 - 1} f(\xi_1, \xi_2) d\xi_1 d\xi_2, \end{aligned} \tag{8.73}$$

where $a = (a_1, a_2), x = (x_1, x_2)$ are points in $\mathbb{R}^2$, and $x_k > a_k$. In case of fractional derivative on only one variable one can set $\alpha_k = 0$ for all other variables.

As shown in Samko *et al.* (1993), the proof of the Picard–Lindelöf theorem in space $L_1[a,b]$ assuming that $f \in L_1[a,b]$ for any y, and also that f is

Lipschitz on the second argument, is constructed based on the equivalence of the problem in Eq. (8.72) and an integral Volterra equation of the second kind

$$u(x) = \sum_{k=1}^{n} \frac{c_k (x-a)^{\alpha-k}}{\Gamma(\alpha-k+1)} + \frac{1}{\Gamma(\alpha)} \int_{\alpha}^{x} (x-\xi)^{\alpha-1} f(\xi, u(\xi)) d\xi, \quad x > a.$$

A similar connection could be established in case when the fractional ODE problem is replace by the fractional PDE problem, again see Samko *et al.* (1993) and references therein.

PART 4
Equities, FX, and Commodities

Chapter 9

Barrier and American Options

In this chapter we develop a semi-analytical solutions for the barrier (perhaps, time-dependent) and American options written on the underlying stock which follows a time-dependent Ornstein-Uhlenbeck (OU) process with a lognormal drift. Semi-analytical means that given the time-dependent interest rate, continuous dividend and volatility functions, one need to solve a linear (for the barrier option) or nonlinear (for the American option) Volterra equation of the second kind (or Fredholm equation of the first kind). After that the option prices in all cases are presented as one-dimensional integrals of combination of the above solutions and Jacobi theta functions. We also demonstrate that computationally our method is more efficient than the backward finite difference method traditionally used for solving these problems, and can also be as efficient as the forward finite difference solver while providing better accuracy and stability.

Below we demonstrate how semi-analytical solutions of these problems can be constructed by using the methods described in Part 3. In particular, we discuss pricing barrier and American options using a time-dependent OU process (Carr and Itkin, 2021); pricing of barrier options in the Hull-White model (Itkin and Muravey, 2020b) and calculate the hitting time density (Alili *et al.*, 2005; Lipton and Kaushansky, 2020a).[1] Also, the method could be used for solving various problems in physics where it was originally developed for the heat equation (Kartashov, 2001; Friedman, 1964.) and references therein.

[1]The corresponding material is an expanded version of those papers.

9.1 Problem for pricing barrier options

As applied to pricing barrier options, our approach to a certain degree is similar to that in Mijatovic (2010), who, however, used a different underlying process (the lognormal model with local spot-dependent volatility, and constant interest rates and dividends, but time-dependent barriers). Therefore, our model is more general in a sense that all parameters of the model are time-dependent, including time-dependent barriers. Also, as compared with (Mijatovic, 2010), we don't use a probabilistic argument, rather a theory of partial differential equations (PDE). At the end we demonstrate that computationally our method is more efficient than the backward finite difference method used to solve these problems, and can also be as efficient as the forward finite difference solver while providing better accuracy and stability.

We start by specifying the dynamics of the underlying spot price S_t to be

$$dS_t = [r(t) - q(t)]S_t dt + \sigma(t) dW_t, \tag{9.1}$$

where now $r(t)$ is the deterministic short interest rate. This model is also known in the financial literature as the Bachelier model. See, e.g., Thomson (2016) for a thorough discussion of pro and contra of this model. One can think about S_t, e.g., as the stock price or the price of some commodity asset. While in the Bachelier model the underlying value could become negative which is not desirable for the stock price, this is fine for commodities under the modern market conditions when the oil prices have been several times observed to be negative, see e.g., CME Clearing (2020). For the sake of certainty, below we will reference S_t as the stock price.

In Eq. (9.1) we don't specify the explicit form of $r(t), q(t), \sigma(t)$ but assume that they are known as a differentiable functions of time $t \in [0, \infty)$. The case of discrete dividends is discussed in Section 9.4.

Further in this section we consider a contingent claim written on the underlying process S_t in Eq. (9.1) which is the Up-and-Out barrier Call option. It is known that by the Feynman-Kac formula (Klebaner, 2005) one can obtain a parabolic (*linear*) PDE whose solution gives the Up-and-Out barrier Call option price $C(S, t)$ conditional on $S_0 = S$, which reads

$$\frac{\partial C}{\partial t} + \frac{1}{2}\sigma^2(t)\frac{\partial^2 C}{\partial S^2} + [r(t) - q(t)]S\frac{\partial C}{\partial S} = r(t)C. \tag{9.2}$$

This equation should be solved subject to the terminal condition at the option maturity $t = T$

$$C(S, T) = (S - K)^+, \tag{9.3}$$

and the boundary conditions

$$C(0,t) = 0, \qquad C(H,t) = 0, \tag{9.4}$$

where $H = H(t)$ is the upper barrier. Note, that for the arithmetic BM process the domain of definition is $S \in (-\infty, H]$, however here we move the boundary condition from minus infinity to zero, see discussion in Itkin and Muravey (2020b) about rigorous boundary conditions for this problem. This is because in practice we can control the left boundary to make the probability of S dropping below 0 rare.

Our goal now is to build a series of transformations to transform Eq. (9.2) to the heat equation. As shown in Chapter 8 in general, if possible, this can be done by using the so-called Cherkasov's condition which guarantees that some process can be transformed to the standard Wiener process.

9.1.1 *Transformation to the heat equation*

To transform the PDE Eq. (9.2) to the heat equation we first make a change of the dependent and independent variables as follows:

$$S \to x/g(t), \qquad C(S,t) \to e^{f(x,t)}u(x,t), \tag{9.5}$$

where new functions $f(x,t), g(t)$ has to be determined in such a way, that the equation for u is the heat equation. This can be done by substituting Eq. (9.5) into Eq. (9.2) and providing some tedious algebra. The result reads

$$f(x,t) = k(t) - \frac{g'(t) + g(t)(r(t) - q(t))}{2g(t)^3\sigma(t)^2}x^2, \tag{9.6}$$

$$k(t) = \frac{1}{2}\log\left(\frac{g(t)}{g(0)}\right) + \frac{1}{2}\int_0^t [3r(s) - q(s)]ds.$$

The function $g(t)$ solves the following ordinary differential equation (ODE)

$$0 = b(t)g(t) - g''(t) + 2g'(t)\frac{\sigma'(t)}{\sigma(t)} + 2\frac{g'(t)^2}{g(t)}, \tag{9.7}$$

$$b(t) = 2(r(t) - q(t))\frac{\sigma'(t)}{\sigma(t)} - \left[(r(t) - q(t))^2 + r'(t) - q'(t)\right].$$

The Eq. (9.7) by substitution

$$g(t) \to e^{\int_0^t w(s)ds} \tag{9.8}$$

can be further transformed to the Riccati equation

$$w'(t) = b(t) + w(t)^2 + 2w(t)\frac{\sigma'(t)}{\sigma(t)}. \tag{9.9}$$

This equation cannot be solved analytically for arbitrary functions $r(t), q(t), \sigma(t)$, but can be efficiently solved numerically. Also, in some cases it can be solved in closed form. For instance, if $|r(t) - q(t)|t = \varepsilon \ll 1$ (which at the current market is a typical case), then $b(t)$ can be reduced to $b(t) = 2(r(t) - q(t))\sigma'(t)/\sigma(t)$. Then assuming in the first approximation on ε

$$|w(t)| \gg |r(t) - q(t)|, \tag{9.10}$$

we obtain the solution

$$w(t) = \frac{\sigma^2(t)}{D - \int_0^t \sigma^2(s)ds}, \tag{9.11}$$

where D is an integration constant. Thus, Eq. (9.10) can be re-written as

$$\mathrm{V}(t)\left[1 + \frac{V}{\bar{V}}\varepsilon\right] \gg D(r(t) - q(t)),$$

where $\mathrm{V}(t) = \sigma^2(t)$ is the normal variance, and $\bar{V}(t) = \frac{1}{t}\int_0^t \sigma^2(s)ds$ is the average normal variance. Thus, our solution in Eq. (9.11) is correct if $\varepsilon \ll 1$ and also $V\varepsilon/\bar{V} \ll 1$, because then D can always be chosen to obey the inequality $\mathrm{V}(t) \gg D(r(t) - q(t))$, $\forall t \in [0, T]$.

With these explicit definitions Eq. (9.2) transforms to the form

$$\frac{1}{2}\sigma(t)^2 e^{2\int_0^t w(s)\,ds}\frac{\partial^2 u}{\partial x^2} + \frac{\partial u}{\partial t} = 0. \tag{9.12}$$

The next step is to make a change of the time variable

$$\tau(t) \to \frac{1}{2}\int_t^T \sigma^2(s)e^{2\int_0^s w(m)\,dm}\,ds, \tag{9.13}$$

so Eq. (9.12) finally takes the form of the heat equation

$$\frac{\partial u}{\partial \tau} = \frac{\partial^2 u}{\partial x^2}. \tag{9.14}$$

The Eq. (9.14) should be solved subject to the terminal condition

$$u(x, 0) = (xe^{-\int_0^T w(s)ds} - K)^+ e^{-f(x,T)}, \tag{9.15}$$

and the boundary conditions

$$u(0, \tau) = 0, \qquad u(y(\tau), \tau) = 0, \qquad y(\tau) = H(t(\tau))g(t(\tau)). \tag{9.16}$$

These conditions directly follow from Eqs. (9.3) and (9.4), while $y(\tau)$ is now a time-dependent upper barrier[2]. The function $t(\tau)$ is the inverse map of Eq. (9.13). It can be computed for any $t \in [0, T]$ by substituting it into Eq. (9.13), then finding the corresponding value of $\tau(t)$, and finally inverting.

[2]Therefore, we can also naturally solve the same problem with the time-dependent upper barrier $H = H(t)$ as this just changes the definition of $y(\tau)$.

9.1.2 *Solution of the barrier pricing problem*

The PDE in Eqs. (9.14)–(9.16) is a parabolic equation whose solution should be found at the domain with moving boundaries. These kind of problems are known in physics for a long time. Similar problems arise in the field of nuclear power engineering and safety of nuclear reactors; in studying combustion in solid-propellant rocket engines; in laser action on solids; in the theory of phase transitions (the Stefan problem and the Verigin problem (in hydromechanics)); in the processes of sublimation in freezing and melting; in the kinetic theory of crystal growth; etc., see Kartashov (1999) and references therein. Analytical solutions of these problems require non-traditional, and sometimes sophisticated methods. Those methods were actively elaborated on by the Russian mathematical school in the 20th century starting from A.V. Luikov, and then by B.Ya. Lyubov, E.M. Kartashov, and many others.

As applied to mathematical finance, one of these methods — the method of heat potentials — was actively utilized by A. Lipton and his co-authors who solved various problems of mathematical finance by using this approach, see Lipton (2001), Lipton and Lopez de Prado (2020a,b) and references therein. Another method that we use here is the method of a generalized integral transform. Below we closely follow (Kartashov, 2001) when give an exposition of the method.

We start by introducing an integral transform of the form

$$\bar{u}(p,\tau) = \int_0^{y(\tau)} u(x,\tau)\sinh(x\sqrt{p})dx, \tag{9.17}$$

where $p = a + i\omega$ is a complex number with $\mathbb{R}(p) \ge \beta > 0$, and $-\frac{\pi}{4} < \arg\left(\sqrt{p}\right) < \frac{\pi}{4}$. Let us multiply both parts of Eq. (9.14) by $\sinh(x\sqrt{p})$ and then integrate on x from zero to $y(\tau)$:

$$\int_0^{y(\tau)} \frac{\partial u}{\partial \tau}\sinh(x\sqrt{p})dx = \int_0^{y(\tau)} \frac{\partial^2 u}{\partial x^2}\sinh(x\sqrt{p})dx.$$

Integrating by parts, we obtain

$$\begin{aligned}
&\frac{\partial}{\partial \tau}\int_0^{y(\tau)} u(x,\tau)\sinh(x\sqrt{p})dx - u(y(\tau),\tau)\sinh(y(\tau)\sqrt{p})y'(\tau)\\
&= \frac{\partial u(x,\tau)}{\partial x}\sinh(x\sqrt{p})\bigg|_0^{y(\tau)} + \sqrt{p}u(x,\tau)\cosh(x\sqrt{p})\bigg|_0^{y(\tau)}\\
&+ p\int_0^{y(\tau)} u(x,\tau)\sinh(x\sqrt{p})dx.
\end{aligned}$$

With allowance for the boundary conditions in Eq. (9.16) and the definition in Eq. (9.17) we obtain the following Cauchy problem

$$\frac{\partial \bar{u}}{\partial \tau} - p\bar{u} = \Psi(\tau)\sinh(y(\tau)\sqrt{p}), \tag{9.18}$$

$$\bar{u}(p,0) = \int_0^{y(0)} u(x,0)\sinh(x\sqrt{p})dx, \qquad \Psi(\tau) = \left.\frac{\partial u(x,\tau)}{\partial x}\right|_{x=y(\tau)}.$$

The Eq. (9.18) can be solved explicitly, assuming that $\Psi(\tau)$ is known. The solution reads

$$\bar{u}e^{-p\tau} = \int_0^{\tau} \Psi(k)e^{-pk}\sinh(y(k)\sqrt{p})dk + \int_0^{y(0)} u(x,0)\sinh(x\sqrt{p})dx. \tag{9.19}$$

As $\mathbb{R}(p) \geq \beta > 0$, and $\bar{u}(x,\tau) < \infty$, the function $\bar{u}e^{-p\tau} \to 0$ at $\tau \to \infty$. Therefore, letting τ tend to ∞, we obtain an equation which makes connection between the moving boundary $y(\tau)$ and $\Psi(\tau)$:

$$\int_0^{\infty} \Psi(\tau)e^{-p\tau}\sinh(y(\tau)\sqrt{p})d\tau = -\int_0^{y(0)} u(x,0)\sinh(x\sqrt{p})dx. \tag{9.20}$$

Using the definitions in Eq. (9.15) and Eq. (9.6), the integral in the RHS of Eq. (9.20) can be represented as

$$\begin{aligned}
F(p) &\equiv -e^{-\int_0^T w(s)ds}\int_{K_1}^{y(0)} (x-K_1)\sinh(x\sqrt{p})e^{k(T)-a(T)x^2}dx \qquad (9.21)\\
&= e^{-\int_0^T w(s)ds}\frac{e^{k(T)-x\left[a(T)x+\sqrt{p}\right]}}{8a^{3/2}(T)}\Bigg\{2\sqrt{a(T)}\left(e^{2\sqrt{p}x}-1\right)\\
&\quad -\sqrt{\pi}e^{\frac{[2a(T)x+\sqrt{p}]^2}{4a(T)}}\Bigg[[\sqrt{p}-2a(T)K_1]\operatorname{erf}\left(\frac{2a(T)x-\sqrt{p}}{2\sqrt{a(T)}}\right)\\
&\quad +[\sqrt{p}+2a(T)K_1]\operatorname{erf}\left(\frac{2a(T)x+\sqrt{p}}{2\sqrt{a(T)}}\right)\Bigg]\Bigg\}\Bigg|_{K_1}^{y(0)},
\end{aligned}$$

$$K_1 = Ke^{\int_0^T w(s)ds}, \quad a(t) = \frac{g'(t)+g(t)(r(t)-q(t))}{2g(t)^3\sigma(t)^2}.$$

Thus, Eq. (9.20) takes the form

$$\int_0^{\infty} \Psi(\tau)e^{-p\tau}\sinh(y(\tau)\sqrt{p})d\tau = F(p), \tag{9.22}$$

where F(p) is known from Eq. (9.21).

The Eq. (9.22) is a linear Fredholm integral equation of the first kind (Polyanin and Manzhirov, 2008). The solution $\Psi(\tau)$ can be found numerically on a grid by solving a system of linear equations. In other words, given functions $r(t), q(t), \sigma(t)$ we can compute first $w(t)$, then $g(t)$, and finally $\tau(t)$ (or $t(\tau)$), thus determining the moving boundary $y(\tau)$. Next we can solve Eq. (9.22) for $\Psi(\tau)$ and substitute it into Eq. (9.19) to obtain the generalized transform of $u(x, \tau)$ in the explicit form. Therefore, if this transform can be inverted back, we solved the problem of pricing Up-and-Out barrier Call options.

9.1.3 *The inverse transform*

In this section the description of inversion is borrowed from Kartashov (2001). Since that book has not been translated into English, we provide a wider exposition of the method. Also, the book contains various typos that are fixed here.

As known from a general theory of the heat equation, the solution of the heat equation $\mathcal{L}u(x,\tau) = 0$, $\mathcal{L} = \partial/\partial\tau - \nu\partial^2/\partial x^2$ at the space domain $0 < x < l$, where l is a constant, can be expressed via Fourier series of the form (Polyanin, 2002)

$$u(x,\tau) = \sum_{n=1}^{\infty} \alpha_n e^{-\nu\gamma_n^2 t} \sin\left(\frac{n\pi x}{l}\right),$$

where $\psi(x) = \sin(n\pi x/l)$ are the eigenfunctions of the heat operator $\mathcal{L}$, and $\gamma_n = n\pi/l$ are its eigenvalues.

Therefore, by analogy we look for the inverse transform of $\bar{u}$, or for the solution of Eq. (9.19) in terms of $u(x, \tau)$, to be a generalized Fourier series of the form (Kartashov, 2001)

$$u(x,\tau) = \sum_{n=1}^{\infty} \alpha_n(\tau) e^{-\left(\frac{n\pi}{y(\tau)}\right)^2 \tau} \sin\left(\frac{n\pi x}{y(\tau)}\right). \tag{9.23}$$

where $\alpha_n(\tau)$ are some functions to be determined. Note, that this definition automatically respects the vanishing boundary conditions for $u(x, \tau)$. We assume that this series converges absolutely and uniformly $\forall x \in [0, y(\tau)]$ for any $\tau > 0$.

Applying this generalized integral transform to both parts of Eq. (9.17) and integrating, we obtain

$$\sum_{n=1}^{\infty} \frac{(-1)^{n+1} n\alpha_n(\tau) e^{-(n\pi/y(\tau))^2\tau}}{p + (n\pi/y(\tau))^2} = \frac{y(\tau)}{\pi \sinh\left(\sqrt{p}\, y(\tau)\right)} \bar{u}(p,\tau). \tag{9.24}$$

The LHS of this equation is regular everywhere except simple poles on the negative semi-axis, see Fig. 9.1

$$p_n = -\left(\frac{n\pi}{y(\tau)}\right)^2, \quad n = 1, 2, ...$$

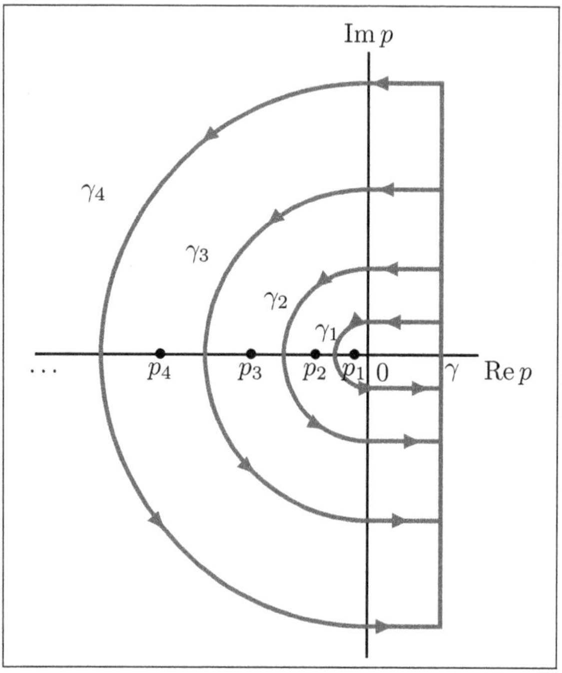

Figure 9.1: Contours of integration γ_n in a complex plane. Here p_n, $n = 1, 2...$ are simple poles of the LHS of the Eq. (9.24).

Let us sequentially integrate both sides of Eq. (9.24) on p along contours $\gamma_1, \gamma_2, \ldots$ The contour γ_n consists of the vertical line $\gamma > 0$, the half-round of radius $R_n = [\pi^2/(2y(\tau)](2n^2 + 2n + 1)$ (the contour γ_n crosses the $Re(p)$ axis in the middle point between p_n and p_{n+1} with the center in the origin), and two horizontal lines $Y = \pm[\pi^2/(2y(\tau))](2n^2 + 2n + 1)$. It means that the circle R_n doesn't hit any pole of the LHS of Eq. (9.24). Then by the Cauchy's residual theorem (Mitrinovic and Keckic, 1984) the integral taken along the contour γ_n is equal to $2\pi\mathrm{i}$ times the sum of residuals of the LHS of Eq. (9.24) that lie inside γ_n.

As poles are simple, and the function under the integral in the LHS of Eq. (9.24) has the form $F_1(p)/F_2(p)$, the residual of such a function is (Mitrinovic and Keckic, 1984)

$$\mathrm{Res}[F_1(p)/F_2(p); p_k] = F_1(p)/F_2'(p)\Big|_{p=p_k}$$

The above analysis is the basis for running a residual machinery to calculate all the coefficients $\alpha_n(\tau)$.

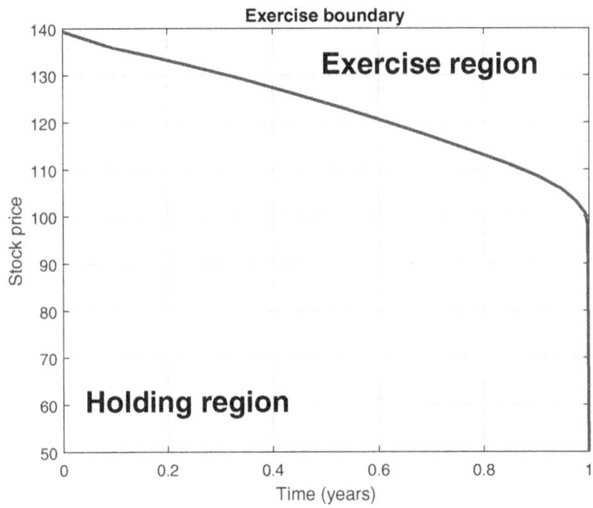

Figure 9.2: Typical exercise boundary for the American Call option.

9.1.3.1 *Residual machinery*

Let us denote via $\mathcal{I}_k$ the following contour integral

$$\mathcal{I}_k = \frac{1}{2\pi \mathrm{i}} \oint_{\gamma_k} \frac{\bar{u}(p,\tau)}{\sinh\left(\sqrt{p}y(\tau)\right)} dp.$$

Below we show that all coefficients α_n, $n = 1, \ldots, \infty$ can be expressed via these integrals.

1. Coefficient $\alpha_1(\tau)$. Integrating Eq. (9.24) along the contour γ_1 gives

$$\alpha_1(\tau) e^{-(\pi/y(\tau))^2 \tau} \oint_{\gamma_1} \frac{1}{p + (\pi/y(\tau))^2} dp + \sum_{n=2}^{\infty} (-1)^{n+1} n \alpha_n(\tau) e^{-(n\pi/y(\tau))^2 \tau} \oint_{\gamma_1} \frac{1}{p + (n\pi/y(\tau))^2} dp = \frac{y(\tau)}{\pi} \oint_{\gamma_1} \frac{\bar{u}(p,\tau) dp}{\sinh\left(\sqrt{p}y(\tau)\right)}.$$

Observe, that

$$\oint_{\gamma_1} \frac{1}{p + (\pi/y(\tau))^2} dp = 2\pi \mathrm{i}, \qquad \oint_{\gamma_1} \frac{1}{p + (n\pi/y(\tau))^2} dp = 0, \quad n \geq 2,$$

where the second result is due to the Cauchy integral theorem (Mitrinovic and Keckic, 1984). Then

$$\alpha_1(\tau, y) = \frac{y(\tau)}{\pi} e^{(\pi/y(\tau))^2 \tau} \mathcal{I}_1. \tag{9.25}$$

2. Coefficient $\alpha_2(\tau)$. By analogy, integrating the second equation of Eq. (9.24) along the contour γ_2 we obtain

$$\begin{aligned}
&\alpha_1(\tau) e^{-(\pi/y(\tau))^2 \tau} \oint_{\gamma_2} \frac{dp}{p + (\pi/y(\tau))^2} - 2\alpha_2(\tau) e^{-(2\pi/y(\tau))^2 \tau} \oint_{\gamma_2} \frac{dp}{p + (2\pi/y(\tau))^2} \\
&\quad + \sum_{n=3}^{\infty} (-1)^{n+1} \alpha_n(\tau) e^{-(n\pi/y(\tau))^2 \tau} \oint_{\gamma_2} \frac{dp}{p + (n\pi/y(\tau))^2} \\
&\quad = \frac{y(\tau)}{\pi} \oint_{\gamma_2} \frac{\bar{u}(p, \tau)}{\sinh\left(\sqrt{p} y(\tau)\right)} dp,
\end{aligned}$$

whence using again the residual theorem and Eq. (9.25) we find

$$\alpha_2(\tau, y) = -\frac{y(\tau)}{2\pi} e^{(2\pi/y(\tau))^2 \tau} \left[\mathcal{I}_2 - \mathcal{I}_1\right].$$

3. Coefficient $\alpha_n(\tau)$. Proceeding in a similar manner, we obtain a general formula for the coefficients α_n, $n \geq 1$

$$\alpha_n(\tau) = (-1)^{n+1} \frac{y(\tau)}{n\pi} e^{(n\pi/y(\tau))^2 \tau} \left[\mathcal{I}_n - (1 - \delta_{n,1}) \mathcal{I}_{n-1}\right], \tag{9.26}$$

where $\delta_{n,1}$ is the Kronecker symbol.

9.1.3.2 *The final solution*

To calculate the integrals in the RHS of Eq. (9.26), we rewrite them in the explicit form by using the solution for $\bar{u}(p, \tau)$ previously found in Eq. (9.19)

$$\begin{aligned}
\mathcal{I}_j = \frac{1}{2\pi \mathrm{i}} \oint_{\gamma_k} \frac{e^{p\tau}}{\sinh(y(\tau)\sqrt{p})} &\left[\int_0^{\tau} \Psi(s) e^{-ps} \sinh(y(s)\sqrt{p}) ds \right. \\
&\left. + \int_0^{y(0)} u(x, 0) \sinh(x\sqrt{p}) dx \right] dp.
\end{aligned}$$

As $\sinh(x)$ is a periodic complex function with the period $\pi k/i$, the RHS of this equation is regular everywhere except simple poles, where $\sinh\left(\sqrt{p} y(\tau)\right)$ vanishes. It is easy to checks that these poles are exactly

$p_i,\ i = 1, \ldots, k$. Therefore, we again can directly apply the Cauchy residual theorem. Computing residuals, after some algebra we obtain

$$\alpha_n(\tau) = \frac{2}{y(\tau)}\Bigg[\int_0^{y(0)} u(x,0)\sin\left(\frac{n\pi x}{y(\tau)}\right)dx \\ + \int_0^{\tau} e^{(n\pi/y(\tau))^2 s}\Psi(s)\sin\left(\frac{n\pi y(s)}{y(\tau)}\right)ds.\Bigg]. \tag{9.27}$$

Thus, from Eq. (9.23) and Eq. (9.27) we find the final solution

$$u(x,\tau) = \frac{2}{y(\tau)}\Bigg[\sum_{n=1}^{\infty} e^{-\left(\frac{n\pi}{y(\tau)}\right)^2\tau}\sin\left(\frac{n\pi x}{y(\tau)}\right)\int_0^{y(0)} u(x,0)\sin\left(\frac{n\pi x}{y(\tau)}\right)dx \\ + \sum_{n=1}^{\infty}\sin\left(\frac{n\pi x}{y(\tau)}\right)\int_0^{\tau} e^{(n\pi/y(\tau))^2(s-\tau)}\Psi(s)\sin\left(\frac{n\pi y(s)}{y(\tau)}\right)ds\Bigg].$$

This can also be re-written as

$$u(x,\tau) = \frac{2}{y(\tau)}\Bigg[\int_0^{y(0)} dz\ u(z,0)\sum_{n=1}^{\infty} e^{-\left(\frac{n\pi}{y(\tau)}\right)^2\tau}\sin\left(\frac{n\pi x}{y(\tau)}\right)\sin\left(\frac{n\pi z}{y(\tau)}\right) \\ + \int_0^{\tau} ds\ \Psi(s)\sum_{n=1}^{\infty} e^{(n\pi/y(\tau))^2(s-\tau)}\sin\left(\frac{n\pi y(s)}{y(\tau)}\right)\sin\left(\frac{n\pi x}{y(\tau)}\right)\Bigg]. \tag{9.28}$$

We proceed with the observation that the sums in Eq. (9.28) could be expressed via Jacobi theta functions of the third kind (Mumford *et al.*, 1983). By definition

$$\theta_3(z,\omega) = 1 + 2\sum_{n=1}^{\infty}\omega^{n^2}\cos(2nz). \tag{9.29}$$

Therefore,

$$\sum_{n=1}^{\infty} e^{-\left(\frac{n\pi}{y(\tau)}\right)^2\tau}\sin\left(\frac{n\pi x}{y(\tau)}\right)\sin\left(\frac{n\pi z}{y(\tau)}\right) = \frac{1}{4}[\theta_3(\phi_-(z),\omega_1) - \theta_3(\phi_+(z),\omega_1)], \\ \sum_{n=1}^{\infty} e^{(n\pi/y(\tau))^2(s-\tau)}\sin\left(\frac{n\pi y(s)}{y(\tau)}\right)\sin\left(\frac{n\pi x}{y(\tau)}\right) \\ = \frac{1}{4}[\theta_3(\phi_-(y(s)),\omega_2) - \theta_3(\phi_+(y(s)),\omega_2)], \\ \omega_1 = e^{-\left(\frac{\pi\sqrt{\tau}}{y(\tau)}\right)^2},\ \omega_2 = e^{-\left(\frac{\pi\sqrt{\tau-s}}{y(\tau)}\right)^2},\ \phi_-(z) = \frac{\pi(x-z)}{2y(\tau)},\ \phi_+(z) = \frac{\pi(x+z)}{2y(\tau)}. \tag{9.30}$$

A well-behaved theta function must have parameter $|\omega| < 1$ (Mumford *et al.*, 1983). This condition holds at any $\tau > 0$.

Thus, Eq. (9.28) transforms to a simpler form

$$
\begin{aligned}
u(x,\tau) = \frac{1}{2y(\tau)} \Bigg[\int_0^{y(0)} dz\ u(z,0)[\theta_3(\phi_-(z),\omega_1) - \theta_3(\phi_+(z),\omega_1)] \qquad (9.31) \\
+ \int_0^{\tau} ds\ \Psi(s)[\theta_3(\phi_-(y(s)),\omega_2) - \theta_3(\phi_+(y(s)),\omega_2)] \Bigg].
\end{aligned}
$$

The RHS of Eq. (9.28) depends on x via functions ϕ_-, ϕ_+. Since the theta function $\theta_3(z,\omega)$ is even in z, the boundary condition at $x = 0$ is satisfied. At $x = y(\tau)$ it is also satisfied as follows from Eq. (9.30) if one reads it from right to left.

The result in Eq. (9.31) to some extent is not a surprise, as it is known that the Jacobi theta function is the fundamental solution of the one-dimensional heat equation with spatially periodic boundary conditions (Ohyama, 1995).

It is also worth mentioning, that Eq. (9.31) can be transformed to the Volterra equation of the second kind for $\Psi(\tau)$ by differentiating both parts on x and then letting $x = y(\tau)$. This equation can be solved instead of Eq. (9.22) which is computationally easier then solving Eq. (9.22), see Carr *et al.* (2020).

9.2 Pricing American options

We recall that an American option is an option that can be exercised at anytime during its life. American options allow option holders to exercise the option at any time prior to and including its maturity date, thus increasing the value of the option to the holder relative to European options, which can only be exercised at maturity. The majority of exchange-traded options are American style. For a more detailed introduction, see Detemple (2006) and Hull (2011).

It is known that pricing American (or Bermudan) options requires solution of a linear complementary problem. Various efficient numerical methods have been proposed for doing that. For instance, when the underlying stock price S_t follows the time-dependent Black-Scholes model these (finite-difference) methods are discussed in Itkin (2017) (see also references therein).

Another approach, elaborated e.g., in Andersen *et al.* (2016) for the Black-Scholes model with constant coefficients, uses a notion of the exercise boundary $S_B(t)$. The boundary is defined in such a way, that, e.g., for the American Put option $P_A(S,t)$ at $S \leq S_B(t)$ it is always optimal to exercise the option, therefore $P_A(S,t) = K - S$. For the complementary domain $S > S_B(t)$ the earlier exercise is not optimal, and in this domain $P_A(S,t)$ obeys the Black-Scholes equation. This domain is called the continuation (holding) region. The problem of pricing American options lies in the fact that $S_B(t)$ is not known in advance. Instead, we only know the price of the American option at the boundary. For instance, for the American Put we have $P_A(S_B(t),t) = K - S_B(t)$, and for the American Call - $C_A(S_B(t),t) = S_B(t) - K$. A typical shape of the exercise boundary for the Call option obtained with the parameters $K = 100, r = 0.05, q = 0.03, \sigma = 0.2$ is presented in Fig. 9.2. The method proposed in Andersen *et al.* (2016) finds $S_B(t)$ by numerically solving an integral (Volterra) equation for $S_B(t)$. The resulting scheme is straightforward to implement and converges at a speed several orders of magnitude faster than existing approaches.

In terms of that paper, the continuation region is a domain with the moving boundary where the option price solves the corresponding PDE. In case of our model in Eq. (9.1) this is the PDE in Eq. (9.2). Therefore, this problem is, by nature, similar to that for the barrier options considered in Section 9.1, but the difference is as follows.

- For the barrier option pricing problem the moving boundary (the time-dependent barrier) is known, as this is stated in Eq. (9.16). But the Option Delta $\frac{\partial u}{\partial x}$ at the boundary $x = y(\tau)$ is not, and should be found by solving the linear Fredholm equation Eq. (9.22). Also, the problem is solved subject to the vanishing condition at the barrier (the moving boundary) for the option value.
- For the American option pricing problem the moving boundary is not known. However, the option Delta $\frac{\partial u}{\partial x}$ at the boundary $x = y(\tau)$ is known (it follows from the conditions $\frac{\partial C_A}{\partial S}|_{S=S_B(t)} = 1$ and $\frac{\partial P_A}{\partial S}|_{S=S_B(t)} = -1$ expressed in variables x and $y(\tau)$ according to their definitions in Section 9.1). Also the boundary condition for the American Call and Put at the exercise boundary (the moving boundary) differs from that for the Up-and-Out barrier option, namely: it is $C_A(S_B(t),t)) = S_B(t) - K$ for the Call, and $P_A(S_B(t),t)) = K - S_B(t)$ for the Put

Because of the similarity of these two problems, it turns out that the American option problem can be solved for the continuation region together with the simultaneous finding of the exercise boundary, by using the same approach that we proposed for solving the barrier option pricing problem. However, due to the highlighted differences, some equations slightly change. Also, it is worth mentioning that a similar approach which uses a method of heat potentials has been developed in Lipton and Kaushansky (2020b).

9.2.1 *Solution for the American Call option price*

Since the PDE we need to solve is the same as in Eq. (9.2), we do same transformations as in Section 9.1, and come up to the same heat equation as in Eq. (9.14). It should be solved subject to the terminal condition

$$u(x,0) = \left(xe^{-\int_0^T w(s)ds} - K\right)^+ e^{-f(x,T)},$$

and the boundary conditions

$$\begin{aligned}
u(0,\tau) &= 0, \qquad u(y(\tau),\tau) \equiv \psi_1(\tau) = y(\tau) - K, \\
\Psi(\tau) &\equiv \left.\frac{\partial u}{\partial x}\right|_{x=y(\tau)} = \frac{e^{-f(y(\tau),t)}}{g^2(t)}\left[1 + a_1(t)y(\tau)(y(\tau)-K)\right], \\
a_1(t) &= \frac{g'(t) + g(t)(r(t)-q(t))}{g(t)\sigma(t)^2},
\end{aligned}$$

where $t = t(\tau)$. We underline once again that the function $y(\tau)$ here is not known yet, while $\Psi(\tau)$ is known. These problems with the free boundaries are also well known in physics.

We proceed by using the same transformation in Eq. (9.17), and by analogy with Eq. (9.18) obtain the following Cauchy problem

$$\begin{aligned}
\frac{\partial \bar{u}}{\partial \tau} - p\bar{u} &= \Psi(\tau)\sinh(x\sqrt{p}) + \psi_1(\tau)\sqrt{p}, \\
\bar{u}(p,0) &= \int_0^{y(0)} u(x,0)\sinh(x\sqrt{p})dx,
\end{aligned}$$

This problem can be solved explicitly to yield (Kartashov, 2001)

$$\begin{aligned}
\bar{u}e^{-p\tau} = {} & \int_0^{\tau} \Psi(\tau)e^{-p\tau}\sinh(y(\tau)\sqrt{p})d\tau + \int_0^{y(0)} u(x,0)\sinh(x\sqrt{p})dx \\
& + \sqrt{p}\int_0^{\tau} e^{-p\tau}\psi_1(\tau)d\tau.
\end{aligned}$$

Accordingly, instead of Eq. (9.20) we obtain

$$\int_0^\infty e^{-p\tau}\left[\Psi(\tau)\frac{\sinh(y(\tau)\sqrt{p})}{\sqrt{p}} + y(\tau)\right]d\tau$$

$$= \frac{K}{p} - \frac{1}{\sqrt{p}} \int_0^{y(0)} u(x,0)\sinh(x\sqrt{p})dx = \frac{K}{p} + \frac{F(p)}{\sqrt{p}}.$$

This is a nonlinear Fredholm equation of the first kind, but now with respect to the function $y(\tau)$. It can also be solved numerically (iteratively).

The next step is to reduce our problem to that with homogeneous boundary conditions. This can be done by change of the dependent variable

$$u(x,\tau) = W(x,\tau) + \Theta(x,\tau), \quad \Theta(x,\tau) = (1 - x/y(\tau)]\psi_1(\tau).$$

The function $W(x,\tau)$ solves the same heat equation with the same terminal condition, and with the homogeneous boundary conditions. Therefore, it can be solved by using the method of generalized integral transform described in Section 9.1.2. The solution reads

$$u(x,\tau) = \Theta(x,\tau) + \sum_{n=1}^{\infty} \alpha_n(\tau) e^{-\left(\frac{n\pi}{y(\tau)}\right)^2 \tau} \sin\left(\frac{n\pi x}{y(\tau)}\right),$$

$$\alpha_n(\tau) = \frac{2}{y(\tau)}\left[\int_0^{y(0)} [u(z,0) - \Theta(z,0)] \sin\left(\frac{n\pi z}{y(\tau)}\right) dz \right.$$
$$\left. + \int_0^{\tau} e^{(n\pi/y(\tau))^2 s} \left[\Psi(s) + \frac{\psi_1(s)}{y(s)}\right] \sin\left(\frac{n\pi y(s)}{y(\tau)}\right) ds\right].$$

Again, using the definition of the Jacobi theta function in Eq. (9.29), this can be finally re-written as

$$u(x,\tau) = \Theta(x,\tau) + \frac{1}{2y(\tau)}\left[\int_0^{y(0)} dz\ [u(z,0) - \Theta(z,0)]\right.$$
$$\times [\theta_3(\phi_-(z),\omega_1) - \theta_3(\phi_+(z),\omega_1)]$$
$$\left. + \int_0^{\tau} ds\ \left[\Psi(s) + \frac{\psi_1(s)}{y(s)}\right] [\theta_3(\phi_-(y(s)),\omega_2) - \theta_3(\phi_+(y(s)),\omega_2)]\right].$$

9.3 Numerical example

To test performance and accuracy of our method in this section we provide a numerical example where a particular time dependence of $r(t), q(t), \sigma(t)$ is chosen as

$$r(t) = r_0 e^{-r_k t}, \qquad q(t) = q_0, \qquad \sigma(t) = \sigma_0 e^{-\sigma_k t}. \tag{9.32}$$

Here $r_0, q_0, \sigma_0, r_k, \sigma_k$ are constants. With this model Eq. (9.9) can be solved analytically to yield

$$w(t) = q_0 - r_0 e^{-r_k t}. \tag{9.33}$$

Accordingly, from Eq. (9.8) we find

$$g(t) = \exp\left[q_0 t + \frac{r_0}{r_k}\left(e^{-r_k t} - 1\right)\right], \tag{9.34}$$

and from Eq. (9.6)

$$f(x,t) = k(t) = \frac{1}{2}\left[\log\left(\frac{g(t)}{g(0)}\right) - q_0 t + 3\frac{r_0}{r_k}\left(1 - e^{-r_k t}\right)\right]. \tag{9.35}$$

The algorithm described in Section 9.1 was implemented in python. We did it for two reasons. First, we found neither any standard implementation of the Jacobi theta functions in Matlab, nor any custom good one. Surprisingly, this is also not a part of numpy or scipy packages in python. However, they are available as a part of the python package mpmath which is a free (BSD licensed) Python library for real and complex floating-point arithmetic with arbitrary precision, see Johansson (2007). It has been developed by Fredrik Johansson since 2007, with help from many contributors.

Also, we didn't find any standard implementation of solver for the Fredholm integral equation of the first kind in both python and Matlab. Therefore, we implemented a Tikhonov regularization method as this is described in Fuhry (2001). In particular, with the model used in this section, the function $F(p)$ reads

$$F(p) = \frac{e^{-k(T)}}{p}\left[-\sqrt{p}(K_1 - y_0)\cosh\left(\sqrt{p}y_0\right) + \sinh\left(K_1\sqrt{p}\right) - \sinh\left(\sqrt{p}y_0\right)\right]. \tag{9.36}$$

The algorithm described in Section 9.1 was implemented in python. Finally to validate the results provided by our method, we implemented a FD solver for pricing Up-and-Out barrier options. This solver is based on the Crank–Nicolson scheme with a few Rannacher first steps, and uses a non-uniform grid, in more detail see, e.g., Itkin (2017). We implemented two solvers: one for the backward PDE, and the other one — for the forward PDE. But logically, since here we solved the backward PDE, it does make sense to compare our method with the backward solver. This implementation has been done in Matlab.

In our particular test we choose parameters of the model as they are presented in Table 9.1.

We recall, that here $\sigma(t)$ is the normal volatility. Therefore, we choose its typical value by multiplying the log-normal volatility by the barrier level.

We run the test for a set of maturities $T \in [1/12, 0.3, 0.5, 1]$ and strikes $K \in [50, 55, 60, 65, 70, 75, 80]$. The Up-and-Out barrier Call option prices computed in such an experiment are presented in Fig. 9.3.

r_0	q_0	σ_0	r_k	σ_k	H	S_0
0.02	0.01	0.5$\cdot H$	0.1	0.2	90	60

Table 9.1: Parameters of the test.

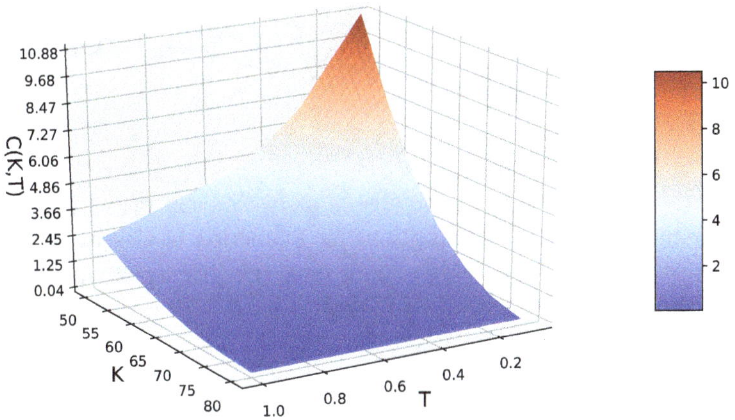

Figure 9.3: Option price of Up-and-Out barrier Call options as a function of the strike K and maturity T computed in the test by using the GIT method.

In Fig. 9.4 the relative difference between the Up-and-Out barrier Call option prices obtained by using our method and the FD solver are presented as a function of the option strike K and maturity T. Here to provide a comparable accuracy we run the FD solver with 101 nodes in space S and the time step $\Delta t = 0.01$.

It can be seen that the quality of the FD solution is not sufficient. Therefore, we reran it by using 201 nodes in space S and the time step $\Delta t = 0.001$. The relative difference between our semi-analytic and the FD solutions in this case are presented in Fig. 9.5.

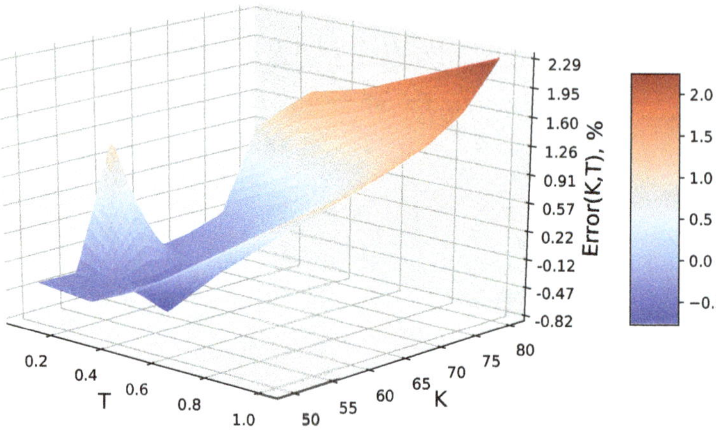

Figure 9.4: The relative difference of two option prices of the Up-and-Out barrier Call options as a function of the strike K and maturity T. The first price is computed by using the GIT method, the other one by an FD solver with 101 nodes in space S and time step $\Delta t = 0.01$.

It can be seen that the agreement between prices obtained by using our method and the FD pricer is good, so the relative difference is about 1%. However, the cost for this improvement of the FD method is speed. In Table 9.2 we compare the elapsed time of both methods. The column "no Ψ" has the following meaning. Since the volatility and the interest rate change with time relatively slow, contribution of the second integral in Eq. (9.31) to the option price is negligible. Therefore, in this particular case we can find the option price by computing only the first integral in Eq. (9.31). Accordingly, we don't need to solve the Fredholm equation Eq. (9.22) that almost halves the elapsed time.

Finally, our tests show that linear algebra in python (numpy) at our machine is about 3 times slower than that in Matlab. Therefore, given the

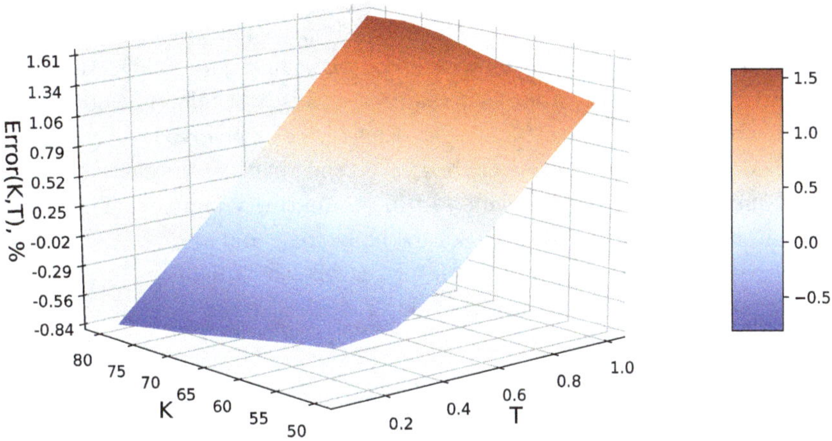

Figure 9.5: The relative difference of two option prices of the Up-and-Out barrier Call options as a function of the strike K and maturity T. The first price is computed by using the GIT method, the other one by an FD solver with 101 nodes in space S and time step $\Delta t = 0.001$.

Test	Elapsed time (s)
Semi-analytic	0.23
Semi-analytic, no Ψ	0.16
FD-101, $\Delta t = 0.01$	0.36
FD-201, $\Delta t = 0.01$	0.65
FD-201, $\Delta t = 0.001$	3.60

Table 9.2: Elapsed time in secs of various tests.

same accuracy, our method is about 30-40 times faster than the backward FD solver.

Of course, the forward FD solver is by an order of magnitude faster than the backward one if we need to simultaneously price multiple options of various strikes and maturities but written on the same underlying. However, for barrier options this approach requires a very careful implementation, which often is not universal and with a lot of tricks involved.

Also, as it can be seen from Fig. 9.3, the relative difference, despite small, increases when the strike and time to maturity increase. This is because in this case the strike is closer to the barrier, and also the probability of reaching the barrier over the life of the option is higher. In this case the FD solver, since the price has a kink at the barrier provides a bigger difference.

9.4 Discussion

Our attention in Section 9.1 was drawn to the Up-and-Out barrier Call option C_{uao}. Obviously, using the barriers parity (Hull, 2011) the price of the Down-and-Out barrier Call option C_{dao} can be found as $C_{dao} = C_{van} - C_{uao}$, where C_{van} is the price of the European vanilla Call option. It is known that the latter is given by the corresponding formula for the process with constant coefficients, where those efficient constant coefficients $\bar{r}, \bar{q}, \bar{\sigma}$ are defined as

$$\bar{r} = \frac{1}{T}\int_0^T r(s)ds, \qquad \bar{q} = \frac{1}{T}\int_0^T q(s)ds, \qquad \bar{\sigma}^2 = \frac{1}{T}\int_0^T \sigma^2(s)ds.$$

Second, we underline that in addition to the model with time-dependent coefficients we also consider the barriers to be some arbitrary functions of time. Our method provides full coverage of this case, while constant barriers are just some particular case of the general solution.

Third, and perhaps the most important point is about computational efficiency of our method. In addition to what was presented in Section 9.3, let's look at this problem from a theoretical pint of view. Suppose the barrier pricing problem is attacked by solving the forward PDE for a set of strikes $K_i,\ i = 1, \dots, \bar{k}$ and a set of maturities $T_j,\ j = 1, \dots, \bar{m}$ numerically by some FD method on a grid with N nodes in the space domain $S \in [0, H]$, and M nodes in the time domain $t \in [0, T_{\bar{m}}]$. Then the complexity of this method is known to be $O(MN + 4N)$. This should be compared with the complexity of our approach.

Let's assume that the Riccati equation in Eq. (9.9) can be solved either analytically, or, at least, approximately, as this is discussed in Section 9.1.1.

Then the first computational step consists in solving the linear Fredholm equation in Eq. (9.22) (or the corresponding Volterra equation). This can be done on a rarefied grid with $M_1 < M$ nodes and complexity $O(M_1^3)$. The intermediate values in t can be found (if necessary) by interpolation with the complexity $O(M_1^2)$. As the integral kernel doesn't depend on strikes K_i, this calculation can be done simultaneously for all strikes still preserving the complexity $O(M_1^3)$.

The final solution of the pricing problem is provided in the form of two integrals in Eq. (9.31). Therefore, if we need the option price at a single value of S_0 (same as when solving the forward PDE), but for all strikes and maturities, the complexity is $O(2\bar{k}L(M_1 + N_1)$, where N_1 is the number of points in the x space, and $O(L)$ is the complexity of computing the Jacobi theta function $\theta_3(z, \omega)$. Normally, $M_1 \le N$, $L \ll N$, $N_1 \ll N$ for the typical values of N in the FD method (about 50–100 or even more). Thus, the total complexity of our method is fully determined by the solution of the Fredholm equation. Therefore, our method is slower than the corresponding FD method if $M_1 > (MN)^{1/3}$. For the American option this situation is worse since instead of solving a linear Fredholm equation we need to solve a nonlinear equation. This can be done iteratively, e.g., using k iterations until the method converges to the given tolerance. Then the total complexity becomes $O(kM_1^3)$. However, our experiments show that using just $M_1 = 10$ points in p space could be sufficient, while further increase of M_1 doesn't change the results.

Also, the accuracy of the method in x can be increased if one uses high-order quadratures for computing the final integrals. For instance, one can use the Simpson instead of the trapezoid rule that doesn't affect the complexity of our method. While increasing the accuracy for the FD method is not easy (i.e., it significantly increases the complexity of the method, e.g., see Itkin (2017)).

Another advantage of our approach also mentioned in Carr *et al.* (2020), is computation of option Greeks. Since the option prices are represented in closed form via integrals, the explicit dependence of prices on the model parameters is available and transparent. Therefore, explicit representations of the option Greeks can be obtained by a simple differentiation under the integrals. This means that the values of Greeks can be calculated simultaneously with the prices almost with no increase in time. This is because differentiation under the integrals slightly changes the integrands, and these changes could be represented as changes in weights of the quadrature scheme used to numerically compute the integrals. Since the major computational time has to be spent for computation of densities which contain special functions, they can be saved during the calculation of the prices, and then reused for computation of Greeks.

It is worth mentioning that our method can also be extended to pricing American options where the underlying pays discrete dividends. Indeed, the constructed analytical solution covers the time interval starting from maturity T and up to the last ex-dividend date t_n using the final payoff as the terminal condition. Then, at t_n we have option prices $u(\tau, x)$ for all x. Due to continuity, shifting the underlying by the dividend amount Δx (so $x \mapsto \bar{x} = x + \Delta x$) and re-interpolating the prices to $u(\tau, \bar{x})$, we obtain the new terminal condition. Then the algorithms continues from t_n to t_{n-1} by replacing T with t_n and the terminal condition at t_n by $u(\tau, \bar{x})$. And so on. This is the same approach which is used when solving this problem using the FD method (Itkin, 2017; Tavella and Randall, 2000).

9.5 Extension to the Black-Scholes model

Despite in this chapter we are mostly dedicated to pricing barrier and American options written on the underlying which follows the OU process (or the Bachelier model), same approach can be developed for pricing same types of options but under the Black-Scholes model (for the model description see Section 2.2.1). That is because by a series of transformations this problem can also be reduced to the heat equation Eq. (9.14) with the initial and boundary conditions as in Eqs. (9.15) and (9.16).

We start by specifying the dynamics of the underlying spot price S_t in the Black-Scholes models

$$
\begin{aligned}
dS_t &= [r(t) - q(t)]S_t dt + \sigma(t) S_t dW_t, \qquad (9.37)\\
S\Big|_{t=0} &= S, \quad (t, S) \in [0, t] \times [0, \infty),
\end{aligned}
$$

which is a geometric Brownian motion process with time-dependent coefficients.

Suppose we want to price a contingent claim written on the underlying process S_t in Eq. (9.37) which is the Up-and-Out barrier Call option. Again, it follows from the Feynman-Kac formula (Klebaner, 2005), that this price $C(S,t)$ solves a linear parabolic PDE conditional on $S_0 = S$, $S \in [0, \infty)$, which reads

$$
\frac{\partial C}{\partial t} + \frac{1}{2}\sigma^2(t) S^2 \frac{\partial^2 C}{\partial S^2} + [r(t) - q(t)] S \frac{\partial C}{\partial S} = r(t) C. \qquad (9.38)
$$

This equation should be solved subject to the terminal condition at the option maturity $t = T$

$$C(S,T) = (S-K)^+, \tag{9.39}$$

and the boundary conditions

$$C(0,t) = 0, \qquad C(H(t),t) = 0, \tag{9.40}$$

where $H(t)$ is the time-dependent upper barrier.

As was claimed in above, by a series of transformations Eq. (9.2) can be reduced to the heat equation Eq. (9.14). Indeed, by the standard change of variables $\log(S/K) \to x$, the Eq. (9.38) can be transformed to

$$\begin{aligned} &\frac{\partial C}{\partial t} + \frac{1}{2}\sigma^2(t)\frac{\partial^2 C}{\partial x^2} + \mu(t)\frac{\partial C}{\partial x} = r(t)C, \\ &\mu(t) = r(t) - q(t) - \frac{1}{2}\sigma^2(t). \end{aligned} \tag{9.41}$$

Further, another change of variables (Polyanin, 2002),

$$C(x,t) = u(z,\tau)e^{\int_t^T r(s)ds}, \quad z = x - \int_t^T \mu(s)ds, \quad \tau = \frac{1}{2}\int_t^T \sigma^2(s)ds, \tag{9.42}$$

transforms Eq. (9.41) to the heat equation

$$\frac{\partial u}{\partial \tau} = \frac{\partial^2 u}{\partial z^2}. \tag{9.43}$$

Equation (9.43) should be solved subject to the terminal condition

$$u(z,0) = Ke^{-\int_t^T r(s)ds}\left(e^{z-\int_t^T \mu(s)ds} - 1\right)^+, \tag{9.44}$$

and the boundary conditions

$$\begin{aligned} &u(0,\tau) = 0, \quad u(y(\tau),\tau) = 0, \\ &y(\tau) = \log\left(\frac{H(t(\tau))}{K}\right) - \int_{t(\tau)}^0 \mu(s)ds. \end{aligned} \tag{9.45}$$

These conditions directly follow from Eqs. (9.39) and (9.40) and changes of variables made. The dependence $t(\tau)$ is an inverse map of the last line in Eq. (9.42), and $y(\tau)$ is a time-dependent upper barrier expressed in new variables.

The problem in Eqs. (9.43)–(9.45) is the same problem as that in Eqs. (9.14)–(9.16). Therefore, they can be solved by using the GIT method described in Section 9.1. Same is true for the American option pricing problem and its solution presented in Section 9.2. Thus, as was stated in above, under the Black-Scholes model these problems can be solved by using the solutions presented in this Chapter for the Bachelier model.

Chapter 10

On the First Hitting Time Density for a Reducible Diffusion Process

Computation of the first hitting time density of a diffusion process is a long-standing problem, which still remains open. Particular examples are the hitting problem of a Wiener process to a curvilinear boundary and the hitting problem of an Ornstein–Uhlenbeck process. An abstract approach applicable to this problem has been found by Fortet (1943); the Fortet's equation can be viewed as a variant of the Einstein-Smoluchowski equation (Einstein, 1905; Von Smoluchowski, 1906). A general overview can be found in Breiman (1967); Horowitz (1985); Borodin and Salminen (2012).

For a standard Wiener process, Shepp (1967) and Novikov (1971) derived the density and moments expansion for the problem with a square-root boundary in terms of special function (parabolic cylinder function and Hermite polynomial) (Daniels, 1996; Novikov *et al.*, 1999; Pötzelberger and Wang, 2001); derived the expression for a linear and piece-wise linear barrier, and as result, developed numerical methods for general curved boundaries by a piece-wise linear approximation of the boundary. Hyer *et al.* (1998) and Avellaneda and Zhu (2001) analyzed the problem with curvilinear boundaries in finance. In a structural default model A. Lipton developed numerical methods to find the default boundary knowing the default probabilities.

Attempts to find an analytical result for an Ornstein–Uhlenbeck have been made since 1998 when (Leblanc and Scaillet, 1998) first derived an analytical formula which contained an error. Two years later, Leblanc *et al.* (2000) published a correction on the paper; unfortunately, the correction was erroneous as well. The authors showed that the density can be expressed as the Laplace transform of a functional of a three-dimensional Bessel bridge. Göing-Jaeschke and Yor (2003) noticed that the authors had incorrectly used a spatial homogeneity property for the three-dimensional

Bessel bridge. It is worth noting that the same conclusion can be obtained by calculating the CDF and showing that it asymptotically exceeds unity for some values of parameters.

In Ricciardi and Sato (1988); Linetsky (2004) and Alili *et al.* (2005) representations are found for the hitting density by using the Laplace transform. Alili *et al.* (2005) gave several representations: a representation in the series of parabolic cylinder functions and its derivatives, an indefinite integral representation via special functions, and a Bessel bridge representation. The first approach is based on inverting the Laplace transform, which is computed analytically. As a result, the authors got the series representation involving parabolic cylinder functions and its derivatives. The second approach is based on the cosine transform and its inverse. The authors got a representation via an indefinite integral involving some special functions and computed it using the trapezoidal rule. The third approach gives a representation of the density via an expectation of a function of the three-dimensional Bessel bridge, which can be computed using the Monte Carlo method. Linetsky (2004) gave an analytical representation via relevant Sturm–Liouville eigenfunction expansions. The coefficients were found as a solution of a nonlinear equation, which involved nonlinear special (Hermite) functions.

In Martin *et al.* (2015) a nonlinear Fokker–Planck equation was solved with steady-state solution by representing it as an infinite product rather than as usual an infinite sum. The PDE, which corresponds to the transition probability of Ornstein–Uhlenbeck process, belongs to the class of equations described in Martin *et al.* (2015). In principle, the results of the paper can be modified to the computation of first time hitting density. An advantage of this method is that it allows quantifying the errors; thereby controlling the number of terms required to reach a given precision.

In Martin *et al.* (2019) a short- and long-time asymptotic expansion are developed for the OU and other mean-reverting processes.

Various numerical approaches for first passage problems of general Gauss–Markov processes are developed in Nardo *et al.* (2001). The authors reduced the problem to an integral equation and solved it numerically. Gobet and Menozzi (2010) developed an improved Euler-type scheme to reduce the bias from hitting the boundary.

The methods described above require substantial numerical computations, and for some of them the convergence rate is unknown. Moreover, most of them are difficult to implement; for example, Cheridito and Xu (2015) preferred the Crank–Nicolson method to other known analytical methods, because it is easier to implement.

Below we describe the solution of this problem presented in Lipton and Kaushansky (2020a). There the authors developed a fast semi-analytical method to compute the first time hitting density of a general diffusion process, which is easy to implement. We give particular examples for the cases of a Wiener and time-dependent Ornstein–Uhlenbeck processes as well as show several financial applications. After an appropriate change of variables, the corresponding PDE becomes a heat equation with a moving boundary. We solve it using the method of heat potentials. As a result, we got recursive formulas to compute the corresponding hitting density. It is also worth to point out that for the case of an Ornstein–Uhlenbeck process, our method is able to deal with time-dependent coefficients, while the methods, which are based on the Laplace transform, are able to deal only with the case of constant coefficients.

10.1 Problem formulation and initial transformations

First hitting time density for a Wiener process. Consider the distribution of the hitting time of a Wiener process with $W_0 = z$ of a given time-dependent barrier $b(t)$

$$s = \inf\{t : W_t = b(t)\}, \quad z \neq b(0).$$

By using the Feynman–Kac formula it is easy to see that the transition probability density solves the problem

$$\begin{aligned} p_t(t, x; z) &= \frac{1}{2} p_{xx}(t, x; z), \\ p(0, x; z) &= \delta(x - z), \qquad p(t, b(t); z) = 0. \end{aligned}$$

Then the distribution in question could be determined as

$$g(t, z) = \frac{\varpi}{2} p_x(t, b(t); z), \tag{10.1}$$

where

$$\varpi = \operatorname{sgn}(z - b(0)).$$

Alternatively, one can solve the corresponding backward problem for the cumulative hitting probability $G(t, T, z)$ which can be represented as

$$G(t, T, z) = \mathbb{E}[\mathbf{1}_{\{\exists s \in [t,T]: W_s = b(s)\}} \mid W_t = z]. \tag{10.2}$$

Again, using the Feynman–Kac formula, one can show that it solves the following problem

$$0 = G_t(t, T, z) + \frac{1}{2} G_{zz}(t, T, z), \tag{10.3}$$

$$0 = G(T, T, z), \qquad 1 = G(t, T, b(t)).$$

Thus, the first hitting density is

$$g(T, z) = G_T(0, T, z).$$

10.2 Main results

In this section we present four main theorems that solve forward and backward problems for the Wiener and general Cherkasov processes.

Theorem 10.1 (Forward problem for Wiener process). *The hitting density Eq. (10.1) can be written as*

$$g(t, z) = \frac{1}{2}\int_0^t \frac{\left(1 - \frac{\Psi(t,t')^2}{(t-t')}\right)\Xi(t, t')\nu(t') - \nu(t)}{\sqrt{2\pi(t-t')^3}}dt' - \frac{1}{\sqrt{2\pi t}}\nu(t) \tag{10.4}$$
$$+ \varpi\left(-b'(t)\nu(t) + \frac{(z - b(t))\exp\left(-\frac{(z-b(t))^2}{2t}\right)}{2\sqrt{2\pi t^3}}\right),$$

where

$$\Psi(t, t') = b(t) - b(t'), \qquad t \geq t',$$
$$\Xi(t, t') = \begin{pmatrix} \exp\left(-\frac{\Psi(t,t')^2}{2(t-t')}\right), & t > t' \\ 1, & t = t', \end{pmatrix}$$

and $\nu(t)$ is the solution of Volterra integral equation of the second kind

$$\nu(t) + \varpi\int_0^t \frac{\Psi(t, t')\Xi(t, t')\nu(t')}{\sqrt{2\pi(t-t')^3}}dt' = -\frac{\exp\left(-\frac{(z-b(t))^2}{2t}\right)}{\sqrt{2\pi t}}. \tag{10.5}$$

Proof. Consider $\varphi > 0$. To find the hitting density, we use the method of heat potentials (see Tikhonov and Samarskii, 1963; Lipton, 2001). As usual, we write

$$p(t, x) = \tilde{H}(t, x, 0, z) + q(t, x),$$

where $\tilde{H}(t, x, 0, z)$ is the standard heat kernel, while $q(t, x)$ solves the IBVP of the form

$$q_t(t, x) = \frac{1}{2}q_{xx}(t, x),$$

$$q(0,x) = 0,$$
$$q(t,b(t)) = -\tilde{H}(t,b(t),0,z).$$

Accordingly,

$$q(t,x) = \int_0^t \frac{(x-b(t'))\exp\left(-\frac{(x-b(t'))^2}{2(t-t')}\right)}{\sqrt{2\pi(t-t')^3}}\nu(t')\,dt',$$

where ν is the solution of the Volterra equation of the second kind,

$$\nu(t) + \int_0^t \frac{\Psi(t,t')\Xi(t,t')\nu(t')}{\sqrt{2\pi(t-t')^3}}dt' + \frac{\exp\left(\left(-\frac{(b(t)-z)^2}{2t}\right)\right)}{\sqrt{2\pi t}} = 0.$$

Assuming that $\nu(t)$ is found, we can proceed as follows:

$$p(t,x) = \frac{\exp\left(-(x-z)^2/(2t)\right)}{\sqrt{2\pi t}} + q(t,x). \tag{10.6}$$

Then, $g(t)$ can be computed using Eq. (10.23) as follows: first, we differentiate Eq. (10.6) to get p_x, then take its limit at $x \to b(t)$. Finally, we use the result in Section 8.1.2 and arrive to the final formula Eq. (10.4). In more detail, this derivation can be found in Lipton and Kaushansky (2020a). □

Theorem 10.2 (Backward problem for Wiener process). *Choose some positive $T > 0$. The CDF Eq. (10.2) can be written as*

$$G(0,T,z) = \varpi \int_0^T \frac{(z-b(T-t'))\exp\left(-\frac{(z-b(T-t'))^2}{2(T-t')}\right)\nu(t',T)}{\sqrt{2\pi(T-t')^3}}dt', \tag{10.7}$$

where

$$\nu(t,T) + \varpi\int_0^t \frac{\Psi(T-t,T-t')\,\Xi(T-t,T-t')\nu(t',T)}{\sqrt{2\pi(t-t')^3}}dt' = 1. \tag{10.8}$$

Proof. Consider $\varphi > 0$. Fix $T > 0$. Consider the change of variables $\tau = T - t$ with $\hat{G}(\tau,T,z) = G(T-t,T,z)$ in Eq. (10.3). Then,

$$\hat{G}_\tau(\tau,T,z) = \frac{1}{2}\hat{G}_{zz}(\tau,T,z), \tag{10.9}$$
$$\hat{G}(0,T,z) = 0, \qquad \hat{G}(\tau,T,b(T-\tau)) = 1.$$

We got that $\hat{G}(t, T, z)$ satisfies the heat equation with moving boundary and zero initial condition. As a result, as before, we are able to apply the method of heat potentials. We get

$$\hat{G}(\tau, T, z) = \int_0^T \frac{(z - b(T - t')) \exp\left(-\frac{(z - b(T-t'))^2}{2(T-t')}\right) \nu(t', T)}{\sqrt{2\pi (T - t')^3}} dt',$$

where

$$\nu(t, T) + \int_0^t \frac{(\Psi(T - t, T - t'))\, \Xi(T - t, T - t')\nu(t', T)}{\sqrt{2\pi (t - t')^3}} dt' = 1.$$

As $G(0, T, z) = \hat{G}(T, T, z)$, we immediately get

$$G(0, T, z) = \int_0^T \frac{(z - b(T - t')) \exp\left(-\frac{(z - b(T-t'))^2}{2(T-t')}\right) \nu(t', T)}{\sqrt{2\pi (T - t')^3}} dt'.$$

□

Remark 10.1. Theorem 10.1 is easier to use because it allows one to calculate $g(t)$ in one go, while Theorem 10.2 requires solving a different equation for every T, except of the trivial case $b(t) \equiv b$. However, Theorem 10.2 is useful if a particular point in time is of interest, say $T \to \infty$.

Next we formulate the result for a general process Eq. (2.46) using the transformations from Section 2.5.4 and the Theorem 10.1.

Theorem 10.3 (Forward problem for a general Cherkasov process). *The hitting density of a general Cherkasov process can be written as*

$$g(t, z) = \frac{d\tilde{t}(t, b(t))}{dt} \tilde{g}(\tilde{t}, \tilde{z}), \tag{10.10}$$

where $\tilde{g}(\tilde{t}, \tilde{z})$ is the hitting density of a Wiener process, $\tilde{t}(t, x)$ is defined in Eq. (2.53), $\tilde{b}(\tilde{t})$ is defined in Eq. (2.57), $\tilde{z}(z)$ is defined in Eq. (2.55), and

$$\frac{d\tilde{t}(t, b(t))}{dt} = \Phi^2(t, b(t)) + 2b'(t) \int_0^t \Phi(s, b(t))\Phi_x(s, b(t))\, ds.$$

Theorem 10.4 (Backward problem for a general Cherkasov process). *Choose some positive $T > 0$. The CDF of the hitting time for Eq. (2.46) can be written as*

$$G(T, z) = \tilde{G}(\tilde{T}, \tilde{z}), \tag{10.11}$$

where $\tilde{G}(\tilde{T}, \tilde{z})$ is the hitting density of the transformed Wiener process.

10.3 Representative examples

10.3.1 *Linear boundary*

This is a classical example which is considered, e.g., in Lipton (2001) among many others. Here, we provide an alternative derivation using the theory we developed before.

When the boundary is linear,

$$b(t) = \mu_0 + \mu_1 t, \quad z > \mu_0,$$

the pair $(\nu(t), g(t))$ can be calculated explicitly by skillfully using a combination of the direct and inverse Laplace transforms. In other cases it has to be calculated numerically.

Since

$$\Psi(t, t') = \mu_1 (t - t'), \quad \Xi(t, t') = \exp\left(-\frac{\mu_1^2 (t - t')}{2}\right),$$

we have

$$\nu(t) + \mu_1 \int_0^t \frac{\exp\left(-\frac{\mu_1^2(t-t')}{2}\right) \nu(t')}{\sqrt{2\pi (t - t')}} dt' = -\frac{\exp\left(-\frac{(z-\mu_0-\mu_1 t)^2}{2t}\right)}{\sqrt{2\pi t}}. \tag{10.12}$$

Accordingly

$$\begin{aligned} g(t) &= -\left(\mu_1 + \frac{1}{\sqrt{2\pi t}}\right) \nu(t) \\ &\quad + \frac{1}{2} \int_0^t \frac{(1 - \mu_1^2 (t - t')) \exp\left(-\frac{\mu_1^2(t-t')}{2}\right) \nu(t') - \nu(t)}{\sqrt{2\pi (t - t')^3}} dt' \\ &\quad + \frac{(z - \mu_0 - \mu_1 t) \exp\left(-\frac{(z-\mu_0-\mu_1 t)^2}{2t}\right)}{2\sqrt{2\pi t^3}}. \end{aligned} \tag{10.13}$$

By rearranging terms and using integration by parts, we can rewrite Eq. (10.13) as follows

$$\begin{aligned} g(t) &= -g_1(t) + g_2(t), \\ g_1(t) &= \int_0^t \frac{\exp\left(-\frac{\mu_1^2(t-t')}{2}\right) \dot{\nu}(t')}{\sqrt{2\pi (t - t')}} dt, \\ g_2(t) &= \frac{(z - \mu_0 + \mu_1 t) \exp\left(-\frac{(z-\mu_0-\mu_1 t)^2}{2t}\right)}{2\sqrt{2\pi t^3}}. \end{aligned} \tag{10.14}$$

Taking the Laplace transform of Eq. (10.12), and the second line of Eq. (10.14) yields

$$\left(1+\frac{\mu_1}{\sqrt{2s+\mu_1^2}}\right)\hat{\nu}(s)=-\frac{\exp\left(-(z-\mu_0)\left(\sqrt{2s+\mu_1^2}-\mu_1\right)\right)}{\sqrt{2s+\mu_1^2}},$$

$$\hat{g}_1(s)=\frac{s\hat{\nu}(s)}{\sqrt{2s+\mu_1^2}}.$$

Accordingly, by taking the inverse Laplace transform, we obtain

$$\begin{aligned}\nu(t)=&-\frac{\exp\left(-\frac{(z-\mu_0-\mu_1 t)^2}{2t}\right)}{\sqrt{2\pi t}}\\&+\mu_1\exp\left(2(z-\mu_0)\mu_1\right)N\left(-\frac{z-\mu_0}{\sqrt{t}}-\mu_1\sqrt{t}\right),\end{aligned}\tag{10.15}$$

$$\begin{aligned}\nu(t)=&-\frac{\exp\left(-\frac{(z-\mu_0-\mu_1 t)^2}{2t}\right)}{\sqrt{2\pi t}}\\&+\mu_1\exp\left(2(z-\mu_0)\mu_1\right)N\left(-\frac{z-\mu_0}{\sqrt{t}}-\mu_1\sqrt{t}\right),\end{aligned}$$

and

$$g_1(t)=-\frac{(z-\mu_0)\exp\left(-\frac{(z-\mu_0-\mu_1 t)^2}{2t}\right)}{2\sqrt{2\pi t^3}}+\frac{\mu_1\exp\left(-\frac{(z-\mu_0-\mu_1 t)^2}{2t}\right)}{2\sqrt{2\pi t}}.$$

Finally, we arrive at

$$g(t)=\frac{(z-\mu_0)\exp\left(-\frac{(z-\mu_0-\mu_1 t)^2}{2t}\right)}{\sqrt{2\pi t^3}},\tag{10.16}$$

as expected. When $\mu_1=0$, Eq. (10.16) trivially follows from the method of images.

10.3.2 *Time-dependent geometric Brownian motion*

Consider a time-dependent geometric Brownian motion. Then

$$\begin{aligned}&\delta(t,x)=\delta(t)x,\ \sigma(t,x)=\sigma(t)x,\ \beta(t,x)=x\ln x,\\&\gamma(t,x)=\left(2\delta(t)-\sigma^2(t)\right)x-2\frac{\sigma_t(t)}{\sigma(t)}x\ln x,\\&P(t,x)=\begin{vmatrix}x\ln x & \left(2\delta(t)-\sigma^2(t)\right)x-2\frac{\sigma_t(t)}{\sigma(t)}x\ln x\\ \ln x+1 & \left(2\delta(t)-\sigma^2(t)\right)-2\frac{\sigma_t(t)}{\sigma(t)}(\ln x+1)\end{vmatrix}=[\sigma^2(t)-2\delta(t)]x,\end{aligned}$$

$$Q(t,x) = \begin{vmatrix} \sigma(t)\,x & \left(2\delta(t) - \sigma^2(t)\right)x - 2\frac{\sigma_t(t)}{\sigma(t)}x\ln x \\ \sigma(t) & \left(2\delta(t) - \sigma^2(t)\right) - 2\frac{\sigma_t(t)}{\sigma(t)}(\ln x + 1) \end{vmatrix} = -2\sigma_t(t)\,x,$$

$$R(t,x) = \begin{vmatrix} \sigma(t)\,x & x\ln x & \left(2\delta(t) - \sigma^2(t)\right)x - 2\frac{\sigma_t(t)}{\sigma(t)}x\ln x \\ \sigma(t) & \ln x + 1 & \left(2\delta(t) - \sigma^2(t)\right) - 2\frac{\sigma_t(t)}{\sigma(t)}(\ln x + 1) \\ 0 & 1/x & -2\frac{\sigma_t(t)}{\sigma(t)x} \end{vmatrix} = 0.$$

Accordingly,

$$\tilde{t} = \int_0^t \exp\left[2\int_0^u \frac{\sigma_t(s)}{\sigma(s)}ds\right]du = \int_0^t \exp\left[2\ln\sigma(u)\right]du = \int_0^t \sigma^2(u)\,du,$$

$$\tilde{x} = \ln(x) - \int_0^t \left(\delta(u) - \frac{1}{2}\sigma^2(u)\right)du,$$

$$\tilde{b}(t) = \ln b(t) - \tilde{\delta}(\tilde{t}) - \frac{\tilde{t}}{2}, \qquad \tilde{z} = \ln z,$$

where

$$\tilde{\delta}(\tilde{t}) = \int_0^t \delta(u)du.$$

Naturally, this is in agreement with Girsanov's theorem.

Thus, using Theorem 10.3, we can express the solution:

$$g(t,z) = \sigma^2(t)\left[\frac{1}{2}\int_0^{\tilde{t}(t)} \frac{\left(1 - \frac{\tilde{\Psi}\left(\tilde{t}(t),\tilde{t}'\right)^2}{\left(\tilde{t}(t)-\tilde{t}'\right)}\right)\tilde{\Xi}\left(\tilde{t}(t),\tilde{t}'\right)\tilde{\nu}\left(\tilde{t}'\right) - \tilde{\nu}\left(\tilde{t}(t)\right)}{\sqrt{2\pi\left(\tilde{t}(t)-\tilde{t}'\right)^3}}d\tilde{t}' \right.$$

$$- \frac{1}{\sqrt{2\pi\tilde{t}(t)}}\tilde{\nu}\left(\tilde{t}(t)\right) + \varpi\left(- \tilde{b}'\left(\tilde{t}(t)\right)\tilde{\nu}\left(\tilde{t}(t)\right) \right. \tag{10.17}$$

$$\left.\left. + \frac{\left(\ln z - \tilde{b}\left(\tilde{t}(t)\right)\right)\exp\left(-\frac{\left(\ln z - \tilde{b}\left(\tilde{t}(t)\right)\right)^2}{2t}\right)}{2\sqrt{2\pi\tilde{t}(t)^3}}\right)\right],$$

where

$$\tilde{b}'(\tilde{t}(t)) = \frac{1}{\sigma^2(t)}\left[\frac{b'(t)}{b(t)} - \delta(t)\right] - \frac{1}{2}.$$

10.3.3 Time-dependent OU process

Consider a time-dependent Ornstein–Uhlenbeck process. Then

$$\delta(t,x) = \lambda(t)(\theta(t) - x), \ \sigma(t,x) = \sigma(t), \ \beta(t,x) = x,$$

$$\gamma(t,x) = 2\lambda(t)(\theta(t) - x) - 2\frac{\sigma_t(t)}{\sigma(t)}x,$$

$$P(t,x) = \begin{vmatrix} x & 2\lambda(t)(\theta(t) - x) - 2\frac{\sigma_t(t)}{\sigma(t)}x \\ 1 & -2\lambda(t) - 2\frac{\sigma_t(t)}{\sigma(t)} \end{vmatrix} = -2\lambda(t)\theta(t),$$

$$Q(t,x) = \begin{vmatrix} \sigma(t) & 2\lambda(t)(\theta(t) - x) - 2\frac{\sigma_t(t)}{\sigma(t)}x \\ 0 & -2\lambda(t) - 2\frac{\sigma_t(t)}{\sigma(t)} \end{vmatrix} = -2\lambda(t)\sigma(t) - 2\sigma_t(t),$$

$$R(t,x) = \begin{vmatrix} \sigma(t) & x & 2\lambda(t)(\theta(t) - x) - 2\frac{\sigma_t(t)}{\sigma(t)}x \\ 0 & 1 & -2\lambda(t) - 2\frac{\sigma_t(t)}{\sigma(t)} \\ 0 & 0 & 0 \end{vmatrix} = 0.$$

Accordingly,

$$\tilde{t} = \int_0^t e^{2\Lambda(u)}\sigma^2(u)\,du,$$

$$\tilde{x} = e^{\Lambda(t)}x - \int_0^t e^{\Lambda(u)}\lambda(u)\theta(u)\,du,$$

$$\tilde{b}(\tilde{t}) = e^{\Lambda(t)}b(t) - \int_0^t e^{\Lambda(u)}\lambda(u)\theta(u)\,du, \quad \tilde{z} = z,$$

where

$$\Lambda(t) = \int_0^t \lambda(u)\,du.$$

Then, using Theorem 10.3, we obtain

$$\begin{aligned} g(t,z) = e^{2\Lambda(t)}\sigma^2(t) &\left[\frac{1}{2}\int_0^{\tilde{t}(t)} \frac{\left(1 - \frac{\tilde{\Psi}(\tilde{t}(t),\tilde{t}')^2}{(\tilde{t}(t)-\tilde{t}')}\right)\tilde{\Xi}(\tilde{t}(t),\tilde{t}')\,\tilde{\nu}(\tilde{t}') - \tilde{\nu}(\tilde{t}(t))}{\sqrt{2\pi(\tilde{t}(t)-\tilde{t}')^3}}d\tilde{t}' \right. \\ &- \frac{1}{\sqrt{2\pi\tilde{t}(t)}}\tilde{\nu}(\tilde{t}(t)) \\ &\left. + \varpi\left(-\tilde{b}'(\tilde{t}(t))\,\tilde{\nu}(\tilde{t}(t)) + \frac{\left(\tilde{z} - \tilde{b}(\tilde{t}(t))\right)\exp\left(-\frac{\left(\tilde{z}-\tilde{b}(\tilde{t}(t))\right)^2}{2t}\right)}{2\sqrt{2\pi\tilde{t}(t)^3}}\right)\right]. \end{aligned}$$

10.4 Alternative transformation for the OU process

Here we develop an alternative transformation.

First, let us transform an OU process with constant coefficients into the standard OU process. We define the standard time-independent OU process as

$$d\bar{x} = -\bar{x}d\bar{t} + dW_{\bar{t}}.$$

Let us consider a typical OU process

$$dX_t = \lambda(\theta - X_t)\,dt + \sigma dW_t, \qquad X_0 = z. \tag{10.18}$$

We wish to calculate the stopping time $s = \inf\{t : X_t = b(t)\}$, where $z > b$. By introducing new variables

$$\bar{t} = \lambda t,\ \bar{X} = \frac{\sqrt{\lambda}}{\sigma}(X - \theta), \quad \bar{z} = \frac{\sqrt{\lambda}}{\sigma}(z - \theta),\ \bar{b}(t) = \frac{\sqrt{\lambda}}{\sigma}(b(t) - \theta)$$

we can rewrite Eq. (10.18) as follows

$$\begin{aligned} d\bar{X}_{\bar{t}} &= -\bar{X}_{\bar{t}}d\bar{t} + dW_{\bar{t}}, \quad \bar{X}_0 = \bar{z}, \\ \bar{s} &= \inf\left\{\bar{t} : \bar{X}_{\bar{t}} = \bar{b}(t)\right\}, \end{aligned}$$

Below bars are omitted for brevity whenever possible.

In this formulation we consider the hitting from above problem. But, by symmetry, one can also consider the hitting from below. It can be formulated as the hitting problem from above with the parameters $\tilde{z} = -z$ and $\tilde{b} = -b$.

Second, consider the time-dependent process

$$dX_t = \lambda(t)\left(\theta(t) - X_t\right)dt + \sigma(t)dW_t, \quad X_0 = z.$$

We transform it into the standard process in two steps. First, let us introduce

$$\bar{x} = p\left(\bar{t}\right)x + q\left(\bar{t}\right),$$

and notice that

$$d\bar{x} = \left(p\lambda\theta + q' + \frac{(p\lambda - p')\,q}{p} - \frac{(p\lambda - p')}{p}y\right)dt + p\sigma dW_t.$$

Now, we impose the following constraints

$$\begin{aligned} \lambda\theta p + q' + \frac{(\lambda p - p')\,q}{p} &= 0, \quad q\,(0) = q_0, \\ \frac{(\lambda p - p')}{\sigma^2 p^3} &= 1, \quad p\,(0) = p_0, \end{aligned}$$

or, equivalently

$$q' = -\sigma^2 p^2 q - \lambda\theta p, \quad q(0) = q_0, \qquad (10.19)$$
$$p' = \lambda p - \sigma^2 p^3, \quad p(0) = p_0.$$

Provided that the system (10.19) is solved, we can introduce

$$\bar{t} = \int_0^t \sigma^2 p^2 dt,$$

and notice that $\bar{x}(\bar{t})$ is governed by the standard OU process. In general, one can solve the system Eq. (10.19) in two steps. First, since the first equation is a linear inhomogeneous equation for q, it can be trivially solved provided that p is known. Second, the second equation is a Bernoulli equation for p, which can be solved via the standard substitution

$$r = p^{-2}, \quad p = r^{-1/2}.$$

The corresponding equation for r is a linear inhomogeneous equation, which can be trivially solved. Indeed,

$$-\frac{r'}{2r^{3/2}} = \frac{\lambda}{r^{1/2}} - \frac{\sigma^2}{r^{3/2}},$$

or, equivalently

$$r' = -2\lambda r + 2\sigma^2.$$

Thus, r, p, q can all be found by quadratures. Finally, $\bar{t}$ can be found by quadrature as well since the corresponding integrand is known

$$\bar{t} = \int_0^t \sigma^2 p^2 dt' = \int_0^t \frac{\sigma^2}{r} dt'.$$

In the following, we omit bar for brevity.

Specifically

$$\begin{aligned} r(t) &= e^{-2(\Lambda(t)-M(t))}, \\ p(t) &= p_0 e^{\Lambda(t)-M(t)}, \\ q(t) &= e^{-M(t)}\left(-p_0\int_0^t \lambda(u)\theta(u)e^{\Lambda(u)}du + q_0\right), \\ \bar{t}(t) &= M(t). \end{aligned} \qquad (10.20)$$

Here

$$\Lambda(t) = \int_0^t \lambda(u)\, du, \qquad (10.21)$$

$$M(t) = \frac{1}{2}\ln\left(2p_0^2\int_0^t \sigma^2(u)\, e^{2\Lambda(u)}du + 1\right).$$

It is easy to see that for constant parameters, expressions Eq. (10.20) produce correct results.

Remark 10.2. It is worth noting that considering $\theta(t)$ as a periodic function, and other parameters as constants, one can reduce an important problem in neuroscience to the standard OU process[1].

Transformation of the standard OU process to a Wiener process (Forward). To calculate the density of the hitting time distribution $g(t,z)$, we need to solve the following forward problem

$$p_t(t,x;z) = p(t,x;z) + xp_x(t,x;z) + \frac{1}{2}p_{xx}(t,x;z), \tag{10.22}$$

$$p(0,x;z) = \delta(x-z), \qquad p(t,b(t);z) = 0.$$

This distribution is given by

$$g(t,z) = \frac{1}{2}\varpi p_x(t,b(t);z). \tag{10.23}$$

For simplicity, consider the case $z > b(0)$. We wish to transform initial boundary value problem (IBVP) Eq. (10.22) into the standard IBVP for a heat equation with a moving boundary. To this end, we introduce new independent and dependent variables as follows

$$q(\tau,\xi) = e^{-t}p(t,x), \quad \tau = e^t\sinh(t), \quad \xi = e^t x. \tag{10.24}$$

This yields

$$\begin{aligned} q_\tau(\tau,\xi) &= \frac{1}{2}q_{\xi\xi}(\tau,\xi), \\ q(0,\xi) &= \delta(\xi - z), \\ q(\tau,\beta(\tau)) &= 0, \\ \beta(\tau) &= \sqrt{2\tau+1}\tilde{b}(t), \end{aligned}$$

where $\tilde{b}(\tau) = b(t(\tau))$. It is clear that $0 \le \tau < \infty$ and $e^t = \sqrt{2\tau+1}$. Then, application of the Theorem 10.1 provides the solution in Eq. (10.4) and Eq. (10.5).

Assuming that the boundary is flat

$$\frac{\beta(\tau)-\beta(\tau')}{\tau-\tau'} = \frac{b\left(\sqrt{2\tau+1}-\sqrt{2\tau'+1}\right)}{\tau-\tau'} = \frac{2b}{\sqrt{2\tau+1}+\sqrt{2\tau'+1}},$$

[1]The solution of this problem will be reported elsewhere.

Eq. (10.5) can be re-written in the form

$$\nu(\tau) + \sqrt{\frac{2}{\pi}} b \int_0^\tau \frac{\exp\left(-b^2 A(\tau,\tau')\right)\nu(\tau')}{\left(\sqrt{2\tau+1}+\sqrt{2\tau'+1}\right)\sqrt{\tau-\tau'}} d\tau' \tag{10.25}$$

$$+ \frac{\exp\left(-\frac{\left(\sqrt{2\tau+1}b-z\right)^2}{2\tau}\right)}{\sqrt{2\pi\tau}} = 0,$$

$$A(\tau,\tau') = \frac{\sqrt{2\tau+1}-\sqrt{2\tau'+1}}{\sqrt{2\tau+1}+\sqrt{2\tau'+1}}.$$

Hence,

$$g(t) = -\frac{(e^t b - z)\exp\left(-\frac{(e^t b-z)^2}{(e^{2t}-1)} + 2t\right)}{\sqrt{\pi\left(e^{2t}-1\right)^3}} - \left(e^t b + \frac{e^{2t}}{\sqrt{\pi\left(e^{2t}-1\right)}}\right)\nu(t)$$

$$+ \frac{e^{2t}}{\sqrt{8\pi}} \int_0^\tau \frac{\left(1-2b^2 A(\tau,\tau')\right)\exp\left(-b^2 A(\tau,\tau')\right)\nu(\tau') - \nu(\tau)}{\sqrt{(\tau-\tau')^3}} d\tau'. \tag{10.26}$$

We can rewrite Eq. (10.25) in an alternative way. Let $\theta = \sqrt{2\tau+1}-1$, $\theta' = \sqrt{2\tau'+1}-1$, $0 \leq \theta' \leq \theta < \infty$. Then

$$\nu(\theta) + \frac{2b}{\sqrt{\pi}} \int_0^\theta \frac{\exp\left(-b^2 \frac{(\theta-\theta')}{(2+\theta+\theta')}\right)(1+\theta')\nu(\theta')}{\sqrt{(2+\theta+\theta')^3(\theta-\theta')}} d\theta' \tag{10.27}$$

$$+ \frac{\exp\left(-\frac{((1+\theta)b-z)^2}{((1+\theta)^2-1)}\right)}{\sqrt{\pi\left((1+\theta)^2-1\right)}} = 0.$$

Or, symbolically

$$\nu(\theta) + \int_0^\theta \frac{\Phi_b^f(\theta,\theta')\nu(\theta')}{\sqrt{\theta-\theta'}} d\theta' + \frac{\exp\left(-\frac{((1+\theta)b-z)^2}{((1+\theta)^2-1)}\right)}{\sqrt{\pi\left((1+\theta)^2-1\right)}} = 0,$$

where

$$\Phi_b^f(\theta,\theta') = \frac{2b}{\sqrt{\pi}} \frac{\exp\left(-b^2 \frac{(\theta-\theta')}{(2+\theta+\theta')}\right)(1+\theta')}{\sqrt{(2+\theta+\theta')^3}}.$$

Accordingly, Eq. (10.26) can be re-written in the form

$$g(t) = -\frac{(e^t b - z)\exp\left(-\frac{(e^t b - z)^2}{(e^{2t}-1)} + 2t\right)}{\sqrt{\pi\left(e^{2t}-1\right)^3}} - \left(e^t b + \frac{e^{2t}}{\sqrt{\pi\left(e^{2t}-1\right)}}\right)\nu(t)$$
$$+\frac{1}{\sqrt{\pi}}e^{2t}\int_0^{\theta}\frac{\left(1-2b^2 B(\theta,\theta')\right)\exp\left(-b^2 B(\theta,\theta')\right)\nu(\theta') - \nu(\theta)}{(1+\theta')^{-1}\sqrt{(2+\theta+\theta')^3(\theta-\theta')^3}}d\theta',$$
$$B(\theta,\theta') = \frac{\theta-\theta'}{2+\theta+\theta'}.$$

An example of thus transformed flat boundary is given in Fig. 10.1.

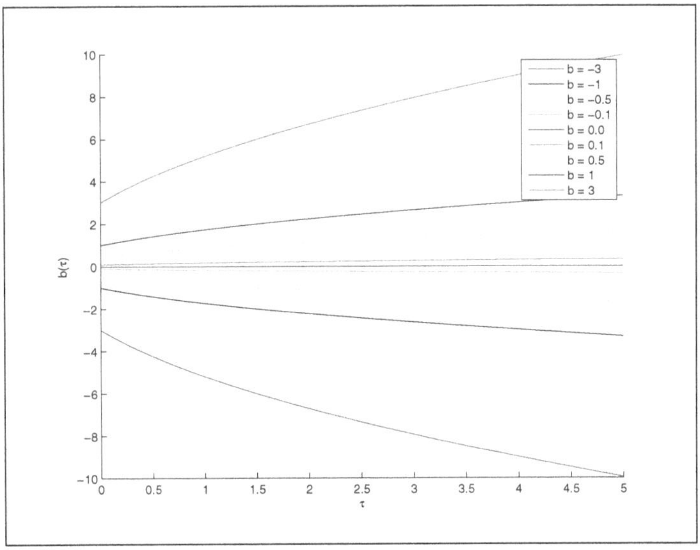

Figure 10.1: Moving boundary $b(\tau)$ for Eq. (10.29).

It is worth noting that the analytical solution is available in two cases:

- $b \equiv 0$: the solution can be trivially found by using the method of images;
- $b(t) = Ae^{-t} + Be^{t}$: by the above transformation the boundary transforms to $2B\tau + A + B$, i.e. the linear boundary which is considered in Section 10.3.1.

Transformation of the standard OU process to a Wiener process (Backward). Alternatively, we can solve the backward problem

$$\begin{aligned} &G_t(t,T,z) = zG_z(t,T,z) - \frac{1}{2}G_{zz}(t,T,z) = 0 \\ &G(T,T,z) = 0, \qquad G(t,T,b(T-t)) = 1. \end{aligned} \tag{10.28}$$

By the same token as before, we introduce

$$\tau = \varpi(t) = e^{-t}\sinh(t),\ \hat{T} = e^{-T}\sinh(T),\ 0 \le \tau < \frac{1}{2},\ x = e^{-t}z,$$

and rewrite IBVP Eq. (10.28) as follows

$$\begin{aligned} G_\tau(\tau,x) &= \frac{1}{2}G_{xx}(\tau,x), \\ G(0,x) = 0, &\qquad G\left(\tau, \beta\left(\hat{T}-\tau\right)\right) = 1, \\ \beta(\tau) &= \sqrt{1-2\tau}\, b(T-t). \end{aligned} \tag{10.29}$$

It is clear that $e^{-t} = \sqrt{1-2\tau}$. Thus, for the backward problem the computational domain is compactified in the τ direction. This fact greatly simplifies numerical evaluation of $G(0,T,z)$, when $T \to \infty$.

Again, application of the Theorem 10.2 provides the solution in the form of Eqs. (10.7) and (10.8).

For the case of flat boundary, $b(t) \equiv b$, we have $\nu(\tau,\hat{T}) \equiv \nu(\tau)$. Accordingly, using the Theorem 10.2 one obtains

$$G(\tau,x) = \int_0^\tau \frac{\left(x - b\sqrt{1-2\tau'}\right)\exp\left(-\frac{\left(x-b\sqrt{1-2\tau'}\right)^2}{2(\tau-\tau')}\right)\nu(\tau')}{\sqrt{2\pi(\tau-\tau')^3}}d\tau', \tag{10.30}$$

Since

$$\frac{b(\tau)-b(\tau')}{\tau-\tau'} = \frac{\left(\sqrt{1-2\tau}-\sqrt{1-2\tau'}\right)b}{\tau-\tau'} = -\frac{2b}{\sqrt{1-2\tau}+\sqrt{1-2\tau'}},$$

we obtain

$$\nu(\tau) - \sqrt{\frac{2}{\pi}}b\int_0^\tau \frac{\exp\left(-b^2\frac{\left(\sqrt{1-2\tau'}-\sqrt{1-2\tau}\right)}{\left(\sqrt{1-2\tau'}+\sqrt{1-2\tau}\right)}\right)\nu(\tau')}{\left(\sqrt{1-2\tau'}+\sqrt{1-2\tau}\right)\sqrt{\tau-\tau'}}d\tau' - 1 = 0. \tag{10.31}$$

Once Eq. (10.31) is solved, $G(\tau,x)$ and $G(t,z)$ can be calculated by virtue of Eq. (10.30) in a straightforward fashion.

It is possible to rewrite Eq. (10.31) in an alternative way. Let $\vartheta = 1-\sqrt{1-2\tau}$, $\vartheta' = 1-\sqrt{1-2\tau'}$, $0 \le \vartheta' \le \vartheta < 1$. Then

$$\nu(\vartheta) - \frac{2b}{\sqrt{\pi}} \int_0^{\vartheta} \frac{\exp\left(-b^2 \frac{(\vartheta-\vartheta')}{(2-\vartheta-\vartheta')}\right)(1-\vartheta')\,\nu(\vartheta')}{\sqrt{(2-\vartheta-\vartheta')^3}\sqrt{\vartheta-\vartheta'}} d\vartheta' - 1 = 0. \quad (10.32)$$

Or, symbolically

$$\nu(\vartheta) - \int_0^{\vartheta} \frac{\Phi_b^b(\vartheta,\vartheta')\nu(\vartheta')}{\sqrt{\vartheta-\vartheta'}} d\vartheta' - 1 = 0,$$

where

$$\Phi_b^b(\vartheta,\vartheta') = \frac{2b}{\sqrt{\pi}} \frac{\exp\left(-b^2 \frac{(\vartheta-\vartheta')}{(2-\vartheta-\vartheta')}\right)(1-\vartheta')}{\sqrt{(2-\vartheta-\vartheta')^3}}.$$

As one would expect,

$$\Phi_b^b(\theta,\theta') = -i\Phi_{ib}^f(-\theta,-\theta').$$

As a result,

$$G(t,z) = 2\int_0^{\vartheta} \frac{(z(1-\vartheta) - b(1-\vartheta'))\exp\left(-\frac{\left(z(1-\vartheta)-b(1-\vartheta')\right)^2}{(\vartheta-\vartheta')(2-\vartheta-\vartheta')}\right)}{\sqrt{\pi(\vartheta-\vartheta')^3(2-\vartheta-\vartheta')^3}} \times (1-\vartheta')\,\nu(\vartheta')\,d\vartheta',$$

where $\vartheta = 1-e^{-t}$.

We show the moving boundary for different values of b in Fig. 10.2.

10.5 Numerical examples

In this section we present results of various test validating our semi-analytical approach described in this chapter. In doing so, to solve the corresponding Volterra equations of the second kind we use numerical methods discussed in Chapter 5.

10.5.1 *Wiener process with linear boundary*

For this problem there exist an analytical solution which is given in Eqs. (10.15) and (10.16). We use the values of the model parameters: $T = 1$, $z = 2$, and $b(t) = 2t + 1$, and compare the numerical results with the corresponding solutions in Eqs. (10.15) and (10.16). The results are presented in Figs. 10.3 and 10.4 It is clear that analytical and numerical results are in good agreement.

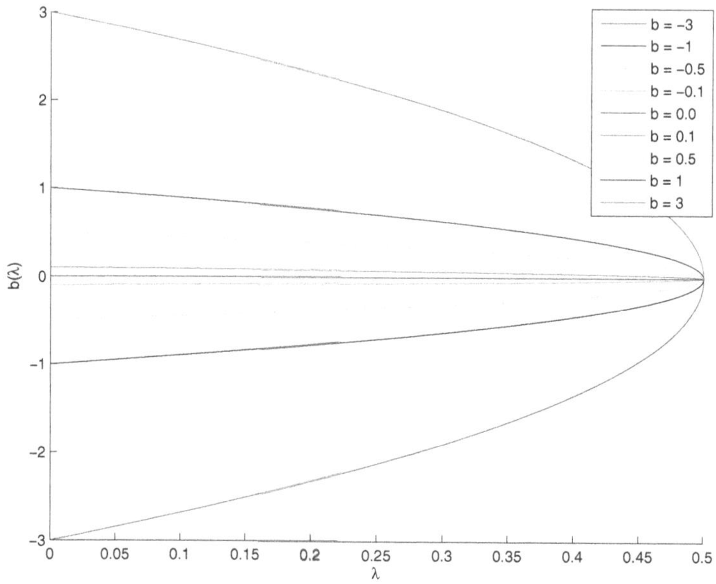

Figure 10.2: Moving boundary $b(\tau)$ for Eq. (10.29).

10.5.2 *Standard OU process with constant boundary*

Again, let us pick the values $T = 2$ and $z = 2$. In Fig. 10.5 we examine the density and cumulative density functions for different values of b and compare them with the analytical solution Eq. (10.21) for $b = 0$. In Fig. 10.6 $G(t, z)$ is shown as a function of both t and z for $b = 1$. In Fig. 10.7 $\nu(\theta)$ is presented for the forward problem, and in Fig. 10.8 - $\nu(\vartheta)$ for the backward problem for various values of b.

To test how good the Abel equation approximates the corresponding Volterra equation, we plot them together for small values of t. In Fig. 10.9 we compare the corresponding forward equations, and in Fig. 10.10 — the backward equations. One can see that up to $t = 0.02$ the solutions are visually indistinguishable.

We also empirically test the convergence rate of our numerical method by implementing it for the forward and backward Abel equations Eqs. (5.9) and (5.12), and comparing them with the analytical solutions Eqs. (5.10) and (5.13). In Fig. 10.11a we get the order of convergence 3.2 for the quadratic interpolation method and order 1 for the trapezoidal method for

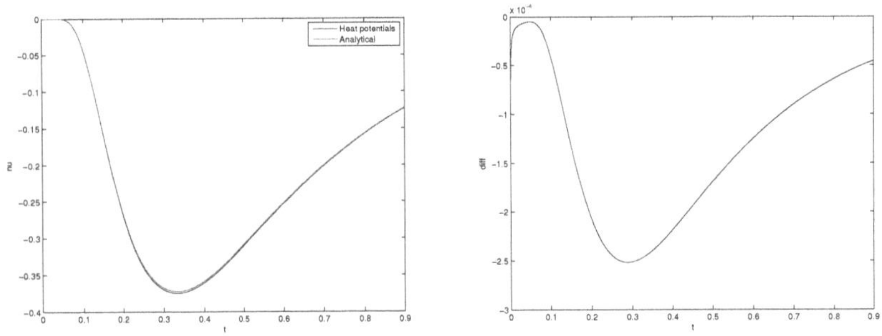

(a) (b)

Figure 10.3: (a) Functions $\nu(t)$ for the Wiener process and linear boundary obtained analytically and by using the HP method; (b) The difference between functions $\nu(t)$ obtained by using the HP method and analytically.

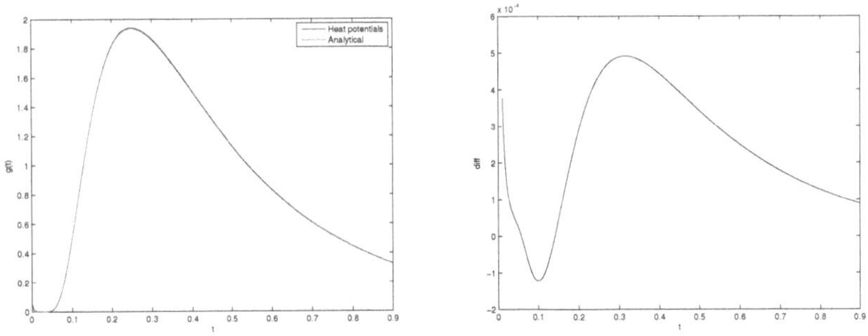

(a) (b)

Figure 10.4: (a) Functions $g(t)$ for the Wiener process and linear boundary obtained analytically and by using the HP method; (b) The difference between functions $g(t)$ obtained by using the HP method and analytically.

the forward equation. It confirms our estimate in Section 5.1. In Fig. 10.11b we get the order 1.5 for the quadratic interpolation method and the order 1 for the trapezoidal method for the backward equation. The convergence order of the quadratic interpolation method for the backward equation

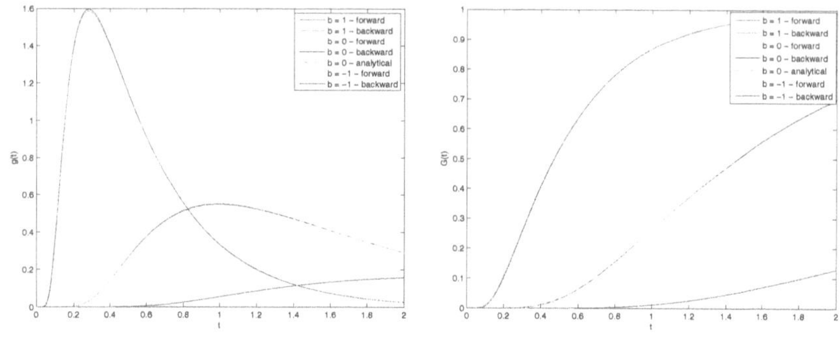

(a) (b)

Figure 10.5: (a) Density function $g(t)$; (b) Cumulative density function $G(t)$.

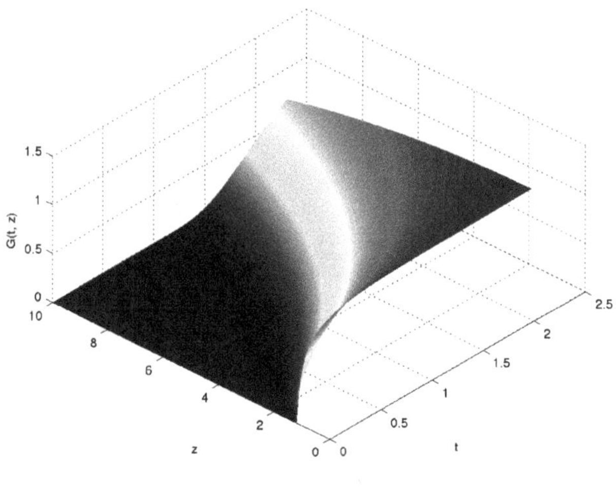

Figure 10.6: $G(t, z)$ as a function of t and z for $b = 1$.

is smaller than the theoretical estimate, because the assumptions on the regularity of the solution are not satisfied. Potentially, a non-uniform grid improves the convergence order.

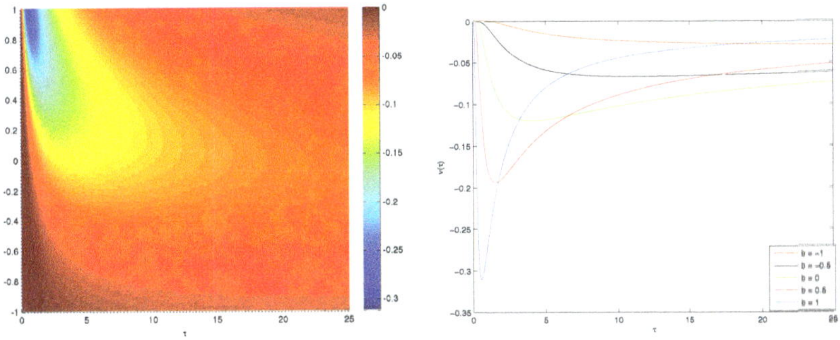

(a) (b)

Figure 10.7: Solution of Eq. (10.27) - $\nu(\theta)$ obtained at $z = 2$: (a) as a function of θ and b; (b) as a function of θ for different values of b.

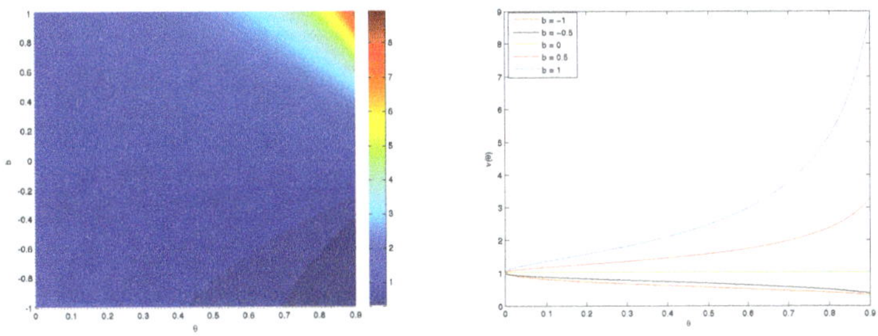

(a) (b)

Figure 10.8: Solution of Eq. (10.32) - $\nu(\vartheta)$: (a) as a function of ϑ and b; (b) as a function of ϑ for different values of b.

10.5.3 *Comparison with other methods*

Here we compare our method with other available alternatives.

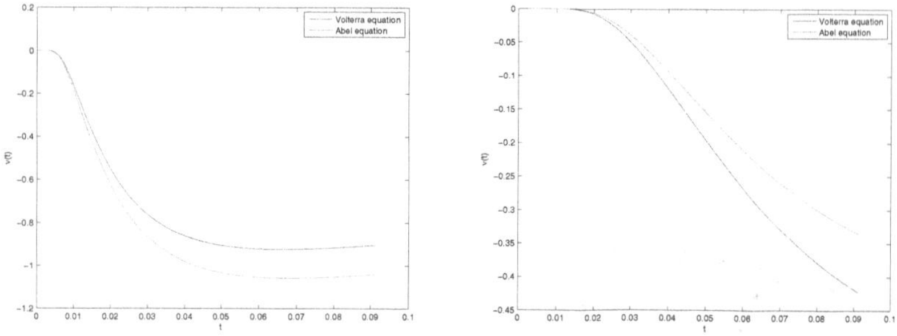

(a) (b)

Figure 10.9: Solution of the forward equation $\nu(t)$ obtained by using the Abel equation approximation and by a numerical solver: (a) $b = -0.5, z = -0.25$; (b) $b = 0.5, z = 1$.

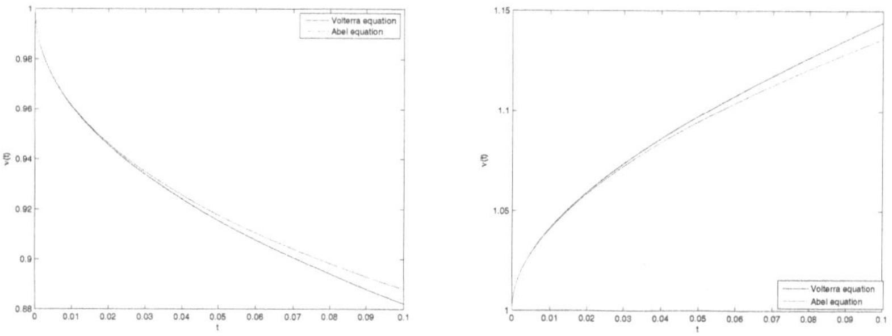

(a) (b)

Figure 10.10: Solution of the forward equation $\nu(t)$ obtained by using the Abel equation approximation and by a numerical solver: (a) $b = -0.5$; (b) $b = 0.5$.

(1) Method of Leblanc *et al.* (2000). We used (3) in Leblanc *et al.* (2000) to compute the hitting density. In our notation it is

$$g(t) = \frac{(z-b)\exp\left(\frac{z^2-b^2+t-(z-b)^2\coth t}{2}\right)}{\sqrt{2\pi\left(\sinh\left(t\right)\right)^3}}$$

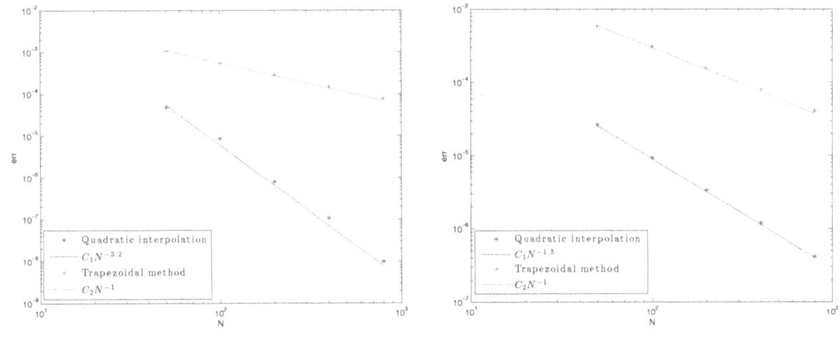

(a) (b)

Figure 10.11: Empirical estimation of the numerical methods from Section 5.1: (a) forward equation; (b) backward equation.

$$= \frac{(z-b)\exp\left(b(z-b) - \frac{e^{-t}(z-b)^2}{2\sinh(t)} + \frac{t}{2}\right)}{\sqrt{2\pi\,(\sinh(t))^3}}.$$

Integrating $g(t)$, we get

$$G(t) = 2e^{b(z-b)} N\left(-\frac{(z-b)e^{-t/2}}{\sqrt{\sinh t}}\right).$$

(2) Method of Alili *et al.* (2005); Linetsky (2004) and Ricciardi and Sato (1988). The method is based on the inversion of the Laplace transform of the hitting density. In our notation, the Laplace transform $u(\Lambda)$ has the form

$$u(\Lambda) = e^{\frac{z^2-b^2}{2}} \frac{D_{-\Lambda}(z\sqrt{2})}{D_{-\Lambda}(b\sqrt{2})}, \tag{10.33}$$

where $D_\nu(x)$ is the parabolic cylinder function. In Fig. 10.12 the function $u(\Lambda)$ is depicted for several values of b.

In Alili *et al.* (2005) a representation of the inverse Laplace transform is found as a series of parabolic cylinder function and its derivatives

$$g(t) = e^{(z^2-b^2)/2} \sum_{j=1}^{\infty} \frac{D_{\nu_{j,b\sqrt{2}}}(z\sqrt{2})}{D'_{\nu_{j,b\sqrt{2}}}(b\sqrt{2})} \exp\left(\nu_{j,b\sqrt{2}}\right),$$

where $D'_\nu(x) = \frac{\partial D}{\partial \nu}(x)$ and $\nu_{j,b}$ the ordered sequence of positive zeros of $\nu \to D_\nu(x)$.

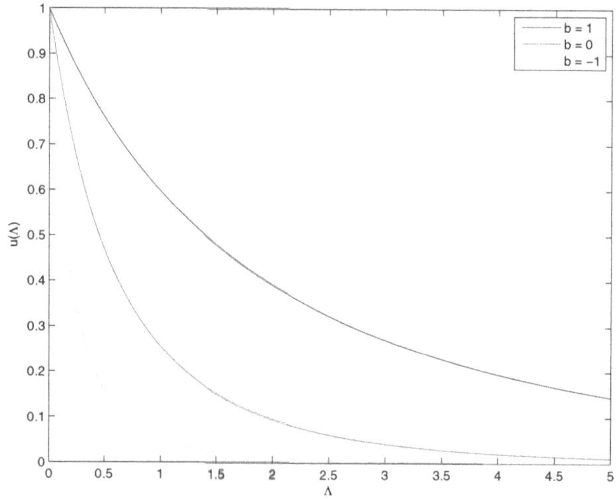

Figure 10.12: The Laplace transform $u(\Lambda)$ for $z = 2$.

Instead, Lipton and Kaushansky (2020a) used the Gaver-Stehfest algorithm (Kuznetsov, 2013; Abate *et al.*, 2000) to compute an inverse Laplace transform as it provides more stable and robust results. Moreover, the representation from Alili *et al.* (2005) works only for $t > t_0$ for some t_0, while the Gaver-Stehfest algorithm works for all positive t. In more detail, see chapter 5.
As an example of using the Gaver-Stehfest algorithm in finance, one can refer to Lipton and Sepp (2011b) and Itkin and Lipton (2018), where it was used for calibration of a local volatility surface.

(3) The Crank–Nicolson method. We solved Eq. (10.28) using a finite difference approach, in particular — the Crank–Nicolson scheme (Duffy, 2013).

We begin by comparing our results with the method in Leblanc *et al.* (2000) to demonstrate that the latter produces wrong results. In Fig. 10.13 we compare two methods for different parameters. We also give the analytical solution, when it is available for $b = 0$. One can observe that only for $b = 0$ the method of Leblanc *et al.* (2000) produces the correct results, while for other parameters they totally differ from those obtained by our method. Moreover in Fig. 10.13c we can see that it gives $G(t) > 1$.

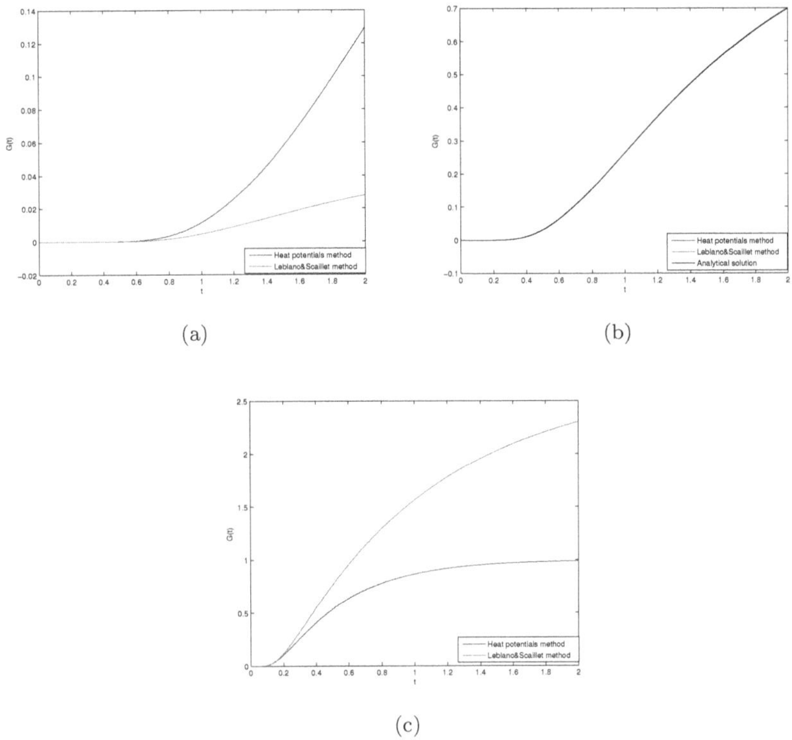

Figure 10.13: Comparison with the method of Leblanc *et al.* (2000): (a) $z = 2, b = -1$; (b) $z = 2, b = 0$; (c) $z = 2, b = -1$.

A closed-form solution is available only when $b = 0$ and is given by Eq. (10.21). The other parameters are taken as before, $T = 2$ and $z = 2$. In Figs. 10.14 and 10.15 we compare our method (forward and backward) with the Crank–Nicolson method, and the method based on the inverse Laplace transform. Note that in this case our method has an advantage, because the Volterra equations Eqs. (10.25) and (10.32) become trivial for $b = 0$. Nevertheless, the comparison is still very useful.

We take $N = 500$ for both forward and backward methods, $h = k = 0.005$ for the Crank–Nicolson method, and $m = 24$ for the Gaver-Stehfest algorithm. We used path library for `Python` (Johansson, 2007) for the implementation of the Gaver–Stehfest algorithm and parabolic cylinder functions with arbitrary precision arithmetics. In our computations we used precision of 100 digits.

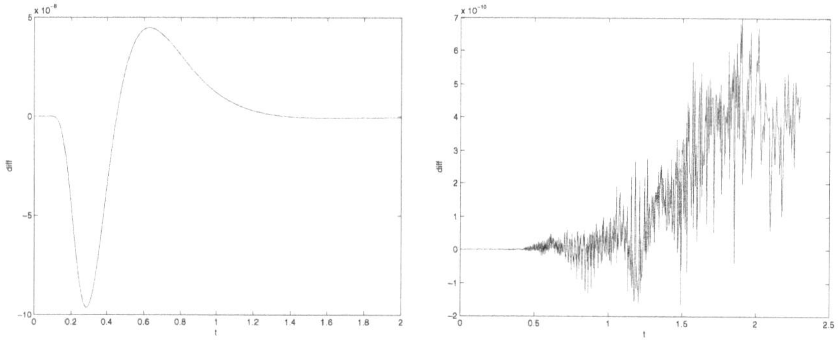

(a) (b)

Figure 10.14: The difference of the numerical and analytical solutions for $b = 0$, (a) forward method; (b) backward method.

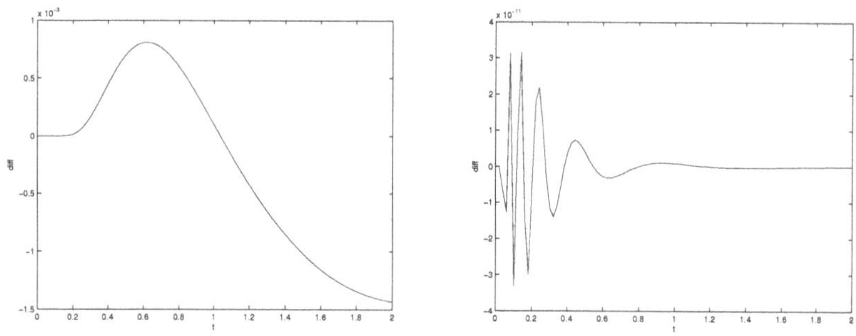

(a) (b)

Figure 10.15: The difference of the numerical and analytical solutions, (a) the Crank–Nicolson method; (b) the Laplace transform method.

Based on the presented results one can see that the methods developed in Lipton and Kaushansky (2020a) and the method based on the inverse Laplace transform significantly outperform the Crank–Nicolson scheme.

In order to compare the solutions for non-trivial parameters, we take the Laplace transform method with a maximum possible precision, and our method for $N = 100$ and $N = 10000$. The first example shows how our results can be comparable to the Laplace transform method for a relatively small value of N, when the computations can be done very fast; the second example demonstrates that our method can obtain a very good precision when N is large. The results are presented in Fig. 10.16.

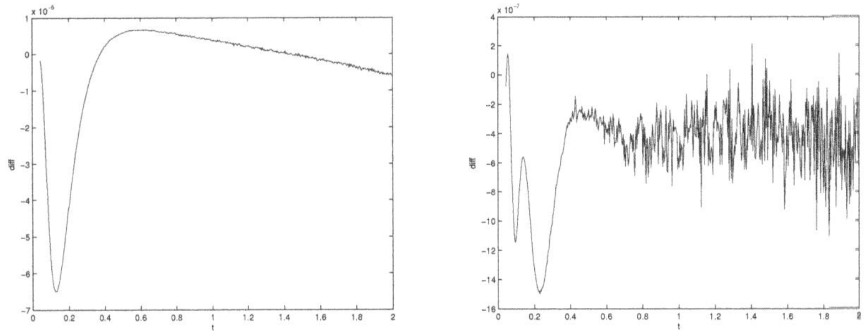

(a) (b)

Figure 10.16: The difference between our method and the Laplace transform method, (a) $N = 100$; (b) $N = 10000$.

At the end of this section we consider an asymptotic behavior of the solution when $t \to \infty$ for the standard OU process. It is clear that for $b = 0$, and hence all $b \geq 0$, $G(t, z) \to 1$, when $t \to \infty$. But it is unclear what happens for a negative b. Let us empirically investigate it by plotting $G(t, z)$ as a function of b for a fixed large value of T and fixed z. In Fig. 10.17a we take $z = 2$ and $T = 500$ and compute $G(T, z)$ for $b \in [-5, 2]$. One can see that $G(t, z)$ still remains close to 1 up to $b = -2$, and then rapidly approaches zero.

Further, we want to explore the asymptotic behavior of $G(t)$ in more detail. In Fig. 10.17b we show the expected hitting time; this quantity is very important for the design of mean-reverting trading strategies since it allows a trader to decide when to go in and out of the trade.

10.6 Final notes

The hitting density of the Wiener and OU processes to a curvilinear boundary has many applications in applied mathematics, especially in mathematical finance. One of them is a problem of pricing barrier options considered in Chapters 9 and 12. Pricing methods for path-dependent options on yields were proposed in Leblanc and Scaillet (1998), where the authors reduced

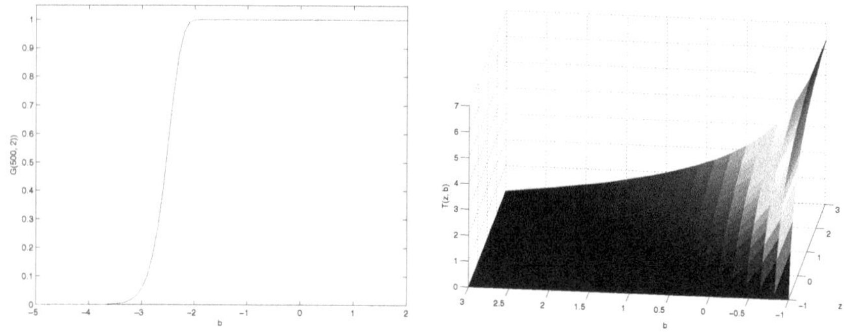

(a) (b)

Figure 10.17: (a) $G(T, z)$ as a function of b with $T = 500$ and $z = 2$; (b) $T(z, b) = \mathbb{E}[s]$, where s is the hitting time.

it to the hitting problem of an OU process. Another important problem is the design of mean-reverting trading strategies and pairs trading, which was considered in Avellaneda and Lee (2010) and Lopez De Prado (2018). Credit risk modeling in a structural framework is another topic, which heavily uses first passage density of various stochastic processes. We refer to Black and Cox (1976a), Hyer *et al.* (1998), Hull and White (2001), Avellaneda and Zhu (2001), Collin-Dufresne and Goldstein (2001), Coculescu *et al.* (2008), Yi (2010), Cheridito and Xu (2015), Lipton and Sepp (2011a) and Lipton and Savescu (2014) just to mention a few. The detailed review can be found in Lipton and Sepp (2011a).

It has also applications in quantitative biology (Smith, 1991) where the hitting time is used for modeling the time between rings of a nerve cell, and numerous other fields.

Various financial applications of the proposed technique are also discussed in Part 6 of this book.

It is worth mentioning that the method, in principle, can be extended to multi-dimensional hitting problems as the heat potentials are also applicable in the multidimensional case (Landis, 1997). A similar Volterra integral equations might be constructed, which would reduce the dimension of the problem.

Chapter 11

Optimal Mean-Reverting Trading Strategies

In this chapter, we closely follow Lipton and Lopez de Prado (2020a,b) who used the method of heat potentials to construct a closed-form solution for optimal mean-reverting trading strategies.

Mean-reverting trading strategies in various contexts have been studied for decades; see, e.g., Vidyamurthy (2004), Gatev *et al.* (2006), Govender (2011), Gregoty *et al.* (2011), Krauss (2015), Leung and Li (2015a), Leung and Li (2015b), Li (2015), Huck and Afawubo (2015), Goncu and Akyldirim (2016), Suzuki (2018). For instance, Elliott *et al.* (2005), explained how mean-reverting processes might be used in pairs trading and developed several methods for parameter estimation. Avellaneda and Lee used mean-reverting processes for pairs trading, and modeled the hitting time to find the exit rule of the trade (Avellaneda and Lee, 2010). Bertram developed some analytic formulae for statistical arbitrage trading where the security price follows an Ornstein–Uhlenbeck (OU) process (Bertram, 2009, 2010). Lindberg and his coauthors model the spread between two assets as an OU process and study the optimal liquidation strategy for an investor who wants to optimize profits over the opportunity cost (Ekstrom *et al.*, 2010, 2011; Larsson *et al.*, 2013; Lindberg, 2014). Lopez De Prado (2018) (Chapter 13) considered trading rules for discrete-time mean-reverting trading strategies and found optimal trading rules using Monte Carlo simulations.

By its very nature, the energy market is particularly well suited to mean-reverting trading strategies, not least because it is affected by strong seasonality. Numerous researchers discuss these strategies (see, e.g., Cummins and Bucca, 2012; Baviera and Baldi, 2019; Liu *et al.*, 2017), among others.

Usually, it is assumed that the stochastic process underlying mean-reverting trading strategies is the standard OU process. However, in practice, jumps play a significant role, as was recognized early on by Norberg (2004). Hence, some attention had been devoted to Lévy-driven OU processes (see, e.g., Larsson *et al.*, 2013; Goncu and Akyldirim, 2016; Endres and Stübinger, 2019). Although Endres and Stübinger (Endres and Stübinger, 2019) present a fairly detailed exposition, their central equation (9) is incorrect because it ignores the possibility of a Lévy-driven OU process to overshoot the chosen boundaries.

We emphasize that most, if not all, analytical results derived by the above authors, are asymptotic and valid for perpetual trading strategies only (see, e.g., Bertram, 2009, 2010; Ekstrom *et al.*, 2010, 2011; Lataillade *et al.*, 2012; Larsson *et al.*, 2013; Lindberg, 2014; Zeng and Lee, 2014; Bai and Wu, 2018). While interesting from a theoretical standpoint, they have limited application in practice. In contrast, our approach deals with finite maturity trading strategies, and, because of that, has immediate applications.

When prices reflect all available information, they oscillate around an equilibrium level. This oscillation is the result of the temporary market impact caused by waves of buyers and sellers. The resulting price behavior can be approximated through an OU process. The parameters of the process might be estimated using historical data.

Market makers provide liquidity in an attempt to monetize this oscillation. They enter a long position when a security is priced below its estimated equilibrium level, and they enter a short position when a security is priced above its estimated equilibrium level. They hold that position until one of three outcomes occurs: (A) they achieve a targeted profit; (B) they experience a maximum tolerated loss; (C) the position is held beyond a maximum tolerated horizon.

All traders are confronted with the problem of defining profit-taking and stop-out levels. More generally, all execution traders acting on behalf of a client must determine at what levels an order must be fulfilled. Lopez De Prado (2018) (Chapter 13) explains how to identify those optimal levels in the sense of maximizing the trader's Sharpe ratio (SR) in the context of OU processes via Monte Carlo experiments. Although the author conjectured the existence of an analytical solution to this problem, he identified it as an open problem. Then in Lipton and Lopez de Prado (2020a,b) the critical question of finding optimal trading rules analytically was answered by using the method of heat potentials. These optimal profit-taking/stop-loss trading rules for mean-reverting trading strategies provide the algorithm that must be followed to exit a position. To put it differently, we find the optimal exit corridor to maximize the SR of the strategy.

The SR is defined as the ratio between the expected returns of an execution algorithm and the standard deviation of the same returns. The returns are computed as the logarithmic ratio between the exit and entry prices, times the sign of the order side (-1 for a sell order, $+1$ for a buy order). Our choice of the SR as an objective function for two reasons: (A) The SR is the most popular criterion for investment efficiency (Bailey and Prado, 2013); (B) The SR can be understood as a t-value of the estimated gains, and modelled accordingly for inferential purposes. The distributional properties of the SR are well-known, and this statistic can be deflated when the assumption of normality is violated (Bailey and Prado, 2014).

Having an analytical estimation of the optimal profit-taking and stop-out levels allows traders to deploy tactical execution algorithms, with maximal expected SR. Rather than deriving an "all-weather" execution algorithm, which supposedly works under every market regime, traders can use our analytical solution for deploying the algorithm that maximizes the SR under the prevailing market regime (Prado, 2019).

11.1 Formulation of the problem

Suppose an investment strategy S invests in $i = 1, ..., I$ opportunities or bets. At each opportunity i, S takes a position of m_i units of security X, where $m_i \in (-\infty, \infty)$. The transaction that entered such opportunity was priced at a value $m_i P_{i,0}$, where $P_{i,0}$ is the average price per unit at which the m_i securities were transacted. As other market participants transact security X, we can mark-to-market (MtM) the value of that opportunity i after t observed transactions as $m_i P_{i,t}$. This represents the value of opportunity i if it were liquidated at the price observed in the market after t transactions. Accordingly, we can compute the MtM profit/loss of opportunity i after t transactions as $\pi_{i,t} = m_i(P_{i,t} - P_{i,0})$.

A standard trading rule provides the logic for exiting opportunity i at $t = T_i$. This opportunity occurs as soon as one of two conditions is verified: (i) $\pi_{i,T_i} \geq \overline{\pi}$, where $\overline{\pi} > 0$ is the profit-taking threshold, or (ii) $\pi_{i,T_i} \leq \underline{\pi}$, where $\underline{\pi} < 0$ is the stop-loss threshold.

Because $\underline{\pi} < \overline{\pi}$, one and only one of the two exit conditions can trigger the exit from opportunity i. Assuming that opportunity i can be exited at T_i, its final profit/loss is π_{i,T_i}. At the onset of each opportunity, the goal is to realize an expected profit

$$\mathbb{E}_0[\pi_{i,T_i}] = m_i(\mathbb{E}_0[P_{i,T_i}] - P_{i,0}), \tag{11.1}$$

where $\mathbb{E}_0[P_{i,T_i}]$ is the forecasted price and $P_{i,0}$ is the entry level of opportunity i.

Now, consider the discrete OU process on a price series $\{P_{i,t}\}$

$$P_{i,t} - \mathbb{E}_0[P_{i,T_i}] = \kappa\left(E_0[P_{i,T_i}] - P_{i,t-1}\right) + \sigma\varepsilon_{i,t}, \tag{11.2}$$

such that the random shocks are IID distributed $\varepsilon_{i,t} \sim \mathcal{N}(0,1)$. The seed value for this process is $P_{i,0}$, the level targeted by opportunity i is $\mathbb{E}_0[P_{i,T_i}]$, and κ determines the speed at which $P_{i,0}$ converges towards $\mathbb{E}_0[P_{i,T_i}]$.

We estimate the input parameters $\{\kappa, \sigma\}$, by stacking the opportunities as

$$\begin{aligned}
X &= \Big(\mathbb{E}_0[P_{0,T_0}] - P_{0,0}, \mathbb{E}_0[P_{0,T_0}] - P_{0,1}, \dots, \mathbb{E}_0[P_{0,T_0}] - P_{0,T-1}, \dots, \\
&\quad \mathbb{E}_0[P_{I,T_I}] - P_{I,0}, \dots, \mathbb{E}_0[P_{I,T_I}] - P_{I,T-1}\Big)^T, \\
Y &= \Big(P_{0,1} - \mathbb{E}_0[P_{0,T_0}], P_{0,2} - \mathbb{E}_0[P_{0,T_0}], \dots, P_{0,T} - \mathbb{E}_0[P_{0,T_0}], \dots, \\
&\quad P_{I,1} - \mathbb{E}_0[P_{I,T_I}], \dots, P_{I,T} - \mathbb{E}_0[P_{I,T_I}]\Big)^T,
\end{aligned}$$

where $(\dots)^T$ denotes vector transposition. Applying ordinary least-square (OLS) on the above equation, we can estimate the original OU parameters as follows

$$\hat{\kappa} = \frac{\operatorname{cov}[Y, X]}{\operatorname{cov}[X, X]}, \quad \hat{\xi} = Y - \hat{\kappa} X, \quad \hat{\sigma} = \sqrt{\operatorname{cov}\left[\hat{\xi}, \hat{\xi}\right]}, \tag{11.3}$$

where, as usual, $\operatorname{cov}[.,.]$ is the covariance operator. We use the above estimations to find optimal stop-loss and take-profit bounds.

11.1.1 *Formulation in terms of HPs*

In this rather technical section, we perform transformations in order to formulate the problem in terms of heat potentials.

Consider a long investment strategy S and suppose profit/loss opportunity is driven by an OU process (see Lopez De Prado, 2018 among many others)

$$dx' = \kappa'\left(\theta' - x'\right) dt' + \sigma' dW_{t'}, \quad x'(0) = 0, \tag{11.4}$$

and a trading rule $R = \{\underline{\pi}', \overline{\pi}', T'\}$, $\underline{\pi}' < 0$, $\overline{\pi}' > 0$. It is important to understand what are the natural units associated with the OU process (11.4). To this end we can use its steady-state. The steady-state expectation of the above process is θ, while its standard deviation is given by $\Omega' = \sigma'/\sqrt{2\kappa'}$.

As usual, an appropriate scaling is helpful to remove superfluous parameters. To this end, we define

$$t = \kappa' t', \quad T = \kappa' T', \quad x = \frac{\sqrt{\kappa'}}{\sigma'} x', \quad \overline{\pi} = \frac{\sqrt{\kappa'}}{\sigma'} \overline{\pi}',$$
$$\underline{\pi} = \frac{\sqrt{\kappa'}}{\sigma'} \underline{\pi}', \quad \theta = \frac{\sqrt{\kappa'}}{\sigma'} \theta', \quad E = \frac{E'}{\sqrt{\kappa'}\sigma'}, \quad F = \frac{F'}{\kappa'\sigma'^2},$$

and get

$$dx_t = (\theta - x)\, dt + dW_t, \tag{11.5}$$

in the domain

$$\underline{\pi} \leq x \leq \overline{\pi}, \quad 0 \leq t \leq T. \tag{11.6}$$

The steady-state distribution has the expectation of θ, and the standard deviation $\Omega = 1/\sqrt{2}$.

According to the trading rule, we exit the trade either when: (A) the price hits $\overline{\pi}$ to take profit; (B) the price hits $\underline{\pi}$ to stop losses; (C) the trade expires at $t = T$. For a short investment strategy, the roles of $\{\underline{\pi}, \overline{\pi}\}$ are reversed - profits equal to $-\underline{\pi}$ are taken when when the price hits $\underline{\pi}$, and losses equal $-\overline{\pi}$ are realized when the price hits $\overline{\pi}$. Given the fact that the reflection $x \to -x$ leaves the initial condition unchanged and transforms the original OU process into the OU process of the form

$$dx = (-\theta - x)\, dt + dW_t, \quad x_0 = 0, \tag{11.7}$$

we can restrict ourselves to the case $\theta \geq 0$. More explicitly and intuitively, we go long when $\theta \geq 0$ and short when $\theta < 0$. Assuming that we know the trading rule $\{\underline{\pi}(\theta, T), \overline{\pi}(\theta, T), T\}$ for $\theta \geq 0$, the corresponding trading rule for $\theta < 0$ has the form

$$\{\underline{\pi}(\theta, T), \overline{\pi}(\theta, T), T\} = \{-\underline{\pi}(-\theta, T), -\overline{\pi}(-\theta, T), T\}. \tag{11.8}$$

Thus, we are interested in the maximization of the SR for nonnegative $\theta \geq 0$. We formulate this mathematically below.

For a given T, we define the stopping time $\iota = \inf\{t : x_t = \overline{\pi} \text{ or } x_t = \underline{\pi} \text{ or } t = T\}$. We wish to determine optimal $\overline{\pi} > 0, \underline{\pi} < 0$, to maximize the SR, which is defined as follows

$$\mathrm{SR} = \frac{\mathbb{E}\{x_\iota/\iota\}}{\sqrt{\mathbb{E}\{x_\iota^2/\iota^2\} - (\mathbb{E}\{x_\iota/\iota\})^2}}. \tag{11.9}$$

We also need to know the expected duration of the trade,

$$\mathrm{DUR} = \mathbb{E}\{\iota\}. \tag{11.10}$$

In order to calculate the corresponding SR and DUR we proceed as follows. We solve three terminal boundary value problems (TBVPs) of the form

$$
\begin{aligned}
&E_t(t,x) + (\theta - x)E_x(t,x) + \frac{1}{2}E_{xx}(t,x) = 0,\\
&E(t,\overline{\pi}) = \frac{\overline{\pi}}{t}, \quad E(t,\underline{\pi}) = \frac{\underline{\pi}}{t}, \quad E(T,x) = \frac{x}{T},\\
&F_t(t,x) + (\theta - x)F_x(t,x) + \frac{1}{2}F_{xx}(t,x) = 0,\\
&F(t,\overline{\pi}) = \frac{\overline{\pi}^2}{t^2}, \quad F(t,\underline{\pi}) = \frac{\underline{\pi}^2}{t^2}, \quad F(T,x) = \frac{x^2}{T^2},
\end{aligned}
$$

and

$$
\begin{aligned}
&G_t(t,x) + (\theta - x)G_x(t,x) + \frac{1}{2}G_{xx}(t,x) = 0,\\
&G(t,\overline{\pi}) = t, \quad G(t,\underline{\pi}) = t, \quad G(T,x) = T.
\end{aligned}
$$

We represent the SR and DUR as

$$
\mathrm{SR} = \frac{E(0,0)}{\sqrt{F(0,0) - (E(0,0))^2}}, \quad \mathrm{DUR} = G(0,0). \tag{11.11}
$$

We wish to use the method of heat potentials to solve the above TBVPs. First, we define $\tau = T - t$ and get initial boundary value problems (IBVPs) of the form

$$
\begin{aligned}
&E_\tau(\tau,x) = (\theta - x)E_x(\tau,x) + \frac{1}{2}E_{xx}(\tau,x),\\
&E(\tau,\overline{\pi}) = \frac{\overline{\pi}}{(T-\tau)}, \quad E(\tau,\underline{\pi}) = \frac{\underline{\pi}}{(T-\tau)}, \quad E(0,x) = \frac{x}{T},\\
&F_\tau(\tau,x) = (\theta - x)F_x(\tau,x) + \frac{1}{2}F_{xx}(\tau,x),\\
&F(\tau,\overline{\pi}) = \frac{\overline{\pi}^2}{(T-\tau)^2}, \quad F(\tau,\underline{\pi}) = \frac{\underline{\pi}^2}{(T-\tau)^2}, \quad F(0,x) = \frac{x^2}{T^2},\\
&G_\tau(\tau,x) = (\theta - x)G_x(\tau,x) + \frac{1}{2}G_{xx}(\tau,x),\\
&G(\tau,\overline{\pi}) = (T-\tau), \quad G(\tau,\underline{\pi}) = (T-\tau), \quad G(0,x) = T,\\
&\mathrm{SR} = \frac{E(T,0)}{\sqrt{F(T,0) - (E(T,0))^2}}, \quad \mathrm{DUR} = G(T,0).
\end{aligned}
$$

Second, we define

$$
\upsilon = \frac{1 - e^{-2\tau}}{2}, \quad \xi = e^{-\tau}(x - \theta), \tag{11.12}
$$

so that

$$\partial_\tau = (1-2\upsilon)\,\partial_\upsilon - \xi\partial_\xi, \quad \partial_x = \sqrt{1-2\upsilon}\partial_\xi. \tag{11.13}$$

Accordingly,

$$\begin{aligned}
E_\upsilon(\upsilon,\xi) &= \frac{1}{2}E_{\xi\xi}(\upsilon,\xi),\\
E\left(\upsilon,\overline{\Pi(\upsilon)}\right) &= \frac{2\overline{\pi}}{\ln\left(\frac{(1-2\upsilon)}{(1-2\Upsilon)}\right)}, \quad E\left(\upsilon,\underline{\Pi(\upsilon)}\right) = \frac{2\underline{\pi}}{\ln\left(\frac{(1-2\upsilon)}{(1-2\Upsilon)}\right)},\\
E(0,\xi) &= -\frac{2(\xi+\theta)}{\ln(1-2\Upsilon)},\\
F_\upsilon(\upsilon,\xi) &= \frac{1}{2}F_{\xi\xi}(\upsilon,\xi),\\
F\left(\upsilon,\overline{\Pi(\upsilon)}\right) &= \frac{4\overline{\pi}^2}{\left(\ln\left(\frac{(1-2\upsilon)}{(1-2\Upsilon)}\right)\right)^2}, \quad F\left(\upsilon,\underline{\Pi(\upsilon)}\right) = \frac{\underline{\pi}^2}{\left(\ln\left(\frac{(1-2\upsilon)}{(1-2\Upsilon)}\right)\right)^2},\\
F(0,\xi) &= \frac{4(\xi+\theta)^2}{(\ln(1-2\Upsilon))^2},\\
G_\upsilon(\upsilon,\xi) &= \frac{1}{2}G_{\xi\xi}(\upsilon,\xi),\\
G\left(\upsilon,\overline{\Pi(\upsilon)}\right) &= \frac{1}{2}\ln\left(\frac{(1-2\upsilon)}{(1-2\Upsilon)}\right), \quad G\left(\upsilon,\underline{\Pi(\upsilon)}\right) = \frac{1}{2}\ln\left(\frac{(1-2\upsilon)}{(1-2\Upsilon)}\right),\\
G(0,\xi) &= -\frac{1}{2}\ln(1-2\Upsilon),\\
\mathrm{SR} &= \frac{E(\Upsilon,\varpi)}{\sqrt{F(\Upsilon,\varpi)-(E(\Upsilon,\varpi))^2}}, \quad \mathrm{DUR} = G(\Upsilon,\varpi).
\end{aligned}$$

Here

$$\Upsilon = \frac{1-e^{-2T}}{2}, \qquad \varpi = -\sqrt{1-2\Upsilon}\theta,$$
$$\overline{\Pi}(\upsilon) = \sqrt{1-2\upsilon}\,(\overline{\pi}-\theta), \quad \underline{\Pi}(\upsilon) = \sqrt{1-2\upsilon}\,(\underline{\pi}-\theta).$$

As usual, we have to account for the initial conditions. To this end, we introduce

$$\begin{aligned}
\hat{E}(\upsilon,\xi) &= E(\upsilon,\xi) + \frac{2(\xi+\theta)}{\ln(1-2\Upsilon)},\\
\hat{F}(\upsilon,\xi) &= F(\upsilon,\xi) - \frac{4\left(\upsilon+(\xi+\theta)^2\right)}{(\ln(1-2\Upsilon))^2},\\
\hat{G}(\upsilon,\xi) &= G(\upsilon,\xi) + \frac{1}{2}\ln(1-2\Upsilon),
\end{aligned}$$

where

$$
\begin{aligned}
\hat{E}_\upsilon(\upsilon,\xi) &= \frac{1}{2}\hat{E}_{\xi\xi}(\upsilon,\xi),\\
\hat{E}(\upsilon,\underline{\Pi}(\upsilon)) &= \underline{e}(\upsilon), \quad \hat{E}\left(\upsilon,\overline{\Pi}(\upsilon)\right) = \overline{e}(\upsilon), \quad \hat{E}(0,\xi) = 0,\\
\hat{F}_\upsilon(\upsilon,\xi) &= \frac{1}{2}\hat{F}_{\xi\xi}(\upsilon,\xi),\\
\hat{F}(\upsilon,\underline{\Pi}(\upsilon)) &= \underline{f}(\upsilon), \quad \hat{F}\left(\upsilon,\overline{\Pi}(\upsilon)\right) = \overline{f}(\upsilon), \quad \hat{F}(0,\xi) = 0,\\
\hat{G}_\upsilon(\upsilon,\xi) &= \frac{1}{2}\hat{G}_{\xi\xi}(\upsilon,\xi),\\
\hat{G}(\upsilon,\underline{\Pi}(\upsilon)) &= \underline{g}(\upsilon), \quad \hat{G}\left(\upsilon,\overline{\Pi}(\upsilon)\right) = \overline{g}(\upsilon), \quad \hat{G}(0,\xi) = 0.
\end{aligned}
$$

Here,

$$
\underline{e}(\upsilon) = \frac{2\underline{\pi}}{\ln\left(\frac{(1-2\upsilon)}{(1-2\Upsilon)}\right)} + \frac{2(\underline{\Pi}(\upsilon)+\theta)}{\ln(1-2\Upsilon)}, \quad \overline{e}(\upsilon) = \frac{2\overline{\pi}}{\ln\left(\frac{(1-2\upsilon)}{(1-2\Upsilon)}\right)} + \frac{2(\overline{\Pi}(\upsilon)+\theta)}{\ln(1-2\Upsilon)}
$$

$$
\underline{f}(\upsilon) = \frac{4\underline{\pi}^2}{\left(\ln\left(\frac{(1-2\upsilon)}{(1-2\Upsilon)}\right)\right)^2} - \frac{4\left(\upsilon+(\underline{\Pi}(\upsilon)+\theta)^2\right)}{(\ln(1-2\Upsilon))^2},
$$

$$
\overline{f}(\upsilon) = \frac{4\overline{\pi}^2}{\left(\ln\left(\frac{(1-2\upsilon)}{(1-2\Upsilon)}\right)\right)^2} - \frac{4\left(\upsilon+(\overline{\Pi}(\upsilon)+\theta)^2\right)}{(\ln(1-2\Upsilon))^2},
$$

$$
\underline{g}(\upsilon) = \frac{1}{2}\ln(1-2\upsilon), \quad \overline{g}(\upsilon) = \frac{1}{2}\ln(1-2\upsilon).
$$

Accordingly,

$$
\mathrm{SR} = \frac{\hat{E}(\Upsilon,\varpi) - \frac{2(\varpi+\theta)}{\ln(1-2\Upsilon)}}{\sqrt{\hat{F}(\Upsilon,\varpi) - \left(\hat{E}(\Upsilon,\varpi)\right)^2 + \frac{4\left(\Upsilon+\ln(1-2\Upsilon)(\varpi+\theta)\hat{E}(\Upsilon,\varpi)\right)}{(\ln(1-2\Upsilon))^2}}}, \tag{11.14}
$$

$$
\mathrm{DUR} = \hat{G}(\Upsilon,\varpi) - \frac{1}{2}\ln(1-2\Upsilon).
$$

After the above transformations are performed, the problem becomes solvable by the method of heat potentials.

11.2 The method of heat potentials

Now we are ready to use the classical method of heat potentials to calculate the SR. Consider $\hat{E}$. We have to solve the following coupled system of Volterra integral equations

$$
\underline{\varepsilon}(\upsilon) + \frac{1}{\sqrt{2\pi}}\int_0^\upsilon \frac{(\underline{\Pi}(\upsilon)-\underline{\Pi}(\zeta))\, e^{-\frac{(\underline{\Pi}(\upsilon)-\underline{\Pi}(\zeta))^2}{2(\upsilon-\zeta)}}}{(\upsilon-\zeta)^{3/2}}\underline{\varepsilon}(\zeta)\, d\zeta \tag{11.15}
$$

$$+\frac{1}{\sqrt{2\pi}}\int_0^{\upsilon}\frac{\left(\underline{\Pi}(\upsilon)-\overline{\Pi}(\zeta)\right)e^{-\frac{\left(\underline{\Pi}(\upsilon)-\overline{\Pi}(\zeta)\right)^2}{2(\upsilon-\zeta)}}}{(\upsilon-\zeta)^{3/2}}\overline{\varepsilon}(\zeta)\,d\zeta=\underline{e}(\upsilon),$$

$$-\overline{\varepsilon}(\upsilon)+\frac{1}{\sqrt{2\pi}}\int_0^{\upsilon}\frac{\left(\overline{\Pi}(\upsilon)-\underline{\Pi}(\zeta)\right)e^{-\frac{\left(\overline{\Pi}(\upsilon)-\underline{\Pi}(\zeta)\right)^2}{2(\upsilon-\zeta)}}}{(\upsilon-\zeta)^{3/2}}\underline{\varepsilon}(\zeta)\,d\zeta$$

$$+\frac{1}{\sqrt{2\pi}}\int_0^{\upsilon}\frac{\left(\overline{\Pi}(\upsilon)-\overline{\Pi}(\zeta)\right)e^{-\frac{\left(\overline{\Pi}(\upsilon)-\overline{\Pi}(\zeta)\right)^2}{2(\upsilon-\zeta)}}}{(\upsilon-\zeta)^{3/2}}\overline{\varepsilon}(\zeta)\,d\zeta=\overline{e}(\upsilon),\qquad(11.16)$$

Once these equations are solved, $\hat{E}(\upsilon,\xi)$ can be written as follows

$$\hat{E}(\upsilon,\xi)=\frac{1}{\sqrt{2\pi}}\int_0^{\upsilon}\frac{\left(\xi-\underline{\Pi}(\zeta)\right)e^{-\frac{\left(\xi-\underline{\Pi}(\zeta)\right)^2}{2(\upsilon-\zeta)}}}{(\upsilon-\zeta)^{3/2}}\underline{\varepsilon}(\zeta)\,d\zeta\qquad(11.17)$$

$$+\frac{1}{\sqrt{2\pi}}\int_0^{\upsilon}\frac{\left(\xi-\overline{\Pi}(\zeta)\right)e^{-\frac{\left(\xi-\overline{\Pi}(\zeta)\right)^2}{2(\upsilon-\zeta)}}}{(\upsilon-\zeta)^{3/2}}\overline{\varepsilon}(\zeta)\,d\zeta.$$

We can find $\hat{F}(\upsilon,\xi)$ by using the same approach by solving the integral equation

$$\underline{\phi}(\upsilon)+\frac{1}{\sqrt{2\pi}}\int_0^{\upsilon}\frac{\left(\underline{\Pi}(\upsilon)-\underline{\Pi}(\zeta)\right)e^{-\frac{\left(\underline{\Pi}(\upsilon)-\underline{\Pi}(\zeta)\right)^2}{2(\upsilon-\zeta)}}}{(\upsilon-\zeta)^{3/2}}\underline{\phi}(\zeta)\,d\zeta\qquad(11.18)$$

$$+\frac{1}{\sqrt{2\pi}}\int_0^{\upsilon}\frac{\left(\underline{\Pi}(\upsilon)-\overline{\Pi}(\zeta)\right)e^{-\frac{\left(\underline{\Pi}(\upsilon)-\overline{\Pi}(\zeta)\right)^2}{2(\upsilon-\zeta)}}}{(\upsilon-\zeta)^{3/2}}\overline{\phi}(\zeta)\,d\zeta=\underline{f}(\upsilon),$$

and

$$-\overline{\phi}(\upsilon)+\frac{1}{\sqrt{2\pi}}\int_0^{\upsilon}\frac{\left(\overline{\Pi}(\upsilon)-\underline{\Pi}(\zeta)\right)e^{-\frac{\left(\overline{\Pi}(\upsilon)-\underline{\Pi}(\zeta)\right)^2}{2(\upsilon-\zeta)}}}{(\upsilon-\zeta)^{3/2}}\underline{\phi}(\zeta)\,d\zeta\qquad(11.19)$$

$$+\frac{1}{\sqrt{2\pi}}\int_0^{\upsilon}\frac{\left(\overline{\Pi}(\upsilon)-\overline{\Pi}(\zeta)\right)e^{-\frac{\left(\overline{\Pi}(\upsilon)-\overline{\Pi}(\zeta)\right)^2}{2(\upsilon-\zeta)}}}{(\upsilon-\zeta)^{3/2}}\overline{\phi}(\zeta)\,d\zeta=\overline{f}(\upsilon).$$

Then the function $\hat{F}(\upsilon,\xi)$ can be found as

$$\hat{F}(\upsilon,\xi)=\frac{1}{\sqrt{2\pi}}\int_0^{\upsilon}\frac{\left(\xi-\underline{\Pi}(\zeta)\right)e^{-\frac{\left(\xi-\underline{\Pi}(\zeta)\right)^2}{2(\upsilon-\zeta)}}}{(\upsilon-\zeta)^{3/2}}\underline{\phi}(\zeta)\,d\zeta\qquad(11.20)$$

$$+\frac{1}{\sqrt{2\pi}}\int_0^{\upsilon}\frac{\left(\xi-\overline{\Pi}(\zeta)\right)e^{-\frac{\left(\xi-\overline{\Pi}(\zeta)\right)^2}{2(\upsilon-\zeta)}}}{(\upsilon-\zeta)^{3/2}}\overline{\phi}(\zeta)\,d\zeta.$$

In particular,

$$\hat{E}(\Upsilon,\varpi)=\frac{1}{\sqrt{2\pi}}\int_0^{\Upsilon}\frac{\left(\varpi-\underline{\Pi}(\zeta)\right)e^{-\frac{(\varpi-\underline{\Pi}(\zeta))^2}{2(\Upsilon-\zeta)}}}{(\Upsilon-\zeta)^{3/2}}\underline{\varepsilon}(\zeta)\,d\zeta \\ +\frac{1}{\sqrt{2\pi}}\int_0^{\Upsilon}\frac{\left(\varpi-\overline{\Pi}(\zeta)\right)e^{-\frac{(\varpi-\overline{\Pi}(\zeta))^2}{2(\Upsilon-\zeta)}}}{(\Upsilon-\zeta)^{3/2}}\overline{\varepsilon}(\zeta)\,d\zeta, \tag{11.21}$$

$$\hat{F}(\Upsilon,\varpi)=\frac{1}{\sqrt{2\pi}}\int_0^{\Upsilon}\frac{\left(\varpi-\underline{\Pi}(\zeta)\right)e^{-\frac{(\varpi-\underline{\Pi}(\zeta))^2}{2(\Upsilon-\zeta)}}}{(\Upsilon-\zeta)^{3/2}}\underline{\phi}(\zeta)\,d\zeta \\ +\frac{1}{\sqrt{2\pi}}\int_0^{\Upsilon}\frac{\left(\varpi-\overline{\Pi}(\zeta)\right)e^{-\frac{(\varpi-\overline{\Pi}(\zeta))^2}{2(\Upsilon-\zeta)}}}{(\Upsilon-\zeta)^{3/2}}\overline{\phi}(\zeta)\,d\zeta. \tag{11.22}$$

It is important to notice that $\left(\underline{\varepsilon}(\zeta),\overline{\varepsilon}(\zeta)\right)$ and $\left(\underline{\phi}(\zeta),\overline{\phi}(\zeta)\right)$ are singular at $\zeta=\Upsilon$. However, due to the dampening impact of the exponents $\exp\left(-\left(\varpi-\overline{\Pi}(\zeta)\right)^2/2(\Upsilon-\zeta)\right)$, the corresponding integrals still converge.

We now know $\hat{E}(\Upsilon,\varpi),\hat{F}(\Upsilon,\varpi)$ and calculate the SR by using Eq. (11.14). Then $\hat{G}(\Upsilon,\varpi)$ and DUR can be calculated in a similar fashion.

11.3 Numerical method

To compute the SR, we need to find $\hat{E}(\Upsilon,\varpi)$ and $\hat{F}(\Upsilon,\varpi)$, and then apply Eq. (11.14). Functions $\hat{E}(\Upsilon,\varpi)$ and $\hat{F}(\Upsilon,\varpi)$ can be computed using Eqs. (11.17) and (11.20) by simple integration with pre-computed $(\underline{\varepsilon},\overline{\varepsilon})$ and $(\underline{\phi},\overline{\phi})$. In this Section, we develop a numerical method to compute these quantities by solving Eqs. (11.15) and (11.18) by extending the methods developed in Lipton and Kaushansky (2020a) and also described in Chapter 5. Here we provide a more detailed exposition of the method as applied to our particular problem in this chapter.[1]

We want to solve a generic system of the form

$$\nu^1(\upsilon)+\int_0^{\upsilon}\frac{K^{1,1}(\upsilon,s)}{\sqrt{\upsilon-s}}\nu^1(s)\,ds+\int_0^{\upsilon}K^{1,2}(\upsilon,s)\nu^2(s)\,ds=\chi^1(\upsilon), \tag{11.23}$$

$$-\nu^2(\upsilon)+\int_0^{\upsilon}K^{2,1}(\upsilon,s)\nu^1(s)\,ds+\int_0^{\upsilon}\frac{K^{2,2}(\upsilon,s)}{\sqrt{\upsilon-s}}\nu^2(s)\,ds=\chi^2(\upsilon),$$

[1] For illustrative purposes we develop a simple scheme based on the trapezoidal rule for Stieltjes integrals.

with respect to variables $(\nu^1(\upsilon), \nu^2(\upsilon))$, where

$$K^{1,1}(\upsilon, s) = \frac{1}{\sqrt{2\pi}} \frac{\underline{\Pi}(\upsilon) - \underline{\Pi}(s)}{\upsilon - s} \exp\left(-\frac{(\underline{\Pi}(\upsilon) - \underline{\Pi}(s))^2}{2(\upsilon - s)}\right),$$

$$K^{1,2}(\upsilon, s) = \frac{1}{\sqrt{2\pi}} \frac{\underline{\Pi}(\upsilon) - \overline{\Pi}(s)}{(\upsilon - s)^{3/2}} \exp\left(-\frac{(\underline{\Pi}(\upsilon) - \overline{\Pi}(s))^2}{2(\upsilon - s)}\right),$$

$$K^{2,1}(\upsilon, s) = \frac{1}{\sqrt{2\pi}} \frac{\overline{\Pi}(\upsilon) - \underline{\Pi}(s)}{(\upsilon - s)^{3/2}} \exp\left(-\frac{(\overline{\Pi}(\upsilon) - \underline{\Pi}(s))^2}{2(\upsilon - s)}\right),$$

$$K^{2,2}(\upsilon, s) = \frac{1}{\sqrt{2\pi}} \frac{\overline{\Pi}(\upsilon) - \overline{\Pi}(s)}{\upsilon - s} \exp\left(-\frac{(\overline{\Pi}(\upsilon) - \overline{\Pi}(s))^2}{2(\upsilon - s)}\right).$$

It is clear that

$$K^{1,1}(\upsilon, \upsilon) = \frac{1}{\sqrt{2\pi}} \lim_{s \to \upsilon} \frac{\underline{\Pi}(\upsilon) - \underline{\Pi}(s)}{\upsilon - s} = \frac{\theta - \underline{\pi}}{\sqrt{2\pi}\sqrt{1 - 2\upsilon}},$$

$$K^{1,2}(\upsilon, \upsilon) = 0, \quad K^{2,1}(\upsilon, \upsilon) = 0,$$

$$K^{2,2}(\upsilon, \upsilon) = \frac{1}{\sqrt{2\pi}} \lim_{s \to \upsilon} \frac{\overline{\Pi}(\upsilon) - \overline{\Pi}(s)}{\upsilon - s} = \frac{\theta - \overline{\pi}}{\sqrt{2\pi}\sqrt{1 - 2\upsilon}}.$$

We can equally rewrite the relevant integrals as Stieltjes integrals

$$\nu^1(\upsilon) - 2\int_0^\upsilon K^{1,1}(\upsilon, s)\nu^1(s)\, d\sqrt{\upsilon - s} + \int_0^\upsilon K^{1,2}(\upsilon, s)\nu^2(s)\, ds = \chi^1(\upsilon),$$

$$-\nu^2(\upsilon) + \int_0^\upsilon K^{2,1}(\upsilon, s)\nu^1(s)\, ds - 2\int_0^\upsilon K^{2,2}(\upsilon, s)\nu^2(s)\, d\sqrt{\upsilon - s} = \chi^2(\upsilon).$$

Consider a grid $0 = \upsilon_0 < \upsilon_1 < \ldots < \upsilon_n = \Upsilon$, and let $\Delta_{k,l} = \upsilon_k - \upsilon_l$. Then, using the trapezoidal rule for approximation of integrals, we get the following approximation of last two equations

$$\nu_k^1 + \sum_{i=1}^k \left[\frac{\left(K_{k,i}^{1,1}\nu_i^1 + K_{k,i-1}^{1,1}\nu_{i-1}^1\right)}{\left(\sqrt{\Delta_{k,i}} + \sqrt{\Delta_{k,i-1}}\right)} \right.$$
$$\left. + \frac{1}{2}\left(K_{k,i}^{1,2}\nu_i^2 + K_{k,i-1}^{1,2}\nu_{i-1}^2\right) \right] \Delta_{i,i-1} = \chi_k^1,$$

$$-\nu_k^2 + \sum_{i=1}^k \left[\frac{1}{2}\left(K_{k,i}^{2,1}\nu_i^1 + K_{k,i-1}^{2,1}\nu_{i-1}^1\right) \right.$$
$$\left. + \frac{\left(K_{k,i}^{2,2}\nu_i^2 + K_{k,i-1}^{2,2}\nu_{i-1}^2\right)}{\left(\sqrt{\Delta_{k,i}} + \sqrt{\Delta_{k,i-1}}\right)} \right] \Delta_{i,i-1} = \chi_k^2,$$

where

$$\nu_i^\alpha = \nu^\alpha(\upsilon_i), \quad \chi_i^\alpha = \chi^\alpha(\upsilon_i), \quad K_{k,j}^{\alpha,\beta} = K^{\alpha,\beta}(\upsilon_k, \upsilon_i), \quad \alpha, \beta = 1, 2.$$

Taking into account that

$$\left(\nu_0^1, \nu_0^2\right) = \left(\chi_0^1, -\chi_0^2\right), \quad \left(\nu_1^1, \nu_1^2\right) = \left(\frac{\chi_1^1}{\left(1 + K_{1,1}^{1,1}\sqrt{\upsilon_1}\right)}, -\frac{\chi_1^2}{\left(1 - K_{1,1}^{2,2}\sqrt{\upsilon_1}\right)}\right), \tag{11.24}$$

and assuming that $\left(\nu_2^1, \nu_2^2\right), \ldots, \left(\nu_{k-1}^1, \nu_{k-1}^2\right)$ have been computed, we can easily find $\left(\nu_k^1, \nu_k^2\right)$

$$\nu_k^1 = \left(1 + K_{k,k}^{1,1}\sqrt{\Delta_{k,k-1}}\right)^{-1}$$
$$\times \left[\chi_k^1 - K_{k,k-1}^{1,1}\nu_{k-1}^1\sqrt{\Delta_{k,k-1}} - \frac{1}{2}K_{k,k-1}^{1,2}\nu_{k-1}^2\Delta_{k,k-1}\right.$$
$$\left. - \sum_{i=1}^{k-1}\left(\frac{\left(K_{k,i}^{1,1}\nu_i^1 + K_{k,i-1}^{1,1}\nu_{i-1}^1\right)}{\left(\sqrt{\Delta_{k,i}} + \sqrt{\Delta_{k,i-1}}\right)} + \frac{1}{2}\left(K_{k,i}^{1,2}\nu_i^2 + K_{k,i-1}^{1,2}\nu_{i-1}^2\right)\right)\Delta_{i,i-1}\right],$$

$$\nu_k^2 = \left(-1 + K_{k,k}^{2,2}\sqrt{\Delta_{k,k-1}}\right)^{-1}$$
$$\times \left[\chi_k^2 - \frac{1}{2}K_{k,k-1}^{2,1}\nu_{k-1}^1\Delta_{k,k-1} - K_{k,k-1}^{2,2}\nu_{k-1}^2\sqrt{\Delta_{k,k-1}}\right.$$
$$\left. - \sum_{i=1}^{k-1}\left(\frac{1}{2}\left(K_{k,i}^{2,1}\nu_i^1 + K_{k,i-1}^{2,1}\nu_{i-1}^1\right) + \frac{\left(K_{k,i}^{2,2}\nu_i^2 + K_{k,i-1}^{2,2}\nu_{i-1}^2\right)}{\left(\sqrt{\Delta_{k,i}} + \sqrt{\Delta_{k,i-1}}\right)}\right)\Delta_{i,i-1}\right].$$

The approximation error of the integrals is of order $O(\Delta^2)$, where $\Delta = \max_i(\Delta_{i,i-1})$. Hence, on uniform grid, the convergence is of order $O(\Delta)$. We emphasize that, due to the nature of $(\underline{e}(\upsilon), \overline{e}(\upsilon))$, etc., it is necessary to use a highly inhomogeneous grid which is concentrated near the right endpoint.

11.3.1 *Computation of the Sharpe ratio*

Once $(\underline{\varepsilon}(\upsilon), \overline{\varepsilon}(\upsilon))$ and $\left(\underline{\phi}(\upsilon), \overline{\phi}(\upsilon)\right)$ are computed, we can approximate $\hat{E}(\upsilon, \xi)$ and $\hat{F}(\upsilon, \xi)$. We interested to compute these functions at one point (Υ, ϖ), which can be done by approximation of the integrals using the trapezoidal rule

$$\hat{E}(\Upsilon, \varpi) = \frac{1}{2}\sum_{i=1}^{k}\left(\underline{w}_{n,i}\underline{\varepsilon}_i + \underline{w}_{n,i-1}\underline{\varepsilon}_{i-1} + \overline{w}_{n,i}\overline{\varepsilon}_i + \overline{w}_{n,i-1}\overline{\varepsilon}_{i-1}\right)\Delta_{i,i-1}, \tag{11.25}$$

and

$$\hat{F}(\Upsilon,\varpi)=\frac{1}{2}\sum_{i=1}^{k}\left(\underline{w}_{n,i}\underline{\phi}_{i}+\underline{w}_{n,i-1}\underline{\phi}_{i-1}+\overline{w}_{n,i}\overline{\phi}_{i}+\overline{w}_{n,i-1}\overline{\phi}_{i-1}\right)\Delta_{i,i-1}. \tag{11.26}$$

The corresponding weights are as follows

$$\begin{aligned}\underline{w}_{n,i}&=\frac{(\varpi-\underline{\Pi}_{i})e^{-\frac{(\varpi-\underline{\Pi}_{i})^{2}}{2\Delta_{n,i}}}}{\sqrt{2\pi}\Delta_{n,i}^{3/2}}, & \overline{w}_{n,i}&=\frac{(\varpi-\overline{\Pi}_{i})e^{-\frac{(\varpi-\overline{\Pi}_{i})^{2}}{2\Delta_{n,i}}}}{\sqrt{2\pi}\Delta_{n,i}^{3/2}}, & &1\le i<n,\\ \underline{w}_{n,i}&=0 & \overline{w}_{n,i}&=0, & &i=n.\end{aligned} \tag{11.27}$$

As a result, we get the following algorithm for the numerical evaluation of the SR.

Algorithm 2 Numerical evaluation of the Sharpe ratio

Step 1	Define a time grid $0=\upsilon_0<\upsilon_1<\ldots<\Upsilon$.
Step 2	Compute $\underline{\epsilon}(\upsilon),\overline{\epsilon}(\upsilon),\underline{\phi}(\upsilon),\overline{\phi}(\upsilon)$ using numerical method in Section 11.3.
Step 3	Compute $\hat{E}(\Upsilon,\varpi)$ by using Eq. (11.25).
Step 4	Compute $\hat{F}(\Upsilon,\varpi)$ by using Eq. (11.26).
Step 5	Compute the Sharpe ratio by using Eq. (11.14).

11.3.2 *Comparison with Monte Carlo simulations*

We compute the SR for various values of $\underline{\pi}$ and $\overline{\pi}$, and as a result show the SR as a function of $(\underline{\pi},\overline{\pi})$. After that one can choose $(\underline{\pi},\overline{\pi})$ in order to maximize the SR.

To be concrete, consider $\theta=1.0,\Upsilon=0.49,T=1.96$. We compare our results with the Monte Carlo method, which simulates the process and compute its expectation and variance (see Lopez De Prado, 2018). First, we compare separately E, $\sigma=\sqrt{F-E^2}$, and G calculated by both methods in Fig. 11.1.

Second, we show the results for the SR itself in Fig. 11.2. We see that the relative difference between the method of heat potentials and the Monte Carlo method is small and mainly comes from the Monte Carlo noise.

11.3.3 *Optimization of the Sharpe ratio*

In this section we solve a problem of finding parameters to maximize the SR by analyzing it as a function of $(\underline{\pi},\overline{\pi})$ for different values of θ and Υ. Two problems are considered

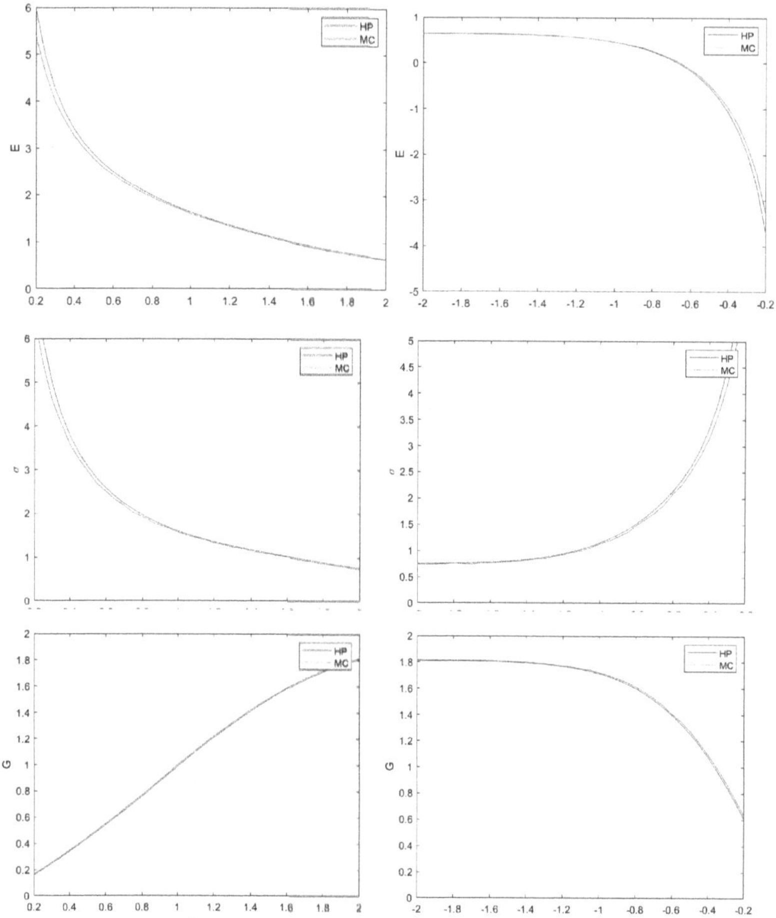

(a) (b)

Figure 11.1: (a) Functions $E, \sigma = \sqrt{F - E^2}$ and G vs $\overline{\pi} \equiv \pi_1$ computed by using the HP and Monte Carlo methods at $\underline{\pi} = -2$; (b) Same quantities as functions of $\underline{\pi} \equiv \pi_0$ for $\overline{\pi} = 1, \theta = 1.0, T = 1.96$.

(A) Fix Υ and maximize the SR over $(\underline{\pi}, \overline{\pi})$;
(B) Maximize the SR over $(\underline{\pi}, \overline{\pi}, \Upsilon)$.

Given that the natural unit $\Omega = 1/\sqrt{2}$, we consider three representative values of θ, namely $\theta = 1$, $\theta = 0.5$, and $\theta = 0$, corresponding to strong and weak mispricing and fair pricing, respectively. We choose three

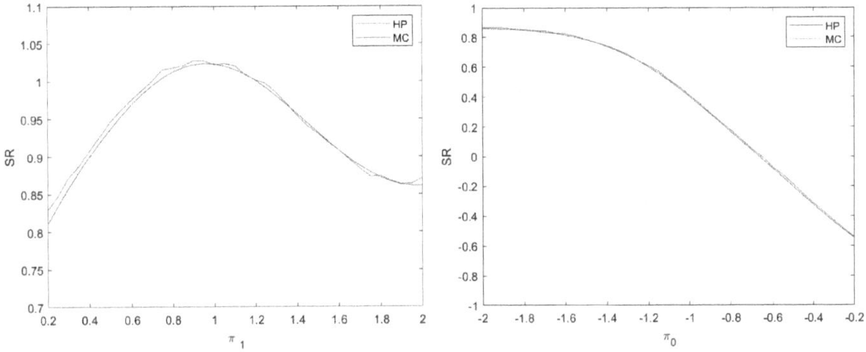

(a) (b)

Figure 11.2: (a) The Sharpe ratio as a function of $\overline{\pi} \equiv \pi_1$ computed by using the HP and Monte Carlo methods for $\underline{\pi} = -1$; (b) Same quantity as a function of $\underline{\pi} \equiv \pi_0$ for $\overline{\pi} = 1, \theta = 1.0, T = 1.96$.

maturities, $\Upsilon = 0.49, 0.4999, 0.499999$ or, equivalently, $T = 1.96, 4.26, 6.56$. For negative θ the corresponding SR can be obtained by reflection if needed.

We show the corresponding SR surfaces in Figs. 11.3–11.5.

$\theta\backslash\Upsilon, T$	0.49, 0.8	0.4999, 1.96	0.499999, 4.26
1.0	$\underline{\pi}^* = -4.0$	$\underline{\pi}^* = -4.0$	$\underline{\pi}^* = -4.0$
	$\overline{\pi}^* = 4.0$	$\overline{\pi}^* = 4.0$	$\overline{\pi}^* = 4.0$
	SR $= 1.2261$	SR $= 1.3824$	SR $= 1.3709$
0.5	$\underline{\pi}^* = -4.0$	$\underline{\pi}^* = -4.0$	$\underline{\pi}^* = -4.0$
	$\overline{\pi}^* = 0.6$	$\overline{\pi}^* = 0.9$	$\overline{\pi}^* = 1.0$
	SR $= 0.8219$	SR $= 0.8792$	SR $= 0.8963$
0.0	$\underline{\pi}^* = -4.0$	$\underline{\pi}^* = -4.0$	$\underline{\pi}^* = -4.0$
	$\overline{\pi}^* = 0.1$	$\overline{\pi}^* = 0.4$	$\overline{\pi}^* = 0.1$
	SR $= 0.7075$	SR $= 0.7139$	SR $= 0.7411$

Table 11.1: The Sharpe ratio maximized over $(\underline{\pi}, \overline{\pi})$ for fixed Υ or T.

The optimal bounds $(\underline{\pi}^*, \overline{\pi}^*)$ are given in Table 11.1. It can be seen that in the case when the original mispricing is strong ($\theta = 1$) it is not optimal to stop the trade early. When the mispricing is weaker ($\theta = 0.5$) or there is no mispricing in the first place ($\theta = 0$) it is not optimal to stop losses, but it might be beneficial to take profits. We emphasize that in practice one needs to use a highly reliable estimation of the OU parameters to be able to use these rules with confidence.

11.4 Traditional approaches

The method of heat potentials boils down to solving a system of Volterra equations of the second kind. However, there are certain quantities of interest that can be calculated directly. To put it into a proper context, in this section we discuss several classical approaches to the problem we are interested in. We emphasize that the method of heat potentials is dramatically different from other method because it allows one to consider strategies with finite duration, say T, whilst, to the best of our knowledge, all other methods are asymptotic in nature and assume that $T \to \infty$.

11.4.1 *Expectation and variance of the trade's duration*

Here we calculate the expected value and the variance of the of duration of a trade, which terminates only when the spread hits one of the barriers, $T = \infty$ (or $\Upsilon = 0.5$). Specifically, we show how to calculate these quantities analytically by solving inhomogeneous linear ordinary differential equations (ODEs).

In the case in question, the second change of variables is not necessary, so that we can concentrate on the following problems

$$G_t^{(1)}(t,x) + (\theta - x) G_x^{(1)}(t,x) + \frac{1}{2} G_{xx}^{(1)}(t,x) = 0, \tag{11.28}$$
$$G^{(2)}(t,\underline{\pi}) = t, \quad G^{(1)}(t,\overline{\pi}) = t,$$

and

$$G_t^{(2)}(t,x) + (\theta - x) G_x^{(2)}(t,x) + \frac{1}{2} G_{xx}^{(2)}(t,x) = 0, \tag{11.29}$$
$$G^{(2)}(t,\underline{\pi}) = t^2, \quad G^{(2)}(t,\overline{\pi}) = t^2,$$

with implicit terminal conditions at $T \to \infty$. The superscripts indicate the first and second moments, respectively.

We start with the expectation. We can represent the solution $G^{(1)}(t,x)$ of Eq. (11.28) in a semi-stationary form

$$G^{(1)}(t,x) = t + g^{(1)}(x)$$

where

$$(\theta - x)\, g_x^{(1)}(x) + \frac{1}{2} g_{xx}^{(1)}(x) = -1, \tag{11.30}$$

and

$$g^{(1)}(\underline{\pi}) = 0, \quad g^{(1)}(\overline{\pi}) = 0. \tag{11.31}$$

The Eq. (11.30) can be solved by the method of variation of constants

$$\begin{aligned} g_x^{(1)}(x) &= a_1 e^{(x-\theta)^2} + \lambda \mathcal{F}(x - \theta), \\ g^{(1)}(x) &= a_0 + a_1 \mathcal{I}(x - \theta) + \lambda \mathcal{G}(x - \theta), \end{aligned} \tag{11.32}$$

where a_0, a_1 are arbitrary constants. Here $\mathcal{D}(x)$ is Dawson's function (Abramowitz and Stegun, 1964), $\mathcal{E}(x)$ is its integral, and $\mathcal{F}(x), \mathcal{G}(x)$ are convenient abbreviations

$$\begin{aligned} \mathcal{I}(x) &= \int_0^x e^{z^2} dz, \quad \mathcal{D}(x) = e^{-x^2} \int_0^x e^{z^2} dz = e^{-x^2} \mathcal{I}(x), \\ \mathcal{E}(x) &= \int_0^x D(z)\, dz, \quad \mathcal{F}(x) = \sqrt{\pi} N\left(\sqrt{2}x\right) e^{x^2}, \\ \mathcal{G}(x) &= \sqrt{\pi} N\left(\sqrt{2}x\right) \mathcal{I}(x) - \mathcal{E}(x). \end{aligned} \tag{11.33}$$

We can use the Taylor series expansion for $\mathcal{G}(x)$ and represent it in the form

$$\mathcal{G}(x) = \frac{1}{4} \sum_{n=1}^{\infty} \frac{\Gamma\left(\frac{n}{2}\right)}{\Gamma(n+1)} (2x)^n, \tag{11.34}$$

see also (Ricciardi and Sato, 1988), where this formula is obtained via the Laplace transform. Taking into account boundary conditions (11.31), we can represent g as follows

$$\begin{aligned} g^{(1)}(x, \underline{\pi}, \overline{\pi}) = 2\Bigg[&\frac{(\mathcal{G}(\overline{\pi} - \theta) - \mathcal{G}(\underline{\pi} - \theta))}{(\mathcal{I}(\overline{\pi} - \theta) - \mathcal{I}(\underline{\pi} - \theta))} (\mathcal{I}(x - \theta) - \mathcal{I}(\underline{\pi} - \theta)) \\ &- (\mathcal{G}(x - \theta) - \mathcal{G}(\underline{\pi} - \theta)) \Bigg]. \end{aligned} \tag{11.35}$$

Finally, the expected duration is given by the following expression:

$$\mathrm{DUR} = g(0). \tag{11.36}$$

We show the expected duration as a function of $\underline{\pi}, \overline{\pi}$ for $\theta = 1$ in Fig. 11.6. There as well as in Figs. 11.3–11.5 $\pi_0 \equiv \underline{\pi}$, $\pi_1 \equiv \overline{\pi}$.

Given the fact that $\Upsilon \to 0.5$ corresponds to $T \to \infty$, we can see from this figure that for sufficiently remote $\underline{\pi}, \overline{\pi}$ the process stays within the range $[\underline{\pi}, \overline{\pi}]$ indefinitely, or, at least, for a very long time.

Now we consider Eq. (11.29) and write

$$G^{(2)}(t,x) = t^2 + tg^{(2,1)}(x) + g^{(2,0)}(x),$$

where

$$\begin{gathered}(\theta - x)\, g_x^{(2,1)}(x) + \frac{1}{2} g_{xx}^{(2,1)}(x) = -2, \\ g^{(2,1)}(\underline{\pi}) = 0, \quad g^{(2,1)}(\overline{\pi}) = 0, \\ (\theta - x)\, g_x^{(2,0)}(x) + \frac{1}{2} g_{xx}^{(2,0)}(x) = -g^{(2,1)}(x), \\ g^{(2,0)}(\underline{\pi}) = 0, \quad g^{(2,0)}(\overline{\pi}) = 0.\end{gathered} \tag{11.37}$$

It is clear that

$$g^{(2,1)}(x) = 2g^{(1)}(x, \underline{\pi}, \overline{\pi}),$$

where $g^{(1)}$ is given by Eq. (11.35).

The Green function $\mathfrak{G}(x,y)$ for the problem in Eq. (11.37) has the form

$$\mathfrak{G}(x,y) = \begin{cases} 2\frac{e^{-(y-\theta)^2}(\mathcal{I}(y-\theta)-\mathcal{I}(\underline{\pi}-\theta))}{(\mathcal{I}(\overline{\pi}-\theta)-\mathcal{I}(\underline{\pi}-\theta))} \left(\mathcal{I}(x-\theta) - \mathcal{I}(\overline{\pi}-\theta)\right), & y \le x \le \overline{\pi}, \\ 2\frac{e^{-(y-\theta)^2}(\mathcal{I}(y-\theta)-\mathcal{I}(\overline{\pi}-\theta))}{(\mathcal{I}(\overline{\pi}-\theta)-\mathcal{I}(\underline{\pi}-\theta))} \left(\mathcal{I}(x-\theta) - \mathcal{I}(\underline{\pi}-\theta)\right), & \underline{\pi} \le x \le y. \end{cases}$$

As usual,

$$g^{(2,0)}(x) = -2\int_{-\infty}^{u} \mathfrak{G}(x,y)\, g^{(1)}(y, \underline{\pi}, \overline{\pi})\, dy. \tag{11.38}$$

The explicit expression for the expected duration given by Eq. (11.36) is interesting in its own right and also can be used for benchmarking solutions obtained via the method of heat potentials.

11.4.2 *Renewal theory approach*

To facilitate the comparison with previously know results, from now on, we assume that $\theta = 0$.

In this subsection, we revisit Bertram's approach (Bertram, 2009, 2010). In a nutshell, Bertram assumes that the underlying OU process, representing portfolio's log-price, is running in perpetuity. He envisions the following investment strategy. When the return process x hits the lower level l, the underlying is bought. When the process x hits the upper level u, the underlying is sold. Thus, the round trip is characterized by two transitions, $x = l \to x = u$, and $x = u \to x = l$; once the round trip is completed, the process starts again.

We can use the same ideas as in Section 11.4.1 to calculate $\mathbb{E}(T)$, $\mathbb{E}(T^2)$, and $\mathbb{V}(T)$ for the hitting time of a given level u, starting at the level $x = l$ by letting $\underline{\pi} \to -\infty, \overline{\pi} = u$

$$\begin{aligned}
\mathbb{E}(T) &= 2(\mathcal{G}(u) - \mathcal{G}(l)), \\
\mathbb{E}(T^2) &= 8(\mathcal{G}(u)(\mathcal{G}(u) - \mathcal{G}(l)) - (\mathcal{J}(u) - \mathcal{J}(l))), \\
\mathbb{V}(T) &= 4\left((\mathcal{G}^2(u) - 2\mathcal{J}(u)) - (\mathcal{G}^2(l) - 2\mathcal{J}(l))\right),
\end{aligned}$$

where

$$\begin{aligned}
\mathcal{J}(x) &= \int_{-\infty}^{x} e^{-y^2}(\mathcal{I}(x) - \mathcal{I}(y))\mathcal{G}(y)\,dy, \\
\mathcal{I}(x) &= \int_{-\infty}^{x} e^{-y^2}\mathcal{G}(y)\,dy - \int_{-\infty}^{x} \mathcal{D}(y)\mathcal{G}(y)\,dy.
\end{aligned}$$

Similarly to Eq. (11.34), we can write

$$\mathcal{J}(x) = \frac{1}{16}\sum_{n=1}^{\infty} \frac{\Gamma\left(\frac{n}{2}\right)\left(\psi\left(\frac{n}{2}\right) + \gamma\right)}{\Gamma(n+1)}(2x)^n,$$

where ψ is the digamma function and γ is the Euler-Mascheroni constant, $\gamma = -\psi(1)$, see also Ricciardi and Sato (1988), where this formula is obtained via the Laplace transform. In summary,

$$\begin{aligned}
\varepsilon(l \to u) &\equiv \mathbb{E}(T) = 2(\mathcal{G}(u) - \mathcal{G}(l)), \\
\vartheta(l \to u) &\equiv \mathbb{V}(T) = 4\left((\mathcal{G}^2(u) - 2\mathcal{J}(u)) - (\mathcal{G}^2(l) - 2\mathcal{J}(l))\right).
\end{aligned}$$

By symmetry,

$$\begin{aligned}
\varepsilon(u \to l) &= \varepsilon(-u \to -l) = 2(\mathcal{G}(-l) - \mathcal{G}(-u)), \\
\vartheta(u \to l) &= \vartheta(-u \to -l) \\
&= 4\left((\mathcal{G}^2(-l) - 2\mathcal{J}(-l)) - (\mathcal{G}^2(-u) - 2\mathcal{J}(-u))\right).
\end{aligned}$$

Finally

$$\begin{aligned}
&\varepsilon(l \to u \to l) \equiv \varepsilon(l \to u) + \varepsilon(u \to l) \\
&\quad = 2(\mathcal{G}(u) - \mathcal{G}(-u) - (\mathcal{G}(l) - \mathcal{G}(-l))) = 2\sqrt{\pi}(\mathcal{I}(u) - \mathcal{I}(l)), \\
&\vartheta(l \to u \to l) \equiv \vartheta(l \to u) + \vartheta(u \to l) = 4\Big[\Big(\mathcal{G}^2(u) - \mathcal{G}^2(-u) \\
&\quad - 2(\mathcal{J}(u) - \mathcal{J}(-u))\Big) - \left(\mathcal{G}^2(l) - \mathcal{G}^2(-l) - 2(\mathcal{J}(l) - \mathcal{J}(-l))\right)\Big] \\
&\quad = 16\left(\left(\mathcal{G}^{(e)}(u)\mathcal{G}^{(o)}(u) - \mathcal{J}^{(o)}(u)\right) - \left(\mathcal{G}^{(e)}(l)\mathcal{G}^{(o)}(l) - \mathcal{J}^{(o)}(l)\right)\right)
\end{aligned}$$

since $\mathcal{G}$, $\mathcal{J}$ are decomposed into the even and odd parts as follows

$$\mathcal{G}(x) = \mathcal{G}^{(e)}(x) + \mathcal{G}^{(o)}(x),$$

$$\mathcal{G}^{(e)}(x) = \sqrt{\pi}\left(N\left(\sqrt{2}x\right) - \frac{1}{2}\right)\mathcal{I}(x) - \mathcal{E}(x),$$

$$\mathcal{G}^{(o)}(x) = \frac{\sqrt{\pi}}{2}\mathcal{I}(x), \mathcal{J}(x) = \mathcal{J}^{(e)}(x) + \mathcal{J}^{(o)}(x),$$

$$\mathcal{J}^{(e)}(x) = \mathcal{I}(x)\int_0^x e^{-y^2}\mathcal{G}^{(e)}(y)\,dy - \int_{-\infty}^0 \mathcal{D}(y)\,\mathcal{G}(y)\,dy - \int_0^x \mathcal{D}(y)\,\mathcal{G}^{(e)}(y)\,dy,$$

$$\mathcal{J}^{(o)}(x) = \mathcal{I}(x)\left(\int_{-\infty}^0 e^{-y^2}\mathcal{G}(y)\,dy + \frac{\sqrt{\pi}}{2}\mathcal{E}(x)\right) - \frac{\sqrt{\pi}}{2}\int_0^x \mathcal{D}(y)\,\mathcal{I}(y)\,dy.$$

Once the requisite quantities are computed, Bertram invokes classical results from renewal theory (Bertram, 2009, 2010).[2] The classical result from renewal theory (see, e.g., Ross, 2010), gives the asymptotic properties of the random variable $M(t,l,u)$ representing the number of round trips on the time interval $[0,t]$

$$M(t,l,u) \sim \mathfrak{N}\left(\frac{t}{\varepsilon(l\to u\to l)}, \frac{\vartheta(l\to u\to l)\,t}{\varepsilon(l\to u\to l)^3}\right), \tag{11.39}$$

where $\mathfrak{N}$ is the normal variable. Given that x represents the log-price of the underlying portfolio, the return and asymptotic Sharpe ratio SR per unit of time are given by

$$r = \frac{(u-l-c)}{\varepsilon(l\to u\to l)}, \quad \mathrm{SR} = \sqrt{\frac{\varepsilon(l\to u\to l)}{\vartheta(l\to u\to l)}\frac{(u-l-c-r_f)}{(u-l-c)}}.$$

Here c, r_f represent transaction fees, and risk-free rate, respectively. Bertram maximizes one of these quantities over the stop-loss/take profit thresholds (l,u).

The main practical problem with this approach is that it assumes stationarity in perpetuity of the underlying process, which is a somewhat questionable assumption. The other problem is that, even for a stationary process, it takes a very long time for the strategy to reach its asymptotic state. The reason why these issues have not been discussed earlier, is that it is very hard to calculate the probability density function (pdf) for the processes $l\to u$, $u\to l$, and the round-trip process $l\to u\to l$. Recently, Lipton and Kaushansky (2020a) proposed a very efficient method for calculating

[2]We note in passing that Bertram uses informal notation, which is dimensionally incorrect, such as $\mathbb{V}(1/T) = \mathbb{V}(T)/\mathbb{E}(T)^3$, and next to impossible to understand, although his final results are correct.

the pdfs for the processes $l \to u$, $u \to l$; the round-trip process $l \to u \to l$ can be analyzed by convolution. We show the corresponding pdfs for a representative choice of l, u, namely $l = -1/\sqrt{2}$, $u = 1/\sqrt{2}$ in Fig. 11.7. This figure clearly demonstrates that a very long right tail characterizes the round-trip process $l \to u \to l$, so that the strategy might never reach its asymptotic limit in practice.

11.4.3 *Perpetual value function*

In this section, we discuss results obtained in Ekstrom *et al.* (2010, 2011); Larsson *et al.* (2013) and Lindberg (2014), and re-derive and improve their findings in a concise semi-analytic fashion.

The stationary problem for determining the value function and the *optimal* take-profit level u for a *given* stop-loss level l (which is determined by the investor's risk appetite) and the time value of money has the form

$$
\begin{gathered}
V_{xx}(x) - 2xV_x(x) - \lambda V = 0, \quad l \le x \le u, \\
V(l) = l, \quad V(u) = u, \quad V'(u) = 1.
\end{gathered} \tag{11.40}
$$

This problem is similar, but by no means identical, to the pricing problem for the perpetual American call option on a dividend-paying stock. Here λ is the non-dimensional discount rate, $\lambda = 2r/\kappa$.

The second-order ordinary differential equation Eq. (11.40) is the well-known Hermite differential equation. Its general solution has the form

$$
V(x) = a_0 M\left(\frac{\lambda}{4}, \frac{1}{2}, x^2\right) + a_1 x M\left(\frac{\lambda+2}{4}, \frac{3}{2}, x^2\right), \tag{11.41}
$$

where $M(a, b, z)$ is the Kummer function (a confluent hypergeometric function of the first kind) (Abramowitz and Stegun, 1964), and a_0, a_1 are arbitrary constants. Boundary conditions Eq. (11.40) yield

$$
\begin{aligned}
a_0 M\left(\frac{\lambda}{4}, \frac{1}{2}, l^2\right) + a_1 l M\left(\frac{\lambda+2}{4}, \frac{3}{2}, l^2\right) &= l, \\
a_0 M\left(\frac{\lambda}{4}, \frac{1}{2}, u^2\right) + a_1 u M\left(\frac{\lambda+2}{4}, \frac{3}{2}, u^2\right) &= u, \\
a_0 \lambda u M\left(\frac{\lambda+4}{4}, \frac{3}{2}, u^2\right) + a_1 \Big[M\left(\frac{\lambda+2}{4}, \frac{3}{2}, u^2\right) & \\
+ \frac{(\lambda+2)}{3} u^2 M\left(\frac{\lambda+6}{4}, \frac{5}{2}, u^2\right) \Big] &= 1.
\end{aligned}
$$

Here we use the identity (Abramowitz and Stegun, 1964)

$$
M_z(a, b, z) = \frac{a}{b} M(a+1, b+1, z),
$$

and eliminate a_0, a_1

$$a_0 = \frac{c_{11}l - c_{01}u}{c_{00}c_{11} - c_{01}c_{10}}, \quad a_1 = \frac{-c_{10}l + c_{00}u}{c_{00}c_{11} - c_{01}c_{10}}, \tag{11.42}$$
$$c_{00} = M\left(\frac{\lambda}{4}, \frac{1}{2}, l^2\right), \quad c_{01} = lM\left(\frac{\lambda+2}{4}, \frac{3}{2}, l^2\right),$$
$$c_{10} = M\left(\frac{\lambda}{4}, \frac{1}{2}, u^2\right), \quad c_{11} = uM\left(\frac{\lambda+2}{4}, \frac{3}{2}, u^2\right).$$

to obtain the following nonlinear algebraic equation for u

$$(c_{11}l - c_{01}u)\lambda uM\left(\frac{\lambda+4}{4}, \frac{3}{2}, u^2\right) + (-c_{10}l + c_{00}u)\left[M\left(\frac{\lambda+2}{4}, \frac{3}{2}, u^2\right)\right.$$
$$\left. + \frac{(\lambda+2)}{3}u^2 M\left(\frac{\lambda+6}{4}, \frac{5}{2}, u^2\right)\right] = (c_{00}c_{11} - c_{01}c_{10}). \tag{11.43}$$

We solve Eq. (11.43) via the Newton-Raphson method.

Once u is found, we use Eqs. (11.41) and (11.42), to construct the value functions $V(x)$. We show $V(x) - x$ and $u(l)$ for several representative values of λ in Fig. 11.8.

The stationary problem for determining the value function and the *optimal* take-profit level U for a *given* stop-loss level L (which is determined by the investor's risk appetite) and the opportunity cost c has the non-dimensional form

$$V_{xx}(x) - 2xV_x(x) = \lambda, \quad l \le x \le u, \tag{11.44}$$
$$V(l) = l, \quad V(u) = u, \quad V_x(u) = 1,$$

where λ is the non-dimensional opportunity cost, $\lambda = 2c/\kappa$. It is easy to show that the general solution of Eq. (11.44) has the form given by Eq. (11.32). As before we get the following set of equations

$$a_0 + a_1\mathcal{I}(l) = l - \lambda\mathcal{G}(l),$$
$$a_0 + a_1\mathcal{I}(u) = u - \lambda\mathcal{G}(u),$$
$$a_1 e^{u^2} = 1 - \lambda\mathcal{F}(u).$$

Accordingly,

$$a_0 = \frac{\mathcal{I}(u)(l - \lambda\mathcal{G}(l)) - \mathcal{I}(l)(u - \lambda\mathcal{G}(u))}{(\mathcal{I}(u) - \mathcal{I}(l))},$$
$$a_1 = \frac{u - \lambda\mathcal{G}(u) - (l - \lambda\mathcal{G}(l))}{(\mathcal{I}(u) - \mathcal{I}(l))},$$
$$\frac{e^{u^2}(u - \lambda\mathcal{G}(u) - (l - \lambda\mathcal{G}(l)))}{(\mathcal{I}(u) - \mathcal{I}(l))} = 1 - \lambda\mathcal{F}(u).$$

In Fig. 11.9a we show $V(x) - x$ for $l = -2.0$ for several representative values of λ, the corresponding optimal values of u are 1.07, 0.74, 0.50. In Fig. 11.9b the optimal boundary $u(l)$ is presented.

It is natural to ask what happens when the underlying mean-reverting process has a jump component, so that

$$dx = -xdt + dW_t + JdP_t,$$

where P_t is a Poisson process with intensity ν, and J is the jump magnitude, which is assumed to be a random variable with density function $\phi(J)$ (see Larsson *et al.*, 2013). Larsson *et al.* (2013) use the finite element method to solve the corresponding free boundary problem. However, if J has a double exponential distribution density function $\phi(J)$,

$$\phi(J) = \kappa e^{-\kappa|J|},$$

or, more generally, a hyper-exponential distribution (see, e.g., Lipton, 2002), the problem can be solved in a semi-analytical fashion.

To this end, it is convenient to write the problem with jumps in terms of $v(x) = V(x) - x$

$$\begin{aligned}
&v''(x) - 2xv'(x) + \omega(I_+(x) + I_-(x) - v(x)) = \lambda + (2+\omega)x,\\
&I_+(x) = \int_0^{x-l} v(x-z)\, e^{-\kappa z} dz = \int_l^x v(z)\, e^{-\kappa(x-z)} dz,\\
&I_-(x) = \int_0^{u-x} v(x+z)\, e^{-\kappa z} dz = \int_x^u v(z)\, e^{-\kappa(z-x)} dz,\\
&\quad v(l) = v(u) = v'(u) = I_+(l) = I_-(u) = 0,
\end{aligned}$$

where $\omega = \kappa\nu$. It can be written as an inhomogeneous system of linear ODEs

$$\begin{aligned}
&v'(x) - w(x) = 0,\\
&w'(x) - 2xw(x) + \omega(I_+(x) + I_-(x) - v(x)) = \lambda + (2+\omega)x,\\
&I_+'(x) - v(x) + \kappa I_+(x) = 0,\\
&I_-(x) + v(x) - \kappa I_-(x) = 0,\\
&\quad v(l) = 0, \quad w(l) = c, \quad I_+(l) = 0, \quad I_-(l) = d.
\end{aligned}$$

This system can be solved by the method of shooting by choosing initial values c, d and the right endpoint of the computational interval b to satisfy the remaining boundary conditions

$$v(u) = w(u) = I_-(u) = 0, \tag{11.45}$$

or in the matrix form

$$\begin{pmatrix} v \\ w \\ I_+ \\ I_- \end{pmatrix}' + \begin{pmatrix} 0 & -1 & 0 & 0 \\ -\omega & -2x & \omega & \omega \\ -1 & 0 & \kappa & 0 \\ 1 & 0 & 0 & -\kappa \end{pmatrix} \begin{pmatrix} v \\ w \\ I_+ \\ I_- \end{pmatrix} = \begin{pmatrix} 0 \\ \lambda + (2+\omega)x \\ 0 \\ 0 \end{pmatrix}$$

In Fig. 11.10a we show the solution vector $(v(x), w(x), I_+(x), I_-(x))$ corresponding to the suboptimal choice of u. The shooting parameters, c, d, are chosen in such a way, that two of the three boundary conditions Eq. (11.45) are satisfied, $v(u) = I_-(u) = 0$. In Fig. 11.10b we show what happens when the upper limit u is chosen optimally, by using the Newton-Raphson method. For $u = 1.18$ all three conditions Eq. (11.45) are met. In Fig. 11.10c we demonstrate the quality of our numerical method by putting $\omega = 0$ and comparing the corresponding numerical solution with the analytical solution given by Eq. (11.32). It can be seen that the agreement is excellent.

In Fig. 11.11a we show $v(x)$ for $l = -2.0$, the corresponding optimal values of u are 1.18, 0.80, 0.55; we show $u(l)$ for several representative values of λ in Fig. 11.11b.

11.4.4 *Linear transaction costs*

Several researchers, including Lataillade *et al.* (2012) concentrated on the critical question on how linear transaction costs affect the profitability of mean-reverting trading strategies. An alternative treatment is given by Martin and Schöneborn (2011), see also Do and Faff (2012). Denuded of all amenities, the approach of Lataillade *et al.* (2012) is almost identical to the method used by Hyer *et al.* (1997), for studying passport options.

Lataillade *et al.* (2012) reduce the problem to solving the following Fredholm integral equation of the second kind

$$g(x) - \int_{-q}^{q} K(x,y)\, g(y)\, dy = f(x), \tag{11.46}$$

where

$$K(x,y) = \frac{\Theta}{\sqrt{\pi(\Theta^2-1)}} \exp\left(-\frac{(\Theta y - x)^2}{(\Theta^2-1)}\right),$$

$$f(x) = x + \Gamma\left(N\left(-\frac{\sqrt{2}(\Theta q - x)}{\sqrt{(\Theta^2-1)}}\right) - N\left(-\frac{\sqrt{2}(\Theta q + x)}{\sqrt{(\Theta^2-1)}}\right)\right).$$

Here $\Theta = e^{\Delta}$, Γ represents transaction cost, while Δ shows how far forward the behavior of the process can be predicted. The trader should not change her position when $-q < x < q$, and go maximally long when $x = q$, and short when $x = -q$. Eq. (11.46) is augmented with the matching condition

$$g(q) = \Gamma. \tag{11.47}$$

While Lataillade *et al.* (2012) use the path integral method to understand the behavior of Eqs. (11.46) and (11.47), we prefer to attack the problem in question directly — by solving the corresponding Fredholm equation. As before, we solve Eq. (11.46) for a given q, and then adjust q by using the Newton-Raphson method until the matching condition Eq. (11.47) is met. We notice in passing that $K(x, y)$ is even, $K(-x, -y) = K(x, y)$, while $f(x)$ is odd, $f(-x) = -f(x)$, so that $g(x)$ is even, $g(-x) = -g(x)$.

Our analysis results in some unexpected findings. Namely, Eqs. (11.46) and (11.47) have multiple solutions. We choose $\Gamma = 0.1$, $\Delta = 1$ and solve the equations in question. It turns out that at least two critical values of q are possible, $q = 0.0561$ and $q = 1.0131$. We show the corresponding solutions in Figs. 11.12a and 11.12b. It can be shown that $g(x)$, which has a single root at $x = 0$ is the solution of interest. With this in mind, we can construct critical boundaries $q(\Delta)$ corresponding to several representative values of Γ. These boundaries are shown in Fig. 11.12c.

To summarize main findings made in Lipton and Lopez de Prado (2020a,b) and presented in this Chapter, first, a method for calibrating the corresponding OU process to market prices is developed . Second, an explicit expression for the SR given by Eq. (11.14) was derived and then maximized with respect to the stop loss/ take profit bounds $(\underline{\pi}, \overline{\pi})$. Third, for three representative values of θ, the SR on a grid of $(\underline{\pi}, \overline{\pi})$ and pre-chosen times values are calculated and graphically summarized in Figs. 11.3–11.5. Next, for each case, optimization is performed and presented $(\underline{\pi}^*, \overline{\pi}^*)$ in Table 11.1. In agreement with intuition, in the case of strong misprising, it is optimal to wait until the trade's expiration without imposing stop losses/take profit bounds. For weaker mispricing, it is not optimal to stop losses, but it might be optimal to take profits early. Still, to be on the safe side, we do recommend imposing stop losses chosen in accordance with one's risk appetite to avoid unpleasant surprises caused by the misspecification of the underlying process.

These rules help liquidity providers to decide how to offer liquidity to the market in the most profitable way, as well as by statistical arbitrage traders to optimally execute their trading strategies.

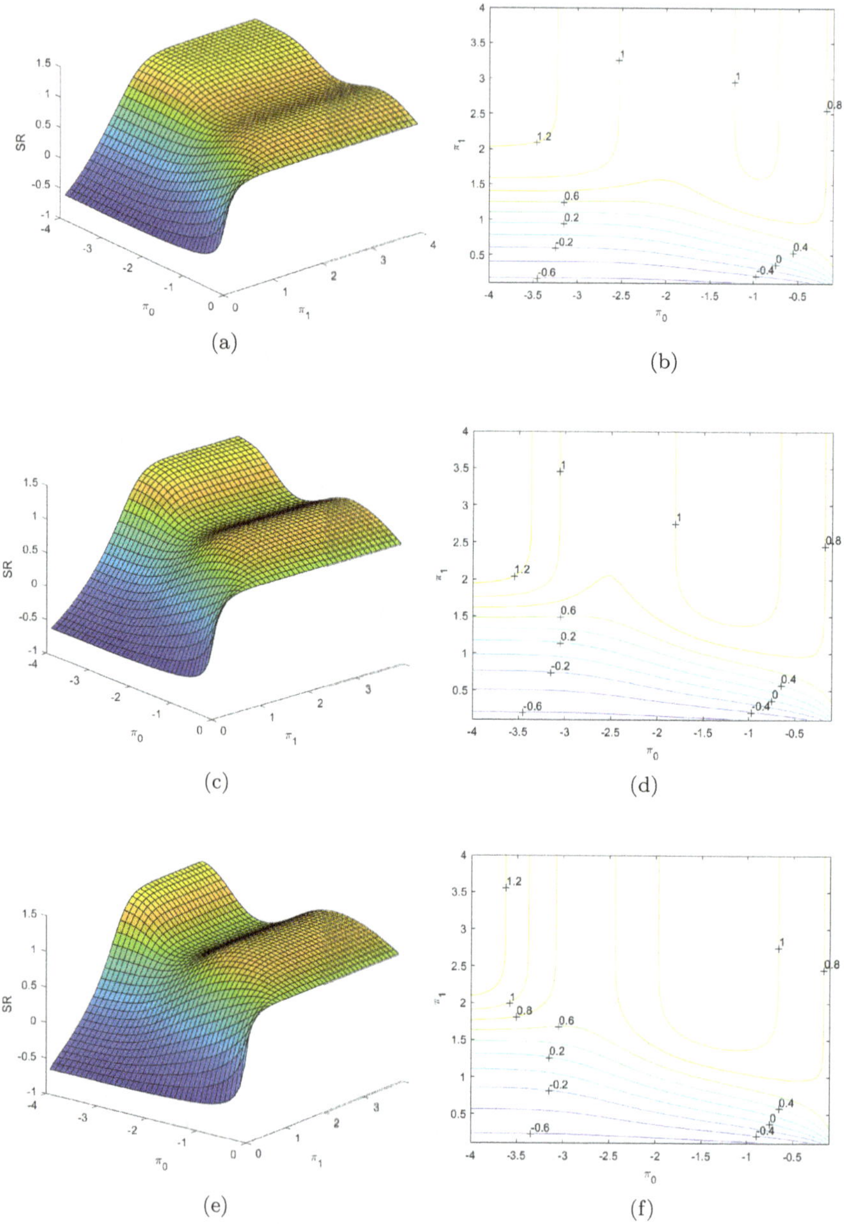

Figure 11.3: The Sharpe ratio as a function of $(\underline{\pi}, \overline{\pi})$ for $\theta = 1.0$; [a,b] - $T = 1.96$, [c,d] - $T = 4.26$, [e,f] - $T = 6.56$.

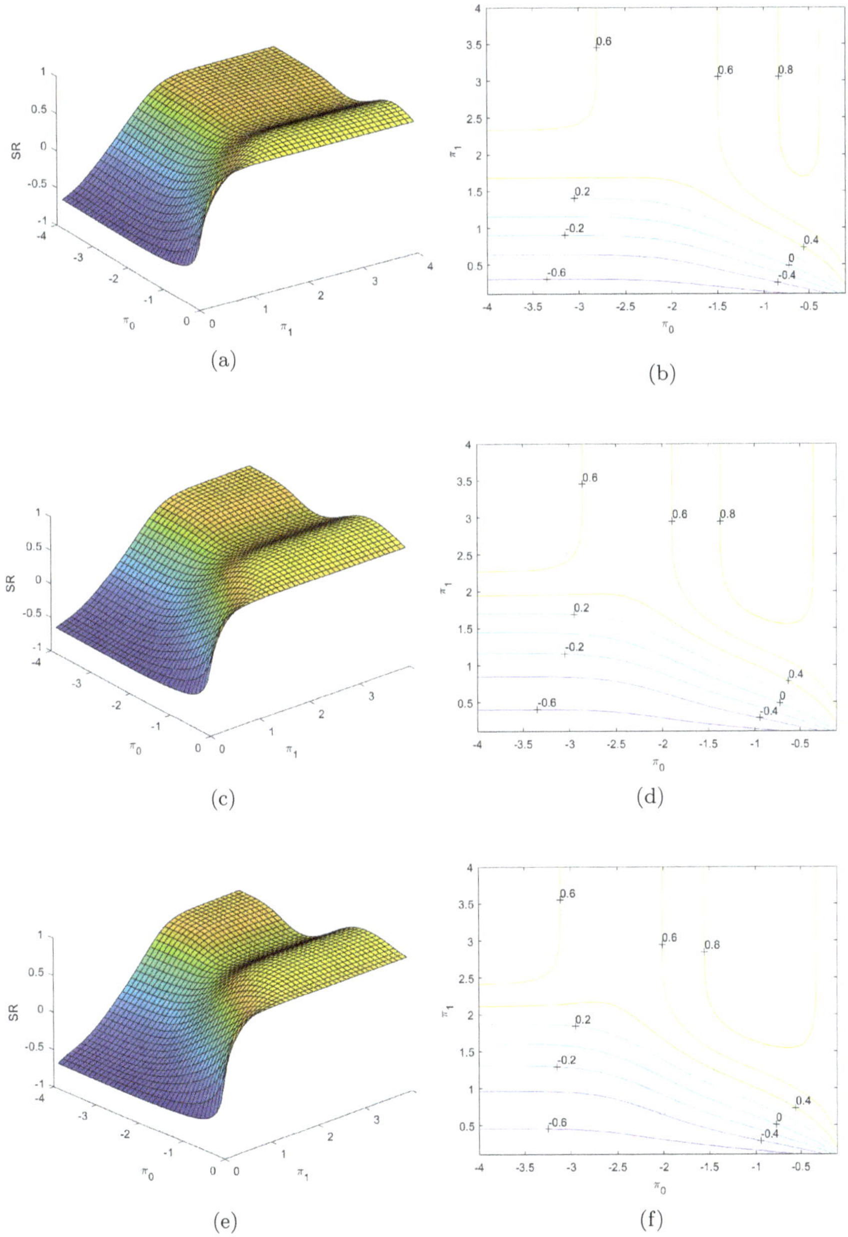

Figure 11.4: The Sharpe ratio as a function of $(\underline{\pi}, \overline{\pi})$ for $\theta = 0.5$; [a,b] - $T = 1.96$, [c,d] - $T = 4.26$, [e,f] - $T = 6.56$.

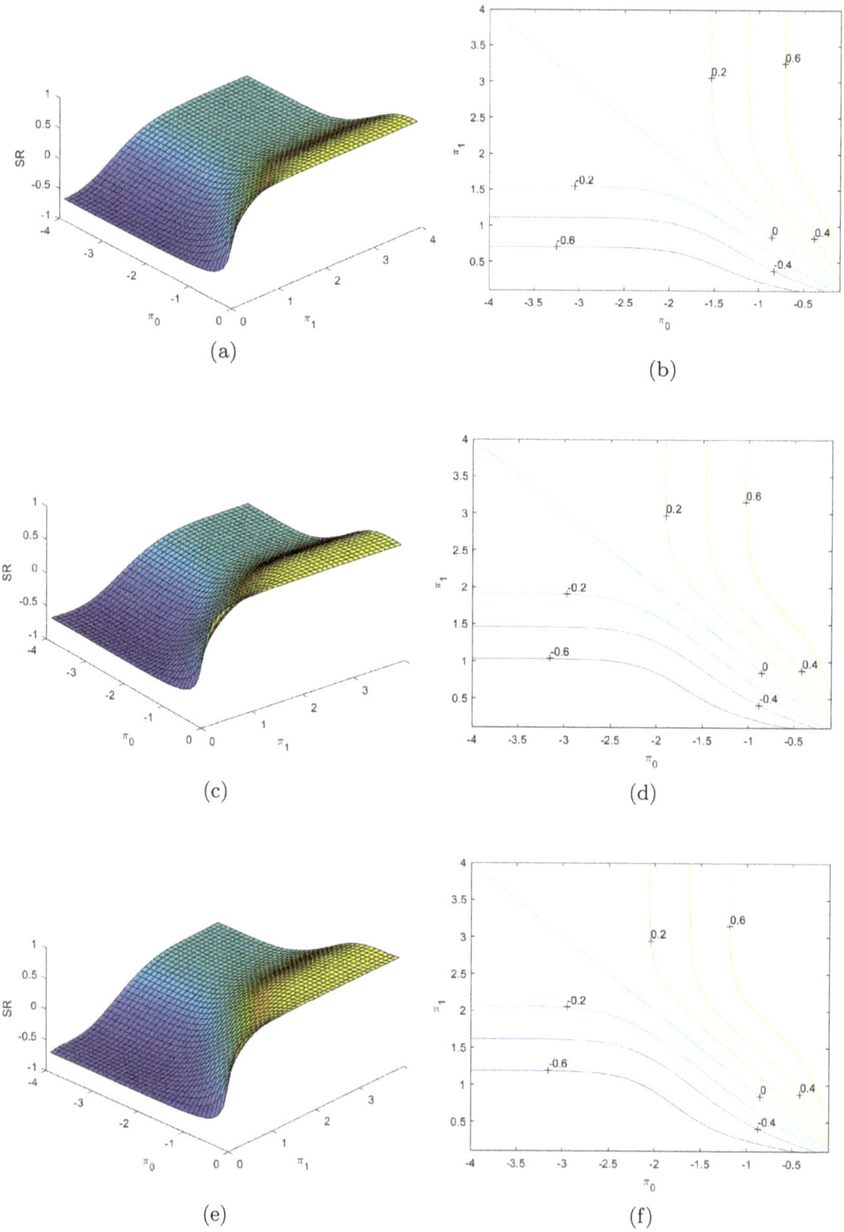

Figure 11.5: The Sharpe ratio as a function of $(\underline{\pi}, \overline{\pi})$ for $\theta = 0.0$; [a,b] - $T = 1.96$, [c,d] - $T = 4.26$, [e,f] - $T = 6.56$.

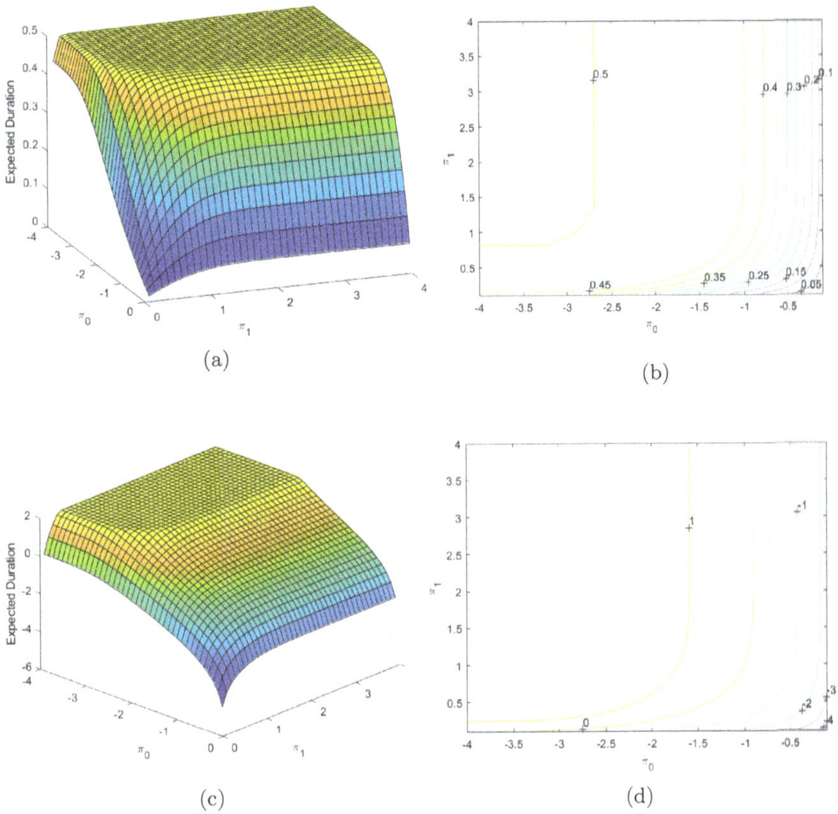

Figure 11.6: The dependence on π_0, π_1 at $\theta = 1.0$ of: (a)-(b) the expected duration $\Upsilon = 1 - \frac{1}{2}e^{-2G}$; (c)-(d) - $\log G$.

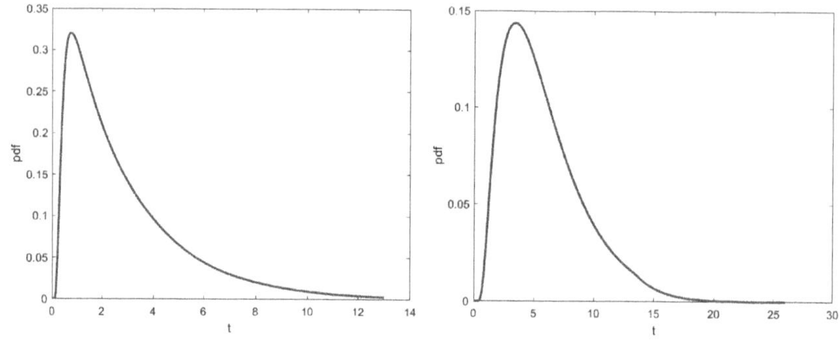

(a) (b)

Figure 11.7: (a) the pdf for the process $l \to u$; (b) the pdf for the round-trip process $l \to u \to l$. The corresponding values $l = -1/\sqrt{2}$, $u = 1/\sqrt{2}$. The choice is made based on the fact that one standard deviation of the stationary OU distribution is $1/\sqrt{2}$.

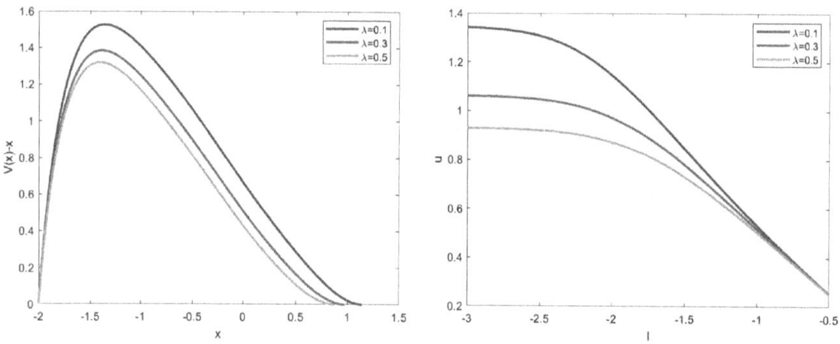

(a) (b)

Figure 11.8: (a) The dimensionless function $V(x)$ for several representative values of λ (the corresponding optimal values of u are 1.15, 0.97, 0.87); (b) the dimensionless optimal take-profit boundary u as a function of the dimensionless stop-loss boundary l.

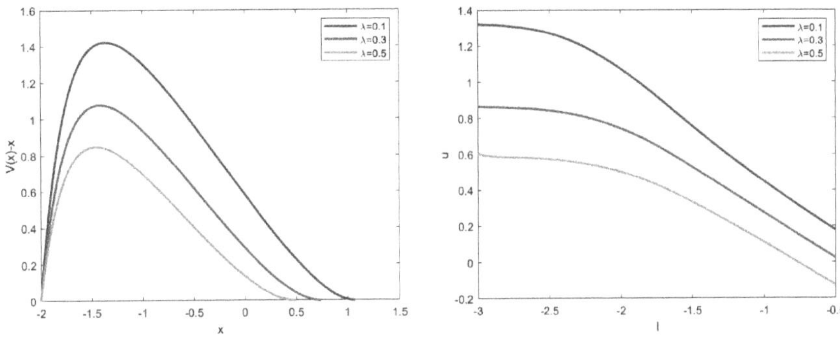

(a) (b)

Figure 11.9: (a) The dimensionless value function $V(x)$ for $l = -2.0$, and several representative values of λ (the corresponding optimal values of u are 1.07, 0.74, 0.50); (b) the dimensionless optimal execution boundary $u(l)$.

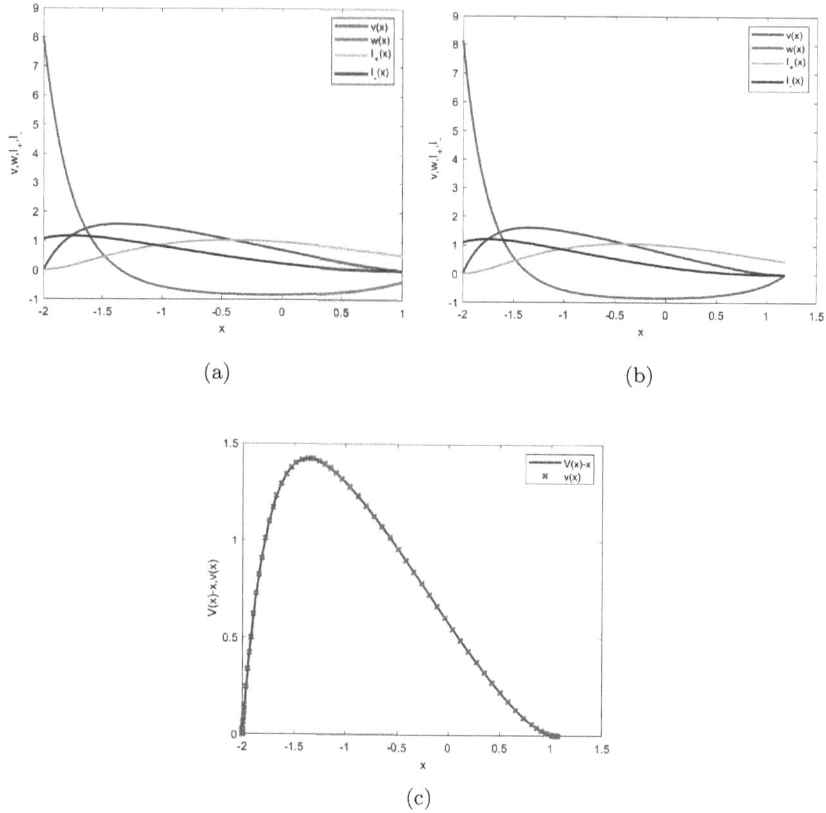

Figure 11.10: (a) The dimensionless function $v(x)$ for $l = -2.0, \lambda = 0.1, \omega = 0.1, \kappa = 1.0$. The value $u = 1$ is taken which is not optimal, and hence, the matching condition is not satisfied; (b) $v(x)$ for the optimal value of $u = 1.18$. Since u is optimal, the matching condition is met, so $w(u) = 0$; (c) - $v(x)$ for $l = -2.0, u = 1.18, \lambda = 0.1, \omega = 0.0, \kappa = 1.0$.

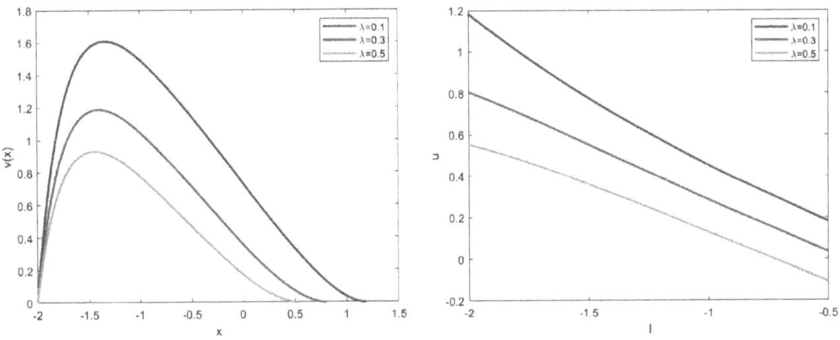

(a) (b)

Figure 11.11: (a) The dimensionless function $v(x)$ for $l = -2.0$, and several representative values of λ (the corresponding optimal values of u are 1.18, 0.81); (b) the dimensionless optimal execution boundary $u(l)$ at $\omega = 0.1, \kappa = 1.0$.

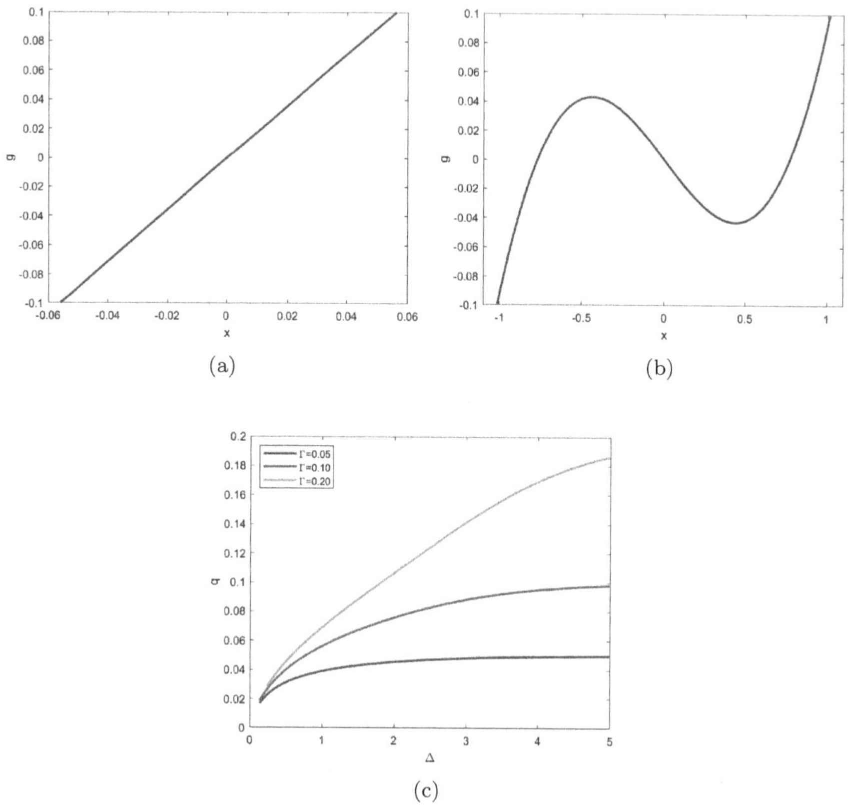

Figure 11.12: (a) $g(x)$ corresponding to the critical value $q = 0.0561$; (b) $g(x)$ for $q = 1.0131$. We can see that in the first case $g(x)$ has a single root at $x = 0$, while in the second case, there are three roots. (c) Critical boundaries $q(\Delta)$ for three representative values of $\Gamma = 0.05, 0.10, 0.20$. It is clear that for larger values of Γ it is beneficial to wait longer before changing one's position.

PART 5

Fixed Income

Chapter 12

Barrier Options in the Hull-White Model

The Hull-White model is a simple and popular one-factor model which is described in Chapter 2. Under the Hull-White model the prices of Zero-coupon bonds and European Vanilla options are known in closed form (Andersen and Piterbarg, 2010). However, for exotic options, e.g., barrier options, these prices are not known yet in closed form. Therefore, various numerical methods are used to obtain such prices, see survey in Kuan and Webber (2003). However, for exotic options, e.g., barrier options, these prices are not known yet in closed form. Therefore, various numerical methods are used to obtain such prices, see survey in Kuan and Webber (2003).

Here we present an analytical solution of this problem, and demonstrate that it significantly accelerates computation of the prices and, accordingly, calibration of the model. First, we solve the problem of pricing Barrier options in semi-closed form and provide the resulting expressions that were not known in the literature until the solution has been found in Itkin and Muravey (2020b). Second, we solve this problem by two methods. The first one is the method of heat potentials (HP). The other method is a method of generalized integral transform (GIT), also known in physics, and introduced to mathematical finance in Carr and Itkin (2021). However, this method solves the problem where the underlying is defined at the domain $S \in [0, y(t)]$ with S being the stock price, and $y(t)$ being the time-dependent barrier. Contrary, in this chapter we consider a complimentary domain $r \in [y(t), \infty)$. The constructed solution can be applied to a wide class of problems, both in physics and finance.

The Hull-White model was introduced in Hull and White (1990), and then named after the authors. The model assumes dynamics of the short interest rate r_t to follow the Ornstein-Uhlenbeck (OU) process with time-dependent coefficients

$$dr_t = \kappa(t)[\theta(t) - r_t]dt + \sigma(t)dW_t, \qquad r_{t=0} = r. \tag{12.1}$$

Here $t \geq 0$ is the time, $\kappa(t) > 0$ is the constant speed of mean-reversion, $\theta(t)$ is the mean-reversion level, $\sigma(t)$ is the volatility of the process, W_t is the standard Brownian motion under the risk-neutral measure. To address calibration to real market rate curves the model could be updated by using a deterministic shift $s(t)$, so $r(t) = s(t) + \bar{r}(t)$ where $\bar{(t)}$ solves Eq. (12.1). This can be easily done using our framework (see Itkin *et al.*, 2020) so here we don't concentrate on that.

In the commodities world this model is known as one-factor Schwartz's model (Schwartz, 1997) for the logarithm of the spot price S. This model is also popular for modeling prices of the plain vanilla and exotic options. In particular, in this Chapter we consider a Down-and-Out barrier option with the time-dependent lower barrier $L(t)$ where the underlying is a zero-coupon bond with maturity S and price $F(r, t, S)$.

We assume that the short interest rate evolves in time as in Eq. (12.1). It is known that $F(r, t, S)$ under a risk-neutral measure solves a linear parabolic partial differential equation (PDE) (Privault, 2012)

$$\frac{\partial F}{\partial t} + \frac{1}{2}\sigma^2(t)\frac{\partial^2 F}{\partial r^2} + \kappa(t)[\theta(t) - r]\frac{\partial F}{\partial r} = rF. \tag{12.2}$$

This equation should be solved subject to the terminal condition

$$F(r, S, S) = 1, \tag{12.3}$$

and the boundary conditions

$$F(0, t, S) = g(t), \qquad F(r, t, S)\Big|_{r\to\infty} = 0, \tag{12.4}$$

where $g(t)$ is some function of the time t. See, for instance (Zhang and Yang, 2017) and references therein.

To find $g(t)$, recall that according to Ekström and Tysk (2011) for single-factor models that predict nonnegative short rates, no second boundary condition is required for Eq. (12.2) if $r_t > 0$, $\forall t \geq 0$ in Eq. (12.1), i.e., the boundary $r_t = 0$ is not attainable. Otherwise, the PDE itself at this point serves as the boundary condition. In particular, as applied to our OU process, it reads

$$\left(\frac{\partial F}{\partial t} + \frac{1}{2}\sigma^2(t)\frac{\partial^2 F}{\partial r^2} + \kappa(t)\theta(t)\frac{\partial F}{\partial r}\right)\Bigg|_{r=0} = 0. \tag{12.5}$$

However, since Eq. (12.1) is the OU process, it allows zero or even negative interest rates, i.e. in this model $r \in \mathbb{R}$. Therefore, the PDE Eq. (12.2) must be solved subject to the second boundary condition either at $r \to -\infty$, or at some artificial left boundary $r = r_{\min} < 0$. It is not obvious, however, how to setup this boundary condition at $r_{\min}$.

On the other hand, since the Hull-White model belongs to the class of affine models (Andersen and Piterbarg, 2010), the solution of Eq. (12.2) can be represented in the form

$$F(r,t,S) = A(t,S)e^{B(t,S)r}. \tag{12.6}$$

Substituting this expression into Eq. (12.2) and separating the terms proportional to r, we obtain two equations to determine $A(t,S), B(t,S)$

$$\begin{aligned} \frac{\partial B(t,S)}{\partial t} &= B(t,S)\kappa(t) + 1, \\ 2\frac{\partial A(t,S)}{\partial t} &= -A(t,S)B(t,S)\left[2\theta(t)\kappa(t) + B(t,S)\sigma(t)^2\right]. \end{aligned} \tag{12.7}$$

To obey the terminal condition Eq. (12.3), the first PDE in Eq. (12.7) should be solved subject to the terminal condition $B(S,S) = 0$, and the second one - to $A(S,S) = 1$. The solution reads

$$\begin{aligned} B(t,S) &= e^{\int_0^t \kappa(x)\,dx} \int_S^t e^{-\int_0^x \kappa(q)\,dq}\,dx, \\ A(t,S) &= \exp\left[-\frac{1}{2}\int_S^t B(x,S)\left(2\theta(x)\kappa(x) + B(x,S)\sigma^2(x)\right)dx\right]. \end{aligned} \tag{12.8}$$

It can be seen that $B(t,S) < 0$ if $t < S$. Therefore, $F(r,t,S) \to 0$ when $r \to \infty$.

12.1 Down-and-Out barrier option

Let us consider a Down-and-Out barrier Call option written on a ZCB as an underlying.[1] It is known that the price of the option $C(r,t)$ written on this bond under a risk-neutral measure solves the PDE (Andersen and Piterbarg, 2010)

$$\frac{\partial C}{\partial t} + \frac{1}{2}\sigma^2(t)\frac{\partial^2 C}{\partial r^2} + \kappa(t)[\theta(t) - r]\frac{\partial C}{\partial r} = rC. \tag{12.9}$$

[1]Other types of underlying can also be handled within the proposed framework, e.g., the LIBOR rate subject to the barrier conditions, etc., by the corresponding modification of Eq. (12.6). Also, given the prices of barrier options on a ZCB, one can evaluate a continuous barrier caplet (flooret).

This PDE should be solved subject to the terminal condition at the option maturity $T \leq S$, and some boundary conditions provided to guarantee a uniqueness of the solution. The terminal condition reads

$$C(r,T) = (F(r,T,S) - K)^+, \tag{12.10}$$

where K is the option strike.

By the contract definition, the lower barrier $L_F(t)$ is set on the Zero coupon bond (ZCB) price. In other words, at the barrier we have the following condition

$$C(r,t) = 0 \quad \text{if } F(r,t,S) = L_F(t). \tag{12.11}$$

Since the ZCB price $F(r,t,S)$ is known in closed form in Eq. (12.6), the above condition can be reformulated in the r domain by solving the equation

$$F(r,t,S) = A(t,S)e^{B(t,S)r} = L_F(t), \tag{12.12}$$

with respect to r. This yields the following equivalent barrier $L(t)$ in the r domain

$$L(t) = \frac{1}{B(t,S)} \log\left(\frac{L_F}{A(t,S)}\right) > 0, \tag{12.13}$$

where it is assumed that $L_F(t) < A(t,S)$. Thus, the boundary condition to Eq. (12.9) now reads

$$C(L(t),t) = 0. \tag{12.14}$$

At the second boundary as $r \to \infty$ the ZCB price tends to zero, see Eq. (12.6), and, therefore, the Call option price tends to zero. Thus, we set

$$C(r,t)\Big|_{r\to\infty} = 0. \tag{12.15}$$

Our goal now is to build a series of transformations to transform Eq. (12.9) to the heat equation.

12.1.1 *Transformation to the heat equation*

To transform the PDE Eq. (12.2) to the heat equation we first rewrite Eq. (12.9) in the form

$$\frac{\partial C}{\partial t} = -\frac{1}{2}\sigma^2(t)\frac{\partial^2 C}{\partial r^2} - [-\kappa(t)r + \kappa(t)\theta(t)]\frac{\partial C}{\partial r} + rC. \tag{12.16}$$

This equation belongs to the type of equations considered in Polyanin (2002), Section 3.8.7. It is shown there that by transformation

$$C(r,t) = \exp[\alpha(t)r + \beta(t)]u(x,\tau), \quad \tau = \phi(t), \quad x = r\psi(t) + \xi(t), \tag{12.17}$$

Eq. (12.16) can be reduced to the heat equation

$$\frac{\partial u}{\partial \tau} = \frac{\partial^2 u}{\partial x^2}. \tag{12.18}$$

Here,

$$\begin{aligned}
\psi(t) &= C_1 \exp\left(\int_0^t \kappa(q)dq\right), \qquad (12.19)\\
\phi(t) &= \frac{1}{2}\int_t^T \sigma^2(q)\psi^2(q)dq + C_2,\\
\alpha(t) &= \psi(t)\int_0^t \frac{1}{\psi(q)}dq + C_3\psi(t),\\
\beta(t) &= -\frac{1}{2}\int_0^t \alpha(q)\left[2\kappa(q)\theta(q) + \sigma^2(q)\alpha(q)\right]dq + C_4,\\
\xi(t) &= -\int_0^t \left[\kappa(q)\theta(q) + \sigma^2(q)\alpha(q)\right]\psi(q)dq + C_5,
\end{aligned}$$

where $C_1, \ldots, C_5$ are some constants. In our case we can choose $C_1 = 1$, $C_2 = C_3 = C_4 = C_5 = 0$.

The Eq. (12.18) should be solved subject to the initial condition

$$\begin{aligned}
u(x,0) &= \exp\left[-\frac{\alpha(T)}{\psi(T)}(x - \xi(T)) - \beta(T)\right]\left(\bar{F}(x,T,S) - K\right)^+, \qquad (12.20)\\
\bar{F}(x,T,S) &= A(T,S)\exp\left[\frac{B(T,S)}{\psi(T)}(x - \xi(T))\right],
\end{aligned}$$

as it follows from Eqs. (12.10), (12.6) and (12.17).

The boundary conditions should be set at the new domain $x \in \Omega$: $[y(\tau), \infty)$, where $y(\tau) = L(t(\tau))\psi(t(\tau)) + \xi(t(\tau))$, and $t(\tau)$ is the inverse map $t \to \tau$. The latter can be found explicitly by solving the second equation in Eq. (12.19)

$$\tau = \frac{1}{2}\int_t^T \sigma^2(q)\psi^2(q)dq, \tag{12.21}$$

with respect to t. Accordingly, the conditions at the boundaries of Ω can be obtained from Eqs. (12.15) and (12.14) and read

$$u(x,\tau)\Big|_{x\to\infty} = 0, \qquad u(y(\tau),\tau) = 0. \tag{12.22}$$

12.1.2 *Solution of the barrier pricing problem*

As applied to equities, the problem of pricing barrier options, where the underlying follows a time-dependent OU process, was considered in Carr and Itkin (2021). There the authors utilized and extended a method of generalized integral transform actively elaborated on by the Russian mathematical school to solve parabolic equations at the domain with moving boundaries (see e.g., Kartashov, 1999) and references therein. However, in Carr and Itkin (2021) and Kartashov (1999) those problems were solved at the domain $x \in [0, y(\tau)]$ while here we have to deal with the infinite domain Ω. For that kind of domains, the above method is not elaborated yet since there is a problem with constructing the inverse transform. Here we have managed to construct this solution (to the best of our knowledge, for the first time) and present it in Section 12.1.3). Some examples where the inverse transform is not necessary can be found in Kartashov (1999, 2001).

But first we will attack this problem by using the method of heat potentials. This method is known in the theory of heat equation for a long time (see, e.g., Tikhonov and Samarskii, 1963; Friedman, 1964.; Kartashov, 2001) and references therein. The first use of this method in mathematical finance is due to Lipton (2002) for pricing path-dependent options with curvilinear barriers, and more recently in Lipton and Kaushansky (2020a), Lipton and Lopez de Prado (2020a,b) (also see references therein).

Recall, that we must solve the heat equation in Eq. (12.18) with the initial condition in Eq. (12.20) and the boundary conditions in Eq. (12.22). Since this problem has an inhomogeneous initial condition, the method of heat potentials cannot be directly applied. Therefore, let us represent $u(x,\tau)$ in the form

$$u(x,\tau) = q(x,\tau) + \frac{1}{2\sqrt{\pi\tau}} \int_{y(0)}^{\infty} u(x',0) e^{-\frac{(x-x')^2}{4\tau}} dx'. \tag{12.23}$$

Here the second term in the RHS is the solution of the heat equation in Eq. (12.18) at the infinite domain with the initial condition in Eq. (12.10). But since $x \in \Omega$, we moved the left boundary from $-\infty$ to $y(0)$.

The function $q(x,\tau)$ solves the problem

$$\begin{aligned} \frac{\partial q(x,\tau)}{\partial \tau} &= \frac{\partial^2 q(x,\tau)}{\partial x^2}, \\ q(x,0) &= 0, \qquad y(0) < x < \infty, \\ q(x,\tau)\Big|_{x\to\infty} &= 0, \qquad q(y(\tau),\tau) = \phi(\tau), \end{aligned} \tag{12.24}$$

$$\phi(\tau) = -\frac{1}{2\sqrt{\pi\tau}} \int_{y(0)}^{\infty} u(x', 0) e^{-\frac{(y(\tau)-x')^2}{4\tau}} dx'.$$

This problem is similar to that in Eqs. (12.18), (12.10) and (12.22), but now with a homogeneous initial condition. Therefore, the solution can be found by the method of heat potentials.

Following the general idea of this method, we are looking for the solution of Eq. (12.24) in the form of a generalized heat potential

$$q(x,\tau) = \frac{1}{4\sqrt{\pi}} \int_0^{\tau} \Psi(k) \frac{x - y(k)}{\sqrt{(\tau - k)^3}} e^{-\frac{(x-y(k))^2}{4(\tau-k)}} dk, \tag{12.25}$$

where $\Psi(k)$ is the heat potential density. It is easy to check that thus defined function $q(x,\tau)$ solves the first line of Eq. (12.24), and satisfies both the initial condition and the vanishing condition at $x \to \infty$. Also, from Eq. (12.25) at the barrier $x = y(\tau)$ we must have (Tikhonov and Samarskii, 1963)

$$2\phi(\tau) = \Psi(\tau) + \frac{1}{2\sqrt{\pi}} \int_0^{\tau} \Psi(k) \frac{y(\tau) - y(k)}{\sqrt{(\tau - k)^3}} e^{-\frac{(y(\tau)-y(k))^2}{4(\tau-k)}} dk, \tag{12.26}$$

since for $x = y(\tau)$ function $q(x,\tau)$ is discontinuous, and its limiting value at $x = y(\tau) + 0$ is equal to $\phi(\tau)$.

The Eq. (12.26) is a linear Volterra equations of the second kind (Polyanin and Manzhirov, 2008). Since $\phi(\tau)$ is a continuously differentiable function, Eq. (12.26) has a unique continuous solution for $\Psi(\tau)$. First, it can be efficiently solved by a standard approach, i.e., the integral in the RHS is approximated using some quadrature rule with N nodes in k, and the solution is obtained at M nodes in τ. This discretization gives rise to a linear equation with respect to the discrete vector of values of $\Psi(\tau)$ with the matrix being lower triangular. Therefore, the solution can be obtained by a simple elimination procedure with quadratic complexity. Alternatively, since the kernel is proportional to Gaussian, the discrete sum approximating the integral can be computed with linear complexity $O(N + M)$ using the Fast Gauss Transform (see e.g., Spivak *et al.*, 2010). Then the final solution can be obtained as this is discussed in Section 12.2.

Second, if in Eq. (12.21) $\tau(0)$ is small, we can approximate a curvilinear boundary $y(\tau)$ by a linear function

$$y(\tau) = a + b\tau, \qquad a = y(0),\ b = \frac{y(\tau(0)) - y(0)}{y(\tau(0))}. \tag{12.27}$$

Then the integral kernel becomes a function of the difference $\tau - k$, and so the integrand is a convolutional function. Thus, Eq. (12.26) can be solved by using the Laplace or G-transforms. For instance, with allowance for Eq. (12.27) let us re-write Eq. (12.26) in the form (Kartashov, 2001)

$$\phi_1(\tau) = \Psi_1(\tau) + \frac{b}{2\sqrt{\pi}} \int_0^\tau \frac{\Psi_1(k)}{\sqrt{\tau - k}} dk, \tag{12.28}$$

$$\phi_1(\tau) = 2\phi(\tau) e^{b^2\tau/4}, \quad \Psi_1(k) = \Psi(k) e^{b^2 k/4}.$$

This is the Abel integral equation of the second kind (Polyanin and Manzhirov, 2008), which can be solved in closed form by using the Laplace transform. Since $\phi(0) = 0$, the solution reads

$$\Psi_1(\tau) = \mathcal{F}(\tau) + \frac{b^2}{4} \int_0^\tau e^{\frac{\tau - k}{4}} \mathcal{F}(k) dk, \qquad \mathcal{F} = \phi_1(\tau) - \frac{b}{2\sqrt{\pi}} \int_0^\tau \frac{\phi_1(k)}{\sqrt{\tau - k}} dk.$$

In case where the linear approximation is too crude, this expression can be used as a smart initial guess for the function $\Psi(\tau)$ which is needed by the iterative numerical method described in above.

Once Eq. (12.26) is solved and the function $\Psi(\tau)$ is found, the final solution reads

$$\begin{aligned} u(x, \tau) &= \frac{1}{4\sqrt{\pi}} \int_0^\tau \Psi(k) \frac{x - y(k)}{\sqrt{(\tau - k)^3}} e^{-\frac{(x - y(k))^2}{4(\tau - k)}} dk \\ &+ \frac{1}{2\sqrt{\pi\tau}} \int_{y(0)}^\infty u(x', 0) e^{-\frac{(x - x')^2}{4\tau}} dx'. \end{aligned} \tag{12.29}$$

The second integral can be further simplified with allowance for the definition of $u(x, 0)$ in Eq. (12.10). This yields

$$\begin{aligned} &\frac{1}{2\sqrt{\pi\tau}} \int_{y(0)}^\infty u(x', 0) e^{-\frac{(x - x')^2}{4\tau}} dx' = \frac{1}{2} e^{-\beta(T)} \\ &\quad \left[e^{A_2 x + A_1} A(T, S) \left(\mathrm{Erf}\left(\frac{x - y(0) + 2\tau A_2}{2\sqrt{\tau}} \right) - \mathrm{Erf}\left(\frac{x - K_1 + 2\tau A_2}{2\sqrt{\tau}} \right) \right) \right. \\ &\quad \left. - K e^{B_2 x + B_1} \left(\mathrm{Erf}\left(\frac{x - y(0) + 2\tau B_2}{2\sqrt{\tau}} \right) - \mathrm{Erf}\left(\frac{x - K_1 + 2\tau B_2}{2\sqrt{\tau}} \right) \right) \right], \\ &A_1 = \frac{A_2}{\psi(T)} [\tau(B(T, S) - \alpha(T)) - \xi(T)\psi(T)], \qquad A_2 = \frac{B(T, S) - \alpha(T)}{\psi(T)}, \\ &B_1 = \frac{B_2}{\psi(T)} [-\tau\alpha(T) - \xi(T)\psi(T)], \qquad B_2 = -\frac{\alpha(T)}{\psi(T)}, \\ &K_1 = \max\left\{ \xi(T) + \frac{\psi(T)}{B(T, S)} \log\left(\frac{K}{A(T, S)} \right), y(0) \right\}. \end{aligned} \tag{12.30}$$

Also, by definition in Eq. (12.24), the function $\phi(\tau)$ can be represented in the same form just by substituting $x = y(\tau)$ and multiplying the result by -1.

12.1.3 *Second solution of the barrier pricing problem*

In this section we solve the same problem but using the method of generalized integral transform. This method was invented by the Russian mathematical school in the 20th century starting from A.V. Luikov, and then by B.Ya. Lyubov, E.M. Kartashov, and some others, see a detailed survey in Kartashov (1999). However, as mentioned in Kartashov (2001), the solution for a semi-infinite domain is not known yet, while some recommendations were given on how one can try to proceed. Therefore, to the best of authors' knowledge, the solution presented in this Section is new and compliments the method of heat potentials presented in Section 12.1.2.

We attack this problem by introducing the following integral transform

$$\bar{u}(p,\tau) = \int_{y(\tau)}^{\infty} u(x,\tau)e^{-\sqrt{p}x}dx, \tag{12.31}$$

where $p = a + i\omega$ is a complex number with $\mathrm{Re}(p) = \beta > 0$, and $-\pi/4 < \arg(\sqrt{p}) < \pi/4$. Let us multiply both parts of Eq. (12.18) by $e^{-x\sqrt{p}}$ and then integrate on x from $y(\tau)$ to ∞:

$$\begin{aligned} &\frac{\partial}{\partial\tau}\int_{y(\tau)}^{\infty} u(x,\tau)e^{-\sqrt{p}x}dx + u(y(\tau),\tau)e^{-\sqrt{p}y(\tau)}y'(\tau) = \\ &= \frac{\partial u}{\partial x}e^{-\sqrt{p}x}\Big|_{x=y(\tau)}^{x\to+\infty} + \sqrt{p}u(x,\tau)e^{-\sqrt{p}x}\Big|_{x=y(\tau)}^{x\to+\infty} + p\int_{y(\tau)}^{\infty} u(x,\tau)e^{-\sqrt{p}x}dx. \end{aligned} \tag{12.32}$$

By taking into account the boundary conditions in Eq. (12.22), Eq. (12.32) can be represented as the following Cauchy problem

$$\begin{aligned} &\frac{d\bar{u}}{d\tau} - p\bar{u} = \Upsilon(\tau)e^{-\sqrt{p}y(\tau)}, \\ &\bar{u}(p,0) = \int_{y(0)}^{\infty} u(x,0)e^{-\sqrt{p}x}dx, \quad \Upsilon(\tau) = -\frac{\partial u(x,\tau)}{\partial x}\Big|_{x=y(\tau)}, \end{aligned} \tag{12.33}$$

where $u(x,0)$ us given in Eq. (12.20). The solution of this problem reads

$$\bar{u}(p,\tau) = e^{p\tau}\int_0^{\tau} \Upsilon(s)e^{-ps}e^{-\sqrt{p}y(s)}ds + e^{p\tau}\int_{y(0)}^{\infty} u(z,0)e^{-\sqrt{p}z}dz. \tag{12.34}$$

By analogy with (Carr and Itkin, 2021), the function $\Upsilon(\tau)$ solves the Fredholm equation of the first type

$$\int_0^{\infty} \Upsilon(\tau)e^{-p\tau-\sqrt{p}\tau}d\tau = F(p), \tag{12.35}$$

with

$$F(p) = -e^{\frac{\alpha(T)\xi(T)}{\psi(T)}-\beta(T)}\left[A(T,S)e^{\frac{B(T,S)\xi(T)}{\psi(T)}}\frac{e^{-f_1(p,T)y_0} - e^{-f_1(p,T)K_1}}{f_1(p,T)}\right. \tag{12.36}$$

$$
\left. - K\frac{e^{-f_2(p,T)y_0} - e^{-f_2(p,T)K_1}}{f_2(p,T)} \right],
$$

$$
f_1(p,T) = \sqrt{p} + \frac{\alpha(T)}{\psi(T)} - \frac{\beta(T)}{\psi(T)}, \quad f_2(p,T) = \sqrt{p} + \frac{\alpha(T)}{\psi(T)},
$$

where K_1 is defined in Eq. (12.30).

To construct the inverse transform, recall that the solution of the heat equation $\mathcal{L}u(x,\tau) = 0$, $\mathcal{L} = \partial/\partial\tau - \partial^2/\partial x^2$ in the half-plane domain $x \in (0,\infty)$ can be expressed via the Fourier sine integral (Cannon and Browder, 1984)

$$
\int_0^\infty \alpha(\xi) e^{-\xi^2\tau} \sin(\xi x) d\xi.
$$

Therefore, by analogy we look for the inverse transform of $\bar{u}$ (or for the solution of Eq. (12.33) in terms of u) to be an oscillatory integral of the form

$$
u(x,\tau) = \int_0^\infty \alpha(\xi,\tau) \sin[\xi(x - y(\tau))] d\xi, \tag{12.37}
$$

where $\alpha(\xi,\tau)$ is some function to be determined. Note, that this definition automatically respects the vanishing boundary conditions for $u(x,\tau)$. We assume that this integral converges absolutely and uniformly $\forall x \in [y(\tau),\infty)$ for any $\tau > 0$.

Applying Eq. (12.31) to both parts of Eq. (12.37) and integrating yields

$$
\begin{aligned}
\bar{u}(p,\tau) &= \int_{y(\tau)}^\infty e^{-\sqrt{p}x} \int_0^\infty \alpha(\xi,\tau) \sin(\xi(x - y(\tau))) d\xi dx \qquad (12.38) \\
&= e^{-\sqrt{p}y(\tau)} \int_0^\infty \alpha(\xi,\tau) \frac{\xi d\xi}{\xi^2 + p}.
\end{aligned}
$$

Replacing $\bar{u}(p,\tau)$ with the solution found in Eq. (12.34), we obtain

$$
\begin{aligned}
\int_0^\infty \alpha(\xi,\tau) \frac{\xi d\xi}{\xi^2 + p} &= \int_0^\tau \Upsilon(s) e^{p(\tau - s)} e^{\sqrt{p}(y(\tau) - y(s))} ds \qquad (12.39) \\
&+ e^{p\tau} \int_{y(0)}^\infty u(z,0) e^{\sqrt{p}(y(\tau) - z)} dz.
\end{aligned}
$$

Now, similar to inverse operator methods, like the inverse Laplace transform, we need an analytic continuation of the transform parameter p into the complex plane. Let us integrate both sides of Eq. (12.39) on p along the so-called keyhole contour presented in Fig. 12.1. In more detail, this contour can be described as follows. It starts with a big symmetric arc Γ around the origin with the radius R; extending to two horizontal line

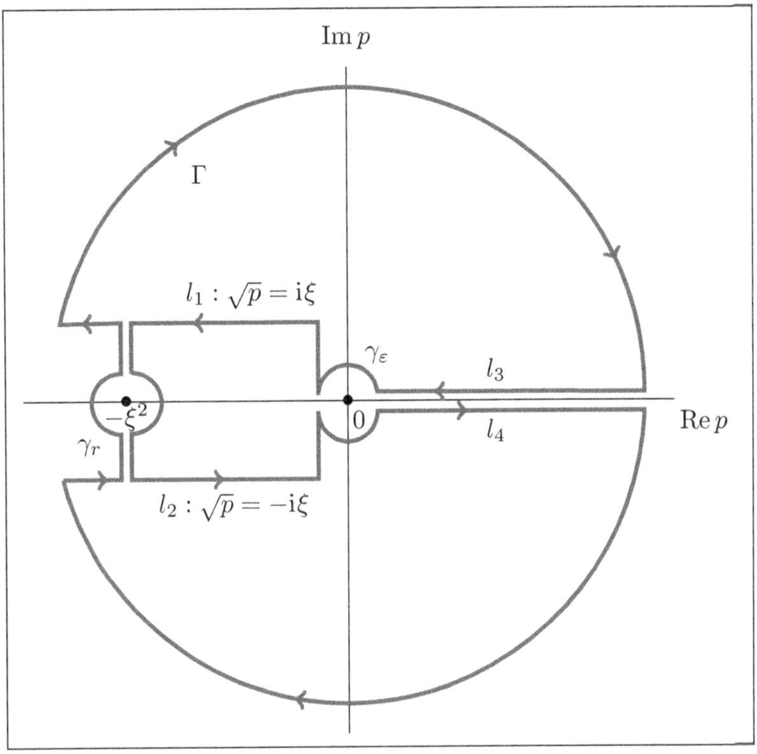

Figure 12.1: Contour of integration of Eq. (12.39) in a complex plane of p.

segments l_3, l_4 (a cut around the line $\operatorname{Im} p = 0, \operatorname{Re} p > 0$); connecting to two small semi-circles γ_ε around the origin with the radius $\varepsilon \ll 1$; then extending to two vertical line segments up to points $\operatorname{Im} p = \pm\xi$; then again two horizontal parallel line segments l_1, l_2 at $\operatorname{Im} p = \pm\xi$, which end points are connected to the arc Γ with a cut at $\operatorname{Im} p = -\xi^2$ (it consists of two vertical line segments and two semi-circles γ_r with the radius ε), such that the whole contour is continuous.

Using a standard technique, we take a limit $\varepsilon \to 0, R \to \infty$, so in this limit the contour takes the form as depicted in Fig, 12.2. It has a horizontal cut along the positive real line with point $p = 0$ excluded from the area inside the contour; another vertical cut at $\operatorname{Re} p = -\xi^2$ with the point $p = -\xi^2$ lying inside the contour; and a branch cut l_1, l_2 of the multivalued function $\sqrt{p}$ at $p = -\xi^2$. Also, in this limit $l_7 \to 0, l_8 \to 0$, but in Fig. 12.2 we left them as it is for a better readability.

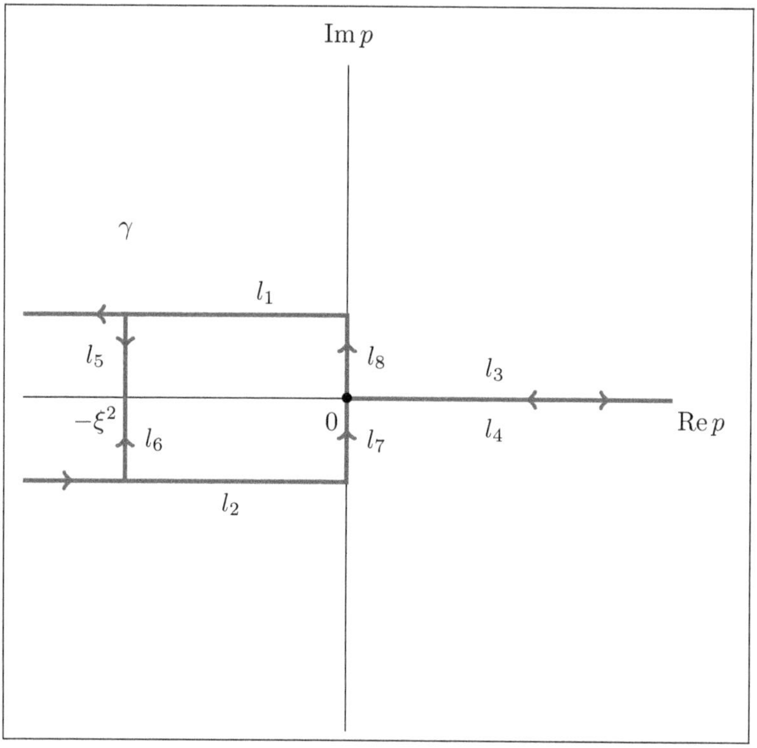

Figure 12.2: Contour of integration γ of Eq. (12.39) in a complex plane of p at $\varepsilon \to 0, R \to \infty$.

Now we are ready to compute the integrals in Eq. (12.39). That one in the LHS is regular everywhere inside this contour except a single pole $p = -\xi^2$. By the residue theorem, we obtain

$$\oint_\gamma \left(\int_0^\infty \alpha(\xi,\tau) \frac{\xi d\xi}{\xi^2 + p} \right) dp = -2\pi i \int_0^\infty \xi \alpha(\xi,\tau) d\xi. \tag{12.40}$$

The integral in the RHS doesn't have any singularity inside the contour γ, however, it has several cuts. As can be easily checked, the integrals along the segments l_3 and l_4 cancel out, as well as those along l_7 and l_8, and those along l_5 and l_6. The integral along the contour Γ tends to zero if $R \to \infty$ due to Jordan's lemma. Hence, the only remaining integrals are those along the horizontal semi-infinite lines l_1 and l_2. They read

$$\begin{aligned}
&\int_{l_1}\left(\int_0^\tau \Upsilon(s)e^{p(\tau-s)}e^{\sqrt{p}(y(\tau)-y(s))}ds + e^{p\tau}\int_{y(0)}^\infty u(z,0)e^{\sqrt{p}(y(\tau)-z)}dz\right)dp \\
&\quad = -2\int_0^\infty \xi\left[\int_0^\tau \Upsilon(s)e^{-\xi^2(\tau-s)}e^{i\xi(y(\tau)-y(s))}ds \right. \qquad (12.41)\\
&\quad \left. + e^{-\xi^2\tau}\int_{y(0)}^\infty u(z,0)e^{i\xi(y(\tau)-z)}dz\right]d\xi, \\
&\int_{l_2}\left(\int_0^\tau \Upsilon(s)e^{p(\tau-s)}e^{\sqrt{p}(y(\tau)-y(s))}ds + e^{p\tau}\int_{y(0)}^\infty u(z,0)e^{\sqrt{p}(y(\tau)-z)}dz\right)dp \\
&\quad = 2\int_0^\infty \xi\left[\int_0^\tau \Upsilon(s)e^{-\xi^2(\tau-s)}e^{-i\xi(y(\tau)-y(s))}ds \right.\\
&\quad \left. + e^{-\xi^2\tau}\int_{y(0)}^\infty u(z,0)e^{-i\xi(y(\tau)-z)}dz\right]d\xi.
\end{aligned}$$

The RHS of Eq. (12.39) is equal to a sum of these integrals and reads

$$\begin{aligned}
4i\int_0^\infty \xi&\left[\int_0^\tau \Upsilon(s)e^{-\xi^2(\tau-s)}\sinh\left(i\xi(y(\tau)-y(s))\right)ds\right.\\
&\left.+ e^{-\xi^2\tau}\int_{y(0)}^\infty u(z,0)\sin\left(\xi(y(\tau)-z\right)dz\right]d\xi.
\end{aligned}$$

Equating the LHS and RHS provides an explicit representation of $\alpha(\xi,\tau)$

$$\begin{aligned}
\alpha(\xi,\tau) = -\frac{2}{\pi}&\left[\int_0^\tau \Upsilon(s)e^{-\xi^2(\tau-s)}\sin\left(\xi(y(\tau)-y(s))\right)ds\right. \qquad (12.42)\\
&\left.+ e^{-\xi^2\tau}\int_{y(0)}^\infty u(z,0)\sin\left(\xi(y(\tau)-z\right)dz\right].
\end{aligned}$$

Finally, we substitute Eq. (12.42) into Eq. (12.39) and take into account the identity (Gradshtein and Ryzhik, 2007)

$$\int_0^\infty e^{-\beta x^2}\sin(ax)\sin(bx)\,dx = \frac{1}{4}\sqrt{\frac{\pi}{\beta}}\left(e^{-\frac{(a-b)^2}{4\beta}} - e^{-\frac{(a+b)^2}{4\beta}}\right), \quad \beta > 0,$$

which yields

$$u(x,\tau) = \frac{1}{2\sqrt{\pi}}\int_0^\tau \frac{\Upsilon(s)}{\sqrt{\tau-s}}\left(e^{-\frac{(x-y(s))^2}{4(\tau-s)}} - e^{-\frac{(x-2y(\tau)+y(s))^2}{4(\tau-s)}}\right)ds \qquad (12.43)$$

$$+ \frac{1}{2\sqrt{\pi\tau}} \int_{y(0)}^{\infty} u(z,0) \left(e^{-\frac{(x-z)^2}{4\tau}} - e^{-\frac{(x-2y(\tau)+z)^2}{4\tau}} \right) dz.$$

Thus, we obtained another representation of the solution which reads a bit different from that in Eq. (12.29), despite the general ansatz looks similar. Indeed, the solution is a sum of two integrals: one with respect to time τ, and the other one with respect to x with the integrand being a product of the initial condition with a Gaussian weight (it can be computed explicitly via Erf functions). The difference can be attributed to the different definitions of function $\Upsilon(k)$, as in Eq. (12.25) it is the heat potential density, while in Eq. (12.33) this is the gradient of the solution at $x = y(\tau)$. The first one is determined by the solution of the Volterra equation of the second kind Eq. (12.26), and the second one - by the solution of the Fredholm equation of the first kind in Eq. (12.35). However, the latter can either be transformed to the Volterra equation of the second kind. For doing that, one needs to differentiate Eq. (12.43) on x, and then let $x = y(\tau)$. This yields

$$\Upsilon(\tau) = \frac{1}{2\sqrt{\pi}} \int_0^{\tau} \Upsilon(s) \frac{y(\tau)-y(s)}{(\tau-s)^{3/2}} e^{-\frac{(y(\tau)-y(s))^2}{4(\tau-s)}} ds \qquad (12.44)$$
$$+ \frac{1}{2\sqrt{\pi\tau^3}} \int_{y(0)}^{\infty} u(z,0)(y(\tau)-z) e^{-\frac{(z-y(\tau))^2}{4\tau}} dz.$$

12.2 Numerical example

To test performance and accuracy of our method we run a test where the explicit form of parameters $\kappa(t), \theta(t), \sigma(t)$ is chosen as

$$\kappa(t) = \kappa_0, \qquad \theta(t) = \theta_0 e^{-\theta_k t}, \qquad \sigma(t) = \sigma_0 e^{-\sigma_k t}, \qquad (12.45)$$

with $\kappa_0, \theta_0, \sigma_0, \theta_k, \sigma_k$ being constants. With these definitions all functions in Eq. (12.19) can be found in closed form.

We approach pricing of Down-and-Out barrier Call option written on a ZCB in two ways. First, we solve the PDE in Eq. (12.9) by using a finite-difference scheme of the second order in space and time. We use the Crank-Nicolson scheme with few first Rannacher steps on a non-uniform grid compressed close to the barrier level at $t = 0$, in more detail, see (Itkin, 2017). Second, we use the method of heat potentials (HP) and solve the same problem as this is described in Section 12.1.2. To solve the Volterra equation in Eq. (12.26) we approximate the kernel on a rectangular grid $M \times M$, and the integral using the trapezoidal rule, which gives rise to the following system of linear equations

$$\|2\phi\| = (I + P)\|\Psi\|. \tag{12.46}$$

Here $\|\Psi\|$ is the vector of discrete values of $\Psi(\tau)$, $\tau \in [0, \tau(0)]$ on a grid with M nodes, $\|\phi\|$ is a similar vector of $\phi(\tau)$, I is the unit $M \times M$ matrix, and P is the $M \times M$ matrix of the kernel values on the same grid. Due to the specific structure of the Gaussian kernel, matrix P is lower triangular. Therefore, solution of Eq. (12.46) can be done with complexity $O(M^2)$.

An important point here is that the kernel (and so the matrix P) doesn't depend of strikes K, but only the function $\phi(\tau)$. Therefore, Eq. (12.46) can be solved simultaneously for all strikes by inverting the matrix $I+P$ with the complexity $O(M^2)$, and then multiplying it by vectors $\|\phi\|_k$, $k = 1, \dots, \bar{k}$, $\bar{k}$ is the total number of strikes. Hence, the total complexity is $O(\bar{k}M^2)$.

We emphasize, that this algorithm of solving the Volterra equation Eq. (12.26) doesn't require iterations, as that which makes use of the FGT. To compare both algorithms, note that complexity of the matrix algorithm is $O(M^2)$ versus $O(2nM)$ - the complexity of the iterative algorithm which requires n iterations to converge. Therefore, if M is relatively small, e.g., $M = 20$, and the number of iterations is near 10, both algorithms are of the same complexity.

For this test parameters of the model are presented in Table 12.1.

r_0	κ_0	θ_0	σ_0	θ_k	σ_k	L_F	S
0.07	1.0	0.08	0.2	0.3	0.2	0.8	7

Table 12.1: Parameters of the test.

We run the test for a set of maturities $T \in [1/12, 0.3, 0.5, 1]$ and strikes $K \in [0.06, 0.08, 0.1, 0.15, 0.2, 0.3]$. The Down-and-Out barrier Call option prices computed in such an experiment are presented in Fig. 12.3.

In Fig. 12.4 the relative differences (in percents) between these prices obtained by using the HP method and the FD solver are presented as a function of the option strike K and maturity T. Here to provide a comparable accuracy we run the FD solver with 200 nodes in space r and 201 steps in time t. Otherwise the quality of the FD solution is poor.

A large difference about 15% at simultaneously high maturities and strikes is due to the very low computed price of the option, which is 1.62 cents for the FD method, and 1.92 cents for the HP method, respectively. Otherwise, at small T and K the difference is about 0.1-2%. Also, since in

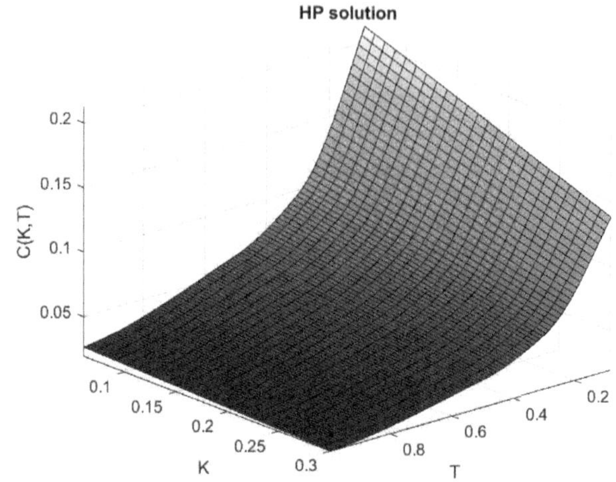

Figure 12.3: Down-and-Out barrier Call option price computed by using the HP method.

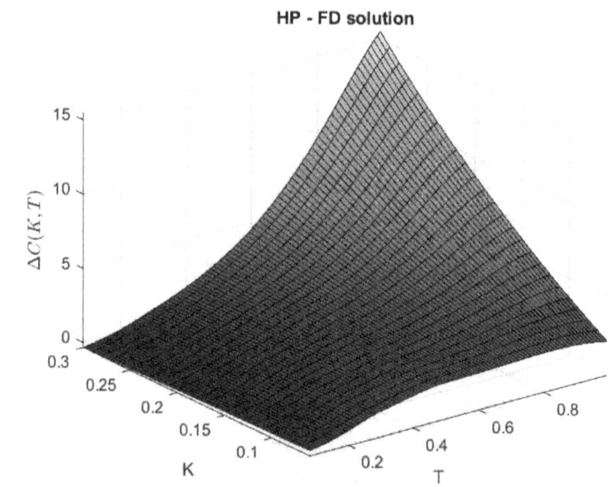

Figure 12.4: The relative difference in % of the Down-and-Out barrier Call option prices computed by using the HP method and the FD solver with 201 nodes in space r and time t.

this test we have chosen a fixed barrier in the ZCB price space, the corresponding barrier in the r space is moving down. Therefore, the computed option price could vanish for strikes close to the barrier at short maturities, but contrary have some positive value for the same strikes at longer maturities.

Obviously, since the FD solver needs a high number of nodes in both space and time to achieve a reasonable accuracy, the cost for this is speed. Suppose that this solver uses N_r nodes in space r, and M_t nodes in time t. Then the total complexity of solving the forward PDE to simultaneously get option prices for all given strikes K_i, $i = 1, \dots, \bar{k}$ and maturities T_j, $j = 1, \dots, \bar{m}$ is $O(N_r \times M_t)$. This should be compared with the complexity of the HP method which is $O(\bar{k}\bar{m}M^2)$. Since, as we saw, $N_r = M_t = 200$, $\bar{k} = 6$, $\bar{m} = 4$, $M = 20$, the HP method should be four times faster than the FD method. In reality, our test with 24 points in the $K \times T$ space shows that the elapsed time of the HP method is 20 mls, while for the FD method this is 146 mls which is 7 times slower. Decreasing the size of the FD grid to 100×100 nodes also decreases the elapsed time to 21 mls, but by the cost of increasing a relative difference up to $\pm 15\%$ for a wide range of maturities and strikes. Thus, overall, the method of HP demonstrates, at least same performance as the forward FD solver.

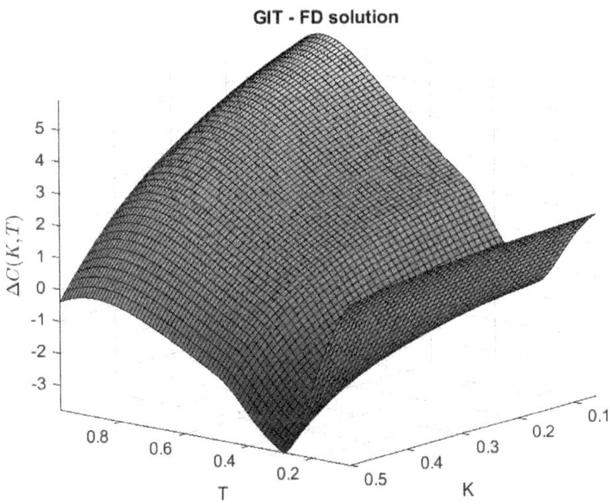

Figure 12.5: The relative difference in % of the Down-and-Out barrier Call option prices computed by using the GIT method and the FD solver.

Similar results obtained by using the GIT method are presented in Fig. 12.5. Computations are done in the same way as for the HP method. The Volterra equation in Eq. (12.44) is solved by discretizing the kernel of the first integral using the trapezoidal rule. The second integral can be computed analytically in terms of Erf functions, similar to how this is done in Eq. (12.30). Actually, first one can compute the second integral in Eq. (12.43), then take the derivative of the result on x, and the let $x = y(\tau)$. Overall, we obtain a system of linear equations with respect to the discrete values of $\Upsilon(\tau)$ which matrix is lower triangular with ones on the main diagonal. Therefore, solving this system is trivial. Once the vector $\Upsilon(\tau)$ is found, it could be substituted into Eq. (12.43). Then the final solution is obtained by computing the first integral since the second integral is known analytically.

As mentioned in Carr and Itkin (2021), the matrix of the linear system doesn't depend on K, just the RHS of the equation. Therefore, the function $\Upsilon(\tau)$ can be found simultaneously for all strikes by using vector elimination (this is equivalent to the solution of the system of linear equations with multiple RHS parts, while the matrix of the system is still lower triangular).

It turns out that the performance of both the GIT and HP methods is same. However, the GIT method produces more accurate results at high strikes and maturities (i.e. where the option price is relatively small) in contrast to the HP method which is more accurate at short maturities and low strikes. This behavior was explained in detail in Carr *et al.* (2020) for the CIR and CEV models. It can be verified by looking at the exponents in Eq. (94) of Carr *et al.* (2020) which are proportional to the time τ. Contrary, when the price is higher (short maturities, low strikes) the GIT method is slightly less accurate than the HP method, as in Eq. (45) of Carr *et al.* (2020) the exponent is inversely proportional to τ.

However, here the situation is a bit different. Indeed, the final PDE in Carr *et al.* (2020) was reduced to the Bessel PDE, why here we work with the heat equation. Therefore, the exponents in both the HP and GIT integrals are inversely proportional to τ. However, the GIT integrals contain a difference of two exponents which becomes small at large τ, in contrast to the HP exponent which tends to 1. Therefore, the convergence properties of two methods are different at large τ.

This situation is well known for the heat equation with constant coefficients. There exist two representation of the solution: one — obtained by using the method of images, and the other one — by the Fourier series. Despite both solutions are equal as the infinite series, their convergence properties are different. Thus, both GIT and HP methods are complementary.

Also, as observed in our experiments, at high maturities and strikes the first integral in Eq. (12.43) is small as compared with the second one. Then the approximate solution is given by the second integral which can be computed analytically. So in this case the whole approximation becomes pure analytical.

12.3 Final notes

In Sections 12.1.2 and 12.1.3 we have constructed semi-closed form solutions for the prices of Down-and-Out barrier Call option C_{dao} where the underlying is a Zero-Coupon Bond, and the interest rate dynamics follows the Hull-White model. Obviously, using the parity for barrier options, the price of the Down-and-In barrier Call option C_{dai} can be found as $C_{dai} = C_{van} - C_{dao}$, where C_{van} is the price of the European vanilla Call option in the Hull-White model. Since this model allows closed-form solutions for European options on Zero-coupon bonds (Andersen and Piterbarg, 2010), our solution also provides a closed form solution for C_{dai}. For the Up-and-Out barrier Call option C_{uao} a simple change of variables $x \to -x$ reduces the pricing problem to that one considered in this Chapter. Accordingly, the price of the Up-and-In barrier Call option can be found by using the barrier parity. The Puts are priced in the same way.

From the computational point of view the proposed solution is very efficient as this is shown in Section 12.2. Using theoretical analysis justified by a test example we conclude that our method is, at least, of the same complexity, or even faster than the forward FD method. On the other hand, our approach provides high accuracy in computing the options prices, as this is regulated by quadrature rule used to discretize the kernel. Therefore, the accuracy of the method in x space can be easily increased by using high order quadratures. For instance, using the Simpson instead of the trapezoid rule doesn't affect the complexity of our method but increase the accuracy, while increasing the accuracy for the FD method is not easy (i.e., it significantly increases the complexity of the method, e.g., see (Itkin, 2017)).

Also, as mentioned in Carr *et al.* (2020), another advantage of the approach advocated here is computation of option Greeks. Since the option prices in both the HP and GIT methods are represented in closed form via integrals, the explicit dependence of prices on the model parameters is available and transparent. Therefore, explicit representations of the option Greeks can be obtained by a simple differentiation under the integrals. This means that the values of Greeks can be calculated simultaneously with the prices almost with no increase in time. This is because differentiation under the integrals slightly changes the integrands, and these changes could be represented as changes in weights of the quadrature scheme used to numerically compute the integrals. Since the major computational time must be spent for computation of densities which contain special functions, they can be saved during the calculation of the prices, and then reused for computation of Greeks.

Chapter 13

Barrier Options in the Time-Dependent CEV and CIR Models

In Part 4 we applied the HP and GIT methods to solve various problems on mathematical finance where the underlying process was the OU process with time-dependent coefficients and boundaries. This methodology can obviously be extended to solve similar problems for other one-factor models. In particular, in this chapter we extend our approach to two additional models: the Cox–Ingersoll–Ross (CIR) model Cox *et al.* (1985), and the time-dependent constant elasticity of variance (CEV) model Cox (1975). Both models are very popular among practitioners and used for pricing various derivatives in asset classes such as equities, fixed income, commodities, FX, etc.

Again, for doing so we develop in parallel two analytic methods. The first method is the method of heat potentials which, as applied to the models considered in this Chapter, is discussed in detail in Section 13.3. We extend this method, since the partial differential equation (PDE) we need to solve cannot be transformed to the heat equation, but rather to the Bessel PDE. This approach is new and has not been developed yet in the literature until (Carr *et al.*, 2020), therefore, in what follows we closely follow that paper. However, once this is done, the same method could be used for solving some other problems implicitly related to pricing of barrier options, e.g., pricing American options (Carr and Itkin, 2021) analyzing the stability of a single bank and a group of banks in the structural default framework, (Kaushansky *et al.*, 2018), calculating the hitting time density (Alili *et al.*, 2005; Lipton and Kaushansky, 2020a), finding an optimal strategy for pairs trading (Lipton and Lopez de Prado, 2020a,b), etc. Also, the method could be used for solving various problems in physics where it was originally developed for the heat equation (Kartashov, 2001; Friedman, 1964.) and references therein.

The other method is the method of generalized integral transform (GIT) considered in chapterchap13 and throughout the whole book. Again, for the Bessel PDE this approach not developed in the literature in full (i.e., up to the final formula) until (Carr *et al.*, 2020), despite some comments on possible ways of achieving this could be found in Kartashov (1999) (but only for the domain $x \in [0, y(t)]$).

Going back to the CIR and CEV models with time-dependent coefficients, the prices of the barrier options in these models are not known in closed form yet. Instead, various numerical methods are used to compute them. This obviously produces a computations burden which could be excessive when these numerical procedures are used as a part of calibration process. Therefore, the closed form solutions discussed in this chapter could be of importance for practitioners. By a semi-closed solution we mean that first, one needs to solve numerically a linear Volterra equation of the second kind, and then the option price is represented as a one-dimensional integral of thus found solution. It is demonstrated in Carr *et al.* (2020) that computationally this new method is more efficient than both the backward and forward finite difference methods while providing better accuracy and stability.

The contribution of Carr *et al.* (2020) to the existing literature is twofold. First, they solve the problem of pricing barrier options in the CIR and CEV models in semi-closed form, and provide the resulting expressions that were not known yet in the literature. Second, they solve these problems by two methods so extending the existing apparatus of mathematical finance.

13.1 The CEV model

The time-dependent constant elasticity of variance (CEV) model is a one-dimensional diffusion process that solves a stochastic differential equation

$$dS_t = \mu(t) S_t dt + \sigma(t) S_t^{\beta+1} dW_t, \qquad S_{t=0} = S_0. \tag{13.1}$$

Here $t \ge 0$ is the time, S_t is the stochastic stock price, $\mu(t)$ is the drift, $\sigma(t)$ is the volatility and β is the elasticity parameter such that $\beta < 1$, $\beta \neq \{0, -1\}$[1], W_t is the standard Brownian motion. It is known, that under risk-neutral measure $\mu(t) = r(t) - q(t)$ where $r(t)$ is the deterministic short interest rate, and $q(t)$ is the continuous dividend. We assume that all

[1] In case $\beta = 0$ this model is the Black-Scholes model, while for $\beta = -1$ this is the Bachelier, or time-dependent Ornstein-Uhlenbeck (OU) model.

parameters of the model are known either as a continuous functions of time $t \in [0, \infty)$, or as a discrete set of N values for some moments t_i, $i = 1, \ldots, N$. This model is discussed in detail in Chapter 2.

For the standard CEV process with constant parameters it is known that change of variable $z_t = 1/(\sigma|\beta|)S_t^{-\beta}$ reduces the CEV process without drift ($\mu = 0$) to the standard Bessel process of order $1/2\beta$ (see Revuz and Yor, 1999; Davydov and Linetsky, 2001). Below we show that a similar connection with the Bessel process can be established for the time-dependent version of the model in Eq. (13.1). Let us consider an Up-and-Out barrier Call option $C(t, S)$ written on the underlying process S_t. By the Feynman–Kac formula this price solves the following partial differential equation (PDE)

$$\frac{\partial C}{\partial t} + \frac{1}{2}\sigma^2(t)S^{2\beta+2}\frac{\partial^2 C}{\partial S^2} + [r(t) - q(t)]S\frac{\partial C}{\partial S} = r(t)C. \tag{13.2}$$

This equation should be solved subject to the terminal condition at the option maturity $t = T$

$$C(T, S) = (S - K)^+, \tag{13.3}$$

where K is the option strike, and the boundary conditions

$$C(t, 0) = 0, \qquad C(t, H(t)) = 0, \tag{13.4}$$

where $H(t)$ is the upper barrier, perhaps time-dependent. Then the following Proposition holds.

Proposition 13.1. *The PDE in Eq. (13.2) can be transformed to*

$$\frac{\partial u}{\partial \tau} = \frac{1}{2}\frac{\partial^2 u}{\partial z^2} + \frac{b}{z}\frac{\partial u}{\partial z}, \tag{13.5}$$

where b is some constant, $u = u(\tau, z)$ is the new dependent variable, and (τ, z) are the new independent variables. The Eq. (13.5) is the PDE associated with the one-dimensional Bessel process (Revuz and Yor, 1999)

$$dX_t = dW_t + \frac{b}{X_t}dt. \tag{13.6}$$

Proof. This transformation can be done in two steps. First, we make a change of variables

$$S = (-x\beta)^{-1/\beta}, \qquad C(t, S) \to u(t, x)e^{\int_0^t r(k)dk}, \qquad \phi = \int_t^T \sigma^2(k)dk. \tag{13.7}$$

This reduces the PDE in Eq. (13.2) to the form

$$\frac{\partial u}{\partial \phi} = \frac{1}{2}\frac{\partial^2 u}{\partial x^2} + \left(xf(t) + \frac{b}{x}\right)\frac{\partial u}{\partial x}, \tag{13.8}$$

$$f(t) = \beta \frac{r(t) - q(t)}{\sigma^2(t)}, \qquad b = \frac{\beta + 1}{2\beta}, \qquad t = t(\phi).$$

The function $t(\phi)$ is the inverse map of the last term in Eq. (13.7). It can be computed for any $t \in [0, T]$ by substituting it into the definition of ϕ, then finding the corresponding value of $\phi(t)$, and finally inverting.

The Eq. (13.8) is a known type of PDE. Therefore, following (Polyanin, 2002) at the second step we make a new change of variables

$$z = xF(\phi), \qquad \tau = \int_0^\phi F^2(k)dk, \qquad F(\phi) = e^{\int_0^\phi f(k)dk}. \tag{13.9}$$

After this change the final PDE takes the form of Eq. (13.5) which finalizes the proof. □

Note, that Carr and Linetsky in Carr and Linetsky (2006) extend the time-dependent CEV model considered here by allowing a jump to zero. They also reduce their stock price process to a time homogeneous Bessel process.

As far as the terminal (now the initial) condition in Eq. (13.3) and the boundary conditions in Eq. (13.4) in the new variables is concerned, we must distinguish two cases, which are determined by the sign of β. If $-1 < \beta < 0$, the domain of definition for z is $z \in [0, y(\tau)]$, where

$$y(\tau) = -\frac{1}{\beta} H^{-\beta} \left(t(\tau)\right) F(\phi(\tau)) > 0. \tag{13.10}$$

Accordingly, the initial condition now reads

$$u(0, z) = e^{-\int_0^T r(k)dk} \left[\left(-\frac{\beta z}{F(\phi(0))} \right)^{-1/\beta} - K \right]^+, \tag{13.11}$$

and the boundary conditions are

$$u(\tau, 0) = u(\tau, y(\tau)) = 0. \tag{13.12}$$

However, if $0 < \beta < 1$, the left boundary goes to $-\infty$. Therefore, in this case it is convenient to redefine $x \to \bar{x} = -x$. This also redefines $z \to \bar{z} = -z$. Then the domain of definition for $\bar{z}$ becomes $\bar{z} \in [y(\tau), \infty)$ where

$$y(\tau) = \frac{1}{\beta} H^{-\beta} \left(t(\tau)\right) F(\phi(\tau)) > 0. \tag{13.13}$$

The initial condition now reads

$$u(0, \bar{z}) = e^{-\int_0^T r(k)dk} \left[\left(\frac{\beta \bar{z}}{F(\phi(0))} \right)^{-1/\beta} - K \right]^+, \tag{13.14}$$

and the boundary conditions are

$$u(\tau, \bar{z})\Big|_{\bar{z}\to\infty} = u(\tau, y(\tau)) = 0. \tag{13.15}$$

Also, in the case $0 < \beta < 1$ the PDE in Eq. (13.5) keeps the same form in the $\bar{z}$ variables. It can be seen, that in this case the Up-and-out barrier options transforms to the Down-and-out counterpart.

We interrupt here dealing with the time-dependent CEV model and postpone construction of the solution of problems in Eqs. (13.5), (13.11), and (13.12) (or Eqs. (13.5), (13.14), and (13.15)) till Section 13.3. Instead, in the next Section we consider the time-dependent CIR model, and barrier options written on a Zero-coupon bond (ZCB) which follows the CIR model. This is done, because, as we show below, the corresponding PDE could also be transformed to Eq. (13.5). Thus, our proposed method could be applied uniformly to both models.

13.2 The CIR model

The Cox–Ingersoll–Ross (CIR) model has been invented in Cox *et al.* (1985) for modeling interest rates. In this model the instantaneous interest rate r_t is stochastic variable which follows the stochastic differential equation (SDE), also named the CIR process. For the time-dependent version of the model this SDE reads

$$dr_t = \kappa(t)[\theta(t) - r_t]dt + \sigma(t)\sqrt{r_t}dW_t, \qquad r_{t=0} = r. \tag{13.16}$$

Here $\kappa(t) > 0$ is the speed of mean-reversion, $\theta(t)$ is the mean-reversion level. The CIR model is an extension of the Hull-White model that we analyzed in Itkin and Muravey (2020b) by making the volatility proportional to $\sqrt{r_t}$. This, on the one hand, allows avoiding the possibility of negative interest rates when the Feller condition $2\kappa(t)\theta(t)/\sigma^2(t) > 1$ is satisfied, while, on the other hand, still preserves tractability of the model (see e.g., Andersen and Piterbarg, 2010) and references therein. Various properties of the CIR model are discussed in Chapter 2.

Since the CIR model belongs to the class of exponentially affine models, the price of the ZCB $F(r, t, S)$ for this model is known in closed form. Here S is the bond expiration time. It is known, that $F(r, t, S)$ under a risk-neutral measure solves a linear parabolic partial differential equation (PDE) (Privault, 2012)

$$\frac{\partial F}{\partial t} + \frac{1}{2}\sigma^2(t) r \frac{\partial^2 F}{\partial r^2} + \kappa(t)[\theta(t) - r]\frac{\partial F}{\partial r} = rF. \tag{13.17}$$

It should be solved subject to the terminal condition

$$F(r, S, S) = 1, \tag{13.18}$$

and the boundary condition

$$F(r, t, S)\Big|_{r\to\infty} = 0. \tag{13.19}$$

The second boundary condition is necessary in case the Feller condition is violated, so the interest rate r_t can hit zero. Otherwise, the PDE in Eq. (13.17) itself at $r = 0$ serves as the second boundary condition.

The ZCB price can be obtained from Eq. (13.17) assuming that the solution is of the form

$$F(r, t, S) = A(t, S)e^{B(t,S)r}. \tag{13.20}$$

Substituting this expression into Eq. (13.2) and separating the terms proportional to r, we obtain two equations to determine $A(t, S), B(t, S)$

$$\begin{aligned}
\frac{\partial B(t, S)}{\partial t} &= 1 + \kappa(t)B(t, S) - \frac{1}{2}\sigma^2(t)B^2(t, S), \\
\frac{\partial A(t, S)}{\partial t} &= -A(t, S)B(t, S)\theta(t)\kappa(t).
\end{aligned} \tag{13.21}$$

To obey the terminal condition Eq. (13.18), the first PDE in Eq. (13.21) should be solved subject to the terminal condition $B(S, S) = 0$, and the second one — to $A(S, S) = 1$.

The first equation in Eq. (13.21) is the Riccati equation. It this general form it cannot be solved analytically for arbitrary functions $\kappa(t), \sigma(t)$, but can be efficiently solved numerically. Also, in some cases it can be solved approximately (asymptotically), see e.g., an example in Carr and Itkin (2021). Once the solution is obtained, the second equation in Eq. (13.21) can be solve analytically to yield

$$A(t, S) = e^{-\int_S^t B(m)\theta(m)\kappa(m)\,dm}. \tag{13.22}$$

When coefficients $\kappa(t), \theta(t), \sigma(t)$ are constants, it is known that the solution $B(t, S)$ can be obtained in closed form and reads (Andersen and Piterbarg, 2010)

$$B(t, S) = -\frac{2[\exp((S - t)h) - 1]}{2h + (\theta + h)[\exp((S - t)h) - 1]}, \qquad h = \sqrt{\theta^2 + 2\sigma^2}. \tag{13.23}$$

Thus, $B(t, S) < 0$ if $t < S$. Therefore, $F(r, t, S) \to 0$ when $r \to \infty$. In other words, the solution in Eq. (13.20) satisfies the boundary condition at $r \to \infty$. In case when all the parameters of the model are deterministic functions of time, and $B(t, S)$ solves the first equation in Eq. (13.21), this also remains to be true. This can be checked as follows. Since $\kappa(t) > 0, \sigma(t) > 0$ and from Eq. (13.20)

$$B(t,S) = \frac{\kappa(t)}{\sigma(t)^2} - \frac{1}{\sigma^2(t)}\sqrt{\kappa(t)^2 + 2[1 - B'(t,S)]}.$$

Therefore, $B(t,S) < 0$ if $B'(t,S) < 1$. On the other hand, if $B(t,S) < 0$ then from Eq. (13.20) $B'(t,S) < 1$. This finalizes the proof.

13.2.1 *Down-and-Out barrier option*

Let us consider a Down-and-Out barrier Call option written on a ZCB. Under a risk-neutral measure the option price $C(t,r)$ solves the same PDE as in Eq. (13.17) (Andersen and Piterbarg, 2010).

$$\frac{\partial C}{\partial t} + \frac{1}{2}\sigma^2(t) r \frac{\partial^2 C}{\partial r^2} + \kappa(t)[\theta(t) - r]\frac{\partial C}{\partial r} = rC. \tag{13.24}$$

The terminal condition at the option maturity $T \le S$ for this PDE reads

$$C(T,r) = (F(r,T,S) - K)^+, \tag{13.25}$$

where K is the option strike.

By a standard contract, the lower barrier $L_F(t)$ (which we assume to be time-dependent as well) is set on the ZCB price, and not on the underlying interest rate r. This means that it can be written in the form

$$C(t,r) = 0 \quad \text{if } F(r,t,S) = L_F(t). \tag{13.26}$$

However, since the ZCB price $F(r,t,S)$ can be expressed in closed form in Eq. (13.20), this condition can be translated into the r domain by solving the equation

$$F(r,t,S) = A(t,S)e^{B(t,S)r} = L_F(t),$$

with respect to r. Denoting the solution of this equation as $L(t)$ we find

$$L(t) = \frac{1}{B(t,S)} \log\left(\frac{L_F(t)}{A(t,S)}\right) > 0, \tag{13.27}$$

where it is assumed that $L_F > A(t,S)$. Accordingly, in the r domain the boundary condition to Eq. (13.24) reads

$$C(t, L(t)) = 0. \tag{13.28}$$

The second boundary can be naturally set at $r \to \infty$. As at $r \to \infty$ the ZCB price tends to zero, see Eq. (13.20), the Call option price also vanishes in this limit. This yields

$$C(t,r)\Big|_{r\to\infty} = 0. \tag{13.29}$$

The PDE in Eq. (13.24) can also be transformed to that for the Bessel process in Eq. (13.6).

Proposition 13.2. *The Eq. (13.24) can be transformed to*

$$\frac{\partial u}{\partial \tau} = \frac{1}{2}\frac{\partial^2 u}{\partial z^2} + \frac{b}{z}\frac{\partial u}{\partial z}, \tag{13.30}$$

where b is some constant, $u = u(\tau, z)$ is the new dependent variable, and (τ, z) are the new independent variables, if

$$\frac{\kappa(t)\theta(t)}{\sigma^2(t)} = \frac{m}{2}, \tag{13.31}$$

where $m \in [0, \infty)$ is some constant. The Eq. (13.30) is the PDE associated with the one-dimensional Bessel process in Eq. (13.6).

Proof. First make a change of variables

$$C(t, r) = u(t, z)e^{a(t)r + \int_0^t a(s)\kappa(s)\theta(s)ds}, \qquad z = g(t)\sqrt{r}, \tag{13.32}$$

$$g(t) = \exp\left[\frac{1}{2}\int_0^t \left(\kappa(s) - a(s)\sigma^2(s)\right) ds\right] \tag{13.33}$$

where $a(t)$ solves the Riccati equation

$$\frac{da(t)}{dt} = -\frac{\sigma^2(t)a^2(t)}{2} + \kappa(t)a(t) + 1. \tag{13.34}$$

This reduces the PDE in Eq. (13.24) to the form

$$\frac{4k(t)\theta(t) - \sigma^2(t)}{2z}\frac{\partial u}{\partial z} + \frac{1}{2}\sigma(t)^2\frac{\partial^2 u}{\partial z^2} + \frac{4}{g^2(t)}\frac{\partial u}{\partial t} = 0. \tag{13.35}$$

Now make a change of time

$$\tau(t) = \frac{1}{4}\int_t^T g^2(s)\sigma^2(s)\, ds, \tag{13.36}$$

which transforms Eq. (13.35) to

$$\left(2\frac{k(t)\theta(t)}{\sigma^2(t)} - \frac{1}{2}\right)\frac{1}{z}\frac{\partial u}{\partial z} + \frac{1}{2}\frac{\partial^2 u}{\partial z^2} = \frac{\partial u}{\partial \tau}, \qquad t = t(\tau). \tag{13.37}$$

The function $t(\tau)$ is the inverse map of Eq. (13.36). It can be computed for any $t \in [0, T]$ by substituting it into the definition of τ, then finding the corresponding value of $\tau(t)$, and inverting.

Finally, as by assumption $k(t)\theta(t)/\sigma^2(t) = m - const$, we set $b = m - 1/2$. Thus, the final PDE takes the form of Eq. (13.5) which finalizes the proof. □

Following from Proposition 13.2, for the time-dependent CIR model the transformation from Eqs. (13.24) to (13.30) cannot be done unconditionally. However, from practitioners' points of view the condition Eq. (13.31) seems not to be too restrictive, Indeed, the model parameters already contain the independent mean-reversion rate $\kappa(t)$ and volatility $\sigma(t)$. Since m is an arbitrary constant, it could be calibrated to the market data together with $\kappa(t)$ and $\sigma(t)$. Therefore, in this form the model should be capable for calibration to the term-structure of interest rates.

Also, according to the change of variables made in Proposition 13.2, the terminal condition Eq. (13.25) in new variables reads

$$u(0,z) = e^{-a(t(0))z^2/g^2(t(0)) - \int_0^T a(s)\kappa(s)\theta(s)ds} \left(A(T,S)e^{B(T,S)z^2/g^2(T)} - K\right)^+. \tag{13.38}$$

And the boundary conditions in Eqs. (13.28) and (13.29) transform to

$$C(\tau, y(\tau)) = 0, \qquad C(\tau, z)\Big|_{z\to\infty} = 0, \tag{13.39}$$
$$y(\tau) = \frac{1}{B(t(\tau), S(\tau))} \log\left(\frac{L_F(t(\tau))}{A(t(\tau), S(\tau))}\right).$$

13.3 The method of Bessel potentials

For convenience of notation, further let us call as *the CEV problem* the PDE in Eq. (13.5) that has to be solved subject to the initial condition in Eq. (13.14) and the boundary conditions in Eq. (13.15). Also, we call *the CIR problem* the PDE in Eq. (13.30) solved subject to the initial condition Eq. (13.38) and the boundary conditions in Eq. (13.39). Both problems can be considered simultaneously, as the PDE in Eq. (13.5) coincides with that in Eq. (13.30), and the boundary conditions in Eq. (13.15) coincide with that in Eq. (13.39). Accordingly, both solutions are defined at the domain $z \in [y(\tau), \infty]$, thought the definitions of $y(\tau)$ in Eqs. (13.13) and (13.39) differ. We will describe our method for this domain in Section 13.3.1. Another type of the CEV problem where $\bar{z} \in [0, y^-(\tau)]$ will be considered separately in Section 13.3.2.

In Carr and Itkin (2021) and Itkin and Muravey (2020b) a similar problem but for the heat equation was solved by using two approaches. The first one is a method of generalized integral transform, actively elaborated on by the Russian mathematical school to solve parabolic equations at the domain with moving boundaries (see Kartashov, 1999, 2001) and references therein. These kind of problems are known in physics for a long time and arise in the field of nuclear power engineering and safety of nuclear reactors; in studying combustion in solid-propellant rocket engines; in the theory of phase transitions (the Stefan problem and the Verigin problem (in hydromechanics)); in the processes of sublimation in freezing and melting; in the kinetic theory of crystal growth; etc. In Carr and Itkin (2021) and Kartashov (1999) this method was applied to the domain $z \in [0, y(\tau)]$, and in Itkin and Muravey (2020b) for the first time the solution was obtained for the semi-infinite domain $z \in [y(\tau), \infty)$. We will further develop this method to solve the CEV and CIR problems in Section 13.4.

The second method used to solve the same problems in Carr and Itkin (2021) and Itkin and Muravey (2020b) is the method of heat potentials (see, e.g., Tikhonov and Samarskii, 1963; Friedman, 1964.; Kartashov, 2001) and references therein. The first use of this method in mathematical finance is due to Lipton (2002) for pricing path-dependent options with curvilinear barriers, and more recently in Lipton and Lopez de Prado (2020a,b) (also see references therein). However, the CEV and CIR problems which we deal with cannot be reduced to the heat equation, but rather to the Bessel PDE. Therefore, in this section we propose generalization of the method for this type of equations. Accordingly, we call this generalization as the method of Bessel potentials.

Note, that the potential method could be constructed for any PDE where the space operator is a linear differential operator with time-independent coefficients. We propose and discuss this generalization of the heat potential method in the Appendix. Thus, the heat and Bessel potentials are just two particular cases of this general scheme.

13.3.1 *Domain* $y(\tau) \le z < \infty$

Since both the CEV and CIR problems have the inhomogeneous initial condition, our first step is to reduce them to the alternative problems with a homogeneous initial condition. Since the Green function of the Bessel equation at the infinite domain is known in closed form (Polyanin, 2002) this can be achieved by representing $u(\tau, z)$ in the form

$$u(\tau, z) = q(\tau, z) + \int_{y(0)}^{\infty} u(0, \xi) q_\tau(z, \zeta, b) d\zeta. \tag{13.40}$$

Here $q_\tau(z, \zeta, b)$ is the fundamental solution (or the transition density, or the Green function) of Eq. (13.30) at the domain $z \in [0, \infty)$. This density can be obtained assumed that the Bessel process stops when it reaches the origin. But since the domain of definition of z is $z \in [y(\tau), \infty)$, we moved the left boundary from 0 to $y(0)$.

By the definition of b in Eq. (13.8), for the CEV model $b = (1+\beta)/(2\beta)$. Since in this case $0 < \beta < 1$, we get $b > 1$. It is known (Lawler, 2018; Linetsky and Mendoza, 2010) that in case $b \geq 1/2$ the density $q_\tau(z, \zeta, b)$ is a good density with no defect of mass, i.e., it integrates into 1. The explicit representation reads (Cox, 1975; Emanuel and Macbeth, 1982)

$$q_\tau(z, \zeta, b) = \frac{\sqrt{z\zeta}}{\tau} \left(\frac{\zeta}{z}\right)^b e^{-\frac{z^2+\zeta^2}{2\tau}} I_{b-1/2}\left(\frac{z\zeta}{\tau}\right). \tag{13.41}$$

Here $I_\nu(x)$ is the modified Bessel function of the first kind (Abramowitz and Stegun, 1964).

For the CIR model $b = m - 1/2$ where $m \in [0, \infty)$ can be found by calibration. Therefore, if $m > 1$ (i.e., if the Feller condition is satisfied, and the process never hits the origin), the Green function $q_\tau(z, \zeta, b)$ is given by Eq. (13.41). Otherwise, $0 < m < 1$ and $-1/2 < b < 1/2$. Then by another change of variables (Polyanin, 2002)

$$w(\tau, z) = z^{2(1-m)} u(\tau, z),$$

the Eq. (13.30) transforms to the same equation with respect to $w(\tau, z)$ but now with $b = (3 - 2m)/2$. Accordingly, since $0 < m < 1$ we have $b > 1/2$, Therefore, again the Green function is represented by Eq. (13.41).

The function $q(x, \tau)$ solves the problem

$$\begin{aligned}
\frac{\partial q(\tau, z)}{\partial \tau} &= \frac{1}{2}\frac{\partial^2 q(\tau, z)}{\partial z^2} + \frac{b}{z}\frac{\partial q(\tau, z)}{\partial z}, \\
q(0, z) &= 0, \qquad y(0) < z < \infty, \\
q(\tau, z)\Big|_{z\to\infty} &= 0, \qquad q(\tau, y(\tau)) = \varsigma(\tau), \\
\varsigma(\tau) &= -\int_{y(0)}^{\infty} u(0, \zeta) q_\tau(y(\tau), \zeta, b) d\zeta.
\end{aligned} \tag{13.42}$$

This problem is like that in Eqs. (13.30), (13.14), (13.15), (13.38) and (13.39)), but now with a homogeneous initial condition. Therefore, following the general idea of the method of heat potentials, we represent the solution in the form of a generalized potential for the Bessel PDE

$$q(\tau, z) = \int_0^\tau \Psi(k) \frac{\partial}{\partial \xi} \left[\frac{\sqrt{z\xi}}{\tau - k}\left(\frac{\xi}{z}\right)^b e^{-\frac{z^2+\xi^2}{2(\tau-k)}} I_{b-1/2}\left(\frac{z\xi}{\tau - k}\right)\right]\Bigg|_{\xi\to y(k)} dk, \tag{13.43}$$

where $\Psi(k)$ is the potential density. It can be seen that $q(\tau, z)$ solves Eq. (13.30) as the derivative of the integral on the upper limit is proportional to the Delta function which vanishes due to $z \neq y(\tau)$. The solution in Eq. (13.43) also satisfies the initial condition at $\tau = 0$, and the vanishing condition at $z \to \infty$. For the large values of argument $z\xi/(\tau - k)$ we propose to use the following approximation :

$$q(\tau, z) \approx \frac{1}{\sqrt{2\pi}} \int_0^\tau \Psi(k) \frac{\partial}{\partial \xi} \left[\frac{1}{\sqrt{\tau - k}} \left(\frac{\xi}{z} \right)^b e^{-\frac{(z-\xi)^2}{2(\tau - k)}} \right] \Bigg|_{\xi \to y(k)} dk \quad (13.44)$$

At the barrier $z = y(\tau)$ function $q(\tau, z)$ is discontinuous. Following a similar approach for the heat potentials method (Tikhonov and Samarskii, 1963), it can be shown that the limiting value of $q(\tau, z)$ at $z = y(\tau) + 0$ is equal to $\varsigma(\tau)$:

$$\varsigma(\tau) = \Psi(\tau) + \int_0^\tau \Psi(k) \frac{\partial}{\partial y(k)} \quad (13.45)$$

$$\times \left[\frac{\sqrt{y(\tau)y(k)}}{\tau - k} \left(\frac{y(k)}{y(\tau)} \right)^b e^{-\frac{y^2(\tau)+y^2(k)}{2(\tau - k)}} I_{b-1/2} \left(\frac{y(\tau)y(k)}{\tau - k} \right) \right] dk.$$

The Eq. (13.45) is a linear Volterra equations of the second kind (Polyanin and Manzhirov, 2008). Since $\varsigma(\tau)$ is a continuously differentiable function, Eq. (13.45) has a unique continuous solution for $\Psi(\tau)$. The Volterra equation can be efficiently solved numerically (see Itkin and Muravey, 2020b) for the discussion on various approached to the numerical solution of this type of equations. In brief, for instance, the integral in the RHS is approximated using some quadrature rule with N nodes in k space, and the solution is obtained at N nodes in the τ space. Thus, obtained matrix equation can be solved with the complexity $O(N^2)$ since the matrix is lower triangular. Since N could be small ($N \approx 20 - 30$), the solution is fast. In more detail we discuss numerical aspects of the solution in Section 13.5.

Once Eq. (13.45) is solved and the function $\Psi(\tau)$ is found, the final solution reads

$$u(\tau, z) = \int_0^\tau \Psi(k) \frac{\partial}{\partial y(k)} \left[\frac{\sqrt{zy(k)}}{\tau - k} \left(\frac{y(k)}{z} \right)^b e^{-\frac{z^2+y^2(k)}{2(\tau - k)}} I_{b-1/2} \left(\frac{zy(k)}{\tau - k} \right) \right] dk$$

$$+ \int_{y(0)}^\infty u(0, \xi) q_\tau(z, \zeta, b) d\zeta. \quad (13.46)$$

13.3.2 *Domain* $0 < z < y(\tau)$

The construction of the solution in this case is similar to that described in the previous Section. Again, to obtain a PDE with a homogeneous initial condition, we represent the solution in the form

$$u(\tau, z) = q(\tau, z) - \varsigma_0(\tau) + \int_0^{y(0)} u(0, \xi) q_\tau(z, \zeta, b) d\zeta, \tag{13.47}$$

$$\varsigma_0(\tau) = -\int_0^{y(0)} u(0, \zeta) q_\tau(0, \zeta, b) d\zeta,$$

where (Abramowitz and Stegun, 1964)

$$q_\tau(0, \zeta, b) = \frac{2^{1/2-b}\zeta^{2b}}{\tau^{b+1/2}\Gamma\left(b + \frac{1}{2}\right)} e^{-\frac{\zeta^2}{2\tau}},$$

and $\Gamma(x)$ is the Euler Gamma function.

Then the function $q(\tau, z)$ solves the problem

$$\begin{aligned} \frac{\partial q(\tau, z)}{\partial \tau} &= \frac{1}{2}\frac{\partial^2 q(\tau, z)}{\partial z^2} + \frac{b}{z}\frac{\partial q(\tau, z)}{\partial z}, \\ q(0, z) &= 0, \qquad 0 < z < y(0), \\ q(\tau, 0) &= 0, \qquad q(\tau, y(\tau)) = \varsigma(\tau) + \varsigma_0(\tau). \end{aligned} \tag{13.48}$$

We search for the solution in the form of the Bessel potential in Eq. (13.43). The potential density $\Psi(\tau)$ solves the following Volterra equation of the second kind

$$\begin{aligned} \varsigma(\tau) + \varsigma_0(\tau) &= \Psi(\tau) + \int_0^\tau \Psi(k) \\ &\times \frac{\partial}{\partial y(k)}\left[\frac{\sqrt{y(\tau)y(k)}}{\tau - k}\left(\frac{y(k)}{y(\tau)}\right)^b e^{-\frac{y^2(\tau)+y^2(k)}{2(\tau-k)}} I_{b-1/2}\left(\frac{y(\tau)y(k)}{\tau - k}\right)\right] dk. \end{aligned} \tag{13.49}$$

Once this function is found, the final solution reads

$$\begin{aligned} u(\tau, z) &= \int_0^\tau \Psi(k)\frac{\partial}{\partial y(k)}\left[\frac{\sqrt{zy(k)}}{\tau - k}\left(\frac{y(k)}{z}\right)^b e^{-\frac{z^2+y^2(k)}{2(\tau-k)}} I_{b-1/2}\left(\frac{zy(k)}{\tau - k}\right)\right] dk \\ &+ \int_0^{y(0)} u(0, \xi) q_\tau(z, \zeta, b) d\zeta. \end{aligned} \tag{13.50}$$

13.3.3 *Double barrier options*

Double barrier options for time-dependent models can be also priced by using the method of potentials. In particular, in Itkin and Muravey (2020b) this is demonstrated for the time-dependent Hull-White model. See also Chapter 18 of this book. Here we use a similar approach and apply the idea proposed in Itkin and Muravey (2020b) to construction of the semi-closed form solutions for double barrier options for the CIR and CEV models.

Let us provide the explicit formulae just for the CEV model as for the CIR model this can be done in the exact same way. Suppose we need the price of a double barrier Call option with the lower barrier $L(t)$ and the upper barrier $H(t) > L(t)$. After doing transformation to the Bessel PDE as this is described in Section 13.1, this implies solving the following problem

$$\begin{aligned} \frac{\partial u}{\partial \tau} &= \frac{1}{2}\frac{\partial^2 u}{\partial z^2} + \frac{b}{z}\frac{\partial u}{\partial z}, \\ u(\tau = 0, z) &= u(0, z), \qquad y(0) < x < h(0), \\ u(y(\tau), \tau) &= u(h(\tau), \tau) = 0, \end{aligned} \tag{13.51}$$

where for $-1 < \beta < 0$

$$y(\tau) = -\frac{1}{\beta} L\left(t(\tau)\right)^{-\beta} F(\phi(\tau)), \; h(\tau) = -\frac{1}{\beta} H\left(t(\tau)\right)^{-\beta} F(\phi(\tau)), \tag{13.52}$$

and for $0 < \beta < 1$

$$y(\tau) = \frac{1}{\beta} L\left(t(\tau)\right)^{-\beta} F(\phi(\tau)), \qquad h(\tau) = \frac{1}{\beta} H\left(t(\tau)\right)^{-\beta} F(\phi(\tau)). \tag{13.53}$$

Thus, in this case the solution is defined at the z-domain with two moving (time-dependent) boundaries.

Since this problem has an inhomogeneous initial condition, the method of potentials cannot be directly applied. Therefore, similar to Eq. (13.40) we represent the solution in the form

$$u(\tau, z) = q(\tau, z) + \int_{y(0)}^{h(0)} u(0, \xi) q_\tau(z, \zeta, b) d\zeta. \tag{13.54}$$

Now the function $q(\tau, z)$ solves a similar problem but with the homogeneous initial condition

$$\begin{aligned} \frac{\partial q}{\partial \tau} &= \frac{1}{2}\frac{\partial^2 q}{\partial z^2} + \frac{b}{z}\frac{\partial q}{\partial z}, \\ q(0, z) &= 0, \qquad y(0) < z < h(0), \\ q(\tau, y(\tau)) &= -\phi_2(\tau), \qquad q(\tau, h(\tau)) = -\psi_2(\tau), \\ \phi_2(\tau) &= -\int_{y(0)}^{h(0)} u(0, \xi) q_\tau(y(\tau), \zeta, b) d\zeta, \\ \psi_2(\tau) &= -\int_{y(0)}^{h(0)} u(0, \xi) q_\tau(h(\tau), \zeta, b) d\zeta, . \end{aligned} \tag{13.55}$$

Based on the method of Itkin and Muravey (2020b), we construct the solution of Eq. (13.55) in the form of a generalized Bessel potential

$$
\begin{aligned}
q(\tau, z) = \int_0^\tau \Bigg\{ & \Psi(k) \frac{\partial}{\partial \xi} \left[\frac{\sqrt{z\xi}}{\tau - k} \left(\frac{\xi}{z}\right)^b e^{-\frac{z^2+\xi^2}{2(\tau-k)}} I_{b-1/2}\left(\frac{z\xi}{\tau-k}\right) \right] \Bigg|_{\xi \to y(k)} \\
& + \Phi(k) \frac{\partial}{\partial \xi} \left[\frac{\sqrt{z\xi}}{\tau - k} \left(\frac{\xi}{z}\right)^b e^{-\frac{z^2+\xi^2}{2(\tau-k)}} I_{b-1/2}\left(\frac{z\xi}{\tau-k}\right) \right] \Bigg|_{\xi \to h(k)} \Bigg\} dk.
\end{aligned}
\tag{13.56}
$$

Here $\Psi(k), \Phi(k)$ are the Bessel potential densities to be determined. Using the boundary conditions in Eq. (13.55), and the fact that the expression in square brackets at $\tau = k$ is the Dirac delta function, one can find that they solve a system of two Volterra equations of the second kind

$$
\begin{aligned}
\phi_2(\tau) = \Psi(\tau) + & \int_0^\tau \\
\times \Bigg\{ & \Psi(k) \frac{\partial}{\partial \xi} \left[\frac{\sqrt{y(\tau)\xi}}{\tau - k} \left(\frac{\xi}{y(\tau)}\right)^b e^{-\frac{y^2(\tau)+\xi^2}{2(\tau-k)}} I_{b-1/2}\left(\frac{y(\tau)\xi}{\tau-k}\right) \right] \Bigg|_{\xi \to y(k)} \\
& + \Phi(k) \frac{\partial}{\partial \xi} \left[\frac{\sqrt{y(\tau)\xi}}{\tau - k} \left(\frac{\xi}{y(\tau)}\right)^b e^{-\frac{y^2(\tau)+\xi^2}{2(\tau-k)}} I_{b-1/2}\left(\frac{y(\tau)\xi}{\tau-k}\right) \right] \Bigg|_{\xi \to h(k)} \Bigg\} dk. \\
\psi_2(\tau) = \Phi(\tau) + & \int_0^\tau \\
\times \Bigg\{ & \Psi(k) \frac{\partial}{\partial \xi} \left[\frac{\sqrt{h(\tau)\xi}}{\tau - k} \left(\frac{\xi}{h(\tau)}\right)^b e^{-\frac{h^2(\tau)+\xi^2}{2(\tau-k)}} I_{b-1/2}\left(\frac{h(\tau)\xi}{\tau-k}\right) \right] \Bigg|_{\xi \to y(k)} \\
& + \Phi(k) \frac{\partial}{\partial \xi} \left[\frac{\sqrt{h(\tau)\xi}}{\tau - k} \left(\frac{\xi}{h(\tau)}\right)^b e^{-\frac{h^2(\tau)+\xi^2}{2(\tau-k)}} I_{b-1/2}\left(\frac{h(\tau)\xi}{\tau-k}\right) \right] \Bigg|_{\xi \to h(k)} \Bigg\} dk.
\end{aligned}
\tag{13.57}
$$

This system can be solved by various numerical methods with complexity $O(N^2)$ (see the discussion after Eq. (13.45)). Once this is done, the solution of the double barrier problem is found.

13.4 The method of generalized integral transform

In this Section we solve the same problem but using the method of generalized integral transform. Note, that so far, the method was elaborated on and used just for getting a semi closed form solution of the heat equation. But in this chapter we deal with the Bessel PDE. In Kartashov (1999) it is proposed to construct the direct integral transform for this equation by using Bessel functions. However, aside from this recommendation no explicit solution has been presented. Moreover, our analysis showed that using the form of the transform proposed in Kartashov (1999) doesn't give rise to the solution, as construction of the inverse transform faces various technical problems.

Therefore, the method presented in this section is (i) completely original and (ii) gives rise to the closed form solution of the problem. In other words, we solve the CIR and CEV problems by using the method of generalized integral transform to the very end, and to the best of the authors knowledge this is done for the first time in the literature. As such, this approach could be also very useful in physics for solving various problems. As mentioned in Kartashov (1999) those problems appear (but not limited to) in the field of nuclear power engineering and safety of nuclear reactors; in studying combustion in solid-propellant rocket engines; in laser action on solids; in the theory of phase transitions (the Stefan problem and the Verigin problem (in hydromechanics)); in the processes of sublimation in freezing and melting; in the kinetic theory of crystal growth; etc. (see Kartashov, 1999 and references therein).

13.4.1 *Domain* $0 < z < y(\tau)$

Recall, that based on the description in Section 13.1, this problem emerges when we consider the CEV problem with $\beta < 0$. Since the Laplace transform of Eq. (13.30) gives rise to the Bessel ODE (Abramowitz and Stegun, 1964), it would be natural seeking for the general integral transform in the class of Bessel functions. Therefore, by analogy with Kartashov (2001) and Carr and Itkin (2021) we introduce the generalized integral transform of the form

$$\bar{u}(\tau, p) = \int_0^{y(\tau)} z^{\nu+1} u(\tau, z) J_{|\nu|}(zp) dz, \tag{13.58}$$

where $p = a + i\omega$ is a complex number, $J_\nu(x)$ is the Bessel function of the first kind, and $\nu = 1/(2\beta) < 0$, since $\beta < 0$. Next, let us multiply both parts of Eq. (13.30) by $z^{\nu+1} J_{|\nu|}(zp)$ and integrate on z from 0 to $y(\tau)$. For the LHS this yields

$$\int_0^{y(\tau)} z^{\nu+1} \frac{\partial u}{\partial \tau} J_{|\nu|}(zp) dz = \frac{\partial \bar{u}}{\partial \tau} - y'(\tau) y^{\nu+1}(\tau) u(\tau, y(\tau)) J_{|\nu|}(y(\tau)p). \tag{13.59}$$

The last term in the RHS of Eq. (13.59) vanishes due to the boundary condition in Eq. (13.12).

Accordingly, for the RHS of Eq. (13.30) we have $b = \nu + 1/2$, and

$$J_1 = \frac{1}{2}\int_0^{y(\tau)} z^{\nu+1}\frac{\partial^2 u}{\partial z^2}J_{|\nu|}(zp)dz = \frac{1}{2}z^{\nu+1}\frac{\partial u}{\partial z}J_{|\nu|}(zp)\Bigg|_0^{y(\tau)} \tag{13.60}$$

$$-\frac{1}{2}u(\tau,z)\frac{\partial}{\partial z}\left(z^{\nu+1}J_{|\nu|}(zp)\right)\Bigg|_0^{y(\tau)} + \frac{1}{2}\int_0^{y(\tau)} u(\tau,z)\frac{\partial^2}{\partial z^2}\left(z^{\nu+1}J_{|\nu|}(zp)\right)dz,$$

$$J_2 = \int_0^{y(\tau)} z^{\nu+1}\frac{\nu+1/2}{z}\frac{\partial u}{\partial z}J_{|\nu|}(zp)dz = (\nu+1/2)z^{\nu}J_{|\nu|}(zp)u(\tau,z)\Bigg|_0^{y(\tau)}$$

$$-(\nu+1/2)\int_0^{y(\tau)} u(\tau,z)\frac{\partial}{\partial z}\left(z^{\nu}J_{|\nu|}(zp)\right)dz.$$

Due to the boundary conditions in Eq. (13.12), the sum $J_1 + J_2$ can be represented as

$$J_1 + J_2 = y^{\nu+1}(\tau)J_{|\nu|}(y(\tau)p)\Psi(\tau) + \frac{1}{2}\int_0^{y(\tau)} u(\tau,z)\Big[\frac{1-2\nu}{z}\frac{\partial}{\partial z}\left(z^{\nu}J_{|\nu|}(zp)\right)$$

$$+\frac{\partial^2}{\partial z^2}\left(z^{\nu}J_{|\nu|}(zp)\right)\Big]zdz, \qquad \Psi(\tau) = \frac{\partial u}{\partial z}\Big|_{z=y(\tau)}. \tag{13.61}$$

It can be checked by using the theory of cylinder functions (Bateman and Erdélyi, 1953) that the Bessel function $J_{|\nu|}(zp)$ also solves the following ordinary differential equation (ODE)

$$\frac{d^2 Z}{dz^2} + \frac{1-2\nu}{z}\frac{dZ}{dz} + p^2 Z = 0, \quad Z(z) = C_1 z^{\nu}J_{|\nu|}(zp) + C_2 z^{\nu}Y_{|\nu|}(pz). \tag{13.62}$$

Here $Y_{|\nu|}(z)$ denotes the Bessel function of the second kind (also known as the Neumann or Weber function) which is linearly independent of $J_{|\nu|}(z)$. Assuming $C_1 = 1$, $C_2 = 0$, from Eqs. (13.59) and (13.60) we obtain the following Cauchy problem for $\bar{u}$

$$\frac{d\bar{u}(\tau,p)}{d\tau} = \frac{1}{2}\left[-p^2\bar{u}(\tau,p) + y^{\nu+1}(\tau)J_{|\nu|}(y(\tau)p)\Psi(\tau)\right], \tag{13.63}$$

$$\bar{u}(p,0) = \int_0^{y(0)} z^{\nu+1}J_{|\nu|}(zp)\,u(0,z)dz.$$

The solution of this problem reads

$$\bar{u} = e^{-p^2\tau/2}\left[\bar{u}(0,p) + \frac{1}{2}\int_0^{\tau} e^{p^2 s/2}\Psi(s)y^{\nu+1}(s)J_{|\nu|}(y(s)p)ds\right]. \tag{13.64}$$

By analogy with (Carr and Itkin, 2021), we can obtain the Fredholm equation of the first type for the function $\Psi(\tau)$. For doing so, let us set $p = i\lambda$, $\lambda \in \mathbb{R}$, tend τ to infinity and apply the formula $J_{|\nu|}(ix) = e^{i\nu\pi/2} I_\nu(x)$ which connects the Bessel function $J_\nu(x)$ with the modified Bessel function $I_\nu(x)$. This yields

$$\int_0^\infty e^{-\lambda^2 s/2} \Psi(s) y^{\nu+1}(s) I_\nu\left(y(s)\lambda\right) ds = -2 \int_0^{y(0)} q^{\nu+1} I_\nu\left(q\lambda\right) u(0,q) dq. \tag{13.65}$$

The solution of this integral equation $\Psi(\tau)$ can be found numerically on a grid by solving a system of linear equations (see e.g., Carr and Itkin, 2021) for the discussion on this subject and a numerical example. Once the function $\Psi(\tau)$ is found, it has to be substituted into Eq. (13.63) to obtain the generalized transform of $u(\tau, z)$ in the explicit form. Then, if this transform can be inverted back, we solved the problem of pricing Up-and-Out barrier Call options for the CEV model with $\beta < 0$.

13.4.1.1 *The inverse transform*

As it was already mentioned, it is reasonable to seek the solution of the CEV problem in the class of the Bessel functions. Therefore, we represent the solution in the form

$$u(\tau, z) = z^{-\nu} \sum_{n=1}^{\infty} \alpha_n(\tau) J_{|\nu|}\left(\frac{\mu_n z}{y(\tau)}\right). \tag{13.66}$$

Here μ_n is an ordered sequence of the positive zeros of the Bessel function $J_{|\nu|}(\mu)$:

$$J_{|\nu|}(\mu_n) = J_{|\nu|}(\mu_m) = 0, \quad \mu_n > \mu_m > 0, \quad n > m.$$

Note, that the definition in Eq. (13.66) automatically respects the vanishing boundary conditions for $u(\tau, z)$. We assume that this series converges absolutely and uniformly $\forall z \in [0, y(\tau)]$ for any $\tau > 0$.

Applying the direct integral transform in Eq. (13.58) to both parts of Eq. (13.66), and using a change of variables $z \to \hat{z} = zy(\tau)$ yields

$$\frac{\bar{u}(\tau, p)}{y^2(\tau)} = \sum_{n=1}^{\infty} \alpha_n(\tau) \int_0^1 \hat{z} J_{|\nu|}\left(\mu_n \hat{z}\right) J_{|\nu|}(p y(\tau)\hat{z}) d\hat{z}. \tag{13.67}$$

The set of functions $J_{|\nu|}(\alpha\hat{z})$ with $\alpha \in \mu_n$, $n = 1, \ldots,$ forms an orthogonal basis in the space $C[0, 1]$ with the scalar product

$$\langle J_{|\nu|}(\alpha z), J_{|\nu|}(\beta z)\rangle = 2\int_0^1 \frac{z J_{|\nu|}(\alpha z) J_{|\nu|}(\beta z) dz}{J_{|\nu|+1}(\alpha) J_{|\nu|+1}(\beta)} = \begin{cases} 1, & \alpha = \beta, \\ 0, & \alpha \neq \beta. \end{cases} \tag{13.68}$$

Therefore, the explicit formula for each coefficient $\alpha_n(\tau)$ is straightforward

$$\alpha_n(\tau) = 2\frac{\bar{u}\left(\tau, \mu_n/y(\tau)\right)}{y^2(\tau)J^2_{|\nu|+1}(\mu_n)}. \tag{13.69}$$

Thus, the final solution for $u(\tau, z)$ reads

$$\begin{aligned} u(\tau, z) &= 2\frac{z^{-\nu}}{y^2(\tau)}\sum_{n=1}^{\infty} \qquad\qquad (13.70)\\ &\times \left[\int_0^{y(0)} u(0,s)s^{\nu+1}e^{-\frac{\mu_n^2}{2y^2(\tau)}\tau}\frac{J_{|\nu|}(\mu_n s/y(\tau))J_{|\nu|}(\mu_n z/y(\tau))}{J^2_{|\nu|+1}(\mu_n)}ds\right.\\ &\left.+\frac{1}{2}\int_0^{\tau} y^{\nu+1}(s)\Psi(s)e^{-\frac{\mu_n^2}{2y^2(\tau)}(\tau-s)}\frac{J_{|\nu|}(\mu_n y(s)/y(\tau))J_{|\nu|}(\mu_n z/y(\tau))}{J^2_{|\nu|+1}(\mu_n)}ds\right]. \end{aligned}$$

This expression can be also re-written in the form

$$\begin{aligned} u(\tau, z) = 2\frac{z^{-2\nu}}{y^2(\tau)}\Bigg[\int_0^{y(0)} s\, u(0,s)\Theta_{|\nu|}\left(\frac{\sqrt{\tau}}{y(\tau)}, \frac{s}{y(\tau)}, \frac{z}{y(\tau)}\right)ds \qquad (13.71)\\ +\frac{1}{2}\int_0^{\tau} y(\tau)\Psi(s)\Theta_{|\nu|}\left(\frac{\sqrt{\tau-s}}{y(\tau)}, \frac{y(s)}{y(\tau)}, \frac{z}{y(\tau)}\right)ds\Bigg], \end{aligned}$$

where we introduced a new function

$$\Theta_{|\nu|}(\theta, x_1, x_2) = \sum_{n=1}^{\infty} e^{-\frac{\mu_n^2\theta^2}{2}}(x_1x_2)^{\nu}\frac{J_{|\nu|}(\mu_n x_1)J_{|\nu|}(\mu_n x_2)}{J^2_{|\nu|+1}(\mu_n)}. \tag{13.72}$$

The function $\Theta_{|\nu|}(\theta, x_1, x_2)$ is an analog of the Jacobi theta function which is a periodic solution of the heat equation. Indeed, in Carr and Itkin (2021) the solution of a similar problem for the time-dependent OU model (so $\beta = -1, \nu = -1/2$ and $|\nu| = 1/2$) with moving boundaries but for the heat equation has been obtained in terms of the theta functions. It can be checked that, if $\nu = 1/2$, we have

$$\begin{aligned} &\Theta_{1/2}\left(\frac{\sqrt{\tau}}{y(\tau)}, \frac{s}{y(\tau)}, \frac{z}{y(\tau)}\right) \qquad\qquad (13.73)\\ &\quad= \frac{y(\tau)}{2}\left[\theta_3\left(e^{-\frac{\pi^2\theta^2}{8}}, \frac{n\pi(s-z)}{4y(\tau)}\right) - \theta_3\left(e^{-\frac{\pi^2\theta^2}{8}}, \frac{n\pi(s+z)}{4y(\tau)}\right)\right], \end{aligned}$$

where $\theta_3(\omega, x)$ is the Jacobi theta function (Mumford *et al.*, 1983). Accordingly, function $\Theta_{|\nu|}(\theta, x_1, x_2)$ is a periodic solution of the Bessel equation.

Alternatively to the Fredholm equation of the first kind in Eq. (13.65) which is ill-posed and requires special methods to solve it (see Carr and Itkin, 2021), we can use a trick proposed in Itkin and Muravey (2020b) and instead derive the Volterra equation of the second kind for the function

$\Psi(\tau)$. For doing that, one needs to differentiate Eq. (13.70) on z, and then let $z = y(\tau)$. This yields

$$\Psi(\tau) = -\frac{2}{y^{3+\nu}(\tau)} \sum_{n=1}^{\infty} (\mu_n + \nu) \left[\int_0^{y(0)} u(0,s) s^{\nu+1} e^{-\frac{\mu_n^2 \tau}{2y^2(\tau)}} \frac{J_{|\nu|}(\mu_n s / y(\tau))}{J_{|\nu|+1}(\mu_n)} ds \right.$$
$$\left. + \frac{1}{2} \int_0^{\tau} y^{\nu+1}(s) \Psi(s) e^{-\frac{\mu_n^2 (\tau - s)}{2y^2(\tau)}} \frac{J_{|\nu|}(\mu_n y(s)/y(\tau))}{J_{|\nu|+1}(\mu_n)} ds \right]. \quad (13.74)$$

This equation has to be solved numerically, again see (Itkin and Muravey, 2020b) for the discussion and examples.

13.4.1.2 *Some approximations*

In some cases, Eq. (13.74) can be solved asymptotically. For instance, one can apply the following approximations

$$J_{|\nu|}^2(z) + J_{|\nu|+1}^2(z) \approx \frac{2}{\pi z}, \; x \gg \nu, \; J_{|\nu|+1}^2(\mu_n) \approx \frac{2}{\pi \mu_n}, \; \mu_n \gg \nu, \quad (13.75)$$
$$J_{|\nu|}(z) \sim \sqrt{\frac{2}{\pi z}} \cos\left(z - \frac{2|\nu|+1}{4}\pi \right), \; z \to \infty,$$
$$\mu_n \approx \pi \left(n + \frac{2|\nu|+1}{4} \right), \; n \to \infty.$$

Then the infinite sum in Eq. (13.70) can be truncated up to keep first N terms. The reminder (the error of this method) reads

$$R(N, \tau, z) = \frac{1}{z^{\nu+1/2} y(\tau)} \sum_{n=N+1}^{\infty} \left\{ \int_0^{y(0)} s^{\nu+1/2} u(0,s) e^{-\frac{\pi^2 \lambda_n^2}{2y^2(\tau)} \tau} \right. \quad (13.76)$$
$$\times \left[\cos\left(\frac{\pi \lambda_n (s+z)}{y(\tau)} - 2\pi\lambda_n \right) + \cos\left(\frac{\pi \lambda_n (s-z)}{y(\tau)} \right) \right] ds$$
$$+ \frac{1}{2} \int_0^{\tau} y^{\nu+1/2}(s) \Psi(s) e^{-\frac{\pi^2 \lambda_n^2}{2y^2(\tau)} (\tau - s)} \left[\cos\left(\frac{\pi \lambda_n (y(s)+z)}{y(\tau)} - 2\pi\lambda_n \right) \right.$$
$$\left. \left. + \cos\left(\frac{\pi \lambda_n (y(s)-z)}{y(\tau)} \right) \right] ds \right\}.$$

Here $\lambda_n = n + (2\nu - 1)/4$

A simple assessment shows that, since $-1 < \beta < 0$, from Eq. (13.8) at typical values of the model parameters we have $f(t) \approx f_1 \ll 1$. Assume that $H = const$. Hence, in Eq. (13.9)

$$\phi = \frac{\log(2\beta f_1 \tau + 1)}{2\beta f_1}, \qquad y(\tau) = -\frac{H^{-\beta}\sqrt{2\beta f_1 \tau + 1}}{\beta}.$$

Then it can be checked that the expression

$$-\frac{\mu_n^2 \tau}{2y^2(\tau)}$$

rapidly drops down for $-1 < \beta < 0$, $\tau > 0$, unless $\tau \ll 1$ and $H \gg 1$. Therefore, in this case just first few terms in the sum in Eq. (13.74) would be a good approximation.

Another approximation can be proposed to compute function $\Theta_{|\nu|}(\theta, x_1, x_2)$ defined in Eq. (13.72). The idea is that, as mentioned in above if $\tau \ll 1$ and $H = O(1)$, the first argument θ of this function is small, $\theta \ll 1$. Then the function $c_0 = e^{-\frac{\mu_n^2\theta^2}{2}}$ is small at large n, and $\frac{\mu_n^2\theta^2}{2}$ is small at small n. Therefore, for small n we represent c_0 by using the Padé approximation $(k, 1)$, and for large n we replace the small values of c_0 with the same Padé approximation. To estimate how accurate is this trick, let us pick, for instance, $k = 2$, so

$$c_0 \approx \frac{1}{1+\frac{x}{2}}\left(1 - \frac{x}{2}\right) + O(x^2), \qquad x = \frac{\mu_n^2\theta^2}{2}. \tag{13.77}$$

Thus, expanding the parenthesis into two terms, the function $(x_1x_2)^{-\nu}\Theta_{|\nu|}(\theta, x_1, x_2)$ can be represented as

$$(x_1x_2)^{-\nu}\Theta_{|\nu|}(\theta, x_1, x_2) = A_1(\theta, x_1, x_2) + A_2(\theta, x_1, x_2). \tag{13.78}$$

Now observe that (Bateman and Erdélyi, 1953)

$$\begin{aligned}
A_1(\theta, x_1, x_2) &= \sum_{n=1}^{\infty} \frac{J_{|\nu|}(\mu_n x_1) J_{|\nu|}(\mu_n x_2)}{J^2_{|\nu|+1}(\mu_n)} \frac{1}{1+\frac{\mu_n^2\theta^2}{4}} \\
&= -\frac{4}{\theta^2}\sum_{n=1}^{\infty} \frac{J_{|\nu|}(\mu_n x_1) J_{|\nu|}(\mu_n x_2)}{J^2_{|\nu|+1}(\mu_n)(\gamma^2 - \mu_n^2)} \\
&= \frac{\pi J_{|\nu|}(x_2\gamma)}{\theta^2 J_{|\nu|}(\gamma)}\left[J_{|\nu|}(x_1\gamma)Y_{|\nu|}(\gamma) - J_{|\nu|}(\gamma)Y_{|\nu|}(x_1\gamma)\right] \\
&= \frac{2 I_{|\nu|}(x_2\bar{\gamma})}{\theta^2 I_{|\nu|}(\bar{\gamma})}\left[K_{|\nu|}(x_1\bar{\gamma})I_{|\nu|}(\bar{\gamma}) - K_{|\nu|}(\bar{\gamma})I_{|\nu|}(x_1\bar{\gamma})\right], \\
\gamma &= 2\mathrm{i}/\theta, \quad \bar{\gamma} = 2/\theta.
\end{aligned} \tag{13.79}$$

Here we again used the formulae

$$J_\nu(ix) = e^{\mathrm{i}\nu\pi/2} I_\nu(x), \qquad Y_\nu(ix) = e^{(\nu+1)\pi\mathrm{i}/2} I_\nu(x) - 2/\pi e^{-\nu\pi\mathrm{i}/2} K_\nu(x),$$

which connect the Bessel functions $J_\nu(x)$ and $Y_\nu(x)$ with the modified Bessel functions $I_\nu(x)$ and $K_\nu(x)$.

To compute the next term in Eq. (13.77)

$$A_2(\theta, x_1, x_2) = -\frac{\theta^2}{4}\sum_{n=1}^{\infty}\frac{J_{|\nu|}(\mu_n x_1)J_{|\nu|}(\mu_n x_2)}{J^2_{|\nu|+1}(\mu_n)}\frac{\mu_n^2}{1+\frac{\mu_n^2\theta^2}{4}} \tag{13.80}$$

let us differentiate the LHS of Eq. (13.79) by θ, so

$$\begin{aligned}
&\frac{\partial}{\partial\theta}\sum_{n=1}^{\infty}\frac{J_{|\nu|}(\mu_n x_1)J_{|\nu|}(\mu_n x_2)}{J^2_{|\nu|+1}(\mu_n)}\frac{1}{1+\frac{\mu_n^2\theta^2}{4}} \\
&\quad = -\frac{\theta}{2}\sum_{n=1}^{\infty}\frac{J_{|\nu|}(\mu_n x_1)J_{|\nu|}(\mu_n x_2)}{J^2_{|\nu|+1}(\mu_n)}\frac{\mu_n^2}{\left(1+\frac{\mu_n^2\theta^2}{4}\right)^2} \\
&\quad \approx -\frac{\theta}{2}\sum_{n=1}^{\infty}\frac{J_{|\nu|}(\mu_n x_1)J_{|\nu|}(\mu_n x_2)}{J^2_{|\nu|+1}(\mu_n)}\frac{\mu_n^2}{1+\frac{\mu_n^2\theta^2}{2}}.
\end{aligned} \tag{13.81}$$

Therefore, the second term in Eq. (13.77) takes the form

$$\begin{aligned}
A_2(\theta, x_1, x_2) &= -\frac{\bar\gamma}{4}\frac{\partial}{\partial\bar\gamma}\left\{\frac{\bar\gamma^2 I_{|\nu|}(x_2\bar\gamma)}{I_{|\nu|}(\bar\gamma)}\left[K_{|\nu|}(x_1\bar\gamma)I_{|\nu|}(\bar\gamma)\right.\right. \\
&\quad \left.\left. - K_{|\nu|}(\bar\gamma)I_{|\nu|}(x_1\bar\gamma)\right]\right\}, \quad \bar\gamma = 2\sqrt{2}/\theta,
\end{aligned} \tag{13.82}$$

$$\frac{\partial I_{|\nu|}(\gamma)}{\partial\gamma} = \frac{1}{2}\left(I_{|\nu|-1}(\gamma) + I_{|\nu|+1}(\gamma)\right),$$

$$\frac{\partial K_{|\nu|}(\gamma)}{\partial\gamma} = \frac{1}{2}\left(K_{|\nu|-1}(\gamma) + K_{|\nu|+1}(\gamma)\right).$$

A similar approximation can be developed for Eq. (13.74) by using the identity (Bateman and Erdélyi, 1953)

$$\sum_{n=1}^{\infty}\frac{\mu_n J_{|\nu|}(\mu_n x)}{(\mu_n^2 - k^2)J_{|\nu|+1}(\mu_n)} = \frac{J_{|\nu|}(kx)}{2J_{|\nu|}(k)}. \tag{13.83}$$

As an example, let us consider the CEV model with constant parameters given in Table 13.1.

r	q	σ	H	N	$x_1/y(\tau)$	$x_2/y(\tau)$
0.02	0.01	0.5	0.2	100	0.5	0.5

Table 13.1: Parameters of the test.

In Fig. 13.1 we present the difference of two values of the function $G_\nu(x_1, x_2) = (x_1x_2)^{-\nu}\Theta_{|\nu|}(\theta, x_1, x_2)$: one computed by the definition in

Eq. (13.72) using the first N terms in the sum; and the other computed by using the approximation in Eq. (13.78). It can be seen that the latter approximation provides an accuracy about 25% except the area where both τ and β are simultaneously kind of large, and hence, the value of $G_\nu(x_1, x_2)$ is very small.

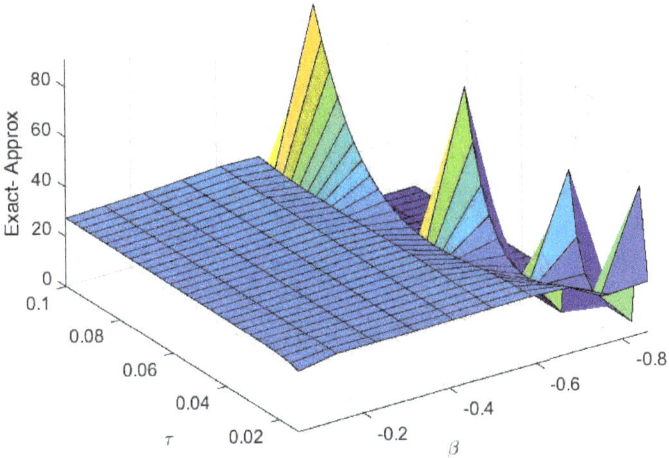

Figure 13.1: Comparison of the exact and approximated values of the function $G_\nu(x_1, x_2)$.

This approach can be further developed by increasing k in the Padé approximations $(k, 1)$, and computing the consecutive terms in the expansion by using the same trick as in above. In other words, to get the next term in the numerator of the Padé approximation we can differentiate the LHS of Eq. (13.79) twice on θ, and then express this derivative via the RHS of Eq. (13.79), etc.

13.4.1.3 *Connection to the first passage time problem*

Let us consider the following stopping moment

$$T_{x_0}^y = \inf\left\{\tau \geq 0, X_\tau \geq y(\tau)\right\},$$

where X_τ is the Bessel process defined by Eq. (13.6) and originated from $X_0 = x_0$. From standard results in probability theory the p.d.f. $\rho_{x_0}^y(\tau)$ of the moment $T^y_{x_0}$ can be found via the Fokker–Planck–Kolmogorov equation associated with the process X_τ:

$$\frac{1}{2}\frac{\partial^2 F_{x_0}(\tau,x)}{\partial x^2} - \frac{b}{x}\frac{\partial F_{x_0}(\tau,x)}{\partial x} + \frac{b}{x^2}F_{x_0}(\tau,x) = \frac{\partial F_{x_0}(\tau,x)}{\partial \tau} \tag{13.84}$$
$$F_{x_0}(\tau, y(\tau)) = 0, \quad F_{x_0}(\tau,0) = \delta(x-x_0).$$

Here $\delta(x-x_0)$ is the Dirac measure at the point x_0. The density $\rho^y_{x_0}(\tau)$ has the following representation:

$$\rho^y_{x_0}(\tau) = \frac{1}{2}\frac{\partial F_{x_0}(\tau,x)}{\partial x}\bigg|_{x=y(\tau)}. \tag{13.85}$$

The solution of the problem Eq. (13.30) and the function $\Psi(\tau)$ can be represented in terms of the function F:

$$u(z,\tau) = z^{-2b}\int_0^{y(0)} s^{2b}u(0,s)F_s(\tau,z)ds, \tag{13.86}$$

$$\Psi(\tau) = 2y(\tau)^{-2b}\int_0^{y(0)} s^{2b}u(0,s)\rho^y_s(\tau)ds.$$

For the boundary moving linearly in time $y(\tau) \approx \alpha + \beta\tau$ it is possible to propose the following approximation for the function $\Psi(\tau)$

$$\Psi(\tau) \approx 2y(\tau)^{-(2\nu+1)}\int_0^{y(0)} s^{2\nu+1}u(s,0)\frac{e^{\frac{\beta}{2\alpha}(\alpha^2-s^2)+\frac{b^2}{2}\tau}}{(\alpha+\beta\tau)^{\nu+2}} \tag{13.87}$$
$$\times \sum_{n=1}^{\infty}\frac{s^{-\nu}\mu_n J_{|\nu|}(\mu_n z/\alpha)}{\alpha^{-2\nu}J_{|\nu|+1}(\mu_n)}e^{-\mu_n^2\frac{\tau}{2\alpha(\alpha+\beta\tau)}}ds.$$

This is due to the fact that the p.d.f $\rho^y_{x_0}(\tau)$ of the first hitting time of the line $a\tau + b$ for the Bessel process is known explicitly (Alili and Patie, 2010).

Note, that a similar problem for the OU process with both constant and time-dependent coefficients has been studied in Lipton and Kaushansky (2020a). The authors consider the first hitting time density to a moving boundary for a diffusion process, which satisfies the Cherkasov condition, and hence, can be reduced to a standard Wiener process. They give two complementary (forward and backward) formulations of this problem and provide semi-analytical solutions for both by using the method of heat potentials.

13.4.2 *Domain $z > y(\tau)$*

To recall, this problem occurs in both the CEV model with $0 < \beta < 1$ and the CIR model. We will construct the solution of this problem by using the

Weber–Orr transform

$$
\begin{aligned}
\bar{u}(\tau, p) &= \int_{y(\tau)}^{\infty} z^{\nu+1} W(\tau, p, z) u(\tau, z) dz, \\
u(\tau, z) &= z^{-\nu} \int_0^{\infty} \frac{p W(\tau, p, z)}{V(\tau, p)} \bar{u}(\tau, p) dp.
\end{aligned} \tag{13.88}
$$

The kernel $W(a, b)$ and the function $V(p)$ are defined as follows (Bateman and Erdélyi, 1953)

$$
\begin{aligned}
W(\tau, a, b) &= J_{|\nu|}(ab) Y_{|\nu|}(ay(\tau)) - Y_{|\nu|}(ab) J_{|\nu|}(ay(\tau)), \\
V(\tau, p) &= {J_{|\nu|}}^2(py(\tau)) + {Y_{|\nu|}}^2(py(\tau)).
\end{aligned} \tag{13.89}
$$

The definitions in Eq. (13.89) are generalizations of the Pythagorean and Angle sum identities for trigonometric functions to the case of cylinder functions $J_{|\nu|}$ and $Y_{|\nu|}$. The functions $W(\tau, a, b)$ as the functions of the second argument a also form an orthogonal basis in the space $C[y(\tau), \infty)$ for all $\tau > 0$.

However, we cannot apply this transform directly to the Bessel equation due to fact that the kernel is time-dependent. Therefore, we propose to represent $\bar{u}$ as the weighted sum of the following transforms

$$
\begin{aligned}
\bar{u}_J(p, \tau) &= \int_{y(\tau)}^{\infty} z^{\nu+1} J_{|\nu|}(zp) u(\tau, z) dz, \\
\bar{u}_Y(p, \tau) &= \int_{y(\tau)}^{\infty} z^{\nu+1} Y_{|\nu|}(zp) u(\tau, z) dz,
\end{aligned} \tag{13.90}
$$

so

$$
\bar{u}\tau, p) = \bar{u}_J(\tau, p) Y_{|\nu|}(y(\tau)p) - \bar{u}_Y(\tau, p) J_{|\nu|}(y(\tau)p) \tag{13.91}
$$

The explicit formulae for $\bar{u}_J$ and $\bar{u}_Y$ read

$$
\bar{u}_J(\tau, p) = e^{-p^2\tau/2} \left[\bar{u}_J(0, p) + \frac{1}{2} \int_0^{\tau} e^{p^2 s/2} \Psi(s) y^{\nu+1}(s) J_{|\nu|}\left(y(s)p\right) ds \right], \tag{13.92}
$$

$$
\bar{u}_Y(\tau, p) = e^{-p^2\tau/2} \left[\bar{u}_Y(0, p) + \frac{1}{2} \int_0^{\tau} e^{p^2 s/2} \Psi(s) y^{\nu+1}(s) Y_{|\nu|}\left(y(s)p\right) ds \right],
$$

where $\Psi(\tau)$ is defined in Eq. (13.61). Using the inversion formula from Eq. (13.88) we immediately get the explicit formula for $u(\tau, z)$:

$$
u(z, \tau) = z^{-\nu} \int_0^{\infty} \int_{y(0)}^{\infty} s^{\nu+1} u(0, s) e^{-\frac{p^2\tau}{2}} \frac{W(\tau, p, z) W(\tau, p, s)}{V(\tau, p)} p \, dp \, ds \tag{13.93}
$$

$$+\frac{z^{-\nu}}{2}\int_0^\infty\int_0^\tau y^{\nu+1}(s)\Psi(s)e^{-\frac{p^2}{2}(\tau-s)}\frac{W(\tau,p,z)W(\tau,p,y(s))}{V(\tau,p)}p\,dp\,ds.$$

By analogy with the previous section, the function $\Psi(\tau)$ solves the Fredholm equation of the first kind

$$\int_0^\infty e^{-\lambda^2 s/2}\Psi(s)y^{\nu+1}(s)K_\nu\left(y(s)\lambda\right)ds = -2\int_{y(0)}^\infty q^{\nu+1}K_\nu\left(q\lambda\right)u(0,q)dq, \tag{13.94}$$

or the Volterra equation of the second kind

$$\Psi(\tau) = y^{-\nu}(\tau)\left[\int_0^\infty\int_{y(0)}^\infty q^{\nu+1}u(0,q)e^{-\frac{p^2\tau}{2}}\frac{Q(\tau,p)W(\tau,p,q)}{V(\tau,p)}dp\,dq\right] \tag{13.95}$$

$$+\frac{1}{2}\int_0^\infty\int_0^\tau y^{\nu+1}(s)\Psi(s)e^{-\frac{p^2}{2}(\tau-s)}\frac{Q(\tau,p)W(\tau,p,y(s))}{V(\tau,p)}dp\,ds.$$

Here

$$Q(\tau,p) = J_{\nu+1}(py(\tau))Y_\nu(py(\tau)) - Y_{\nu+1}(py(\tau))J_\nu(py(\tau)). \tag{13.96}$$

In some cases, Eq. (13.93) can be solved asymptotically. For instance, one can apply the following approximations

$$J_{|\nu|}{}^2(z) + Y_{|\nu|}{}^2(z) \approx \frac{2}{\pi z}\sum_{k=0}^\infty\frac{(2k-1)!!}{2^k z^{2k}}\frac{\Gamma(|\nu|+k+1/2)}{k!\Gamma(|\nu|-k+1/2)}, \tag{13.97}$$

$$J_{|\nu|}{}^2(z) + Y_{|\nu|}{}^2(z) \sim \frac{2}{\pi z}, \quad z\to\infty,$$

$$Y_{|\nu|}(z) \sim \sqrt{\frac{2}{\pi z}}\sin\left(z - \frac{2|\nu|+1}{4}\right), \quad z\to\infty,$$

$$W(\tau,a,b) \sim \frac{2\sin\left(a[b-y(\tau)]\right)}{\pi a\sqrt{by(\tau)}}, \quad a\to\infty.$$

Then, the outer integral in Eq. (13.93) can be truncated from above and approximated by the integral over the domain $[0,P]$, $0<P<\infty$. The reminder (the error of this method) reads

$$R(P,\tau,z) = z^{-\nu-1/2}\int_P^\infty dp \tag{13.98}$$

$$\left\{\int_{y(0)}^\infty s^{\nu+1/2}u(0,s)e^{-\frac{p^2\tau}{2}}\sin\left(p[z-y(\tau)]\right)\sin\left(p[s-y(\tau)]\right)ds\right.$$

$$+\frac{1}{2}\int_0^{\tau} y^{\nu+1/2}(s)\Psi(s)e^{-\frac{p^2}{2}(\tau-s)}\sin\left(p[z-y(\tau)]\right)$$
$$\times \sin\left(p[y(s)-y(\tau)]\right)ds\Bigg\}.$$

Introducing the new function $\Upsilon(P,t,\eta)$

$$\Upsilon(P,t,\eta)=e^{-\frac{\eta^2}{2t}}\left[\operatorname{erfc}\left(\frac{Pt+i\eta}{\sqrt{2t}}\right)+\operatorname{erfc}\left(\frac{Pt-i\eta}{\sqrt{2t}}\right)\right],$$

and taking into account the identity

$$\int_P^{\infty} e^{-\frac{p^2 t}{2}}\cos(\eta p)\,dp=\frac{1}{2}\sqrt{\frac{\pi}{2t}}\Upsilon(P,t,\eta),$$

we obtain the following explicit representation for $R(P,\tau,z)$

$$R(P,\tau,z)=\frac{z^{-\nu-1/2}\sqrt{\pi}}{4\sqrt{2}}\Bigg\{\int_{y(0)}^{\infty}\frac{s^{\nu+1/2}u(0,s)}{\sqrt{\tau}}\Big[\Upsilon(P,\tau,z-s)$$
$$-\Upsilon(P,\tau,z+s-2y(\tau))\Big]ds+\frac{1}{2}\int_0^{\tau}\frac{y^{\nu+1/2}(s)\Psi(s)}{\sqrt{\tau-s}}$$
$$\times\Big[\Upsilon(P,\tau-s,z-y(s))-\Upsilon(P,\tau-s,z+y(s)-2y(\tau))\Big]ds\Bigg\}.$$

Next, assuming that there exist $M_1, M_2 = const$ such that

$$\int_{y(0)}^{\infty} q^{\nu+1/2}u(0,q)dq\le M_1,\qquad \int_{y(0)}^{\infty} q^{2\nu+1}u(0,q)dq\le M_2,$$

we can set the upper limit of integration P such that $|R(P,\tau,z)|<\epsilon$ for any $\epsilon>0$. Indeed, using the following inequalities (see Eq. (13.86)):

$$\frac{1}{\sqrt{2t}}\Upsilon(P,t,\eta)\le\operatorname{erfc}(P),\quad \Psi(\tau)\le 2y(\tau)^{-(1+2\nu)}\int_{y(0)}^{\infty} q^{2\nu+1}u(0,q)dq,$$

we obtain

$$|R(P,\tau,z)|\le\frac{z^{-\nu-1/2}\sqrt{\pi}\operatorname{erfc}(P)}{2}\Bigg\{\int_{y(0)}^{\infty} q^{\nu+1/2}u(0,q)dq$$
$$+\int_0^{\tau} y^{-\nu-1/2}(s)ds\int_{y(0)}^{\infty} q^{2\nu+1}u(0,q)dq\Bigg\}.$$

Since the second integral is bounded for any τ, we obtain the following inequality:

$$|R(P,\tau,z)|\le\frac{z^{-\nu-1/2}\sqrt{\pi}\operatorname{erfc}(P)}{2}\left(M_1+M_2M_3(\tau)\right),$$
$$M_3(\tau)=\int_0^{\tau} y^{-\nu-1/2}(s)ds.$$

13.5 Numerical experiments

Similar to Itkin and Muravey (2020b), to check performance and accuracy of the proposed methods we construct the following test. Let us consider an Up-and-Out Barrier Call option written on the underlying stock which follows the CEV process with $0 < \beta < 1$. This case is described at the end of Section 13.1. To recall, after the change of variables proposed in that Section is done, the problems is transformed to the solution of the Bessel PDE at the domain $\bar{z} \in [y(\tau), \infty)$ with the boundary and initial conditions given in Eqs. (13.15) and (13.14).[2]

In this test we use the following explicit form of parameters $r(t), q(t), \sigma(t)$

$$r(t) = r_0 - r_k(a+t), \; q(t) = q_0 - q_k(a+t), \; \sigma(t) = \sigma_0\sqrt{a+t}, \tag{13.99}$$

where $r_0, q_0, \sigma_0, r_k, q_k$ are constants. We also assume $r_0 = q_0$, and $H-const$. With these definitions one can find

$$\begin{aligned} \phi(t) &= -\frac{1}{2}\sigma_0^2(t-T)(2a+t+T), \\ \phi(\tau) &= \frac{\sigma_0^2}{2\beta(q_k - r_k)}\log\left(\frac{\sigma_0^2 + 2\beta(q_k - r_k)\tau}{\sigma_0^2}\right), \\ F(\phi) &= \frac{\sqrt{2\beta\tau(q_k - r_k) + \sigma_0^2}}{\sigma_0}, \qquad y(\tau) = F(\phi)H^{-\beta}/\beta. \end{aligned} \tag{13.100}$$

We approach pricing the Up-and-Out barrier Call option in the CEV model twofold. First, as a benchmark we solve the PDE in Eq. (13.2) by using a finite-difference (FD) scheme of the second order in space and time. We use the Crank-Nicolson scheme with few first Rannacher steps on a non-uniform grid compressed close to the barrier level (see Itkin, 2017). Accordingly, our domain in S space is $S \in [0, H]$.[3]

Alternatively, we apply the method of Bessel potentials (BP) developed in Section 13.3.1 to solve the Bessel PDE Eq. (13.5). For doing so, first we solve the Volterra equation in Eq. (13.45) where the kernel is approximated on a rectangular grid $M \times M$, and the integral is computed using

[2] Hence, in new variables the Up-and-Out option transforms to the Down-and-Out option.

[3] A similar approach can be developed for the Down-and-Out options with $S \in [L, \infty)$. Then instead of truncating the infinite semi-interval, one can transform it to the fixed interval $[-1, 1)$ or to $[0, 1)$ and solve a modified PDE on the new interval. The boundary behavior of the solution can be obtained using Fichera theory and/or Greeen's integral formula, see Wilmott *et al.* (2014) where this was done in many cases and proved that this approach works well.

the trapezoidal rule. This implies solving the following system of linear equations

$$\|\varsigma\| = (I + P)\|\Psi\|. \tag{13.101}$$

Here $\|\Psi\|$ is the vector of discrete values of $\Psi(\tau)$, $\tau \in \left[0, \tau(t)\Big|_{t=0}\right]$ on a grid with M nodes, $\|\varsigma\|$ is a similar vector of $\varsigma(\tau)$, I is the unit $M \times M$ matrix, and P is the $M \times M$ matrix of the kernel values on the same grid. Note, that the matrix P is lower triangular. Therefore, solution of Eq. (13.101) can be done with complexity $O(M^2)$.

As the kernel (and so the matrix P) doesn't depend of strikes K, but only the function $\varsigma(\tau)$, Eq. (13.101) can be solved simultaneously for all strikes by inverting the matrix $I + P$ with the complexity $O(M^2)$, and then multiplying it by vectors $\|\varsigma\|_k$, $k = 1, \dots, \bar{k}$, $\bar{k}$ is the total number of strikes. Therefore, the total complexity of this step remains $O(\bar{k}M^2)$, but this operation, however, can be vectorized in k. The Volterra equation could also been solved by iterative methods, but with almost the same complexity, see discussion in Itkin and Muravey (2020b).

r_0	q_0	σ_0	r_k	q_k	σ_k	a	H	S
0.01	0.01	0.3	0.01	0.005	0.2	1.0	100	70

Table 13.2: Parameters of the test.

The model parameters for this test parameters are presented in Table 13.2. We run the test for a set of maturities $T \in [1/12, 0.3, 0.5, 1]$ and strikes $K \in [59, 64, 69, 74, 79, 84]$. The Up-and-Out barrier Call option prices computed in such an experiment are presented in Table 13.3.

13.5.1 *Comparison with the BP method*

The same results computed by using the BP method are displayed in Fig. 13.2 for the option prices. Also, in Fig. 13.3 the percentage difference between the prices obtained by using the BP and FD methods is presented as a function of the option strike K and maturity T. Here to provide a comparable accuracy we run the FD solver with 101 nodes in space S and 100 steps in time t. Otherwise the quality of the FD solution is not sufficient.

	BP				FD				Difference %			
K\T	0.0833	0.3	0.5	1.0	0.0833	0.3	0.5	1.0	0.0833	0.3	0.5	1.0
59	9.3192	3.3642	1.6845	0.4976	9.2924	3.3554	1.6884	0.5175	0.2876	0.2604	-0.2321	-3.9899
64	6.2167	2.1795	1.0671	0.3038	6.2025	2.1831	1.0793	0.3252	0.2286	-0.1654	-1.1438	-7.0291
69	3.8402	1.3219	0.6339	0.1731	3.8341	1.3319	0.6494	0.1931	0.1597	-0.7624	-2.4444	-11.5443
74	2.1608	0.7350	0.3450	0.0891	2.1605	0.7477	0.3606	0.1061	0.0118	-1.7293	-4.5240	-19.1029
79	1.0746	0.3612	0.1652	0.0388	1.0775	0.3736	0.1787	0.0522	-0.2700	-3.4102	-8.1779	-34.6241
84	0.4448	0.1462	0.0641	0.0121	0.4484	0.1561	0.0743	0.0216	-0.7971	-6.7649	-15.9277	-78.1360

Table 13.3: Up-and-Out barrier Call option prices computed by using the BP and FD methods.

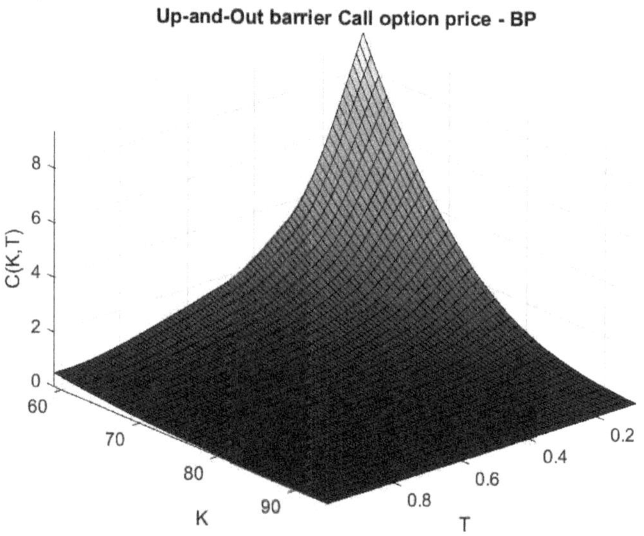

Figure 13.2: Up-and-Out barrier Call option price computed by using the BP method.

It can be seen that the agreement of both methods is good (less than 1%) if the option price is not too small which happens when the strike K is close to the barrier or at high maturities. In this case, as this is seen from Table 13.3, the relative difference becomes large, but the absolute difference of two methods is about one cent, which is almost insignificant. Obviously, such cases are a challenge for any FD method, as at $t = T$ there is a jump in the initial condition at the boundary, and the first derivative of the solution doesn't exists in this point.

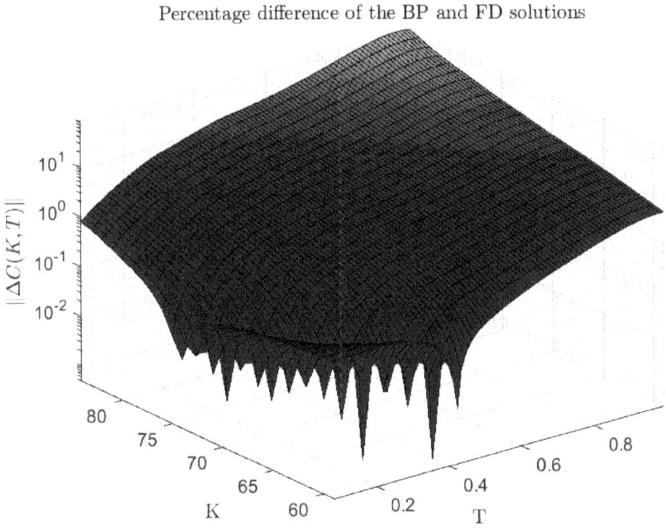

Figure 13.3: Percentage difference of Up-and-Out barrier Call option prices computed by using the BP and FD methods.

As far as performance of both methods is concerned, to decrease the elapsed time for the FD method instead of Eq. (13.2) we solve the corresponding forward PDE. Therefore, prices of all options for a given set of strikes and maturities could be obtained within one sweep. This also requires $\bar{m} \times \bar{k}$ integrations of the product of thus found density function with the payoff function, where $\bar{m}$ is the total number of maturities, and $\bar{k}$ is the total number of strikes. In this test the elapsed time for the FD method is, on average, 140 msec.

For the BP method, since the expression for $\varsigma(\tau)$ in Eq. (13.42) is not known in closed form, we compute this integral numerically by using the Simpson quadratures. Nevertheless, to make the results accurate, we need to increase the number of the grid nodes M. As compared with Itkin and Muravey (2020b), where a similar expression for the Hull-White model could be computed in closed form, and $M = 20$ provided a sufficient accuracy, here we need to take $M = 100$. Nevertheless, the elapsed time, on average, is 70 msec, i.e. twice faster than the FD method for the forward equation.

Decreasing M almost doesn't impact the accuracy of the method at large values of the options prices, while slightly deteriorates the quality of the feed at large maturities and large strikes. Changing M from 100 to 70 drops down the elapsed time to 50 msec. However, for the FD scheme decreasing the grid to 50×50 drops down the elapsed time to 30 msec while almost twice increasing the error for small maturities. Overall, we can conclude that the method of BP demonstrates, at least, same performance as the forward FD solver.

13.5.2 *Comparison with the GIT method*

Here we solve the same problem by using the GIT method developed in Section 13.3.1. Our numerical scheme is similar to that for the BP method: first we solve the Volterra equation Eq. (13.95) and then compute the value of the integrals in Eq. (13.93) using a trapezoidal rule. The inner integrals in the first summand in Eqs. (13.95) and (13.93) can be computed explicitly for the payoff Eq. (13.11) via the following formulas (Gradshtein and Ryzhik, 2007)

$$
\begin{aligned}
&\int_0^1 x^{\nu+1} J_\nu(ax)dx = a^{-1} J_{\nu+1}(a), && \Re(\nu) > -1,\\
&\int_0^1 x^{1-\nu} J_\nu(ax)dx = \frac{a^{\nu-2}}{2^{\nu-1}\Gamma(\nu)} - a^{-1} J_{\nu-1}(a), && \Re(\nu) < 1,\\
&\int_0^1 x^{\nu+1} Y_\nu(ax)dx = a^{-1} Y_{\nu+1}(a) + 2^{\nu+1} a^{-\nu-2}\Gamma(\nu+1), && \Re(\nu) > -1,\\
&\int_0^1 x^{1-\nu} Y_\nu(ax)dx = \frac{a^{\nu-2}\cot(\nu\pi)}{2^{\nu-1}\Gamma(\nu)} - a^{-1} Y_{\nu-1}(a), && \Re(\nu) < 1.
\end{aligned}
$$

However, for the second summands we have to numerically compute two-dimensional integrals containing special functions, see Eq. (13.93) and the definition of W, V in Eq. (13.88).

We run the same test described in above, and the results of this numerical experiment are presented in Table 13.4 and also in Fig. 13.3 which depicts the percentage difference between the prices obtained by using the GIT and FD methods. For this test we use $M = 10$ steps in time. This algorithm was implemented in python.

It can be seen that this method produces very accurate results at high strikes and maturities (i.e., where the option price is relatively small) in contrast to the BP method. This can be verified by looking at the exponents in Eq. (13.93) which are proportional to the time τ. Contrary, when the price is higher (short maturities, low strikes) the GIT method is slightly less

	GIT				FD				Difference %			
K\T	**0.0833**	**0.3**	**0.5**	**1.0**	**0.0833**	**0.3**	**0.5**	**1.0**	**0.0833**	**0.3**	**0.5**	**1.0**
59	9.3617	3.3532	1.6865	0.5158	9.2924	3.3554	1.6884	0.5175	0.71108	-0.2191	-0.3634	-0.7888
64	6.2737	2.1817	1.0781	0.3241	6.2025	2.1831	1.0793	0.3252	1.1101	-0.2149	-0.3654	-0.7997
69	3.90483	1.3315	0.6488	0.1924	3.8341	1.3319	0.6494	0.1931	1.7729	-0.20803	-0.3556	-0.8132
74	2.2194	0.7475	0.3604	0.1057	2.1605	0.7477	0.3606	0.1061	2.6313	-0.1879	-0.3318	-0.8992
79	1.1171	0.3736	0.1786	0.0519	1.0775	0.3736	0.1787	0.0522	3.5572	-0.1327	-0.3129	-1.0134
84	0.4692	0.1563	0.0742	0.0214	0.4484	0.1561	0.0743	0.0216	4.4718	0.0044	-0.2848	-1.1541

Table 13.4: Up-and-Out barrier Call option prices computed by using the GIT and FD methods.

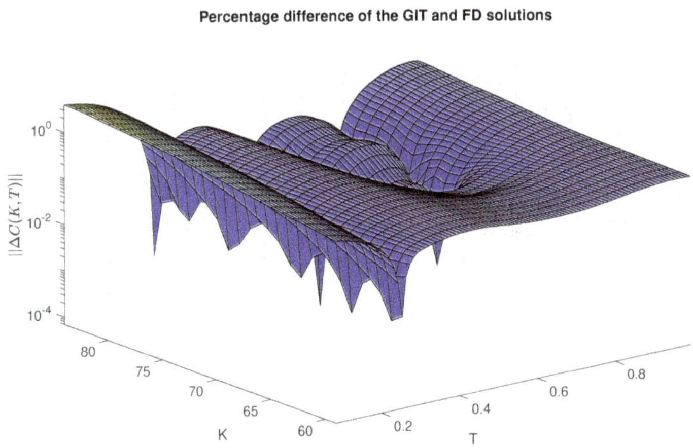

Figure 13.4: Percentage difference of Up-and-Out barrier Call option prices computed by using the FD and GIT methods.

accurate than the BP method, as in Eq. (13.45) the exponent is inversely proportional to τ. Obviously, the accuracy of the GIT method increases when M increases.

This situation is well investigated for the heat equation with constant coefficients. As applied to pricing double barrier options, it is described in Lipton (2002). There exist two representation of the solution: one — obtained by using the method of images, and the other one — by the Fourier series. Despite both solutions are equal as the infinite series, their convergence properties are different. In particular, the Fourier method is superior when the difference between the upper H and lower L barriers is small. and the time is relatively large. And the image expansion should be used otherwise.

Here we come to a similar principle for the time-dependent problems, and not just for the heat equation but also for the Bessel one. Thus, it is important that both the BP and GIT methods don't duplicate but rather compliment each other.

The speed of our python implementation is a bit slower than that for BP (approx. 0.158 sec). However, the latter method was implemented in Matlab. It is known that linear algebra in python (numpy) is almost 3 times slower than that in Matlab. Therefore, performance of both the BP and GIT methods is roughly same.

In addition, one can find that the main computational time is spent by a lot of calls to the routine computing the values of the integrands. The integrand in Eq. (13.93) is a four-dimensional function of z, p, $y(\tau)$ and $y(s)$. Let us denote this functions as $Z(z, p, v, w)$, where $v = y(\tau)$ and $w = y(s)$. We can gain the speed by the following trick: first pre-compute the values of Z on a regular four-dimensional grid, and then design this computational routine as interpolation.

It is also worth mentioning that in many situations the parameters of the model are such that the boundary $y(\tau)$ changes slowly with time, i.e. $y(\tau)$ is almost const. Then the first integral in Eq. (13.93) is a good approximation of the price. Accordingly, one don't need to solve the Volterra equation Eq. (13.95) that makes the algorithm about 2.5 faster.

13.6 Discussion

We have already mentioned that both the method of heat potentials (which in our case is extended to the method of Bessel potentials) and the method of generalized integral transforms were first developed in physics (see Kartashov, 1999, 2001; Tikhonov and Samarskii, 1963; Friedman, 1964.) and references therein. This was done to solve various problems of heat and mass transfer which are widely present in physics, chemistry, energetic, nuclear engineering, geology and many other areas of science and engineering. It turns out that many of those problems mathematically can be formulated in terms of stationary and non-stationary heat transfer. This includes such problems as diffusion, sedimentation, viscosity flows accompanied by various kinetic processes, astronomy, atomic physics, absorbtion, combustion, phase transitions and many others.

As an example in this Section we consider the Stefan problem which is a particular kind of the boundary value problem for the heat equation adapted to the case in which a phase boundary can move with time. It was introduced by I. Stefan in 1889 (see a detailed review in Lyubov (1978). The classical Stefan problem describes the temperature distribution in a homogeneous medium undergoing a phase change, for example ice passing to water. This is accomplished by solving the heat equation imposing the initial temperature distribution on the whole medium, and a particular boundary condition, the Stefan condition, on the evolving boundary between its two phases. Note that this evolving boundary is an unknown hyper-surface; hence, Stefan problems are examples of free boundary problems. However, a temperature gradient at this boundary is supposed to be known.

As such, treating it in financial terms, one can immediately recognize this as the pricing problem for the American option where the exercise boundary is also a free boundary, i.e., is not known. However, the option Delta $\frac{\partial u}{\partial z}$ at the boundary $z = y(\tau)$ is known (it follows from the conditions $\frac{\partial C_A}{\partial S}|_{S=S_B(t)} = 1$ and $\frac{\partial P_A}{\partial S}|_{S=S_B(t)} = -1$. Also the boundary condition for the American Call and Put at the exercise boundary (the moving boundary) is set as $C_A(S_B(t), t)) = S_B(t) - K$ for the Call, and $P_A(S_B(t), t)) = K - S_B(t)$ for the Put. This problem was solved in Carr and Itkin (2021) by using the method of generalized integral transform for the time-dependent OU model, and in Lipton and Kaushansky (2020b) for the Black-Scholes model with constant coefficients by using the method of heat potentials.

On contrary, we want to emphasize that many of the problems considered in above have not been solved yet in physics, and are mentioned in Kartashov (2001) as yet unsolved problems. Therefore, the results obtained in Carr and Itkin (2021) and presented in this Chapter also make a contribution to physics as they can be easily re-formulated in terms of the above mentioned physics problems.

Another connection of our results with physics is about the first passage time (FPT) problem considered in Section 13.4.1.3. As mentioned in Ding and Rangarajan (2004) (see also references therein), the FPT problem finds applications in many areas of science and engineering. A sampling of these applications includes (but not limited to)

- statistical physics (study of anomalous diffusion)
- neuroscience (analysis of neuron firing models)
- civil and mechanical engineering (analysis of structural failure)
- chemical physics (study of noise assisted potential barrier crossings)

- hydrology (optimal design of dams)
- imaging (study of image blurring due to hand jitter)

In particular, in Redner (2001) the author analyses the fundamental connection between the first-passage properties of diffusion and electrostatics. Basic questions of first passage include where a diffusing particle is absorbed on a boundary and when does this absorption event occur. These are time-integrated attributes, obtained by integration of a time-dependent observable over all time. For example, to determine when a particle is absorbed, we should compute the first-passage probability to the boundary and then integrate over all time to obtain the eventual hitting probability. However, it is more elegant to reverse the order of calculation and first integrate the equation of motion over time and then compute the outgoing flux at the boundary. This first step transforms the diffusion equation to the simpler Laplace equation. Then, in computing the flux, the exit probability is just the electric field at the boundary point. Thus, there is a complete correspondence between a first-passage problem and an electrostatic problem in the same geometry. This mapping is simple yet powerful, and can be adapted to compute related time-integrated properties, such as the splitting probabilities and the moments of the exit time.

Other connections to physics problems such as kinetics of spin systems, first passage in composite and fluctuating systems, hydrodynamic transport. reaction-rate theory, etc. could also be found in Redner (2001).

With the hope that we managed to convince the reader about a strong connection between the financial problems considered in this chapter and physics, we stop here this excursus into a wonderful world of physics leaving curious readers to extend it themselves.

Now, let us mention that in Sections 13.1 and 13.2 we constructed semi-closed form solutions for the prices of Up-and-Out barrier Call option C_{uao}. Despite same could be done for the Down-and-Out options, alternatively we can use the parity for barrier options (Hull, 2011). Then the price of the Down-and-Out barrier Call option C_{dao} can be found as $C_{dao} = C_v - C_{uao}$, where C_v is the price of the European vanilla Call option. For the models considered in above the latter is known in closed form (Andersen and Piterbarg, 2010). The double barrier case was also considered in Section 13.3.3.

Despite in our test we assumed the barrier H to be constant in time, the whole framework is developed for the general case where the barrier is some arbitrary function of time.

From the computational point of view the proposed solution is very efficient as this is shown in Section 13.5. Using theoretical analysis justified by a test example we conclude that our method is, at least, of the same complexity, or even faster than the forward FD method. On the other hand, our approach provides high accuracy in computing the options prices, as this is regulated by the order of a quadrature rule used to discretize the kernel. Therefore, the accuracy of the method in z space can be easily increased by using high order quadratures. On the other hand doing same for the FD method is not easy (i.e., it significantly increases the complexity of the method, e.g., see Itkin, 2017).

Another advantage of the approach advocated in Carr and Itkin (2021) is computation of option Greeks. Since the option prices in both the BP and GIT methods are represented in closed form via integrals, the explicit dependence of prices on the model parameters is available and transparent. Therefore, explicit representations of the option Greeks can be obtained by a simple differentiation under the integrals. This means that the values of Greeks can be calculated simultaneously with the prices almost with no increase in time. This is because differentiation under the integrals slightly changes the integrands, and these changes could be represented as changes in weights of the quadrature scheme used to numerically compute the integrals. Since the major computational time has to be spent for computation of densities which contain special functions, they can be saved during the calculation of the prices, and then reused for computation of Greeks.

Note, that the FD method also provides the values of Delta, Gamma and Theta on the FD grid, while, for instance, for Vega one need to bump the model volatility and rerun the whole scheme. But for the BP and GIT methods computation of Delta or Vega is done uniformly. Also, the ability of fast computation of Greeks is important for model calibration. Therefore, one can efficiently calibrate the CIR and CEV models to the market data by using the BP and GIT methods, since the semi-explicit nature of the final expressions allows quasi-analytical formulae for the gradient of the loss function.

Chapter 14

Barrier Options in the BK and Verhulst Models

In this chapter we continue construction of semi-analytical solutions for problems with moving boundaries, now by considering the Black-Karasinski (BK) and Verhulst models. As we have already argued in Chapter 2, some of the most popular short-term interest rates models have to be revisited and modified to better reflect current market conditions. In particular, this is true for the popular Black-Karasinski model, which is widely used by practitioners for modeling interest rates, credit, and commodities. Our adjustment made in Chapter 8 gave rise to the stochastic Verhulst model, which is well-known in the population dynamics and epidemiology as a logistic model. We also demonstrated that the Verhulst model's dynamics are well suited to the current economic environment and the Fed's actions.

Here, we proceed by deriving an integral equations for the zero-coupon bond prices for both the BK and Verhulst (MBK) models. For the BK model for small maturities up to 2 years, we solve the corresponding integral equation by using the reduced differential transform (RDTM) method which is discussed in Chapter 5. For the Verhulst PDE, under some mild assumptions, we find the closed-form solution. Numerical examples show that computationally our approach is more efficient than the standard finite difference method.

14.1 Integral equation for the ZCB price in the BK model

To remind, in Section 2.3.3 we showed that the short rate $\bar{r}_t = (r_t - s(t))/R$ in the BK model solves the following stochastic differential equation (SDE)

$$d\bar{r}_t = [k\theta(t) + \frac{1}{2}\sigma(t)^2 - k\log\bar{r}_t]\bar{r}_t dt + \sigma(t)\bar{r}_t dW_t. \tag{14.1}$$

It is known that, written in terms of variable $z = \log(\bar{r})$, the corresponding PDE for the ZCB price $F(t, z)$ reads (Andersen and Piterbarg, 2010)

$$0 = \frac{\partial F}{\partial t} + \frac{1}{2}\sigma^2(t)\frac{\partial^2 F}{\partial z^2} + \kappa(t)[\theta(t) - z]\frac{\partial F}{\partial z} - [s(t) + Re^z]V, \tag{14.2}$$

This equation should be solved subject to the terminal and boundary conditions (Andersen and Piterbarg, 2010) (see also discussion in Carr and Itkin, 2021)

$$F(T, z) = 1, \qquad F(t, z)\Big|_{z\to\infty} = 0. \tag{14.3}$$

To proceed, let us make the change of variables

$$\begin{aligned} x &= a(t)z + b(t), \qquad \tau = \frac{1}{2}\int_\nu^t \sigma(m)^2 e^{2\int_\nu^m \kappa(m)\,dm}\,dm, \qquad (14.4)\\ v(\tau, x) &= e^{-\int_\nu^t s(t)\,dt} F(t, z),\\ a(t) &= e^{\int_\nu^t \kappa(m)\,dm}, \quad b(t) = -\int_\nu^t \theta(m)\kappa(m)e^{\int_\nu^m \kappa(m)\,dm}\,dm, \end{aligned}$$

where $\nu = const$. As is discussed below, this constant can be chosen to simplify the final expressions. With this change Eq. (14.2) can be transformed to

$$\begin{aligned} 0 &= \frac{\partial v}{\partial \tau} + \frac{\partial^2 v}{\partial x^2} - \beta(t(\tau))e^{\bar{a}(t(\tau))x}v, \qquad (14.5)\\ \beta(t) &= \frac{2R}{\sigma(t)^2}e^{-2\int_\nu^t \kappa(m)\,dm - \frac{b(t)}{a(t)}}, \quad \bar{a}(t) = \frac{1}{a(t)},\\ v(\tau_0, x) &= e^{-\int_\nu^T s(t)dt}, \qquad \tau_0 = \frac{1}{2}\int_\nu^T \sigma(m)^2 e^{2\int_\nu^m \kappa(m)\,dm}\,dm. \end{aligned}$$

Next, we apply the Fourier transform

$$w(\omega, \tau) = \int_{-\infty}^{\infty} v(\tau, x)e^{\mathrm{i}\omega x}dx,$$

to Eq. (14.5) to get the following problem

$$w_\tau - \omega^2 w = \beta(\tau)\int_{-\infty}^{\infty} v(\tau, x)e^{\bar{a}(t)x}e^{\mathrm{i}\omega x}dx, \qquad w(\tau_0, \omega) = 2\pi\delta(\omega), \tag{14.6}$$

where $\delta(\omega)$ is the Dirac delta function. The solution of Eq. (14.6) reads

$$\begin{aligned} w &= e^{\omega^2(\tau-\tau_0)}w(\tau_0, \omega) \qquad (14.7)\\ &+ \int_{\tau_0}^{\tau} e^{\omega^2(\tau-k)}\beta(k)\int_{-\infty}^{\infty} F(k, \zeta)e^{\bar{a}(k)\zeta - \int_{\nu(\tau)}^{t(\tau)} s(t)\,dt}e^{\mathrm{i}\omega\zeta}\,d\zeta\,dk. \end{aligned}$$

Applying the inverse transform

$$v(\tau, x) = \frac{1}{2\pi}\int_{-\infty}^{\infty} w(\tau, \omega)e^{-\mathrm{i}\omega x}d\omega$$

we obtain

$$\begin{aligned}
v(\tau,x) &= 1 + \frac{1}{2\pi}\int_{-\infty}^{\infty}\int_{\tau_0}^{\tau} e^{\omega^2(\tau-k)}\beta(k)\int_{-\infty}^{\infty} v(k,\zeta)e^{\bar{a}(k)\zeta}e^{\mathrm{i}\omega(\zeta-x)}d\zeta\, dk\, d\omega \\
&= 1 - \frac{1}{2\sqrt{\pi}}\int_{-\infty}^{\infty}\int_{\tau_0}^{\tau} \frac{\beta(k)e^{\zeta\bar{a}(k)}}{\sqrt{\tau-k}} e^{-\frac{(x-\zeta)^2}{4(\tau-k)}} v(k,\zeta)d\zeta dk, \qquad (14.8)
\end{aligned}$$

where in the last line the change of variables $k \to -k$, $\tau \to -\tau$ was made.

Thus, the ZCB price solves the following two-dimensional Volterra integral equation of the second kind

$$v(\tau,x) = 1 - \frac{1}{2\sqrt{\pi}}\int_{-\infty}^{\infty}\int_{\tau_0}^{\tau} \frac{\beta(t(k))e^{\bar{a}(t(k))\zeta}}{\sqrt{\tau-k}} e^{-\frac{(x-\zeta)^2}{4(\tau-k)}} v(k,\zeta)d\zeta dk. \qquad (14.9)$$

It can also be obtained by applying the Duhamel's principle to Eq. (14.5).

14.2 Integral equation for the ZCB price in the Verhulst model

Again, we remind that in Section 2.3.4 we considered any contingent claim written on the r_t as the underlying (for instance, the price $F(\bar{r},t,T)$ of a Zero-coupon bond (ZCB) with maturity T) where the interest rate r_t follows the Verhulst model. By using Itô's lemma and the Feynman–Kac formula it is shown that the price of such a claim solves the following partial differential equation

$$\begin{aligned}
0 &= \frac{\partial F}{\partial t} + \frac{1}{2}\sigma^2(t)\bar{r}^2\frac{\partial^2 F}{\partial \bar{r}^2} + \kappa(t)\bar{r}[\tilde{\theta}(t) - \bar{r}]\frac{\partial F}{\partial \bar{r}} - (s(t) + R\bar{r})F, \qquad (14.10) \\
\bar{r}_t &= \frac{r_t - s(t)}{r_0 - s(0)} = e^{z_t}, \qquad \tilde{\theta}(t) = \bar{\theta}(t) + \frac{\sigma^2(t)}{2\kappa(t)}.
\end{aligned}$$

This equation should be solved subject to the same terminal and boundary conditions as in Eq. (2.39)

$$F(T,\bar{r}) = 1, \qquad F(t,\bar{r})\Big|_{\bar{r}\to\infty} = 0. \qquad (14.11)$$

As this is discussed in Chapter 2, since $\bar{r} \in (0,\infty)$, i.e., the boundary $\bar{r} = 0$ is not attainable, Eq. (14.10) doesn't need the boundary condition at the left boundary $\bar{r} \to 0$, as this is discussed in Itkin and Muravey (2020b) with a reference to Fichera theory (in other words, the PDE itself with substituted $\bar{r} = 0$ serves as the boundary condition). If, however, the boundary $\bar{r} = 0$ is attainable, the boundary condition at this point should be set as in Eq. (14.35).

With this in mind, we find the value of the ZCB by deriving and solving a Volterra integral equation of the second kind. We proceed with the elimination of the squared term in the drift in Eq. (14.10) via the following change of variables

$$F(t,\bar{r}) = e^{\frac{\kappa(t)}{\sigma^2(t)}\bar{r}+\int_T^t s(k)dk} W(t,\bar{r}). \tag{14.12}$$

This yields

$$\begin{aligned}
0 &= \frac{\partial W}{\partial t} + \frac{1}{2}\sigma^2(t)\bar{r}^2\frac{\partial^2 W}{\partial \bar{r}^2} + \kappa(t)\tilde{\theta}(t)\bar{r}\frac{\partial W}{\partial \bar{r}} + \left[\gamma(t)\bar{r} - \frac{\kappa^2(t)}{2\sigma^2(t)}\bar{r}^2\right] W,\\
\gamma(t) &= -R + \frac{\kappa^2(t)\tilde{\theta}(t)}{\sigma^2(t)} + \frac{d}{dt}\left(\frac{\kappa(t)}{\sigma^2(t)}\right),\\
W(T,\bar{r}) &= e^{-\frac{\kappa(T)}{\sigma^2(T)}\bar{r}}, \qquad W(T,\bar{r})\Big|_{\bar{r}\to\infty} = 0.
\end{aligned} \tag{14.13}$$

Another change of variables

$$\begin{aligned}
W(t,\bar{r}) &= u(\tau,y), \qquad y = \log\bar{r} + \int_0^t \left[\frac{\sigma^2(k)}{2} - \kappa(k)\tilde{\theta}(k)\right] dk,\\
\tau &= \frac{1}{2}\int_t^T \sigma^2(k)dk
\end{aligned} \tag{14.14}$$

transforms this PDE into the following one

$$\begin{aligned}
\frac{\partial u}{\partial \tau} &= \frac{\partial^2 u}{\partial y^2} + \left[\alpha(\tau)e^y + \beta(\tau)e^{2y}\right] u,\\
\alpha(\tau) &= \frac{2\gamma(t(\tau))}{\sigma^2(t(\tau))} e^{-\int_0^t\left[\frac{\sigma^2(k)}{2} - \kappa(k)\tilde{\theta}(k)\right]dk},\\
\beta(\tau) &= -\frac{\kappa^2(t(\tau))}{\sigma^4(t(\tau))} e^{-\int_0^t\left[\sigma^2(k) - 2\kappa(k)\tilde{\theta}(k)\right]dk}.
\end{aligned} \tag{14.15}$$

Indeed, it follows from a series of equalities

$$\begin{aligned}
\bar{r} &= e^y e^{-\int_0^t\left[\frac{\sigma^2(k)}{2} - \kappa(k)\tilde{\theta}(k)\right]dk},\\
\frac{\partial W}{\partial t} &= \frac{\partial u}{\partial \tau}\frac{\partial \tau}{\partial t} + \frac{\partial u}{\partial y}\frac{\partial y}{\partial t} = -\frac{\sigma^2(t)}{2}\frac{\partial u}{\partial \tau} + \left(\frac{\sigma^2(t)}{2} - \kappa(t)\tilde{\theta}(t)\right)\frac{\partial u}{\partial y},\\
\frac{\partial W}{\partial \bar{r}} &= \frac{\partial u}{\partial y}\frac{\partial y}{\partial \bar{r}} = \frac{1}{\bar{r}}\frac{\partial y}{\partial \bar{r}}, \quad \frac{\partial^2 W}{\partial \bar{r}^2} = \frac{\partial^2 u}{\partial y^2}\left(\frac{\partial y}{\partial \bar{r}}\right)^2 + \frac{\partial u}{\partial y}\frac{\partial^2 y}{\partial \bar{r}^2} = \frac{1}{\bar{r}^2}\frac{\partial^2 u}{\partial y^2} - \frac{1}{\bar{r}^2}\frac{\partial u}{\partial y}.
\end{aligned}$$

It is worth mentioning that, by the change of variables $y \to -y$ and $\tau \to -i\tau$, this PDE transforms into the time-dependent Schrödinger equation with the unsteady Morse potential.

This PDE in Eq. (14.15) should be solved subject to the initial and boundary conditions (see discussion in (Itkin and Muravey, 2020b))

$$u(0,y) = \exp[-B(T)e^{y-A(T)}], \qquad u(\tau,y)\Big|_{y\to\infty} = 0, \tag{14.16}$$

$$B(T) = \frac{\kappa(T)}{\sigma^2(T)}, \qquad A(T) = \int_0^T \left[\frac{\sigma^2(k)}{2} - \kappa(k)\tilde{\theta}(k)\right] dk.$$

Since $u(0,y)$ is continuous for $y \in (-\infty,\infty)$, $u(0,y)$ can be represented as

$$u(0,y) = \frac{1}{2} - \frac{1}{2}(H(y) - H(-y)) + g(y), \qquad g(\infty) = g(-\infty) = 0,$$

where $H(y)$ is the Heaviside step function. The Fourier transforms of all parts in the RHS of this equation exist implying that it also exists for $u(0,y)$. Therefore, applying the Fourier transform

$$\bar{u}(\tau,\omega) = \int_{-\infty}^{\infty} u(\tau,y)e^{-i\omega y}dy, \tag{14.17}$$

to both parts of Eq. (14.15) yields the ordinary differential equation

$$\frac{d\bar{u}}{d\tau} + \omega^2\bar{u} = g(\tau,\omega), \tag{14.18}$$

$$g(\tau,\omega) = \int_{-\infty}^{\infty} u(\tau,y)\left[\alpha(\tau)e^{y} + \beta(\tau)e^{2y}\right]e^{-i\omega y}dy,$$

$$\bar{u}(0,\omega) = \int_{-\infty}^{\infty} u(0,y)e^{-i\omega y}dy.$$

The solution of this problem can be written as

$$\bar{u}(\tau,\omega) = e^{-\omega^2\tau}\bar{u}(0,\omega) + \int_0^{\tau} e^{-\omega^2(\tau-s)}g(s,\omega)ds \tag{14.19}$$

Now, applying the inversion formula

$$u(\tau,y) = \frac{1}{2\pi}\int_{-\infty}^{\infty} \bar{u}(\tau,\omega)e^{i\omega y}d\omega,$$

we obtain the following representation for $u(\tau,y)$

$$u(\tau,y) = \frac{1}{2\pi}\int_{-\infty}^{\infty}\left[e^{-\omega^2\tau}\bar{u}(0,\omega) + \int_0^{\tau} e^{-\omega^2(\tau-s)}g(s,\omega)ds\right]e^{i\omega y}d\omega \tag{14.20}$$

Substituting the explicit representations for $\bar{u}(0,\omega)$ and $g(\tau,\omega)$ into Eq. (14.20), and taking into account that the function $e^{-\omega^2\tau}$ is an even function, we obtain

$$u(\tau,y) = \frac{1}{\pi}\int_0^{\infty}\left\{e^{-\omega^2\tau}\int_{-\infty}^{\infty} u(0,\xi)\cos[\omega(y-\xi)]d\xi\right. \tag{14.21}$$

$$+\int_0^\tau ds\, e^{-\omega^2(\tau-s)} \int_{-\infty}^{\infty} u(\tau,\xi) \left[\alpha(s)e^{\xi} + \beta(s)e^{2\xi}\right] \cos[\omega(y-\xi)]d\xi \Bigg\} d\omega.$$

Applying the identity (Gradshtein and Ryzhik, 2007)

$$\int_{-\infty}^{\infty} e^{-\beta x^2} \cos(bx)dx = \sqrt{\frac{\pi}{\beta}} \exp\left(-\frac{b^2}{4\beta}\right),$$

and changing the order of integration, we get the integral equation for $u(\tau, y)$

$$u(\tau,y) = \frac{1}{2\sqrt{\pi}} \Bigg\{ \int_{-\infty}^{\infty} \frac{u(0,\xi)}{\sqrt{\tau}} e^{-\frac{(y-\xi)^2}{4\tau}} d\xi \tag{14.22}$$

$$+ \int_{-\infty}^{\infty} d\xi \int_0^\tau \frac{u(k,\xi)}{\sqrt{\tau-k}} e^{-\frac{(y-\xi)^2}{4(\tau-k)}} \left[\alpha(k)e^{\xi} + \beta(k)e^{2\xi}\right] dk \Bigg\}.$$

This is a two-dimensional Volterra equation of the second kind (see Lipton, 2001; Lipton and Kaushansky, 2020b; Itkin and Muravey, 2020b; Carr *et al.*, 2020) for the discussion. It is also straightforward to show that this equation is a direct consequence of the Duhamel's principle applied to inhomogeneous PDE.

As far as the numerical solution of Eq. (14.22) is concerned, the simplest scheme would be Picard iterations. First, we can change the order of integration, so the integral in time becomes the outer one. It can be approximated by using, e.g. the trapezoidal rule. Then the coefficients of integration can be computed by using Improved Fast Gauss Transform with the complexity $O(N+n)$, N is the number of grid points in y and n - in ξ. And computing the outer integral requires $O(M)$ operations. Thus, the speed of the method is same as of a FD scheme with the second order of approximation in space and time (e.g., the Crank-Nicolson one). However, using higher order Simpson quadratures can reduce M to $\sqrt{M}$ providing same accuracy, while doing same for the FD method is not trivial. We discuss this in more detail below in the chapter.

14.3 A closed-form solution for the ZCB price in the Verhulst model

Here, we show that for some dependencies between the parameters of the model, the Cauchy problem Eq. (14.13) can be solved explicitly in terms of the confluent hypergeometric function (Abramowitz and Stegun, 1964).

We start with the following change of variables

$$x = \phi(t)\bar{r}, \qquad \phi(t) = e^{\int_0^t \left[C_\alpha \sigma^2(k) - \kappa(k)\tilde{\theta}(k)\right]dk}, \tag{14.23}$$

where C_α is a constant. This change of variables yields

$$0 = \frac{\partial W}{\partial t} + \frac{1}{2}\sigma^2(t)x^2\frac{\partial^2 W}{\partial x^2} + C_\alpha\sigma^2(t)x\frac{\partial W}{\partial x} + \left[\frac{\gamma(t)}{\phi(t)}x - \frac{\kappa^2(t)}{2\sigma^2(t)\phi^2(t)}x^2\right]W$$

$$W(T,x) = e^{-\frac{\kappa(T)}{\sigma^2(T)\phi(T)}x}, \qquad W(T,x)\Big|_{x\to\infty} = 0. \tag{14.24}$$

We assume $\kappa(t)$, $\tilde{\theta}(t)$ and $\sigma(t)$ satisfy the following conditions

$$C_\gamma = \frac{\gamma(t)}{\kappa(t)}, \qquad C_\sigma = \frac{\sigma^2(t)\phi(t)}{2\kappa(t)}, \tag{14.25}$$

where C_γ and $C_\sigma > 0$ are some constants. Using the definitions of $\gamma(t), \phi(t)$, and some algebra, one can show that under these conditions, we have

$$C_\gamma = C_\alpha - \frac{R}{\kappa(t)}, \quad C_\sigma = \frac{\sigma^2(0)}{2\kappa(t)}, \tag{14.26}$$
$$\tilde{\theta}(t) = \frac{C_\alpha\sigma^2(t)}{\kappa(t)} + \frac{2}{\kappa(t)}\frac{\sigma'(t)}{\sigma(t)}, \quad \phi(t) = \frac{\sigma^2(0)}{\sigma^2(t)}.$$

The first equality implies that in this case the mean reversion rate $\kappa(t) = \kappa = const$, and C_γ depends on C_α and κ, while C_σ depends on κ and $\sigma(0)$. It means that we can calibrate the model as follows. First, we calibrate the volatility term structure to the market, together with the constant mean reversion rate of κ and the constant C_α. Second, we determine the time-dependent mean-reversion level $\tilde{\theta}(t)$ by using Eq. (14.26). Thus, in this version of the model, we have three calibration parameters: two of them - κ and C_α are constants, and the normal volatility $\sigma(t)$ is time-dependent. In other words, this enables capturing the volatility term-structure of the market which seems to be the most important property, while assuming a constant mean reversion speed is not too restrictive. The time-dependence of the mean reversion level, however, is fully defined by $\sigma(t)$ and is corrected by another calibrated constant C_α. So this seems to be a weak side of the model.

Now, applying another change of variables to Eq. (14.24)

$$w = \frac{x}{C_\sigma}, \ \tau = \frac{1}{2}\int_t^T \sigma^2(k)\,dk, \ W(t,x) = w^{-C_\alpha}e^{C_\alpha(1-C_\alpha)\tau}u(\tau,w), \tag{14.27}$$

we obtain the PDE with the time-homogeneous coefficients

$$0 = \frac{\partial^2 u}{\partial w^2} + \left[-\frac{1}{4} + \frac{C_\gamma}{w}\right]u - \frac{1}{w^2}\frac{\partial u}{\partial \tau}. \tag{14.28}$$

This PDE should be solved subject to the initial and boundary conditions

$$u(0,w) = w^{C_\alpha} e^{-\frac{\kappa(T)}{\sigma^2(T)\phi(T)} C_\sigma w} = w^{C_\alpha} e^{-w/2}, \ u(\tau,w)\Big|_{w\to\infty} = 0. \qquad (14.29)$$

Applying the Laplace transform

$$\bar{u}(\lambda,w) = \int_0^\infty e^{-\lambda\tau} u(w,\tau) d\tau. \qquad (14.30)$$

to Eq. (14.28) and introducing $\mu = \sqrt{\lambda + 1/4}$, we obtain the following inhomogeneous ordinary differential equation

$$\frac{d^2\bar{u}}{dw^2} + \left[-\frac{1}{4} + \frac{C_\gamma}{w} + \frac{1/4-\mu^2}{w^2}\right]\bar{u} = -\frac{u(0,w)}{w^2}, \ \bar{u}(\lambda,w)\Big|_{w\to\infty} = 0. \ (14.31)$$

The corresponding homogeneous Eq. (14.31) is a Whittaker equation, which has two linearly independent solutions (the Whittaker functions) $M_{C_\gamma,\mu}(w)$ and $W_{C_\gamma,\mu}(w)$ (Abramowitz and Stegun, 1964). A general solution of the problem Eq. (14.31) reads

$$\begin{aligned}
\bar{u}(\lambda,w) &= C_1 M_{C_\gamma,\mu}(w) + C_2 W_{C_\gamma,\mu}(w) \qquad (14.32)\\
&+ \frac{\Gamma(1/2 - C_\gamma + \mu)}{\Gamma(1+2\mu)}\Big(W_{C_\gamma,\mu}(w)\int_0^w \frac{u(0,\zeta)}{\zeta^2} M_{C_\gamma,\mu}(\zeta)d\zeta \\
&- M_{C_\gamma,\mu}(w)\int_0^w \frac{u(0,\zeta)}{\zeta^2} W_{C_\gamma,\mu}(\zeta)d\zeta\Big)\\
&= C_3 M_{C_\gamma,\mu}(w) + C_2 W_{C_\gamma,\mu}(w)\\
&+ \frac{\Gamma(1/2 - C_\gamma + \mu)}{\Gamma(1+2\mu)}\Big(W_{C_\gamma,\mu}(w)\int_0^w \frac{u(0,\zeta)}{\zeta^2} M_{C_\gamma,\mu}(\zeta)d\zeta \\
&+ M_{C_\gamma,\mu}(w)\int_w^\infty \frac{u(0,\zeta)}{\zeta^2} W_{C_\gamma,\mu}(\zeta)d\zeta\Big)\\
C_3 &= C_1 + \int_0^\infty \frac{u(0,\zeta)}{\zeta^2} W_{C_\gamma,\mu}(\zeta)d\zeta.
\end{aligned}$$

Using the asymptotic expressions of the Whittaker functions (Abramowitz and Stegun, 1964) at $w \to 0$

$$\begin{aligned}
M_{\kappa,\mu}(w) &= w^{\mu+1/2}(1 + O(w)), \qquad (14.33)\\
W_{\kappa,\mu}(w) &= \frac{\Gamma(2\mu)}{\Gamma(1/2+\mu-\kappa)} w^{1/2-\mu} + O\left(w^{3/2-\Re(\mu)}\right),
\end{aligned}$$

and at $w \to \infty$

$$\begin{aligned}
M_{\kappa,\mu}(w) &\sim \frac{\Gamma(1+2\mu)}{\Gamma(1/2+\mu-\kappa)} e^{w/2} w^{-\kappa}, \qquad (14.34)\\
W_{\kappa,\mu}(w) &\sim e^{-w/2} w^{\kappa},
\end{aligned}$$

and the boundary condition in Eq. (14.31), we can set $C_3 = C_2 = 0$. Here $\Re(x)$ denotes the real part of x.

Since the integrands in Eq. (14.32) have singularities at the points $w = 0$ and $w \to \infty$ we need to check that both functions $u(\tau, w), W(t, x)$ are regular at these points. Applying Eq. (14.33) and L'Hôspital's rule yields

$$\begin{aligned}
\lim_{w\to 0} \frac{\bar{u}(\lambda, w)}{w^{C_\alpha}} &= \lim_{w\to 0} \left[\frac{1}{2\mu} \frac{\int_0^w u(0,\zeta)\zeta^{-2} M_{C_\gamma,\mu}(\zeta)d\zeta}{w^{C_\alpha+\mu-1/2}} \right. \\
&\quad \left. + \frac{\Gamma(1/2 - C_\gamma + \mu)}{\Gamma(1+2\mu)} \frac{\int_w^\infty u(0,\zeta)\zeta^{-2} W_{C_\gamma,\mu}(\zeta)d\zeta}{w^{C_\alpha-\mu-1/2}} \right] \\
&= \frac{1}{2\mu} \lim_{w\to 0} \left[\frac{u(0,w)w^{-C_\alpha}}{C_\alpha + \mu - 1/2} - \frac{u(0,w)w^{-C_\alpha}}{C_\alpha - \mu - 1/2} \right] \\
&= \frac{1}{(\mu + C_\alpha - 1/2)(\mu - C_\alpha + 1/2)} \lim_{w\to 0} e^{-w/2} \\
&= \frac{1}{\mu^2 - 1/4 - C_\alpha^2 + C_\alpha} = \frac{1}{\lambda - C_\alpha^2 + C_\alpha}
\end{aligned}$$

$$\begin{aligned}
\lim_{w\to\infty} \frac{\bar{u}(\lambda, w)}{w^{C_\alpha}} &= \lim_{w\to\infty} \left[\frac{\Gamma(1/2 - C_\gamma + \mu)}{\Gamma(1+2\mu)} \frac{\int_0^w u(0,\zeta)\zeta^{-2} M_{C_\gamma,\mu}(\zeta)d\zeta}{e^{w/2} w^{C_\alpha - C_\gamma}} \right. \\
&\quad \left. + \frac{\int_w^\infty u(0,\zeta)\zeta^{-2} W_{C_\gamma,\mu}(\zeta)d\zeta}{e^{-w/2} w^{C_\alpha + C_\gamma}} \right] \\
&= \lim_{w\to\infty} \left[\frac{e^{w/2} w^{-C_\gamma - 2} u(0,w)}{e^{w/2} w^{C_\alpha - C_\gamma} (1/2 + (C_\alpha - C_\gamma)/w)} \right. \\
&\quad \left. - \frac{e^{-w/2} w^{C_\gamma - 2} u(0,w)}{e^{-w/2} w^{C_\alpha + C_\gamma} (-1/2 + (C_\alpha + C_\gamma)/w)} \right] \\
&= \lim_{w\to\infty} \left[\frac{u(0,\omega)w^{-C_\alpha}}{w/2 + w^2(C_\alpha - C\gamma)} - \frac{u(0,\omega)w^{-C_\alpha}}{-w/2 + w^2(C_\alpha + C\gamma)} \right] \\
&= \lim_{w\to\infty} \left[\frac{e^{-w/2}}{w/2 + w^2(C_\alpha - C\gamma)} - \frac{e^{-w/2}}{-w/2 + w^2(C_\alpha + C\gamma)} \right] = 0.
\end{aligned}$$

And inverting the Laplace transform we obtain

$$\lim_{w\to 0} \frac{u(\tau, w)}{w^{C_\alpha}} = e^{-C_\alpha(1-C_\alpha)\tau}, \quad \lim_{w\to\infty} \frac{u(\tau, w)}{w^{C_\alpha}} = 0.$$

Returning back to the original variables yields

$$\lim_{x\to 0} W(t,x) = 1, \quad \lim_{x\to\infty} W(t,x) = 0.$$

Accordingly, since $\bar{r} \to 0$ implies $x \to 0$, this yields

$$F(\tau, \bar{r})|_{\bar{r}\to 0} \to e^{\int_T^{t(\tau)} s(k)dk}. \tag{14.35}$$

At $t = T$ (or $\tau = 0$) this limit is consistent with the terminal condition in Eq. (14.11).

Expression Eq. (14.32) can be further simplified by using the formula (Gradshtein and Ryzhik, 2007)

$$\begin{aligned}
&\int_0^\infty e^{-\frac{1}{2}(a_1+a_2)t\cosh(k)} \coth^{2\nu}\left(\frac{k}{2}\right) I_{2\mu}\left(t\sqrt{a_1 a_2}\sinh(k)\right) dk \qquad (14.36)\\
&\quad = \frac{\Gamma(1/2+\mu-\nu)}{t\sqrt{a_1 a_2}\Gamma(1+2\mu)} W_{\nu,\mu}(a_1 t) M_{\nu,\mu}(a_2 t),\\
&\quad \Re(1/2+\mu-\nu) > 0, \qquad \Re(\mu) > 0, \qquad a_1 > a_2,
\end{aligned}$$

where $I_\nu(w)$ is the modified Bessel function. Therefore, setting $t = 1, a_1 = w, a_2 = \xi$, and perceiving that Eq. (14.36) is symmetric with respect to a_1 and a_2 (so that the integrals on ξ in Eq. (14.32) are complimentary and sum up to a single integral from 0 to infinity), we obtain

$$\begin{aligned}
\bar{u}(\lambda, w) = &\int_0^\infty \int_0^\infty u(0,\zeta)\sqrt{\frac{w}{\zeta^3}} e^{-\frac{1}{2}(w+\zeta)\cosh(k)} \coth^{2C_\gamma}\left(\frac{k}{2}\right) \qquad (14.37)\\
&\times I_{2\mu}\left(\sqrt{w\zeta}\sinh(k)\right) dk\, d\zeta.
\end{aligned}$$

Using the inversion formula for the Laplace transform, we get $u(\tau, w)$ of the form

$$\begin{aligned}
u(\tau, w) = &\frac{1}{2\pi \mathrm{i}} \int_{\gamma_*} e^{\lambda\tau} \int_0^\infty \int_0^\infty u(0,\zeta)\sqrt{\frac{w}{\zeta^3}} e^{-\frac{1}{2}(w+\zeta)\cosh(k)} \coth^{2C_\gamma}\left(\frac{k}{2}\right)\\
&\times I_{2\mu}\left(\sqrt{w\zeta}\sinh(k)\right) dk\, d\zeta\, d\lambda. \qquad (14.38)
\end{aligned}$$

Here γ_* denotes any vertical line $\Re(\lambda) = \gamma_*$ in the complex plane such that all singularities of the integrand in Eq. (14.38) lie to the left of this line.

Applying another identity (Gradshtein and Ryzhik, 2007)

$$\begin{aligned}
&\int_0^\infty x^{\nu-\frac{1}{2}} e^{-\alpha x} I_{2\mu}\left(2\beta\sqrt{x}\right) dx = \frac{\Gamma(\nu+\mu+1/2)}{\Gamma(2\mu+1)} \beta^{-1} e^{\frac{\beta^2}{2\alpha}} \alpha^{-\nu} M_{-\nu,\mu}\left(\frac{\beta^2}{\alpha}\right),\\
&\qquad \Re\left(\mu+\nu+\frac{1}{2}\right) > 0,
\end{aligned}$$

to the internal integral of ζ, we obtain the following representation for $u(\tau, w)$

$$u(\tau, w) = \frac{e^{-w/2}}{2\pi \mathrm{i}} \int_{\gamma_*} e^{\lambda\tau} \int_0^\infty \frac{2}{\sinh(k)} \left\{ \Upsilon(\lambda) e^{-\frac{w}{4}(\cosh(k)-1)} \coth^{2C_\gamma}\left(\frac{k}{2}\right) \right. \tag{14.39}$$

$$\times \left(\frac{1+\cosh(k)}{2}\right)^{1-C_\alpha} M_{1-C_\alpha,\sqrt{\lambda+1/4}}\left(\frac{w}{2}(\cosh(k)-1)\right)\Bigg\} dk\, d\lambda,$$

$$\Upsilon(\lambda) = \frac{\Gamma\left(\sqrt{\lambda+1/4}+C_\alpha-1/2\right)}{\Gamma\left(2\sqrt{\lambda+1/4}+1\right)}.$$

By changing the variable of integration k in Eq. (14.39) as

$$2\log\left[\tanh\left(\frac{k}{2}\right)\right] = -\nu, \quad \frac{2dk}{\sinh(k)} = -d\nu, \quad \frac{\cosh(k)-1}{2} = \frac{1}{e^\nu - 1},$$

we get

$$u(\tau,w) = \frac{e^{-w/2}}{2\pi \mathrm{i}}\int_{\gamma_*} e^{\lambda\tau}\Upsilon(\lambda)\int_0^\infty e^{-\frac{w}{2(e^\nu-1)}+\nu C_\gamma} M_{1-C_\alpha,\sqrt{\lambda+1/4}}\left(\frac{w}{e^\nu-1}\right) \tag{14.40}$$

$$\times\left(\frac{e^\nu}{e^\nu-1}\right)^{1-C_\alpha} d\nu d\lambda.$$

14.3.1 *Residuals and branching points*

The integral in Eq. (14.40) has poles at the points $\lambda = \lambda_k$

$$\sqrt{\lambda_k + \frac{1}{4}} = \mu_k = \frac{1}{2} - C_\alpha - k, \quad k = 0, \dots, \left\lfloor \frac{1}{2} - C_\alpha \right\rfloor, \tag{14.41}$$

since the Gamma function in the numerator of $\Upsilon(\lambda)$ turns to complex infinity when its argument is a non-positive integer or zero. The number of poles depends on the value of C_α : if $C_\alpha \geq 1/2$, the integrand function has no poles, if $C_\alpha < 1/2$ the number of the poles is $K+1$, where $K = \lfloor \frac{1}{2} - C_\alpha \rfloor$, and $\lfloor x \rfloor$ is the floor of x. The function $\sqrt{\lambda + 1/4}$ is a multivalued function of λ, i.e., the point $\lambda = -1/4$ is a branching point. Therefore, let us construct the contour of integration in Eq. (14.38) as the so-called keyhole contour presented in Fig. 14.1. In more detail, this contour can be described as follows. It starts with a vertical line γ_* extending to the two big symmetric arcs Γ_1 and Γ_2 around the origin with the radius R_1; connecting to two horizontal parallel lines segments l_1, l_2 at $\mathrm{Im}(\sqrt{\lambda+1/4}) = \pm\omega$; then extending to two vertical line segments l_3, l_4 which end points are connected to the semi-arc γ_ε with the radius ε around the point $\mathrm{Re}(\lambda) = -1/4$. Using a standard technique, we take the limit $\varepsilon \to 0, R_1 \to \infty$, so in this limit the integrals along the lines l_3 and l_4 are cancelled out. The integral along the contours Γ_1 and Γ_2 tends to zero if $R_1 \to \infty$ due to Jordan's lemma.

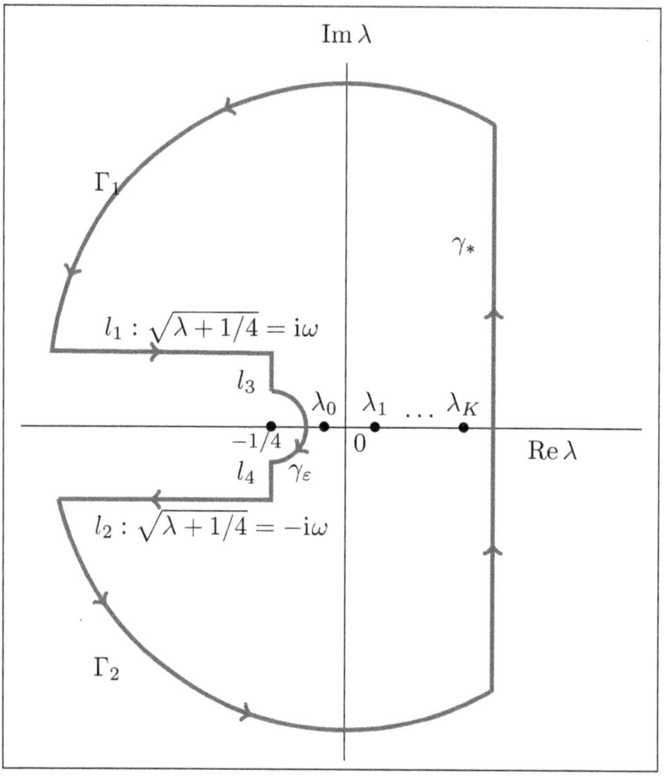

Figure 14.1: Contour of integration of Eq. (14.38) in the complex λ plane with poles at $\lambda_0, \lambda_1, \ldots, \lambda_K$. This picture corresponds to $C_\alpha > 0$, and $C_\alpha < 1/2$. If $C_\alpha < 0$, all poles are positive.

Hence, according to the Cauchy residue theorem, the sum of integral along the vertical line γ_* and two integrals along the horizontal semi-infinite lines l_1 and l_2 is equal to the sum of residuals.

Let us define the sum of residuals as $\mathcal{R}(\tau, z)$ and the sum of integrals along the lines l_1 and l_2 with a negative sign as $\mathcal{I}(\tau, z)$. We explicitly compute them in the next section.

14.3.2 *Calculation of residuals*

Using the well-known expressions for the poles of the Gamma function, and the connection formula for the Whittaker functions $M_{k,\mu}(z)$ and $W_{k,\mu}(z)$

(Abramowitz and Stegun, 1964)

$$\operatorname*{Res}_{x=-n}\Gamma(x) = \frac{(-1)^n}{n!}, \; W_{\frac{1}{2}a+\frac{1}{2}+n,\frac{1}{2}a}(z) = (-1)^n \frac{\Gamma(a+1+n)}{\Gamma(a+1)} M_{\frac{1}{2}a+\frac{1}{2}+n,\frac{1}{2}a}(z),$$

we obtain

$$\begin{aligned}
&\operatorname*{Res}_{\lambda=\lambda_k}\left\{ e^{\lambda\tau}\Upsilon(\lambda) M_{1-C_\alpha,\sqrt{\lambda+1/4}}\left(\frac{w}{e^\nu - 1}\right)\right\} \\
&\quad = \frac{2\mu_k}{k!}\frac{e^{\left(\mu_k^2-\frac{1}{4}\right)\tau}}{\Gamma\left(\frac{3}{2}-C_\alpha+\mu_k\right)} W_{1-C_\alpha,\mu_k}\left(\frac{w}{e^\nu-1}\right).
\end{aligned} \tag{14.42}$$

Thus, the sum of the residuals after the substitutions $1/(e^\nu - 1) \mapsto \varsigma$, $\mu_k = \frac{1}{2} - k - C_\alpha$ reads

$$\begin{aligned}
\mathcal{R}(\tau, w) &= e^{-w/2}\sum_{k=0}^{K}\frac{2\mu_k}{k!}\frac{e^{\left(\mu_k^2-\frac{1}{4}\right)\tau}}{\Gamma\left(\frac{3}{2}-C_\alpha+\mu_k\right)} \\
&\quad \times \int_0^\infty W_{1-C_\alpha,\mu_k}(w\varsigma)\, e^{-\frac{w\varsigma}{2}}(1+\varsigma)^{C_\gamma - C_\alpha}\varsigma^{-1-C_\gamma}\, d\varsigma.
\end{aligned} \tag{14.43}$$

The integrals of ς can be computed analytically

$$\begin{aligned}
&\int_0^\infty W_{1-C_\alpha,\mu}(w\varsigma)\, e^{-\frac{w\varsigma}{2}}(1+\varsigma)^{C_\gamma - C_\alpha}\varsigma^{-1-C_\gamma}\, d\varsigma \\
&\quad = w^{\frac{1}{2}+\mu}\frac{\Gamma^{+}_{C_\gamma}(\mathrm{i}\mu)\,\Gamma^{-}_{C_\gamma}(\mathrm{i}\mu)}{\Gamma(C_\alpha - C_\gamma)} U\left(\frac{1}{2}+\mu - C_\gamma, 2\mu+1, w\right), \quad w > 0, \\
&\Gamma^{\pm}_{y}(\omega) = \Gamma\left(\frac{1}{2} - y \pm \mathrm{i}\omega\right), \; \Re(C_\gamma) < \frac{1}{2} + \Re(\mu), \; \Re(C_\gamma) < \frac{1}{2} - \Re(\mu).
\end{aligned} \tag{14.44}$$

Here $U(a, b, x)$ is a Kummer confluent hypergeometric function, (Abramowitz and Stegun, 1964). One can check that the conditions on C_γ in Eq. (14.44) are automatically satisfied for any μ_k. Using the relation between the Kummer and Whittaker functions

$$W_{k,\mu}(x) = e^{-\frac{z}{2}} z^{1/2+\mu} U\left(\frac{1}{2}+\mu-k, 1+2\mu, x\right), \qquad W_{k,\mu}(x) = W_{k,-\mu}(x),$$

we finally obtain from Eq. (14.43), Eq. (14.44)

$$\mathcal{I}(\tau, w) = \sum_{k=0}^{K}\frac{e^{\left(\mu_k^2-\frac{1}{4}\right)\tau}}{k!}\frac{2\mu_k}{\Gamma\left(\frac{3}{2}-C_\alpha+\mu_k\right)}\frac{\Gamma^{+}_{C_\gamma}(\mathrm{i}\mu_k)\,\Gamma^{-}_{C_\gamma}(\mathrm{i}\mu_k)}{\Gamma(C_\alpha - C_\gamma)} W_{C_\gamma,\mu_k}(w). \tag{14.45}$$

14.3.3 *Calculation of the integrals at different branches*

The integrals along the lines l_1 and l_2 read

$$\begin{aligned}&\int_{l_1} e^{\lambda\tau}\Upsilon(\lambda)M_{1-C_\alpha,\sqrt{\lambda+1/4}}\left(\frac{w}{e^\nu-1}\right)d\lambda\\&\quad=2e^{-\tau/4}\int_0^\infty \omega e^{-\omega^2\tau}\frac{\Gamma\left(\mathrm{i}\omega+C_\alpha-1/2\right)}{\Gamma\left(2\mathrm{i}\omega+1\right)}M_{1-C_\alpha,\mathrm{i}\omega}\left(\frac{w}{e^\nu-1}\right)d\omega,\\&\int_{l_2} e^{\lambda\tau}\Upsilon(\lambda)M_{1-C_\alpha,\sqrt{\lambda+1/4}}\left(\frac{w}{e^\nu-1}\right)d\lambda\\&\quad=-2e^{-\tau/4}\int_0^\infty \omega e^{-\omega^2\tau}\frac{\Gamma\left(-\mathrm{i}\omega+C_\alpha-1/2\right)}{\Gamma\left(-2\mathrm{i}\omega+1\right)}M_{1-C_\alpha,-\mathrm{i}\omega}\left(\frac{w}{e^\nu-1}\right)d\omega.\end{aligned}$$

Applying the connection formula between the Whittaker functions $M_{\nu,\mu}(w)$ and $W_{\nu,\mu}(w)$, and the Euler's reflection formulas (Abramowitz and Stegun, 1964)

$$\begin{aligned}W_{\lambda,\mu}(w)&=\frac{\Gamma(-2\mu)}{\Gamma(1/2-\mu-\lambda)}M_{\lambda,\mu}(w)+\frac{\Gamma(2\mu)}{\Gamma(1/2+\mu-\lambda)}M_{\lambda,-\mu}(w)\\\Gamma(w)\Gamma(1-w)&=\frac{\pi}{\sin(\pi w)},\qquad \Gamma(1/2+w)\Gamma(1/2-w)=\frac{\pi}{\cos(\pi w)}\end{aligned}$$

we get the following expression for the sum of the integrals along the lines l_1 and l_2

$$\begin{aligned}&\int_{l_1+l_2} e^{\lambda\tau}\Upsilon(\lambda)M_{1-C_\alpha,\sqrt{\lambda+1/4}}\left(\frac{w}{e^\nu-1}\right)d\lambda\\&\quad=2e^{-\tau/4}\int_0^\infty\frac{\omega e^{-\omega^2\tau}}{2\mathrm{i}\omega}\left\{\frac{\Gamma\left(\mathrm{i}\omega+C_\alpha-1/2\right)}{\Gamma\left(2\mathrm{i}\omega\right)}M_{1-C_\alpha,\mathrm{i}\omega}\left(\frac{w}{e^\nu-1}\right)\right.\\&\quad\left.+\frac{\Gamma\left(-\mathrm{i}\omega+C_\alpha-1/2\right)}{\Gamma\left(-2\mathrm{i}\omega\right)}M_{1-C_\alpha,-\mathrm{i}\omega}\left(\frac{w}{e^\nu-1}\right)\right\}d\omega\\&\quad=2e^{-\tau/4}\int_0^\infty\frac{\omega e^{-\omega^2\tau}}{2\mathrm{i}\omega}\frac{\Gamma^+_{1-C_\alpha}\left(\omega\right)\Gamma^-_{1-C_\alpha}\left(\omega\right)}{\Gamma\left(2\mathrm{i}\omega\right)\Gamma\left(-2\mathrm{i}\omega\right)}W_{1-C_\alpha,\mathrm{i}\omega}\left(\frac{w}{e^\nu-1}\right)d\omega\\&\quad=-\frac{2\mathrm{i}}{\pi}e^{-\tau/4}\int_0^\infty \omega e^{-\omega^2\tau}\sinh\left(2\pi\omega\right)\\&\quad\times\Gamma^+_{1-C_\alpha}\left(\omega\right)\Gamma^-_{1-C_\alpha}\left(\omega\right)W_{1-C_\alpha,\mathrm{i}\omega}\left(\frac{w}{e^\nu-1}\right)d\omega.\end{aligned}$$

Substituting this expression into Eq. (14.40), we obtain

$$\mathcal{I}(\tau,w)=\frac{e^{-w/2-\tau/4}}{\pi^2}\int_0^\infty\int_0^\infty \omega e^{-\omega^2\tau}\sinh\left(2\pi\omega\right)\Gamma^+_{1-C_\alpha}\left(\omega\right)\Gamma^-_{1-C_\alpha}\left(\omega\right)$$

$$\times W_{1-C_\alpha,\mathrm{i}\omega}\left(\frac{w}{e^\nu-1}\right)e^{-\frac{w}{2(e^\nu-1)}}\frac{e^{\nu(1-C_\alpha+C_\gamma)}}{(e^\nu-1)^{1-C_\alpha}}d\nu\, d\omega \tag{14.46}$$

$$= \frac{1}{\pi^2\Gamma(C_\alpha-C_\gamma)}\int_0^\infty \omega e^{-(\omega^2+1/4)\tau}\sinh(2\pi\omega)\,\Gamma^-_{C_\gamma}(\omega)\Gamma^+_{C_\gamma}(\omega)$$

$$\times \Gamma^-_{1-C_\alpha}(\omega)\Gamma^+_{1-C_\alpha}(\omega)W_{C_\gamma,\mathrm{i}\omega}(w)\, d\omega.$$

14.3.4 *A closed-form solution for the ZCB price*

Combining Eq. (14.46) and Eq. (14.45), we obtain $u(\tau, w)$ in closed-form

$$u(\tau,w) = \sum_{k=0}^{K}\frac{e^{\left(\mu_k^2-\frac{1}{4}\right)\tau}}{k!}\frac{2\mu_k}{\Gamma\left(\frac{3}{2}-C_\alpha+\mu_k\right)}\frac{\Gamma^-_{C_\gamma}(\mathrm{i}\mu_k)\,\Gamma^+_{C_\gamma}(\mathrm{i}\mu_k)}{\Gamma(C_\alpha-C_\gamma)}W_{C_\gamma,\mu_k}(w)$$

$$+\frac{1}{\pi^2\Gamma(C_\alpha-C_\gamma)}\int_0^\infty \omega e^{-(\omega^2+1/4)\tau}\sinh(2\pi\omega) \tag{14.47}$$

$$\times \Gamma^-_{C_\gamma}(\omega)\Gamma^+_{C_\gamma}(\omega)\Gamma^-_{1-C_\alpha}(\omega)\Gamma^+_{1-C_\alpha}(\omega)W_{C_\gamma,\mathrm{i}\omega}(w)\, d\omega.$$

Since $u(0, w) = w^{C_\alpha}e^{-w/2}$, we obtain a previously unknown identity

$$\frac{1}{\pi^2}\int_0^\infty \omega\sinh(2\pi\omega)\,\Gamma^-_{C_\gamma}(\omega)\Gamma^+_{C_\gamma}(\omega)\Gamma^-_{1-C_\alpha}(\omega)\Gamma^+_{1-C_\alpha}(\omega)W_{C_\gamma,\mathrm{i}\omega}(w)\, d\omega$$

$$= w^{C_\alpha}e^{-w/2}\Gamma(C_\alpha-C_\gamma) \tag{14.48}$$

$$-\sum_{k=0}^{K}\frac{1}{k!}\frac{2\mu_k}{\Gamma\left(\frac{3}{2}-C_\alpha+\mu_k\right)}\Gamma^-_{C_\gamma}(\mathrm{i}\mu_k)\,\Gamma^+_{C_\gamma}(\mathrm{i}\mu_k)\,W_{C_\gamma,\mu_k}(w),$$

which could be verified by numerical integration. Thus, Eq. (14.26), Eq. (14.47) can also be represented in the form

$$F(\tau,w)e^{C_a(C_a-1)\tau-\int_T^{t(\tau)}s(k)dk} = 1 + e^{w/2}w^{-C_\alpha} \tag{14.49}$$

$$\times\left(\sum_{k=0}^{K}\frac{2\mu_k\left[e^{\left(\mu_k^2-\frac{1}{4}\right)\tau}-1\right]}{k!\Gamma\left(\frac{3}{2}-C_\alpha+\mu_k\right)}\frac{\Gamma^-_{C_\gamma}(\mathrm{i}\mu_k)\,\Gamma^+_{C_\gamma}(\mathrm{i}\mu_k)}{\Gamma(R/\kappa)}W_{C_\gamma,\mu_k}(w)\right.$$

$$+\frac{1}{\pi^2\Gamma(R/\kappa)}\int_0^\infty \omega\left[e^{-(\omega^2+1/4)\tau}-1\right]\sinh(2\pi\omega)\,\Gamma^-_{C_\gamma}(\omega)\Gamma^+_{C_\gamma}(\omega)$$

$$\left.\times \Gamma^-_{1-C_\alpha}(\omega)\Gamma^+_{1-C_\alpha}(\omega)W_{C_\gamma,\mathrm{i}\omega}(w)\, d\omega\right).$$

Also, it is known that $\Gamma(a + b\mathrm{i})\Gamma(a - b\mathrm{i}) \in \mathbb{R}$ (Cohen Jr., 1940). Since (Temme, 1978)

$$W_{k,\mu}\left(\frac{1}{2}w^2\right) = 2^{-k}\sqrt{w}\sum_{n=0}^{\infty}\frac{1}{n!}\frac{\Gamma(\frac{1}{2}+2\mu+n)}{\Gamma(\frac{1}{2}+2\mu-n)}\frac{D_{2k-n-1/2}(w)}{(2w)^n},$$

$$D_\nu(x) = 2^{\nu/2} U\left(-\frac{1}{2}\nu, \frac{1}{2}, \frac{1}{2}x^2\right),$$

and denoting $\mu = \mathrm{i}\omega$ we have

$$\frac{\Gamma\left(\frac{1}{2}+2\mu+n\right)}{\Gamma\left(\frac{1}{2}+2\mu-n\right)} = \Gamma\left(\frac{1}{2}+2\mu+n\right)\frac{\cos(\pi(2\mu-n))}{\pi}\Gamma\left(\frac{1}{2}-2\mu+n\right) \in \mathbb{R}.$$

Hence $W_{C_\gamma,\mathrm{i}\omega}(w) \in \mathbb{R}$. Thus, in Eq. (14.47)

$$\Gamma^-_{C_\gamma}(\omega)\Gamma^+_{C_\gamma}(\omega)\Gamma^-_{1-C_\alpha}(\omega)\Gamma^+_{1-C_\alpha}(\omega)W_{C_\gamma,\mathrm{i}\omega}(w) \in \mathbb{R}.$$

14.4 Numerical examples

To validate our analytical solution, we compute the ZCB prices by using numerical integration of Eq. (14.49), where the explicit form of the parameter $\sigma(t)$ is

$$\sigma^2(t) = \sigma_a + \frac{\sigma_b}{t+\sigma_c}, \tag{14.50}$$

where $\sigma_a, \sigma_b, \sigma_c$ are constants. We also assume that $s(t) = 0$, and so $R = r_0$. With these assumptions, we have

$$\begin{aligned}\tilde{\theta} &= \frac{C_\alpha\left(\sigma_a(\sigma_c+t)+\sigma_b\right)}{\kappa(\sigma_c+t)} - \frac{\sigma_b}{k(\sigma_c+t)[\sigma_a(\sigma_c+t)+\sigma_b]}, \\ \tau &= \frac{1}{2}\left[\sigma_a t + \sigma_b \log\left(\frac{t+\sigma_c}{\sigma_c}\right)\right].\end{aligned} \tag{14.51}$$

Since we are interested in the positive values of $\tilde{\theta}(t)$, it implies $\sigma_b < 0$.

We present the model parameters for this test in Table 14.1 (the parameters are artificial and not obtained by calibration to the market). We run the test for a set of maturities $T \in [1/12, 0.3, 0.5, 1, 2, 5, 10, 20, 30, 50]$ years. We show the ZCB prices computed in our numerical experiment in Fig 14.2. As a benchmark, we use the numerical solution of Eq. (14.10) obtained by the FD method (Itkin, 2017), and the solution of Eq. (2.37) obtained by Monte Carlo. To accelerate the FD solution, instead of Eq. (14.10), we solve the forward equation for the density and then find the prices by integrating the payoff with the corresponding density. The FD solver runs on a non-uniform grid with 100 nodes in space w and 200 steps in time t. However, for long maturities, more nodes in space might be necessary (see Itkin, 2017) in more detail. The Monte Carlo method uses 500,000 paths and 500 steps in time (so for $T = 50$ years there is some bias in the results due to a small number of time steps). We perform all calculations in Matlab.

r_0	κ_0	σ_a	σ_b	σ_c	C_α
0.03	2.0	0.64	-1.0	5.0	0.3

Table 14.1: Parameters of the test.

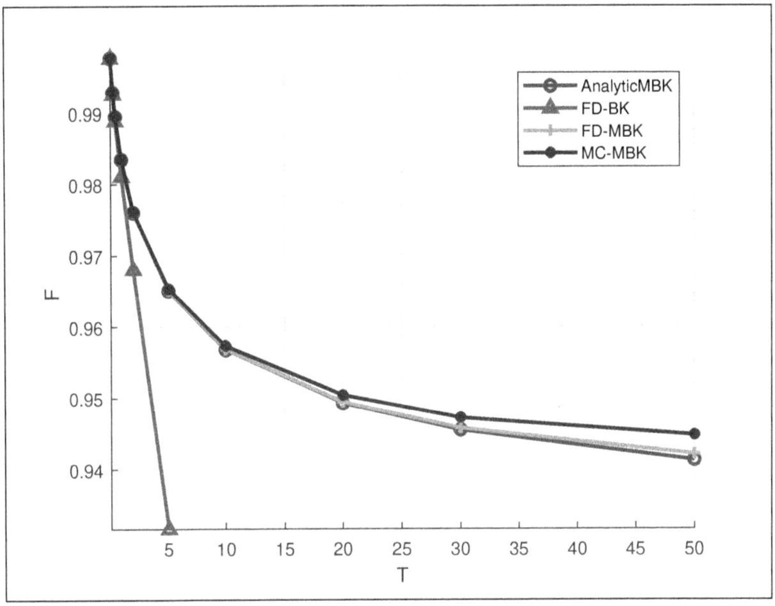

Figure 14.2: ZCB prices computed in the test by using Analytic, FD and Monte Carlo solutions.

	T, years									
Rel error	0.0833	0.3	0.5	1	2	5	10	20	30	50
Anal - FD	-0.0385	-0.0625	-0.0713	-0.0929	-0.2377	-0.2818	-0.5767	-1.4387	-2.8153	-8.7802
Anal - MC	-0.0844	-0.2447	-0.2803	-0.6167	-1.1802	-2.5238	-5.8173	-11.7015	-18.0272	-36.9372

Table 14.2: Relative error in bps between Analytic, FD and MC solutions in the test.

The results obtained by all methods coincide with high accuracy. We compare the corresponding relative errors of those results in Table 14.2. Also, we compute the ZCB prices in the BK model for small T, so that $e^{z_t} \approx 1 + z_t$. We choose $\bar{\theta}(t)_{BK} = \bar{\theta}(t)_{MBK} - 1$. We show the result in

Fig 14.2 as well. It can be seen that the BK ZCB prices agree with the corresponding MBK ZCB prices for $T < 3$, which is due to the fact that z is small.

As far as the performance of the methods is concerned, the elapsed time for computing all 10 ZCB bond prices by using the FD methods is 130 msec. For the analytics, since the only term under the integral in Eq. (14.49), which depends on τ, is $e^{-(\omega^2+1/4)\tau}$, the other terms, including complex-valued Gamma and Whittaker functions, can be computed just once at the beginning of the script and then re-used. We calculate the integral by using the Simpson rule with 75 nodes; the elapsed time for getting all 10 ZCB prices is 55 msec. We note that internal Matlab implementation of the Whittaker functions is a bit slow as it relies on Simulink to compute them. In other programming languages, e.g., C++ or python this is not an issue. Anyway, the performance of our method is on par with that of the forward FD solver, while the accuracy is higher. Moreover, we can further improve the accuracy of the integration by using higher-order quadratures while keeping the elapsed time similar. At the same time, for the FD method, this is problematic (but perhaps can be done by using Radial Basis Functions methods).

14.4.1 *Solving Eq. (14.9) by using RDTM*

To test the RDTM as applied to our problem, we solve Eq. (14.9) by using the modified RDTM described in Section 5.3. Here we use the following explicit form for $\kappa(t), \tilde{\theta}(t), \sigma(t)$

$$\kappa(t) = \kappa_0, \qquad \tilde{\theta}(t) = \theta_0 e^{\theta_1 t}, \qquad \sigma(t) = \sigma_0 e^{-\sigma_1 t}, \tag{14.52}$$

where $\kappa_0, \theta_0, \sigma_0, \theta_1, \sigma_1$ are constants. We also assume $s(t) = 0$, and $R = 1$. With these definitions, we have

$$\begin{aligned}
a(t) &= e^{-k(t-\nu)}, \quad b(t) = \left(e^{\theta_1 \nu} - e^{t(\kappa_0+\theta_1)-\kappa_0\nu}\right)\frac{\kappa_0\theta_0}{\kappa_0+\theta_0}, \qquad (14.53)\\
\beta(t) &= \frac{2}{\sigma_0^2}\exp\left[\frac{2\theta_0}{\kappa_0+\theta_1}\left(e^{\theta_1 t} - e^{\nu(\theta_1+\kappa_0)-\kappa_0 t}\right) + 2t(\sigma_1-\kappa_0) + 2\kappa_0\nu\right],\\
t(\tau) &= \frac{\log\left(e^{-2\sigma_1\nu} - \frac{2\tau(\kappa_0-2\sigma_1)}{\sigma_0^2}\right) + \kappa_0\nu}{\kappa_0 - 2\sigma_1}.
\end{aligned}$$

We present the model parameters for this test in Table 14.3. We run the test for a set of maturities $T \in [1/12, 0.3, 0.5, 1, 2, 5]$. We show the corresponding ZCB prices in Table 14.4. As a benchmark, we use the

r_0	κ_0	θ_0	σ_0	θ_1	σ_1
0.01	1.0	0.05	0.5	0.2	0.2

Table 14.3: Parameters of the test.

	ZCB price, $					
T	**0.0833**	**0.3**	**0.5**	**1.0**	**2.0**	**5.0**
FD	0.9990	0.9938	0.9839	0.9216	0.6201	0.0444
RDTM-1	0.9990	0.9950	0.9887	0.9617	0.5857	0.2477
RDTM-2	0.9990	0.9949	0.9883	0.9587	0.6186	0.6087
	Difference with the FD method, %					
T	**0.0833**	**0.3**	**0.5**	**1.0**	**2.0**	**5.0**
FD	0.0000	0.0000	0.0000	0.0000	0.0000	0.0000
RDTM-1	0.0047	0.1142	0.4907	4.3563	-5.5489	457.6831
RDTM-2	0.0042	0.1039	0.4494	4.0263	-0.2491	1270.3588

Table 14.4: Prices of ZCB bonds with different maturities computed by using the modified RDTM with 1 and 2 terms, and the FD difference method.

solution of Eq. (14.2) obtained by using the FD method described above. The modified RDTM provides reasonable accuracy for small maturities (up to 2 years), while for $T > 2$, one or two terms in the expansion are insufficient to get the correct price. Therefore, for larger T, the Eq. (14.9) has to either be solved numerically or more terms should be taken in the RDTM.

Note, that there are at least two choices for ν in Eq. (14.4): $\nu = 0$ and $\nu = T$. We found that for small T the choice $\nu = 0$ provides slightly better results, while for $T > 1$ it is better to use $\nu = T$.

Also, note that this method, in some sense, is similar to that in Capriotti and Stehlikova (2014). However, we solve an integral equation instead of a PDE in Capriotti and Stehlikova (2014). Besides, there is a difference in parametrization, since we assume that all the parameters are time-dependent. In contrast, in Capriotti and Stehlikova (2014), all model parameters are constant.

As far as the performance of the RDTM is concerned, we compared it with the performance of the FD method applied to the forward equation (the forward analog of Eq. (14.2)). The elapsed time of getting the ZCB prices by solving such the equation is 40 msec while using the RDTM even with the numerical computation of all integrals in Eq. (5.24) takes 13 msec. Therefore, this method allows fast calculation of ZCB prices in the time-dependent BK model for $T < 2$.

PART 6

Credit and Miscellaneous Problems

Chapter 15

Calibrating the Default Boundary to a Constant Default Intensity

In this chapter we discuss calibration of default boundary modelled by a structural default model to a constant default intensity. The original, and straightforward, structural default model was introduced by Merton (1974), who assumed that default could happen only at debt maturity. His model was extended by Black and Cox (1976a), who considered the default, which can happen at any time by introducing flat default boundary representing debt covenants. Numerous authors expanded their model including (Hyer *et al.*, 1998; Hull and White, 2001; Avellaneda and Zhu, 2001), who considered a curvilinear boundary whose shape can be calibrated to the market default probability. One of the major unsolved issues with the above model was articulated by Hyer *et al.* (1998), who pointed out that, unless the shape of the default boundary is very carefully chosen, the probability of short-term default is too low. This issue was addressed by several authors, including Finkelstein and Lardy (2001), Hilberink and Rogers (2002) and Lipton (2002), who proposed to introduce jumps and or uncertainty to increase this probability. We show below that it is possible to calibrate the default boundary in such a way that constant default intensity can be matched. We emphasize that the direct problem — calculating the default probability given the boundary — is linear (albeit relatively involved), while the inverse problem — finding the boundary given the default probability — is nonlinear (and hence even more involved). Additional details are given in Lipton and Kaushansky (2020b)[1]

Here we wish to find the boundary for a structural default model corresponding to a constant default intensity η. We denote the corresponding default probability by

$$\pi(t) = 1 - e^{-\eta t}. \tag{15.1}$$

[1]The material of this chapter is an abridge version of Lipton and Kaushansky (2020b).

Let us introduce the time τ, such that default is impossible for $t < \tau$. Thus the default boundary starts at $t = \tau$. The idea is to calculate the corresponding boundary $b(t; \tau, \eta)$, provided it exists, and then let $\tau \to 0$. It is clear that at time $t = \tau - 0$, the transition probability is

$$p(\tau, x) = H(\tau, x), \tag{15.2}$$

where H is the heat kernel

$$H(\tau, x) = \frac{e^{-\frac{x^2}{2\tau}}}{\sqrt{2\pi\tau}}. \tag{15.3}$$

At time $t = \tau$, the first possibility of default occurs. For $t > \tau$ the transition probability satisfies the following Fokker–Planck problem

$$\begin{gathered}\frac{\partial}{\partial t}p(t,x) = \frac{1}{2}\frac{\partial^2}{\partial x^2}p(t,x), \quad b(t) \leq x < \infty, \\ p(\tau, x) = H(\tau, x), \quad p(t, b(t)) = 0, \quad p(t, x \to \infty) \to 0.\end{gathered} \tag{15.4}$$

The default probability density $g(t)$ is given by

$$g(t) = \frac{1}{2}\frac{\partial}{\partial x}p(t, b(t)), \tag{15.5}$$

or, alternatively

$$\pi(t) = 1 - \int_{b(t)}^{\infty} p(t,x)\,dx, \quad g(t) = \frac{d\pi(t)}{dt}. \tag{15.6}$$

15.1 Governing system of integral equations and its solutions

We split p as follows:

$$p(t,x) = q(t,x) + r(t,x), \tag{15.7}$$

where $q(t,x)$ solves the problem

$$\frac{\partial}{\partial t}q(t,x) = \frac{1}{2}\frac{\partial^2}{\partial x^2}q(t,x), \quad -\infty < x < \infty, \tag{15.8}$$

$$q(\tau, x) = H(\tau, x)\,\Theta(x - b(\tau)), \quad q(t, x \to -\infty) \to 0, \quad q(t, x \to \infty) \to 0, \tag{15.9}$$

and $r(t,x)$ solve the problem

$$\begin{gathered}\frac{\partial}{\partial t}r(t,x) = \frac{1}{2}\frac{\partial^2}{\partial x^2}r(t,x), \quad b(t) \leq x < \infty, \\ r(\tau, x) = 0, \quad r(t, b(t)) = -q(t, b(t)), \quad r(t, x \to \infty) \to 0.\end{gathered} \tag{15.10}$$

Here $\Theta(x)$ is the Heaviside function. Solving Eq. (15.8) as a convolution of heat kernel with the initial condition, we get

$$q(t,x)=\frac{e^{-\frac{x^2}{2t}}}{\sqrt{2\pi t}}N\left(\frac{\frac{ux}{t}-b(\tau)}{\sqrt{u}}\right), \tag{15.11}$$

where $u=(t-\tau)\tau/t$ (see Lipton and Kaushansky, 2020b). Thus

$$\begin{aligned}g(t)=\frac{1}{2}\frac{\partial}{\partial x}r(t,b(t))-\frac{H(t,b(t))}{2t}\Bigg[b(t)\,N\left(\frac{\frac{ub(t)}{t}-b(\tau)}{\sqrt{u}}\right)\\ -uH\left(u,\frac{ub(t)}{t}-b(\tau)\right)\Bigg].\end{aligned} \tag{15.12}$$

Accordingly, as per discussion in Section 8.1.2, we need to solve the following system of integral equations:

$$\begin{aligned}&\nu(t)+\int_\tau^t\frac{\Theta(t,t')\,\Xi(t,t')\,\nu(t')}{\sqrt{2\pi(t-t')}}dt'+H(t,b(t))\,N\left(\frac{ub(t)-tb(\tau)}{t\sqrt{u}}\right)=0,\\ &\eta e^{-\eta t}+\left(\frac{1}{\sqrt{2\pi t}}+\frac{db(t)}{dt}\right)\nu(t)+\frac{1}{2}\int_\tau^t\frac{\Phi(t,t')+\Theta^2(t,t')\,\Xi(t,t')\,\nu(t')}{\sqrt{2\pi(t-t')}}dt'\\ &\quad+\frac{H(t,b(t))}{2t}\left(b(t)\,N\left(\frac{\frac{ub(t)}{t}-b(\tau)}{\sqrt{u}}\right)-uH\left(u,\frac{ub(t)}{t}-b(\tau)\right)\right)\\ &\quad=0.\end{aligned} \tag{15.13}$$

Alternatively, we can rewrite Eq. (15.13) in an integrated form

$$\begin{aligned}&\nu(t)+\int_\tau^t\frac{\Theta(t,t')\,\Xi(t,t')\,\nu(t')}{\sqrt{2\pi(t-t')}}dt'+H(t,b(t))\,N\left(\frac{ub(t)-tb(\tau)}{t\sqrt{u}}\right)=0,\\ &1-e^{-\eta t}+\int_\tau^t\frac{\Xi(t,t')\,\nu(t')}{\sqrt{2\pi(t-t')}}dt'-N\left(\frac{b(t)}{\sqrt{t}}\right)-N\left(\frac{\sqrt{t}b(\tau)}{\sqrt{u(u+t)}}\right)\\ &\quad+BVN\left(\frac{\sqrt{t}b(\tau)}{\sqrt{u(u+t)}},\frac{b(t)}{\sqrt{t}};\sqrt{\frac{u}{u+t}}\right)=0,\end{aligned} \tag{15.14}$$

where $BVN(\cdot,\cdot;\cdot)$ is the bivariate normal distribution.

Now, recall that the default probability has the form

$$\pi(t)=1-e^{-\eta t}. \tag{15.15}$$

The barrier has to start at $\tau=\hat{\tau}$, $\hat{\tau}\to 0$, and there should be no barrier before that. We wish to find $b(\hat{\tau})$ such that

$$\pi(\hat{\tau})=1-\int_{b(\hat{\tau})}^{\infty}\frac{\exp\left(-\frac{x^2}{2\hat{\tau}}\right)}{\sqrt{2\pi\hat{\tau}}}dx=1-N\left(-\frac{b(\hat{\tau})}{\sqrt{\hat{\tau}}}\right)=1-e^{-\eta\hat{\tau}}. \tag{15.16}$$

Thus,

$$N\left(-\frac{b(\hat{\tau})}{\sqrt{\hat{\tau}}}\right)=e^{-\eta\hat{\tau}},\tag{15.17}$$

and

$$b(\hat{\tau})=-\sqrt{\hat{\tau}}N^{-1}\left(e^{-\eta\hat{\tau}}\right).\tag{15.18}$$

Now,

$$N^{-1}(y)\underset{y\to 1}{\sim}\sqrt{2f(\eta)},\tag{15.19}$$

where

$$\begin{aligned}\eta&=-\ln\left(2\sqrt{\pi}\left(1-y\right)\right),\\ f(\eta)&=\eta-\frac{\ln\eta}{2}+\frac{\ln\eta-2}{4\eta}+\frac{(\ln\eta)^2-6\ln\eta+14}{16\eta^2}.\end{aligned}\tag{15.20}$$

Hence,

$$b(\hat{\tau})=-\sqrt{2\hat{\tau}f\left(-\ln\left(2\sqrt{\pi}\left(1-e^{-\eta\hat{\tau}}\right)\right)\right)}\approx-\sqrt{2\hat{\tau}\ln\left(\frac{1}{2\sqrt{\pi}\eta\hat{\tau}}\right)}.\tag{15.21}$$

The Eqs. (15.13) and (15.14) can be solve numerically by using the methods described in Chapter 5. In more detail, see Lipton and Kaushansky (2020b). By using these methods we compute the default boundaries and calibrate them to several representative values of η. The results are shown in Fig. 15.1.

Also, Fig. 15.2 demonstrates that solutions of Eqs. (15.13) and (15.14) coincide modulo numerical error.

These results allow formulation of the following conjecture

Conjecture 15.1. *For a given time interval $I^{(T)}=[0,T]$ there exists a parameter interval $I^{(\eta)}(T)=[0,\eta^*(T)]$ such that for any $\eta\in I^{(\eta)}(T)$ the default boundary $b(t)$ can be calibrated to the default intensity η. We can construct the corresponding boundary as follows*

$$b(t;\eta)=\lim_{\tau\to 0}b(t;\tau,\eta),\quad 0<t\leq T,\tag{15.22}$$

where $b(t;\tau,\eta)$ is found by solving either Eq. (15.13) or Eq. (15.14).

We illustrate this conjecture in Fig. 15.3.

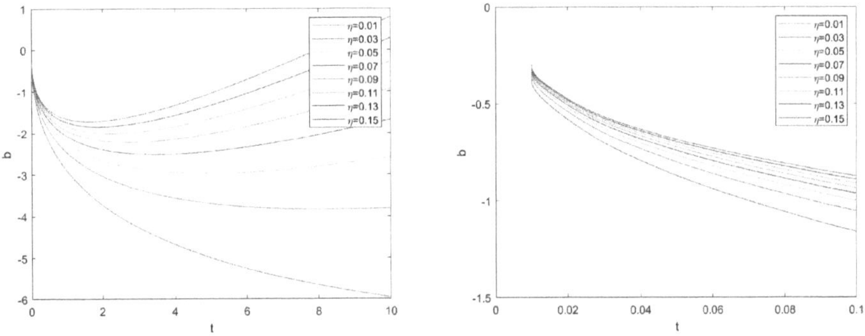

(a) (b)

Figure 15.1: The default boundaries for several representative values of the default intensity η and $\tau = 0.01$: (a) $0.01 < t < 10.0$ to capture their overall behavior; (b) $0.01 < t < 0.1$, so that small features can be shown.

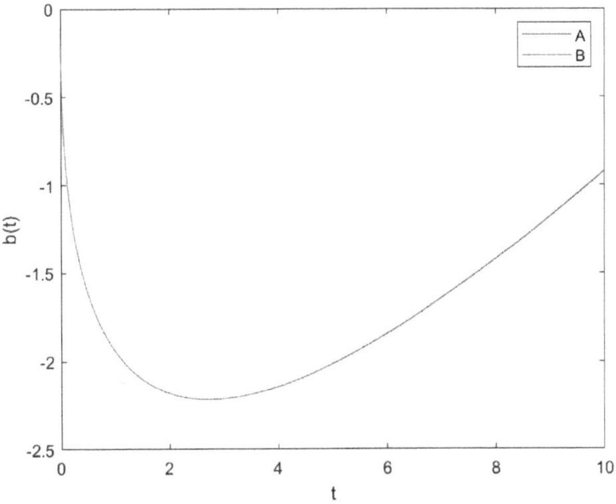

Figure 15.2: With $\eta = 0.09$, the graph shows that boundaries calculated by solving Eqs. (15.13) and (15.14) coincide modulo numerical errors.

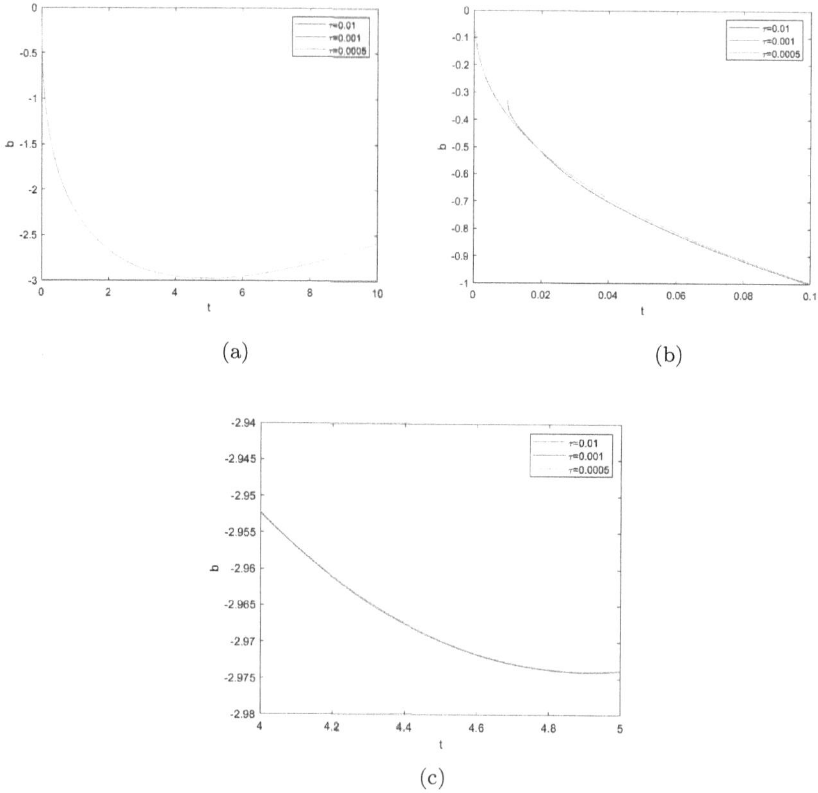

Figure 15.3: To illustrate our main conjecture numerically we construct three boundaries corresponding to $\tau = 0.01, 0.001, 0.0005$, respectively by using the default intensity $\eta = 0.05$. It is clear that after a short initial period, these boundaries begin overlapping.

Chapter 16

McKean-Vlasov Equation with Feedback Through Hitting a Boundary

In this chapter, we derive semi-analytical solutions for the density of interacting particles hitting a boundary.[1] Models of this type have recently arisen as models for the "integrate and fire" behavior of neuronal networks and systemic default risk in networks of interconnected banks. In more detail these problems are discussed in Chapters 11 and 17, respectively. The equation is also closely related to the (supercooled) Stefan problem for heat transfer in the presence of a phase transition, also see Chapter 17.

McKean-Vlasov type equations were introduced by Kac (1956) as a stochastic toy model for the Vlasov kinetic equation of plasma, with a detailed study by McKean (1966). In recent years, mean-field problems, including McKean-Vlasov type equations, have become a trendy topic in applied mathematics from both a theoretical and practical perspective. Different versions of such problems have been used in mathematical finance. For example, in portfolio optimization (Borkar and Suresh Kumar, 2010 consider optimal allocation into sectors for a large number of stocks) and in-game theory (e.g., Huang *et al.*, 2006 discuss an agent's optimal behavior concerning a mass effect).

There are established simulation methods for typical McKean-Vlasov equations (see Bossy and Talay, 1997; Antonelli *et al.*, 2002) and, more recently, several authors have analyzed multilevel and multi-index schemes (see Ricketson, 2015; Szpruch *et al.*, 2017; Haji-Ali and Tempone, 2018) and importance sampling (Reis *et al.*, 2018).

However, none of these methods cover the models described above due to the feedback's singular, path-dependent nature. Here, we consider the Hambly *et al.* (2018) version for simplicity. For this model, Kaushansky and Reisinger (2019) proposed an Euler-type particle method and proved

[1]The material is a slightly streamlined version of Lipton *et al.* (2019).

convergence with order 1/2 in the timestep for a sufficiently small time interval in the regular regime, which can be improved to 1 using Brownian bridges. This chapter shows how to solve these equations by reformulation first as a non-linear free boundary problem similar to the classical Stefan problem and then as a system of two coupled Volterra equations.

First, for *given* drift term from the mean-field interaction, we formulate the problem as diffusion problem on a semi-infinite domain with curvilinear boundary, and its solution represented semi-analytically by the method of heat potentials described in Chapter 8. Second, expressing the interaction term by the solution from the first step results in two coupled Volterra equations. For early applications of heat potentials to versions of the Stefan problem, see already, e.g., Rubinstein (1971). We solve these singular Volterra equations by either an expansion method for the small interaction parameter or by discretization and Newton-Raphson iteration.

A direct PDE expansion to the Stefan problem for both small and large nonlinearity parameters was derived in Addison *et al.* (2006). More recently, an expansion for a particular McKean-Vlasov equation with mean-field interaction through the drift was studied in Gobet and Pagliarani (2018), which performs an iterative two-step procedure decoupling the non-linearity arising from the dependence of the drift on the law of the process from the standard dependence on the state-variables. Our treatment differs not only in the solution approach taken but also in the considered mean-field interaction (through hitting times rather than the drift's expectation) and the expansion parameter (for small drift interaction rather than small volatility).

To assess the accuracy of the (first-order) perturbation solution and illustrate the behavior for strong interactions, where the expansion breaks down, we describe a simple numerical algorithm for solving the relevant Volterra equation. We refer the reader to the large and well-established body of literature on more advanced numerical methods for Volterra equations, see Chapter 5.

16.1 Mean-field limit for large banking system

Following Itkin and Lipton (2015, 2017) and Lipton (2016), we consider a system of N banks with external and mutual assets and liabilities. We denote by L_i the external liabilities for bank i and by L_{ij} the liability from bank i to bank j.

Assume that the dynamics of bank i's total external assets is governed by

$$\frac{dA_t^i}{A_t^i} = \mu\, dt + \sigma_i\, dW_t^i,$$

where μ is the growth rate, W^i are independent standard Brownian motions for $1 \le i \le n$, and the liabilities, both external L_i and mutual L_{ij}, are constant.

We assume that bank i defaults when its assets fall below a certain threshold determined by its liabilities, namely at time $\tau_i = \inf\{t : A_t^i \le \Lambda_t^i\}$, where Λ^i is a default boundary which we now work out. Following Section 6 of Lipton (2016), we choose

$$\Lambda_0^i = R_i \left(L_i + \sum_{j \neq i} L_{ij} \right) - \sum_{j \neq i} L_{ji},$$

where $R_i \in (0,1)$ is the recovery rate of bank i, i.e., bank i defaults if the recovery value of its liabilities, external and to other banks, exceeds the sum of its assets, external and from other banks.

Liabilities and recovery rates are assumed constant in time. Hence, the default boundary remains constant until some bank defaults. If bank k defaults at time t, bank i has to pay all its mutual liabilities L_{ik}, while k returns only a portion of liabilities $R_k L_{ki}$. This can be viewed as a jump in external liabilities from $L_i \to L_i + L_{ik} - R_k L_{ki}$. Hence, the default boundary of bank i becomes

$$\Lambda_t^i = R_i \left([L_i + L_{ik} - R_k L_{ki}] + \sum_{j \neq i,k} L_{ij} \right) - \sum_{j \neq i,k} L_{ji}.$$

As a result, we obtain

$$\Delta \Lambda_t^i = (1 - R_i R_k) L_{ki}.$$

For simplicity, we assume that the banks have the same parameters, i.e., $\sigma_i = \sigma$ and $R_i = R$, for some μ, σ, R, and $L_i = L$ and $L_{ij} = \frac{\gamma}{N}$, both constant, respectively, for some $L, \gamma > 0$. This implies that the asset value processes are exchangeable, and we have $\Lambda_0^i = \Lambda_0$ for some Λ_0, which allows us to take a large pool limit.

At first glance, these assumptions might seem very strong because the banking system is highly heterogeneous. However, these assumptions are a meaningful approximation to large pools of small banks and allow to demonstrate how even the default of small banks can lead to systemic events.

A possible extension in the presence of more prominent, systemically important banks might be to consider several clusters, where banks are similar within a set.

Under the assumptions above, we can write Λ_t^i as

$$\Lambda_t^i = \Lambda_0^i + \frac{\gamma}{N}\sum_{k\neq i}(1-R^2)\mathbf{1}_{\{\tau_k\leq t\}}.$$

It is more convenient to introduce the distance to default $Y_t^i = \log(A_t^i/\Lambda_t^i)/\sigma$, then

$$Y_t^i = Y_0^i + \frac{1}{\sigma}(\mu-\sigma^2/2)t + W_t^i - \frac{1}{\sigma}\log\left[1+\frac{\gamma}{N}\sum_{k\neq i}(1-R^2)\frac{1}{\Lambda_0}\mathbf{1}_{\{\tau_k\leq t\}}\right].$$

Using the approximation $\log(1+x)\approx x$ for small x (i.e., assuming only a small proportion of the banks have yet defaulted), we get for $t<\tau_i$

$$\begin{aligned} Y_t^i &= Y_t^0 + \frac{1}{\sigma}(\mu-\sigma^2/2)t + W_t^i - \frac{\gamma(1-R^2)}{\sigma\Lambda_0}L_t^N, \\ L_t^N &= \frac{1}{N}\sum_k \mathbf{1}_{\{\tau_k\leq t\}}. \end{aligned}$$

For simplicity, we restrict ourselves to the particular case $\mu-\sigma^2/2=0$ (in any case, this term will be small for realistic parameter values and not have any qualitative impact on the results). Then, using propagation of chaos as in Delarue *et al.* (2015), one can obtain that in the limit for $N\to\infty$ all Y_t^i have the same distribution as Y given by[2]

$$\begin{aligned} Y_t &= Y_0 + W_t - \alpha L_t, \\ L_t &= \mathbb{P}(\tau\leq t), \\ \tau &= \inf\{t\in[0,T]: Y_t\leq 0\}, \quad \alpha = \frac{\gamma(1-R^2)}{\sigma\Lambda_0}. \end{aligned} \tag{16.1}$$

To study the effect of the approximation $\log(1+x)\approx x$ above, we compute L_t using particle system simulations (as in Kaushansky and Reisinger (2019) for both αL_t and $-\log(1+\alpha L_t)$. We conclude from Fig. 16.1 that for small α, the results are almost indistinguishable. In contrast, for larger α, the losses accumulate slightly faster for the linearised version, as is to be expected from the expressions. Note that the effect is compensated for if

[2]Note the slight ambiguity between the liabilities above and the loss function below, which are both denoted by L. It should be clear from the context and indices applied which one is being referred to, hence consistently with the literature, we keep the notation.

this model is used to calibrate α to credit market data. As the linear loss term coincides with the model in Hambly *et al.* (2018), we will use that model subsequently.

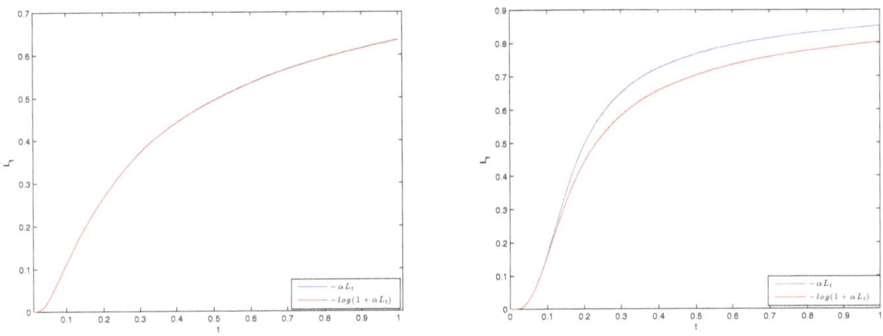

(a) (b)

Figure 16.1: Comparison of L_t computed for $Y_t = Y_0 + W_t - \alpha L_t$ and $Y_0 + W_t - \log(1+\alpha L_t)$, with $z = 0.5$: (a) $\alpha = 0.1$, (b) $\alpha = 0.7$.

The derivation above from first principles allows us to estimate economically meaningful values of α (see also Bujok and Reisinger, 2012) for estimating the initial values Y_0^i from credit spreads. According to David and Lehar (2014), on average, the fraction of interbank liabilities compared to total liabilities is 12% for the EU, 8% for Canada, and 4.5% for the US.

Consider, for example, the EU area. In our notation, $\sum_{j \neq i} L_{ij} \approx \gamma \approx \frac{0.12}{1-0.12} L \approx 0.14L$, where $L_i = L$ are the external liabilities for each bank, assumed identical. We can write α as

$$\alpha \approx \frac{1}{\sigma} \frac{(1-R^2)\gamma}{RL - (1-R)\gamma} \approx \frac{1}{\sigma} \frac{(1-R^2)0.14}{R - (1-R)0.14}.$$

The typical volatility of assets varies from 1% to 8%, which can be confirmed by calibration of the one-dimensional model (see Lipton and Sepp, 2009). Even for a conservative case, when the recovery rate is close to 1, we get a significant value of α. To be precise, for $R = 0.9$ and $\sigma = 0.08$, we get $\alpha \approx 0.3$. On the other hand, for typical recovery rates of $R \approx 0.4$ and volatility at the lower end, one can easily get $\alpha > 5$.

The impact of α on the loss process is discussed in the next section.

16.2 Transition density and known regularity results

In Hambly *et al.* (2018) and Kaushansky and Reisinger (2019) the variable Y_0 is considered a random variable from a given distribution with a known density. For simplicity, let us assume in some of our derivations that $Y_0 = z$ a.s. for some $z > 0$, by taking A_0^i and Λ_0^i the same for all i. If necessary, the results can be extended by making A_0^i random and drawn from the same distribution.

Writing the increasing process L as $\alpha L_t = -\int_0^t \mu(t')\,dt'$ for some negative μ, $p(t,x)$, the Radon-Nikodym density of the distribution with respect to Lebesgue measure of the stopped process $Y_{t\wedge T}$, satisfies the Kolmogorov forward equation

$$
\begin{aligned}
&p_t\,(t,x;z) = -\mu(t)\,p_x\,(t,x;z) + \frac{1}{2}p_{xx}\,(t,x;z)\,, \\
&p\,(0,x;z) = \delta\,(x-z)\,, \quad p\,(t,0;z) = 0,
\end{aligned} \tag{16.2}
$$

where the subscripts t and x denote partial derivatives. Given the continuous differentiability assumption of L, we can consider classical solutions of Eq. (16.2).

Using the relation $L_t = 1 - \int_0^\infty p(t,x;z)\,dx$, we can express μ in terms of p by

$$
\begin{aligned}
g(t;z) &= \frac{\mathrm{d}L_t}{\mathrm{d}t} = -\int_0^\infty p_t\,dx = \mu(t)\int_0^\infty p_x\,dx - \frac{1}{2}\int_0^\infty p_{xx}\,dx \\
&= \frac{1}{2}p_x(t,0;z),
\end{aligned} \tag{16.3}
$$

where we have used the PDE Eq. (16.2) as well as the condition $p(t,0;z) = 0$ and

$$
\lim_{x\to\infty} p(t,x;z) = \lim_{x\to\infty} p_x(t,x;z) = 0.
$$

Hence, Eq. (16.2) can be written in the self-consistent form

$$
\begin{aligned}
&p_t\,(t,x;z) = \frac{\alpha}{2}p_x(t,0;z)\,p_x\,(t,x;z) + \frac{1}{2}p_{xx}\,(t,x;z)\,, \\
&p\,(0,x;z) = \delta\,(x-z)\,, \quad p\,(t,0;z) = 0.
\end{aligned} \tag{16.4}
$$

From the second equation in Eq. (16.1), g is also the density of the first passage time of Y. A result from Peskir (2002) for the first-passage problem of Brownian motion, applied to $Y_t - L_t$ (Hambly *et al.*, 2018) gives the following Volterra equation for L

$$
\Phi\left(-\frac{z-\alpha L_t}{\sqrt{t}}\right) = \int_0^t \Phi\left(\alpha\frac{L_t - L_s}{\sqrt{t-s}}\right)\,dL_s,
$$

where $\Phi(x)$ is the CDF of the standard normal distribution.

In contrast to this equation, we will derive a coupled system of Volterra equations that give both p and g (i.e., not the cumulative distribution). These equations turn to appear in a more standard form without the non-linearity Φ on the left-hand side and integration over L on the right-hand side. Hence, numerically, they are more amenable.

In Fig. 16.2(a), we plot the loss function $t \to L_t$, computed by the methods described in the remainder of this Chapter, for different values of α, where α measures the interconnectivity of the banking system.

The losses increase dramatically because of interbank liabilities, which may lead to systemic events, here for α larger than around 0.9. Hereby the rate of losses increases to infinity, as seen in Fig. 16.2(b) from the large gradient $p_x(t, 0)$ for t immediately before the blow-up, and then triggers a jump in L_t. Nadtochiy and Shkolnikov (2017) first give a rigorous mathematical characterization of this type of behavior in their model, and Theorem 1.1 in Hambly *et al.* (2018) shows the necessity of such "blow-ups" for large enough α, depending on the initial distribution (see also Fasano *et al.*, 1990) for this property phrased in terms of the Stefan problem). Additionally, Theorem 1.1 in Delarue *et al.* (2017) gives a detailed characterization of the regularity of the solution and conditions for jumps.

We note that the derivative of L_t' may explode without necessarily triggering a jump, and instead, the PDE solution immediately returns to the smooth case.

Here we consider the solution of this problem at an initial interval where the loss function is differentiable. It is well known that if

$$\inf\{x \geq 0 : m(0, \alpha x) < x\} = 0,$$

where m is the initial measure, a unique classical solution (with continuously differentiable L_t) exists for small times under mild regularity assumptions on the initial density (see Fasano *et al.*, 1983).

In the rest of the chapter, we will often take the initial distribution as a delta function centred at some point z, and assume that continuous differentiability of L still holds (possibly without a control on the derivative) in this case. Then μ is continuous and the solutions to Eq. (16.2) are classical, so that the derivations of the integral equations will be justified.

Lastly, a comparison of the data for the loss function with $\alpha = 0.1$ and $\alpha = 0.7$ in Fig. 16.2(a), computed by numerical solution (see Section 16.4.3) of the integral equations derived by the techniques in Section 16.4, with the data in Fig. 16.1 (blue curves) for the same values of α and computed by

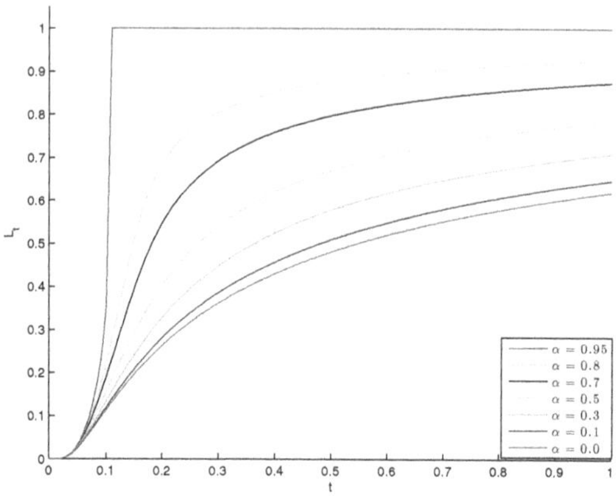

(a)

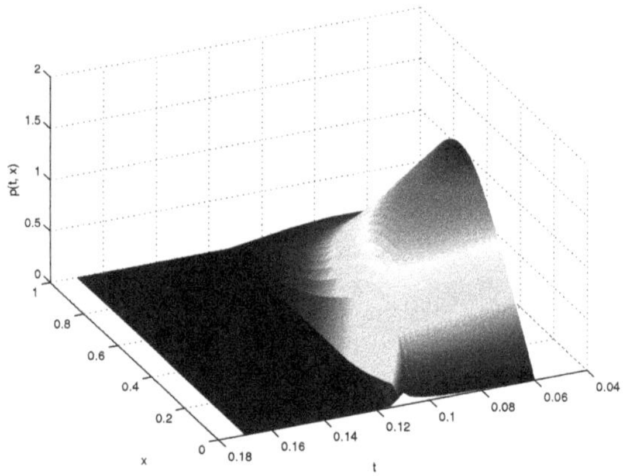

(b)

Figure 16.2: (a) L_t for different values of α and $z = 0.5$. (b) The density $p(t, x)$ for $\alpha = 0.95$.

the particle method in Kaushansky and Reisinger (2019) applied directly to the McKean-Vlasov SDE, shows that they differ only by numerical errors. This provides an empirical verification of the correctness of the derivations.

16.3 The method of heat potentials

We first derive an expression for $p(t, x; z)$ using the method of heat potentials with an unknown weight function which one can find by solving a Volterra equation of the second kind. Next, we differentiate the expression for $p(t, x; z)$ with respect to x and take its limit to 0 in order to find the first passage density of Y in Eq. (16.1) (or, equivalently, g in Eq. (16.3). Below we omit z when possible.

16.3.1 *Semi-closed formula for the transition density*

Consider the case of a continuous and bounded drift $\mu(t)$ on $[0, T^*]$. Let $M(t) = \int_0^t \mu(t')\, dt'$, which is a differentiable function on $[0, T^*]$. The change of variables $\tilde{p}(t, y) = p(t, -M(t))$ yields the following Cauchy-Dirichlet problem:

$$\begin{aligned} \tilde{p}_t(t, y) &= \frac{1}{2}\tilde{p}_{yy}(t, y), \qquad y > -M(t),\, t > 0 \\ \tilde{p}(0, y) &= \delta(y - z), \quad \tilde{p}(t, -M(t)) = 0. \end{aligned} \tag{16.5}$$

With the boundary condition $\tilde{p}_y(t, -M(t)) = -2/\alpha\, (\mathrm{d}M/\mathrm{d}t)(t)$, this is the supercooled Stefan problem (see, e.g., Fasano *et al.*, 1990; Delarue *et al.*, 2017) and also Chapter 17 of this book.

In the following, we omit tilde for brevity. Let us split p from the last equation into

$$p(t, y) = q(t, y) + u(t, y),$$

where u is solution of

$$\begin{aligned} u_t(t, y) &= \frac{1}{2}u_{yy}(t, y), \qquad y \in \mathbb{R},\ t > 0, \\ u(0, y) &= \delta(x - z). \end{aligned} \tag{16.6}$$

The solution of this equation is clearly the heat kernel

$$u(t, y) = \frac{\exp\left(-\frac{(y-z)^2}{2t}\right)}{\sqrt{2\pi t}}.$$

The corresponding problem for q has the form

$$\begin{aligned} q_t(t, y) &= \frac{1}{2}q_{yy}(t, y), \\ q(0, y) &= 0, \quad q(t, -M(t)) = -u(t, -M(t)). \end{aligned}$$

We use the method of heat potentials described in detail in Chapter 8. Thereby, we represent q in the form

$$q(t,y)=\int_0^t \frac{(y+M(t'))\exp\left(-\frac{(y+M(t'))^2}{2(t-t')}\right)}{\sqrt{2\pi(t-t')^3}}\nu(t')\,dt',$$

where ν is a suitable weight which will be determined to match the boundary condition. Assuming that $\nu(t)$ is known, we can revert to the original variables and get

$$p(t,x)=\int_0^t \frac{(x-\Psi(t,t'))\exp\left(-\frac{(x-\Psi(t,t'))^2}{2(t-t')}\right)}{\sqrt{2\pi(t-t')^3}}\nu(t')\,dt' + \frac{\exp\left(-\frac{(x-M(t)-z)^2}{2t}\right)}{\sqrt{2\pi t}}, \tag{16.7}$$

where $\Psi(t,t')=M(t)-M(t')$.

By construction, p in Eq. (16.7) satisfies the first two equations in Eq. (16.5). We also need to satisfy the boundary condition in Eq. (16.5). Using the definition of $\Psi(t,t')$, it is easy to see that the limiting value of the first integral in Eq. (16.7) at $x\to 0$ coincides with the limiting value of the function $\phi(t)$ discussed in Section 8.1.1. Therefore, using that result, we immediately have

$$\lim_{x\to 0}\int_0^t \frac{(x-\Psi(t,t'))\exp\left(-\frac{(x-\Psi(t,t'))^2}{2(t-t')}\right)}{\sqrt{2\pi(t-t')^3}}\nu(t')\,dt' = \nu(t)-\int_0^t \frac{\Psi(t,t')\,\Xi(t,t')}{\sqrt{2\pi(t-t')^3}}\nu(t')\,dt'.$$

The requirement $\lim_{x\to 0}p(t,x)=0$ thus leads to the following Volterra integral equation of the second kind for ν,

$$\nu(t)-\int_0^t \frac{\Psi(t,t')\,\Xi(t,t')}{\sqrt{2\pi(t-t')^3}}\nu(t')\,dt' + \frac{\exp\left(-\frac{(M(t)+z)^2}{2t}\right)}{\sqrt{2\pi t}}=0, \tag{16.8}$$

where

$$\Xi(t,t')=\exp\left(-\frac{\Psi(t,t')^2}{2(t-t')}\right),\quad t\neq t',$$

$$\Xi(t,t)=1.$$

16.3.2 *Computation of loss rate over the boundary*

In this section, we compute $g(t)$ from Eq. (16.3), suppressing z, by first differentiating Eq. (16.7),

$$p_x(t,x) = \int_0^t \left(1 - \frac{(x-\Psi(t,t'))^2}{(t-t')}\right) \frac{\exp\left(-\frac{(x-\Psi(t,t'))^2}{2(t-t')}\right)}{\sqrt{2\pi(t-t')^3}} \nu(t')\,dt' \quad (16.9)$$
$$- (x - M(t) - z) \frac{\exp\left(-\frac{(x-M(t)-z)^2}{2t}\right)}{\sqrt{2\pi t^3}}.$$

Again, using the definition of $\Psi(t,t')$, it is easy to see that the limiting value of the first integral in Eq. (16.9) at $x \to 0$ coincides with the limiting value of the function $\psi(t)$ discussed in Section 8.1.2. Therefore, using that result, we obtain

$$\lim_{x\to 0} \int_0^t \left(1 - \frac{(x-\Psi(t,t'))^2}{(t-t')}\right) \frac{\exp\left(-\frac{(x-\Psi(t,t'))^2}{2(t-t')}\right)}{\sqrt{2\pi(t-t')^3}} \nu(t')\,dt'$$
$$= 2\left(M'(t) - \frac{1}{\sqrt{2\pi t}}\right)\nu(t)$$
$$+ \int_0^t \frac{\left(\left(1 - \frac{\Psi(t,t')^2}{(t-t')}\right)\Xi(t,t')\nu(t') - \nu(t)\right)}{\sqrt{2\pi(t-t')^3}}\,dt.$$

Accordingly,

$$g(t) = \left(M'(t) - \frac{1}{\sqrt{2\pi t}}\right)\nu(t) \quad (16.10)$$
$$+ \frac{1}{2}\int_0^t \frac{\left(\left(1 - \frac{\Psi(t,t')^2}{(t-t')}\right)\Xi(t,t')\nu(t') - \nu(t)\right)}{\sqrt{2\pi(t-t')^3}}\,dt'$$
$$+ \frac{(M(t)+z)\exp\left(-\frac{(M(t)+z)^2}{2t}\right)}{2\sqrt{2\pi t^3}}.$$

16.3.3 *Direct computation of loss rate*

Alternatively, we can represent $g(t)$ using Eq. (16.3) by

$$g(t) = -\frac{d}{dt}\int_0^\infty p(t,x)\,dx,$$

so that

$$
\begin{aligned}
g(t) &= -\frac{d}{dt}\int_0^t \left(\int_0^\infty \frac{(x - \Psi(t,t')) \exp\left(-\frac{(x-\Psi(t,t'))^2}{2(t-t')}\right)}{\sqrt{2\pi (t-t')^3}} dx \right) \nu(t')\, dt' \\
&\quad - \frac{d}{dt}\int_0^\infty \frac{\exp\left(-\frac{(x-M(t)-z)^2}{2t}\right)}{\sqrt{2\pi t}} dx \qquad (16.11) \\
&= -\frac{d}{dt}\int_0^t \left(\int_{-\Psi(t,t')}^\infty \frac{\xi \exp\left(-\frac{\xi^2}{2(t-t')}\right)}{\sqrt{2\pi (t-t')^3}} d\xi \right) \nu(t')\, dt' \\
&\quad - \frac{d}{dt}\int_{-\frac{(M(t)+z)}{\sqrt{t}}}^\infty \frac{\exp\left(-\frac{\xi^2}{2}\right)}{\sqrt{2\pi}} d\xi \\
&= -\frac{d}{dt}\int_0^t \frac{\Xi(t,t')}{\sqrt{2\pi(t-t')}} \nu(t')\, dt' \\
&\quad - \left(M'(t) - \frac{(M(t)+z)}{2t} \right) \frac{\exp\left(-\frac{(M(t)+z)^2}{2t}\right)}{\sqrt{2\pi t}}.
\end{aligned}
$$

As an aside, we can verify easily by direct computation that the two expressions for g agree. We use the result provided by Lemma 8.1 together with Eq. (16.8) and rewrite the second term in the form

$$
\frac{\exp\left(-\frac{(M(t)+z)^2}{2t}\right)}{\sqrt{2\pi t}} = -\nu(t) + \int_0^t \frac{\Psi(t,t')\,\Xi(t,t')}{\sqrt{2\pi(t-t')^3}} \nu(t')\, dt'.
$$

Using the first equality in the lemma, we arrive at the following expression

$$
\begin{aligned}
g(t) &= \left(M'(t) - \frac{1}{\sqrt{2\pi t}} \right) \nu(t) - \frac{1}{2}\int_0^t \frac{dt'}{\sqrt{2\pi(t-t')^3}} \\
&\quad \times \left[\nu(t) - \left(\Xi(t,t') - 2\left(M'(t)\Psi(t,t')\,\Xi(t,t') + (t-t')\,\Xi_t(t,t')\right)\right)\nu(t')\right] \\
&\quad + \frac{(M(t)+z)\exp\left(-\frac{(M(t)+z)^2}{2t}\right)}{2\sqrt{2\pi t^3}}.
\end{aligned}
$$

We notice that

$$
\Xi_t(t,t') = \left(-\frac{M'(t)\,\Psi(t,t')}{(t-t')} + \frac{\Psi(t,t')^2}{2(t-t')^2} \right) \Xi(t,t'),
$$

so that

$$\Xi(t,t') - 2\left(M'(t)\Psi(t,t')\Xi(t,t') + (t-t')\Xi_t(t,t')\right)$$
$$= \left(1 - \frac{\Psi(t,t')^2}{(t-t')}\right)\Xi(t,t'),$$

from which Eq. (16.10) follows.

16.4 Solution of the McKean-Vlasov equation

Now, for the McKean-Vlasov equation Eq. (16.4), we obtain from Eqs. (16.7), (16.8) and (16.10) the following.

Proposition 16.1. *If the loss function L is continuously differentiable on $[0,t]$, then*

$$p(t,x) = \int_0^t \frac{(x - \Psi(t,t'))\exp\left(-\frac{(x-\Psi(t,t'))^2}{2(t-t')}\right)}{\sqrt{2\pi(t-t')^3}}\nu(t')\,dt'$$
$$+ \frac{\exp\left(-\frac{(x-M(t)-z)^2}{2t}\right)}{\sqrt{2\pi t}}.$$

Here ν and g satisfy the system of integral equations

$$\nu(t) + \int_0^t \frac{\alpha\Omega(t,t')\exp\left(-\frac{\alpha^2\Omega(t,t')^2}{2(t-t')}\right)}{\sqrt{2\pi(t-t')^3}}\nu(t')\,dt' + \frac{\exp\left(-\frac{(\alpha\Omega(t,0)-z)^2}{2t}\right)}{\sqrt{2\pi t}}$$
$$= 0, \tag{16.12}$$

$$g(t) + \left(\alpha g(t) + \frac{1}{\sqrt{2\pi t}}\right)\nu(t) + \frac{(\alpha\Omega(t,0)-z)\exp\left(-\frac{(\alpha\Omega(t,0)-z)^2}{2t}\right)}{2\sqrt{2\pi t^3}}$$
$$+ \frac{1}{2}\int_0^t \frac{\left(\nu(t) - \left(1 - \frac{\alpha^2\Omega(t,t')^2}{(t-t')}\right)\exp\left(-\frac{\alpha^2\Omega(t,t')^2}{2(t-t')}\right)\nu(t')\right)}{\sqrt{2\pi(t-t')^3}}\,dt' = 0,$$

where M, Ψ and Ω are given by

$$M(t) = -\alpha\int_0^t g(t')\,dt', \quad \Psi(t,t') = M(t) - M(t'), \tag{16.13}$$

$$\Omega(t,t') = \int_{t'}^t g(t'')\,dt''.$$

Moreover, $L_t = -M(t)/\alpha$.

In Section 16.4.1, we give the explicit solution for special cases, in particular when there is no feedback, $\alpha = 0$, $M(t) = 0$. In general, only approximations to the solution can be found. We give an asymptotic and a numerical approach in the remainder of this section.

16.4.1 *Particular cases*

For illustration, we work out the solution for two particular cases which are also accessible by the standard reflection principle for Brownian motion or method of images for parabolic equations.

$\boldsymbol{M(t) = 0}$. When $M(t) = 0$, we get

$$\nu(t) + \frac{\exp\left(-\frac{z^2}{2t}\right)}{\sqrt{2\pi t}} = 0,$$

$$g(t) + \frac{1}{\sqrt{2\pi t}}\nu(t) + \frac{1}{2}\int_0^t \frac{(\nu(t) - \nu(t'))}{\sqrt{2\pi(t-t')^3}}dt' - \frac{z\exp\left(-\frac{z^2}{2t}\right)}{2\sqrt{2\pi t^3}} = 0.$$

Integration by parts of the second equation and use of the first yields

$$\begin{aligned} g(t) &= -\int_0^t \frac{\dot{\nu}(t')}{\sqrt{2\pi(t-t')}}dt' + \frac{z\exp\left(-\frac{z^2}{2t}\right)}{2\sqrt{2\pi t^3}} \\ &= \frac{z\exp\left(-\frac{z^2}{2t}\right)}{2\sqrt{2\pi t^3}} + \frac{z\exp\left(-\frac{z^2}{2t}\right)}{2\sqrt{2\pi t^3}} = \frac{z\exp\left(-\frac{z^2}{2t}\right)}{\sqrt{2\pi t^3}}, \end{aligned} \tag{16.14}$$

as expected.

$\boldsymbol{M(t) = \mu t}$. When $M(t) = \mu t$, we get

$$\nu(t) - \mu\int_0^t \frac{\exp\left(-\frac{\mu^2(t-t')}{2}\right)}{\sqrt{2\pi(t-t')}}\nu(t')dt' + \frac{\exp\left(-\frac{(\mu t+z)^2}{2t}\right)}{\sqrt{2\pi t}} = 0, \tag{16.15}$$

$$\begin{aligned} g(t) = {}& \frac{1}{2\sqrt{2\pi}}\int_0^t \frac{1}{\sqrt{(t-t')^3}}\left(\exp\left(-\frac{\mu^2(t-t')}{2}\right)\nu(t') - \nu(t)\right)dt' \\ &- \frac{\mu^2}{2\sqrt{2\pi}}\int_0^t \frac{1}{\sqrt{t-t'}}\exp\left(-\frac{\mu^2(t-t')}{2}\right)\dot{\nu}(t')\,dt' + \left(\mu - \frac{1}{\sqrt{2\pi t}}\right)\nu(t) \\ &+ \frac{(\mu t + z)\exp\left(-\frac{(\mu t+z)^2}{2t}\right)}{\sqrt{2\pi t^3}}. \end{aligned}$$

Taking the Laplace transform of the first equation in Eq. (16.15), we have

$$\hat{\nu}(s) - \mu\hat{\nu}(s)\frac{1}{\sqrt{2s+\mu^2}} + \frac{\exp(-z\sqrt{2s+\mu^2}-\mu z)}{\sqrt{2s+\mu^2}} = 0.$$

Hence,

$$\hat{\nu}(s) = -\frac{\exp(-z\sqrt{2s+\mu^2}-\mu z)}{\sqrt{2s+\mu^2}-\mu}.$$

The inverse Laplace transform of $\hat{\nu}(s)$ can be found analytically, but we do not need it to compute $g(t)$. Consider the second equation of Eq. (16.15). The first integral can be rewritten as

$$\begin{aligned}
&\frac{1}{2\sqrt{2\pi}}\int_0^t \frac{1}{\sqrt{(t-t')^3}}\left(\exp\left(-\frac{\mu^2(t-t')}{2}\right)\nu(t') - \nu(t)\right)dt' \\
&\quad= \frac{1}{\sqrt{2\pi}}\int_0^t \left(\exp\left(-\frac{\mu^2(t-t')}{2}\right)\nu(t') - \nu(t)\right) d\left(\frac{1}{\sqrt{(t-t')}}\right) \\
&\quad= -\frac{1}{\sqrt{2\pi t}}\left(\exp\left(-\frac{\mu^2 t}{2}\right)\nu(0) - \nu(t)\right) \\
&\quad\quad - \frac{1}{\sqrt{2\pi}}\int_0^t \frac{1}{\sqrt{t-t'}}\exp\left(-\frac{\mu^2(t-t')}{2}\right)\dot{\nu}(t')\,dt' \\
&\quad\quad - \frac{\mu^2}{2\sqrt{2\pi}}\int_0^t \frac{1}{\sqrt{(t-t')}}\exp\left(-\frac{\mu^2(t-t')}{2}\right)\nu(t')\,dt'.
\end{aligned}$$

Thus,

$$\begin{aligned}
g(t) &= \mu\nu(t) - \frac{1}{\sqrt{2\pi t}}\exp\left(-\frac{\mu^2 t}{2}\right)\nu(0) \\
&\quad - \frac{1}{\sqrt{2\pi}}\int_0^t \frac{1}{\sqrt{t-t'}}\exp\left(-\frac{\mu^2(t-t')}{2}\right)\dot{\nu}(t')\,dt' \\
&\quad - \frac{\mu^2}{\sqrt{2\pi}}\int_0^t \frac{1}{\sqrt{(t-t')}}\exp\left(-\frac{\mu^2(t-t')}{2}\right)\nu(t')\,dt' \\
&\quad + \frac{(\mu t + z)\exp\left(-\frac{(\mu t+z)^2}{2t}\right)}{2\sqrt{2\pi t^3}}.
\end{aligned}$$

Taking the Laplace transform of the last equation, we get

$$\hat{g}(s) = \mu\hat{\nu}(s) - \frac{1}{\sqrt{2s+\mu^2}}\nu(0) - \frac{1}{\sqrt{2s+\mu^2}}(s\hat{\nu}(s) - \nu(0))$$

$$
\begin{aligned}
& -\frac{\mu^2}{\sqrt{2s+\mu^2}}\hat{\nu}(s)+\left(\frac{1}{2}+\frac{\mu}{2\sqrt{2s+\mu^2}}\right)\exp(-z\sqrt{2s+\mu^2}+\mu z) \\
& =\left(-\frac{\mu}{\sqrt{2s+\mu^2}-\mu}+\frac{s+\mu^2}{\sqrt{2s+\mu^2}}\frac{1}{\sqrt{2s+\mu^2}-\mu^2}\right. \\
& \left.+\frac{1}{2}+\frac{\mu}{2\sqrt{2s+\mu^2}}\right)\exp(-z\sqrt{2s+\mu^2}+\mu z) \\
& =\exp(-z\sqrt{2s+\mu^2}+\mu z).
\end{aligned}
$$

The inverse Laplace transform yields the final expression for $g(t)$

$$
g(t)=\frac{z\exp\left(-\frac{(z-\mu t)^2}{2t}\right)}{\sqrt{2\pi t^3}},
$$

as expected.

16.4.2 *Perturbation solution*

We formally expand the solution of Eq. (16.12) into series of α powers

$$
(\nu(t), g(t)) = (\nu_0(t), g_0(t)) + \sum_{n=1}^{\infty} \alpha^n (\nu_n(t), g_n(t)), \tag{16.16}
$$

which we will truncate after the first two terms. This will give an analytical expression which can be expected to be a good approximation for small values of α.

By plugging the formal expansion for (ν, g) into Eq. (16.12) and collecting the terms of the same order in α, we get the following systems for $(\nu_0(t), g_0(t))$ and $(\nu_1(t), g_1(t))$

$$
\nu_0(t)+\frac{\exp\left(-\frac{z^2}{2t}\right)}{\sqrt{2\pi t}}=0,
$$

$$
g_0(t)+\frac{1}{\sqrt{2\pi t}}\nu_0(t)+\frac{1}{2}\int_0^t\frac{(\nu_0(t)-\nu_0(t'))}{\sqrt{2\pi(t-t')^3}}dt'-\frac{z\exp\left(-\frac{z^2}{2t}\right)}{2\sqrt{2\pi t^3}}=0,
$$

$$
\nu_1(t)+\int_0^t\frac{\Omega_0(t,t')}{\sqrt{2\pi(t-t')^3}}\nu_0(t')\,dt'+\frac{z\Omega_0(t,0)\exp\left(-\frac{z^2}{2t}\right)}{\sqrt{2\pi t^3}}=0,
$$

$$
g_1(t)+g_0(t)\nu_0(t)+\frac{1}{\sqrt{2\pi t}}\nu_1(t)+\frac{1}{2}\int_0^t\frac{(\nu_1(t)-\nu_1(t'))}{\sqrt{2\pi(t-t')^3}}dt'
$$

$$+\left(1-\frac{z^2}{t}\right)\frac{\Omega_0\left(t,0\right)\exp\left(-\frac{z^2}{2t}\right)}{2\sqrt{2\pi t^3}}=0,$$

where $\Omega_0\left(t,t'\right)=\int_{t'}^{t}g_0\left(t''\right)dt''$.

The equations for g_0 and g_1 can be written as

$$g_0\left(t\right)+\int_0^t\frac{\dot{\nu}_0\left(t'\right)}{\sqrt{2\pi\left(t-t'\right)^3}}dt'-\frac{z\exp\left(-\frac{z^2}{2t}\right)}{2\sqrt{2\pi t^3}}=0,$$

$$g_1\left(t\right)+g_0\left(t\right)\nu_0\left(t\right)+\int_0^t\frac{\dot{\nu}_1\left(t'\right)}{\sqrt{2\pi\left(t-t'\right)}}dt'$$

$$+\left(1-\frac{z^2}{t}\right)\frac{\Omega_0\left(t,0\right)\exp\left(-\frac{z^2}{2t}\right)}{2\sqrt{2\pi t^3}}=0.$$

Thus, using the results in 16.4.1 for $\alpha=0$,

$$\nu_0\left(t\right)=-\frac{\exp\left(-\frac{z^2}{2t}\right)}{\sqrt{2\pi t}},\qquad g_0\left(t\right)=\frac{z\exp\left(-\frac{z^2}{2t}\right)}{\sqrt{2\pi t^3}}\tag{16.17}$$

$$\Omega_0\left(t,t'\right)=2\left(\Phi\left(\frac{z}{\sqrt{t'}}\right)-\Phi\left(\frac{z}{\sqrt{t}}\right)\right),$$

$$\frac{\partial}{\partial t}\Omega_0\left(t,t'\right)=\frac{z\exp\left(-\frac{z^2}{2t}\right)}{\sqrt{2\pi t^3}}=g_0(t),\qquad\frac{\partial}{\partial t'}\Omega_0\left(t,t'\right)=-g_0(t').$$

Next,

$$\nu_1\left(t\right)=-\int_0^t\frac{\Omega_0\left(t,t'\right)}{\sqrt{2\pi\left(t-t'\right)^3}}\nu_0\left(t'\right)dt'-\frac{z\Omega_0\left(t,0\right)\exp\left(-\frac{z^2}{2t}\right)}{\sqrt{2\pi t^3}}\tag{16.18}$$

$$=2\int_0^t\frac{\left(\Phi\left(\frac{z}{\sqrt{t}}\right)-\Phi\left(\frac{z}{\sqrt{t'}}\right)\right)}{\sqrt{2\pi\left(t-t'\right)^3}}\frac{\exp\left(-\frac{z^2}{2t'}\right)}{\sqrt{2\pi t'}}dt'$$

$$+\frac{2z\left(\Phi\left(\frac{z}{\sqrt{t}}\right)-1\right)\exp\left(-\frac{z^2}{2t}\right)}{\sqrt{2\pi t^3}},$$

$$g_1\left(t\right)=-g_0\left(t\right)\nu_0\left(t\right)-\int_0^t\frac{\dot{\nu}_1\left(t'\right)}{\sqrt{2\pi\left(t-t'\right)}}dt'-\left(1-\frac{z^2}{t}\right)\frac{\Omega_0\left(t,0\right)\exp\left(-\frac{z^2}{2t}\right)}{2\sqrt{2\pi t^3}},$$

where

$$\Omega_0\left(t,0\right)=2\left(1-\Phi\left(\frac{z}{\sqrt{t}}\right)\right).$$

We can re-write $\nu_1(t)$ in the form

$$\nu_1(t) = -\int_0^t \frac{\omega_0(t,t')}{\sqrt{2\pi(t-t')}} g_0(t')\,\nu_0(t')\,dt' - \frac{z\Omega_0(t,0)\exp\left(-\frac{z^2}{2t}\right)}{\sqrt{2\pi t^3}},$$

with

$$\begin{aligned}\omega_0(t,t') &= \frac{\Omega_0(t,t')}{g_0(t')(t-t')} \\ &= -\frac{2\left(\Phi\left(\frac{z}{\sqrt{t}}\right)-\Phi\left(\frac{z}{\sqrt{t'}}\right)\right)\sqrt{2\pi t'^3}\exp\left(\frac{z^2}{2t'}\right)}{z(t-t')}, \qquad t\neq t', \\ \omega_0(t,t) &= 1,\end{aligned}$$

and obtain an expression for $\dot{\nu}_1$

$$\dot{\nu}_1(t) = -\frac{d}{dt}\int_0^t \frac{\omega_0(t,t')}{\sqrt{2\pi(t-t')}} g_0(t')\nu_0(t')\,dt' - \frac{d}{dt}\left(\frac{z\Omega_0(t,0)\exp\left(-\frac{z^2}{2t}\right)}{\sqrt{2\pi t^3}}\right). \tag{16.19}$$

To compute the first term of Eq. (16.19), we use the second equality in Lemma 8.1 with $\nu(t)=\nu_0(t)g_0(t)$ and $\Xi(t,t')=\omega_0(t,t')$, to get

$$\begin{aligned}\dot{\nu}_1(t) &= -\int_0^t \frac{\left(\left(\omega_0(t,t')-2(t-t')\frac{\partial}{\partial t}\omega_0(t,t')\right)g_0(t')\,\nu_0(t')\right)_{t'}}{\sqrt{2\pi(t-t')}}\,dt' \qquad (16.20)\\ &\quad -\frac{d}{dt}\left(\frac{z\Omega_0(t,0)\exp\left(-\frac{z^2}{2t}\right)}{\sqrt{2\pi t^3}}\right) \\ &= -\int_0^t \frac{\left(\left(\frac{3\Omega_0(t,t')}{(t-t')}-2\frac{\partial}{\partial t}\Omega_0(t,t')\right)\nu_0(t')\right)_{t'}}{\sqrt{2\pi(t-t')}}\,dt' \\ &\quad -\frac{d}{dt}\left(\frac{z\Omega_0(t,0)\exp\left(-\frac{z^2}{2t}\right)}{\sqrt{2\pi t^3}}\right) \\ &= \int_0^t \frac{\left(\frac{3g_0(t')}{(t-t')}-3\frac{\Omega_0(t,t')}{(t-t')^2}\right)\nu_0(t')-\left(3\frac{\Omega_0(t,t')}{(t-t')}-2g_0(t)\right)\dot{\nu}_0(t')}{\sqrt{2\pi(t-t')}}\,dt' \\ &\quad -\frac{z^2\exp\left(-\frac{z^2}{t}\right)}{2\pi t^3}+2z\left(1-\Phi\left(\frac{z}{\sqrt{t}}\right)\right)\left(3-\frac{z^2}{t}\right)\frac{\exp\left(-\frac{z^2}{2t}\right)}{2\sqrt{2\pi t^5}}.\end{aligned}$$

Substituting this in the second equation Eq. (16.18) yields an expression for $g_1(t)$, which can be evaluated by numerical integration.

In summary, in the formal expansions Eq. (16.16), ν_0 and g_0 are given by Eq. (16.17), and ν_1 and g_1 are given by Eq. (16.18), where $\dot{\nu}_1$ can be evaluated (without differentiation) by Eq. (16.20).

Finally, to evaluate the complexity of the computation of $g_1(t)$, the most expensive expression to compute, we consider numerical quadrature with N points. First, we precompute $\dot{\nu}_1(t)$ using Eq. (16.20); it can be done in $O(N^2)$ operations. Then, we can compute $g_1(t)$ using the second equation in Eq. (16.18) with precomputed $\dot{\nu}_1(t)$ again in $O(N^2)$. Thus, the total complexity of the perturbation method is $O(N^2)$.

16.4.3 *Numerical solution*

In this section, we present (without convergence analysis) a simple method for the numerical approximation of the solution to the coupled Volterra equations Eq. (16.12). We note that Volterra equations and their numerical solution are a well-established research field; see Chapter 5 and references therein. Also, in Noble (1969) possible instabilities of multi-step methods in the presence of weak singularities are discussed.

Several papers propose higher-order methods and collocation techniques to improve the convergence and treat instabilities. For example, Brunner (1985) proved the convergence for a polynomial spline collocation method with quadratures. Subsequently Kolk *et al.* (2009) and Kolk and Pedas (2009, 2013) used a piecewise polynomial collocation method to solve a Volterra equation with weak singularity to derive optimal global convergence estimates and a local super-convergence result. An alternative is to consider a particular functional basis, such as Chebyshev polynomials and Bernstein polynomials (see Maleknejad *et al.*, 2007 and Maleknejad *et al.*, 2011, respectively). In both cases, the approximation produces a system of linear or nonlinear algebraic equations. Hairer *et al.* (1985) developed a method based on fast Fourier transform to reduce the number of kernel evaluations on an N-point grid from $O(N^2)$ to $O(N(\log N)^2)$.

For simplicity, we use trapezoidal quadrature, with a special treatment of the interval containing the singularity, to obtain the numerical solution recursively. We divide the interval $[0,T]$ into equally spaced subintervals of length Δ and discretize Eq. (16.12) appropriately. To this end, we assume that ν and g are piecewise linear with $\nu(l\Delta) = \nu_l$ and $g(l\Delta) = g_l$, so that on the interval $[(l-1)\Delta, l\Delta]$ we have

$$\nu(t) = \frac{\nu_{l-1}(l\Delta - t) + \nu_l(t-(l-1)\Delta)}{\Delta} = \nu_l - \frac{(\nu_l - \nu_{l-1})}{\Delta}(l\Delta - t),$$

$$g(t) = \frac{g_{l-1}(l\Delta - t) + g_l(t - (l-1)\Delta)}{\Delta} = g_l - \frac{(g_l - g_{l-1})}{\Delta}(l\Delta - t).$$

Accordingly,

$$\Omega_{nl} = \int_{l\Delta}^{n\Delta} g(t'')\,dt' = \frac{\Delta}{2}(g_l + 2g_{l+1} + \dots + 2g_{n-1} + g_n),$$

$$\Omega_{nl} = \Omega_{(n-1)l} + \frac{\Delta}{2}(g_n + g_{n-1}).$$

Inserting in Eq. (16.12), the discretized system of equations has the form

$$\nu_n + \alpha\sum_{l=1}^{n}\mathcal{I}_l^n + \frac{\exp\left(-\frac{(\alpha\Omega_{n0}-z)^2}{2n\Delta}\right)}{\sqrt{2\pi n\Delta}} = 0, \tag{16.21}$$

$$g_n + \left(\alpha g_n + \frac{1}{\sqrt{2\pi n\Delta}}\right)\nu_n + \frac{1}{2}\sum_{l=1}^{n}\mathcal{J}_l^n + \frac{(\alpha\Omega_{n0} - z)\exp\left(-\frac{(\alpha\Omega_{n0}-z)^2}{2n\Delta}\right)}{2\sqrt{2\pi n^3\Delta^3}} = 0.$$

For a given n, all the relevant integrals $\mathcal{I}_l, \mathcal{J}_l, 1 \le l < n$, can be approximated by using the trapezoidal rule (or via more accurate composite formulas, if necessary). Accordingly,

$$\mathcal{I}_l^n = \int_{(l-1)\Delta}^{l\Delta} \frac{\Omega(n\Delta, t')\exp\left(-\frac{\alpha^2\Omega(n\Delta,t')^2}{2(n\Delta-t')}\right)\nu(t')}{\sqrt{2\pi(n\Delta - t')^3}}\,dt' \tag{16.22}$$

$$\approx \frac{1}{\sqrt{8\pi\Delta}}\left(\frac{\Omega_{nl}\exp\left(-\frac{\alpha^2\Omega_{nl}^2}{2(n-l)\Delta}\right)\nu_l}{(n-l)^{3/2}} + \frac{\Omega_{n(l-1)}\exp\left(-\frac{\alpha^2\Omega_{n(l-1)}^2}{2(n-l+1)\Delta}\right)\nu_{l-1}}{(n-l+1)^{3/2}}\right),$$

and

$$\mathcal{J}_l^n = \int_{(l-1)\Delta}^{l\Delta} \frac{\left(\nu_n - \left(1 - \frac{\alpha^2\Omega(n\Delta,t')^2}{(n\Delta-t')}\right)\exp\left(-\frac{\alpha^2\Omega(n\Delta,t')^2}{2(n\Delta-t')}\right)\nu(t')\right)}{\sqrt{2\pi(n\Delta - t')^3}}\,dt'$$

$$\approx \frac{1}{\sqrt{8\pi\Delta}}\left(\frac{\left(\nu_n - \left(1 - \frac{\alpha^2\Omega_{nl}^2}{(n-l)\Delta}\right)\exp\left(-\frac{\alpha^2\Omega_{nl}^2}{2(n-l)\Delta}\right)\nu_l\right)}{(n-l)^{3/2}}\right. \tag{16.23}$$

$$\left. + \frac{\left(\nu_n - \left(1 - \frac{\alpha^2 \Omega^2_{n(l-1)}}{(n-l+1)\Delta}\right) \exp\left(-\frac{\alpha^2 \Omega^2_{n(l-1)}}{2(n-l+1)\Delta}\right) \nu_{l-1}\right)}{(n-l+1)^{3/2}} \right).$$

However, the last two integrals, $\mathcal{I}_n^n, \mathcal{J}_n^n$, require special care, because they have weak singularities. Consider the integral $\mathcal{I}_n$, which has the form

$$\mathcal{I}_n^n = \int_{(n-1)\Delta}^{n\Delta} \frac{\Omega(n\Delta, t') \exp\left(-\frac{\alpha^2 \Omega(n\Delta, t')^2}{2(n\Delta - t')}\right)}{\sqrt{2\pi (n\Delta - t')^3}} \nu(t')\, dt'.$$

In view of our piecewise linearity assumption, we have

$$\Omega(n\Delta, t') = g_n \tau - \frac{g_n - g_{n-1}}{2\Delta} \tau^2, \tag{16.24}$$

where $\tau = n\Delta - t'$. Accordingly,

$$\mathcal{I}_n^n = \int_0^\Delta \frac{d\tau}{2\sqrt{2\pi\tau}} \left(g_n - \frac{g_n - g_{n-1}}{2\Delta}\tau\right) \exp\left(-\frac{\alpha^2 \left(g_n - \frac{(g_n - g_{n-1})}{2\Delta}\tau\right)^2 \tau}{2}\right)$$
$$\times \left(\nu_n - \frac{(\nu_n - \nu_{n-1})}{\Delta}\tau\right).$$

A standard change of variables $\tau = u^2$ yields

$$\mathcal{I}_n^n = \frac{2}{\sqrt{2\pi}} \int_0^{\sqrt{\Delta}} \left(g_n - \frac{(g_n - g_{n-1})}{2\Delta} u^2\right) \tag{16.25}$$
$$\times \exp\left(-\frac{\alpha^2 \left(g_n - \frac{(g_n - g_{n-1})}{2\Delta} u^2\right)^2 u^2}{2}\right) \left(\nu_n - \frac{(\nu_n - \nu_{n-1})}{\Delta} u^2\right) du.$$

The latter integral is now non-singular and can be approximated by using the trapezoidal rule

$$\mathcal{I}_n^n \approx \frac{1}{\sqrt{2\pi\Delta}} \left(\Delta g_n \nu_n + \Gamma_n \exp\left(-\frac{\alpha^2 \Gamma_n^2}{2\Delta}\right) \nu_{n-1}\right) \tag{16.26}$$
$$= \sqrt{\frac{\Delta}{2\pi}} g_n \nu_n + \frac{\Gamma_n \exp\left(-\frac{\alpha^2 \Gamma_n^2}{2\Delta}\right)}{\sqrt{2\pi\Delta}} \nu_{n-1},$$

where

$$\Gamma_n = \frac{\Delta}{2}\left(g_n + g_{n-1}\right).$$

Similarly,

$$\mathcal{J}_n^n = \int_{(n-1)\Delta}^{n\Delta} \frac{\left(\nu_n - \exp\left(-\frac{\alpha^2\Omega(n\Delta,t')^2}{2(n\Delta-t')}\right)\nu(t')\right)}{\sqrt{2\pi(n\Delta-t')^3}}dt' \quad (16.27)$$

$$+\alpha^2\int_{(n-1)\Delta}^{n\Delta}\frac{\Omega(n\Delta,t')^2\exp\left(-\frac{\alpha^2\Omega(n\Delta,t')^2}{2(n\Delta-t')}\right)\nu(t')}{(n\Delta-t')\sqrt{2\pi(n\Delta-t')^3}}dt'$$

$$= \mathcal{J}_n^{n,1} + \alpha^2\mathcal{J}_n^{n,2},$$

for $l = 1, \ldots, n-1$.

Computing the integrals in $\mathcal{J}_n^{n,1}$ and $\mathcal{J}_n^{n,2}$, we have

$$\mathcal{J}_n^{n,1} = \int_0^{\Delta}\frac{\left(\nu_n - \exp\left(-\frac{\alpha^2\left(g_n - \frac{(g_n-g_{n-1})}{2\Delta}\tau\right)^2\tau}{2}\right)\left(\nu_n - \frac{(\nu_n-\nu_{n-1})}{\Delta}\tau\right)\right)}{\sqrt{2\pi\tau^3}}d\tau \quad (16.28)$$

$$= \int_0^{\Delta}\frac{\left(\left(1-\exp\left(-\frac{\alpha^2\left(g_n-\frac{(g_n-g_{n-1})}{2\Delta}\tau\right)^2\tau}{2}\right)\right)\nu_n\right)}{\sqrt{2\pi\tau^3}}d\tau$$

$$+\frac{(\nu_n-\nu_{n-1})}{\Delta}\int_0^{\Delta}\frac{\exp\left(-\frac{\alpha^2\left(g_n-\frac{(g_n-g_{n-1})}{2\Delta}\tau\right)^2\tau}{2}\right)}{\sqrt{2\pi\tau}}d\tau$$

$$\approx \frac{\alpha^2\nu_n}{2}\int_0^{\Delta}\frac{\left(g_n-\frac{(g_n-g_{n-1})}{2\Delta}\tau\right)^2}{\sqrt{2\pi\tau}}d\tau \quad (16.29)$$

$$+\frac{(\nu_n-\nu_{n-1})}{\Delta}\int_0^{\Delta}\frac{\exp\left(-\frac{\alpha^2\left(g_n-\frac{(g_n-g_{n-1})}{2\Delta}\tau\right)^2\tau}{2}\right)}{\sqrt{2\pi\tau}}d\tau$$

$$= \frac{\alpha^2\nu_n}{\sqrt{2\pi}}\int_0^{\sqrt{\Delta}}\left(g_n - \frac{(g_n-g_{n-1})}{2\Delta}u^2\right)^2du$$

$$+\frac{2(\nu_n-\nu_{n-1})}{\sqrt{2\pi}\Delta}\int_0^{\sqrt{\Delta}}\exp\left(-\frac{\alpha^2\left(g_n-\frac{(g_n-g_{n-1})}{2\Delta}u^2\right)^2u^2}{2}\right)du$$

$$\approx \frac{\alpha^2}{2}\frac{1}{\sqrt{2\pi\Delta^3}}\left(\Delta^2 g_n^2 + \Gamma_n^2\right)\nu_n + \frac{1}{\sqrt{2\pi\Delta}}\left(1+\exp\left(-\frac{\alpha^2\Gamma_n^2}{2\Delta}\right)\right)(\nu_n - \nu_{n-1}).$$

and

$$\mathcal{J}_n^{n,2} = \int_0^\Delta \frac{d\tau}{\sqrt{2\pi\tau}}\left(g_n - \frac{g_n - g_{n-1}}{2\Delta}\tau\right)^2 \exp\left(-\frac{\alpha^2\left(g_n - \frac{(g_n-g_{n-1})}{2\Delta}\tau\right)^2\tau}{2}\right) \times \left(\nu_n - \frac{(\nu_n-\nu_{n-1})}{\Delta}\tau\right) \tag{16.30}$$

$$= \frac{2}{\sqrt{2\pi}}\int_0^{\sqrt{\Delta}}\left(g_n - \frac{(g_n-g_{n-1})}{2\Delta}u^2\right)^2 \times \exp\left(-\frac{\alpha^2\left(g_n - \frac{(g_n-g_{n-1})}{2\Delta}u^2\right)^2 u^2}{2}\right)\left(\nu_n - \frac{(\nu_n-\nu_{n-1})}{\Delta}u^2\right)du$$

$$\approx \frac{1}{\sqrt{2\pi\Delta^3}}\left(\Delta^2 g_n^2\nu_n + \Gamma_n^2\exp\left(-\frac{\alpha^2\Gamma_n^2}{2\Delta}\right)\nu_{n-1}\right).$$

Thus,

$$\mathcal{J}_n^n \approx \left(\frac{\left(1+\exp\left(-\frac{\alpha^2\Gamma_n^2}{2\Delta}\right)\right)}{\sqrt{2\pi\Delta}} + \frac{\alpha^2\left(\frac{3}{2}\Delta^2 g_n^2 + \Gamma_n^2\right)}{\sqrt{2\pi\Delta^3}}\right)\nu_n - \left(\frac{\left(1+\exp\left(-\frac{\alpha^2\Gamma_n^2}{2\Delta}\right)\right)}{\sqrt{2\pi\Delta}} - \frac{\alpha^2\Gamma_n^2\exp\left(-\frac{\alpha^2\Gamma_n^2}{2\Delta}\right)}{\sqrt{2\pi\Delta^3}}\right)\nu_{n-1}. \tag{16.31}$$

By using Eq. (16.24) we can represent expressions (16.22) and (16.23) in a recurrent form, neglecting now quadrature errors:

$$\mathcal{I}_l^n = \frac{1}{\sqrt{8\pi\Delta}}\left(\frac{\Omega_{nl}\exp\left(-\frac{\alpha^2\Omega_{nl}^2}{2(n-l)\Delta}\right)\nu_l}{(n-l)^{3/2}} + \frac{\Omega_{n(l-1)}\exp\left(-\frac{\alpha^2\Omega_{n(l-1)}^2}{2(n-l+1)\Delta}\right)\nu_{l-1}}{(n-l+1)^{3/2}}\right)$$

$$= \mathbb{I}_l^n\left(g_n \middle| \nu_{l-1}, \nu_l, g_{l-1}, \ldots, g_{n-1}\right),$$

$$\mathcal{J}_l^n = \left(\frac{1}{\sqrt{8\pi\Delta}(n-l)^{3/2}} + \frac{1}{\sqrt{8\pi\Delta}(n-l+1)^{3/2}}\right)\nu_n - \frac{\left(1-\frac{\alpha^2\Omega_{nl}^2}{(n-l)\Delta}\right)}{\sqrt{8\pi\Delta}(n-l)^{3/2}}\exp\left(-\frac{\alpha^2\Omega_{nl}^2}{2(n-l)\Delta}\right)\nu_l - \frac{\left(1-\frac{\alpha^2\Omega_{n(l-1)}^2}{(n-l+1)\Delta}\right)}{\sqrt{8\pi\Delta}(n-l+1)^{3/2}} \tag{16.32}$$

$$\times \exp\left(-\frac{\alpha^2\Omega_{n(l-1)}^2}{2(n-l+1)\Delta}\right)\nu_{l-1} = \mathbb{J}_l^n(\nu_n, g_n|\nu_{l-1}, \nu_l, g_{l-1}, ..., g_{n-1})$$

$$= \mathbb{U}_l^n \nu_n + \mathbb{V}_l^n(g_n|\nu_{l-1}, \nu_l, g_{l-1}, ..., g_{n-1}).$$

By the same token, $\mathcal{I}_n^n$, $\mathcal{J}_n^n$ given by Eqs. (16.26) and (16.27) can be written in the form

$$\mathcal{I}_n^n = \mathbb{I}_n^n(\nu_n, g_n|\nu_{n-1}, g_{n-1}) = \mathbb{A}_n^n(g_n|g_{n-1})\nu_n + \mathbb{B}_n^n(g_n|\nu_{n-1}, g_{n-1}),$$

$$\mathcal{J}_n^n = \mathbb{J}_n^n(\nu_n, g_n|\nu_{n-1}, g_{n-1}) = \mathbb{U}_n^n(g_n|g_{n-1})\nu_n + \mathbb{V}_n^n(g_n|\nu_{n-1}, g_{n-1}).$$

In view of the above, system Eq. (16.21) can be written as follows

$$\nu_n + \alpha\sum_{l=1}^{n-1}\mathbb{I}_l^n(g_n) + \alpha\mathbb{I}_n^n(\nu_n, g_n) + \frac{\exp\left(-\frac{(\alpha\Omega_{n0}-z)^2}{2n\Delta}\right)}{\sqrt{2\pi n\Delta}} = 0, \tag{16.33}$$

$$g_n + \left(\alpha g_n + \frac{1}{\sqrt{2\pi n\Delta}}\right)\nu_n + \frac{1}{2}\sum_{l=1}^{n-1}\mathbb{J}_l^n(\nu_n, g_n) + \frac{1}{2}\mathbb{J}_n^n(\nu_n, g_n)$$

$$+ \frac{(\alpha\Omega_{n0} - z)\exp\left(-\frac{(\alpha\Omega_{n0}-z)^2}{2n\Delta}\right)}{2\sqrt{2\pi n^3\Delta^3}} = 0,$$

where we suppress explicit dependencies on $g_1, ..., g_{n-1}, \nu_1, ..., \nu_{n-1}$ for brevity provided that $g_1, ..., g_{n-1}, \nu_1, ..., \nu_{n-1}$ are given. This system can be solved by using the Newton-Raphson method, say. As a result, the new pair (ν_n, g_n) can be found and the recurrence advanced by one more step as required.

If so desired, system Eq. (16.33) can be further simplified. We notice that the dependence on ν_n is linear, eliminate ν_n in favor of g_n from the first equation,

$$\nu_n = -\frac{\alpha\left(\sum_{l=1}^{n-1}\mathbb{I}_l^n(g_n) + \mathbb{B}_n^n(g_n)\right) + \frac{\exp\left(-\frac{(\alpha\Omega_{n0}-z)^2}{2n\Delta}\right)}{\sqrt{2\pi n\Delta}}}{(1 + \alpha\mathbb{A}_n^n(g_n))},$$

and substitute this expression in the second equation, obtaining a scalar recursive nonlinear equation of the form

$$g_n + \frac{1}{2}\left(\sum_{l=1}^{n-1}\mathbb{V}_l^n(g_n) + \mathbb{V}_n^n(g_n)\right) \tag{16.34}$$

$$-\frac{\left(\alpha g_n + \frac{1}{\sqrt{2\pi n\Delta}} + \frac{1}{2}\left(\sum_{l=1}^{n-1}\mathbb{U}_l^n + \mathbb{U}_n^n(g_n)\right)\right)}{(1 + \alpha\mathbb{A}_n^n(g_n))}$$

$$\times \left(\alpha \left(\sum_{l=1}^{n-1} \mathbb{I}_l^n (g_n) + \mathbb{B}_n^n (g_n) \right) + \frac{\exp\left(-\frac{(\alpha\Omega_{n0}-z)^2}{2n\Delta}\right)}{\sqrt{2\pi n\Delta}} \right)$$

$$+ \frac{(\alpha\Omega_{n0} - z)\exp\left(-\frac{(\alpha\Omega_{n0}-z)^2}{2n\Delta}\right)}{2\sqrt{2\pi n^3\Delta^3}} = 0.$$

16.5 Numerical tests and results

In this section, we first analyze the convergence (order) of the numerical method, then test the accuracy of the first order expansion against the numerical solution, and finally perform parameter studies (in α) to investigate the influence of the mean-field interaction on the behavior of the solution.

To demonstrate the performance of the discretization scheme, we compare the solution with Eq. (16.14), the analytical solution, in the case $\alpha = 0$.

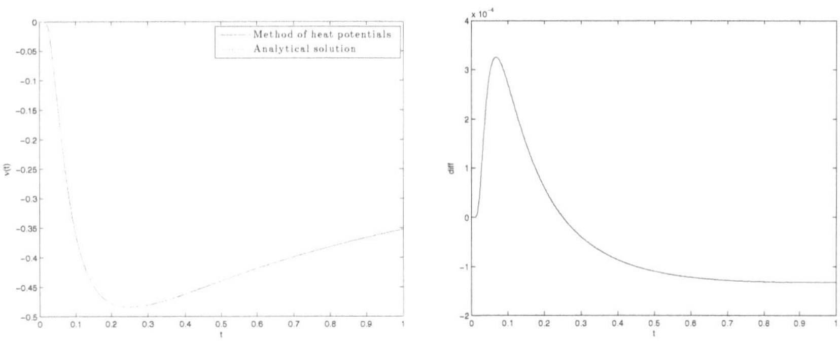

(a) (b)

Figure 16.3: $\nu(t)$: (a) Numerical and analytical solutions for $\alpha = 0, N = 1000$ (they are visually indistinguishible). (b) The difference with the exact solution for $\alpha = 0$.

For $\alpha > 0$, no closed-form solution is available, and we therefore use the Euler timestepping particle method from Kaushansky and Reisinger (2019) with sufficiently many particles and timesteps as benchmark. We illustrate the difference between our method and Kaushansky and Reisinger (2019) in Fig. 16.6.

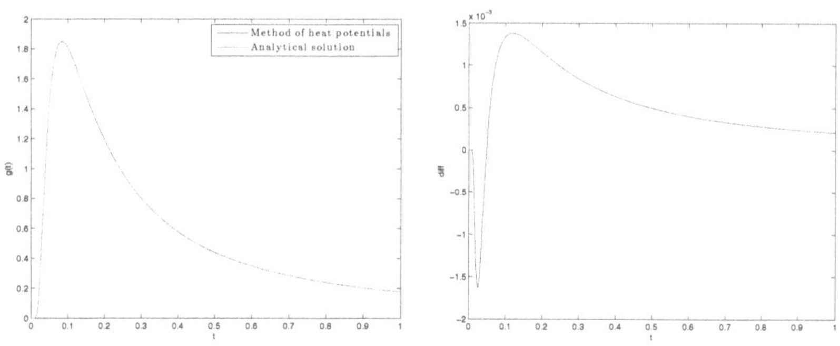

(a) (b)

Figure 16.4: $g(t)$: (a) Numerical and analytical solutions for $\alpha = 0, N = 1000$ (they are visually indistinguishible). (b) The difference with the exact solution for $\alpha = 0$.

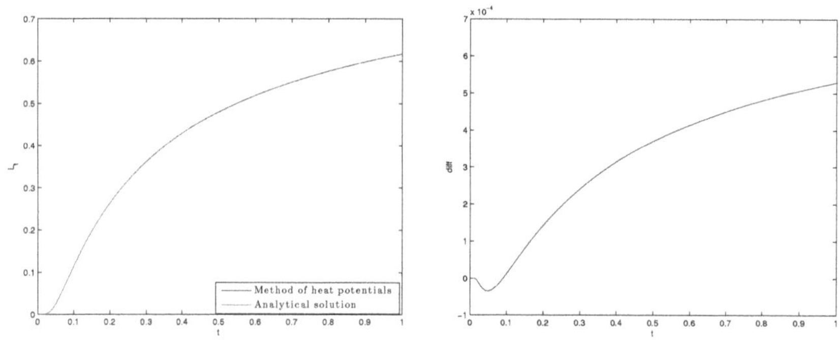

(a) (b)

Figure 16.5: L_t: (a) Numerical and analytical solutions for $\alpha = 0, N = 1000$ (they are visually indistinguishible). (b) The difference with the exact solution for $\alpha = 0$.

We now analyze the convergence order of the discretization scheme for the Volterra equation empirically. With N time steps, the error of the approximation Eq. (16.34) is expected to be $O(N^{-1})$, because the trapezoidal integration in Eqs. (16.25), (16.28), and (16.30) is on intervals $(0, \sqrt{\Delta})$,

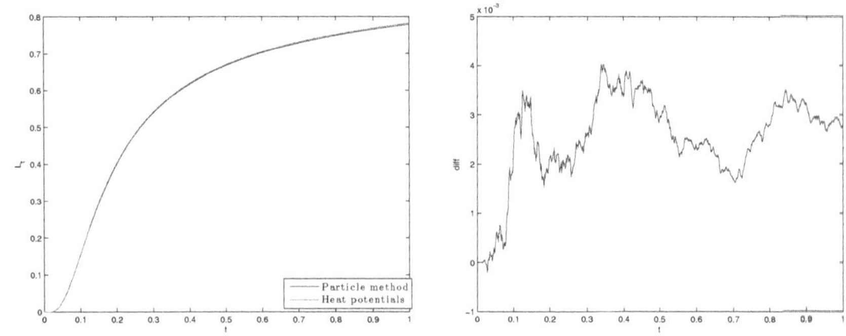

(a) (b)

Figure 16.6: L_t at $\alpha = 0.5$ obtained by numerically solving the Volterra equations ($N = 1000$) and by the particle method in Kaushansky and Reisinger (2019).

and the result is divided by $\sqrt{\Delta}$ after that. We empirically confirm this in Fig. 16.7.

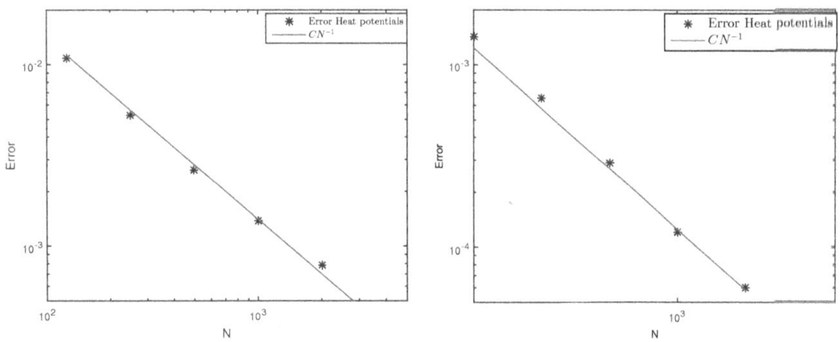

(a) (b)

Figure 16.7: Error of the maximum norm for the numerical method in Section 16.4.3: (a) in g vs the exact solution for $\alpha = 0$; (b) in L vs the simulation solution for $\alpha = 0.5$.

The complexity of our method is $O(N^2)$. Hence, in order to achieve precision ε, we need $O(\varepsilon^{-2})$ operations. In comparison, the particle method with Brownian bridge in Kaushansky and Reisinger (2019) requires $O(\varepsilon^{-3})$ operations. The latter could be improved to $O(\varepsilon^{-2})$ by multilevel simulation, but equally a higher order method for the Volterra equation would bring the complexity down. Another advantage of the method above is that we automatically get directly the derivative g of the loss function, which is harder to obtain in Kaushansky and Reisinger (2019) because of Monte Carlo noise.

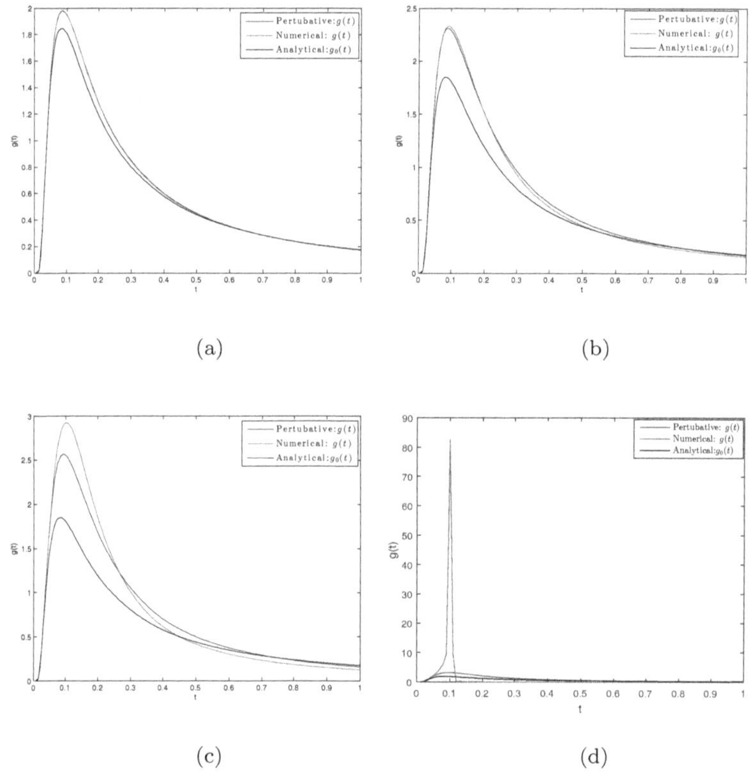

Figure 16.8: Comparison of the numerical and perturbative methods for different values of α: (a) $\alpha = 0.1$; (b) $\alpha = 0.3$; (c) $\alpha = 0.5$; (d) $\alpha = 1$. For visibility, we set $N = 100$ for the numerical method in the last plot.

Next, we compare the numerical and perturbation solutions described above. We fix $T = 1$, $z = 0.5$, and choose $N = 1000$, the number of grid points, sufficiently large so that the numerical error is negligible. In Fig. 16.8 we plot g, the hitting time density, computed with numerical and perturbation methods as well as $g_0(t)$, the solution for $\alpha = 0$, to measure the impact of the nonlinear term, for different values of α. For $\alpha = 0.1$, the two solutions are visually indistinguishable; for $\alpha = 0.3$ there is small but visible difference between the solutions, which increases further for $\alpha = 0.5$ and arises from the higher order terms.[3] For $\alpha = 1$, where the numerical solution shows a jump in the loss function at around $t = 0.1$, the first order expansion approximation breaks down.

16.5.1 *Parameter studies*

We now assess the impact of the mean-field interaction by varying the parameter α. We fix $T = 1, z = 0.5$, and choose $N = 1000$, the number of the grid points. Fig. 16.9 demonstrates the behavior of $\nu(t)$ and $g(t)$ for different values of α, starting with $\alpha = 0$; for L_t, including a case with discontinuity, see already Fig. 16.2.

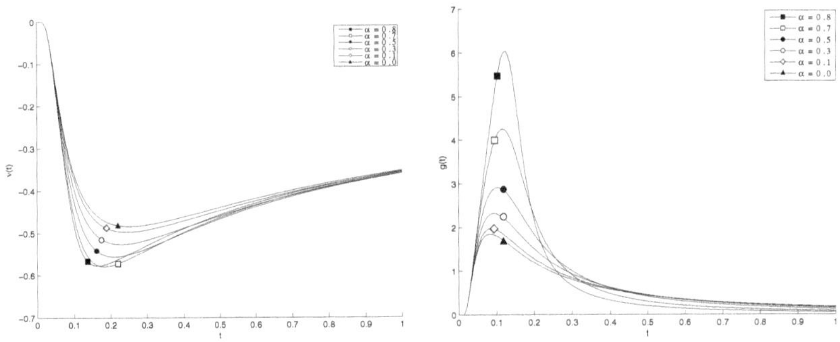

(a) (b)

Figure 16.9: The dependence on α of: (a) $\nu(t)$; (b) $g(t)$. Both functions are computed by solving Eq. (16.12) via numerical method.

To illustrate the impact of the interaction further, we consider the expectation and variance of the default time depending on α. Since the expectation is infinite, we restrict it to the interval $(0, T)$, i.e. consider

[3]In our implementation of the perturbation solution, we also perform a scaling to ensure the correct cumulative density at T (which in practice is unknown) to improve the results slightly.

$\mathbb{E}[\tau|\tau < T]$ and $\mathbb{V}[\tau|\tau < T]$. These expectations must be finite for any fixed T, and go to infinity when $T \to \infty$. The conditional density is then $p_{\tau|\tau<T}(t) = \frac{p_\tau(t)}{\int_0^T p_\tau(s)\,ds}$ for $t \in [0,T]$. Using this, one can easily evaluate the moments numerically. We present the results in Fig. 16.10. As expected, we observe that the expected default time and its variance become smaller with increasing of α, and grow with increasing T. In Fig. 16.11 we show cumulative loss probability for several representative values of α. A striking feature of this figure is the "phase transition" occurring at $\alpha \approx 1.0$, when default after a finite time becomes inevitable. By contrast, for $\alpha = 0$, the default probability reaches unity only asymptotically when $t \to \infty$.

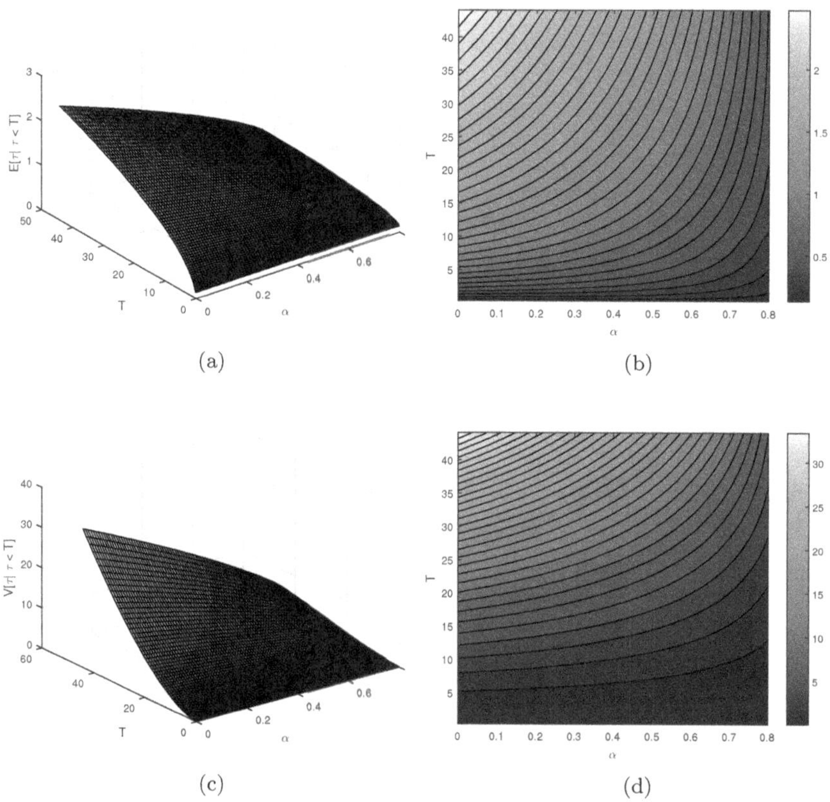

Figure 16.10: (a) and (b) $\mathbb{E}[\tau|\tau < T]$; (c) and (d) $\mathbb{V}[\tau|\tau < T]$ for different values of α and T.

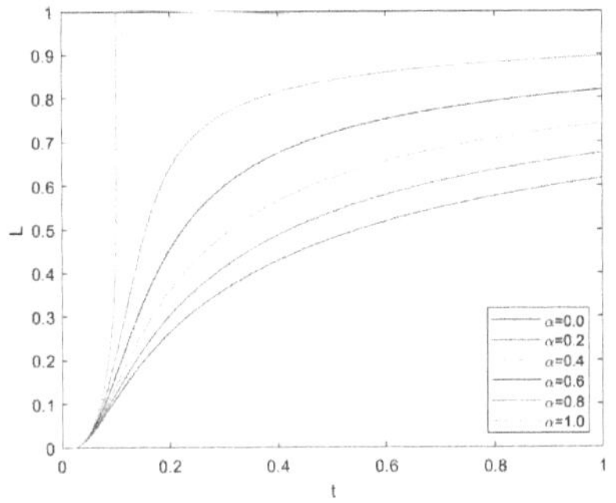

Figure 16.11: Variation of the loss probability $L(t)$ at the initial position $z = 0.5$ with α which characterizes the strength of interbank interactions.

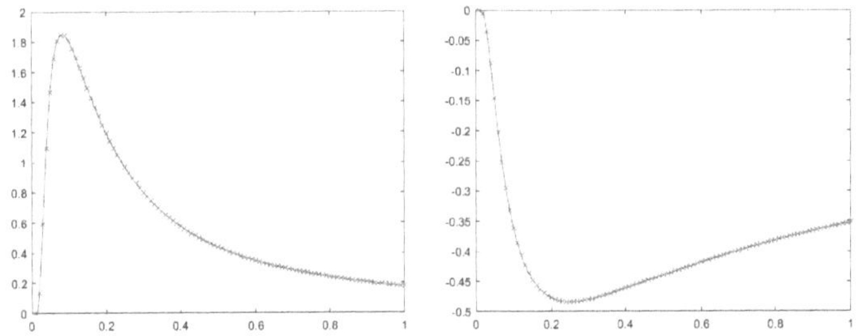

(a) (b)

Figure 16.12: (a) $\mu(t)$; (b) $\nu(t)$ — calculated numerically and analytically at $z = 0.5$, $\alpha = 0$.

We notice that for $\alpha = 0$, $\mu(t)$, $\nu(t)$ can be calculated analytically. For benchmarking purposes, we compare numerical and analytical results in Figs. 16.12a and 16.12b. As usual, the efficiency of the Newton-Raphson method is nothing short of miraculous.

In Fig. 16.13 we represent shifted probability density surfaces $p(t, x - z; z)$ for the values of α used in Fig. 16.11. The shift is made in order to make the connection with the results in Chapter 15 more transparent; after this shift all the processes start at 0 and the boundaries are given by $b = -0.5$.

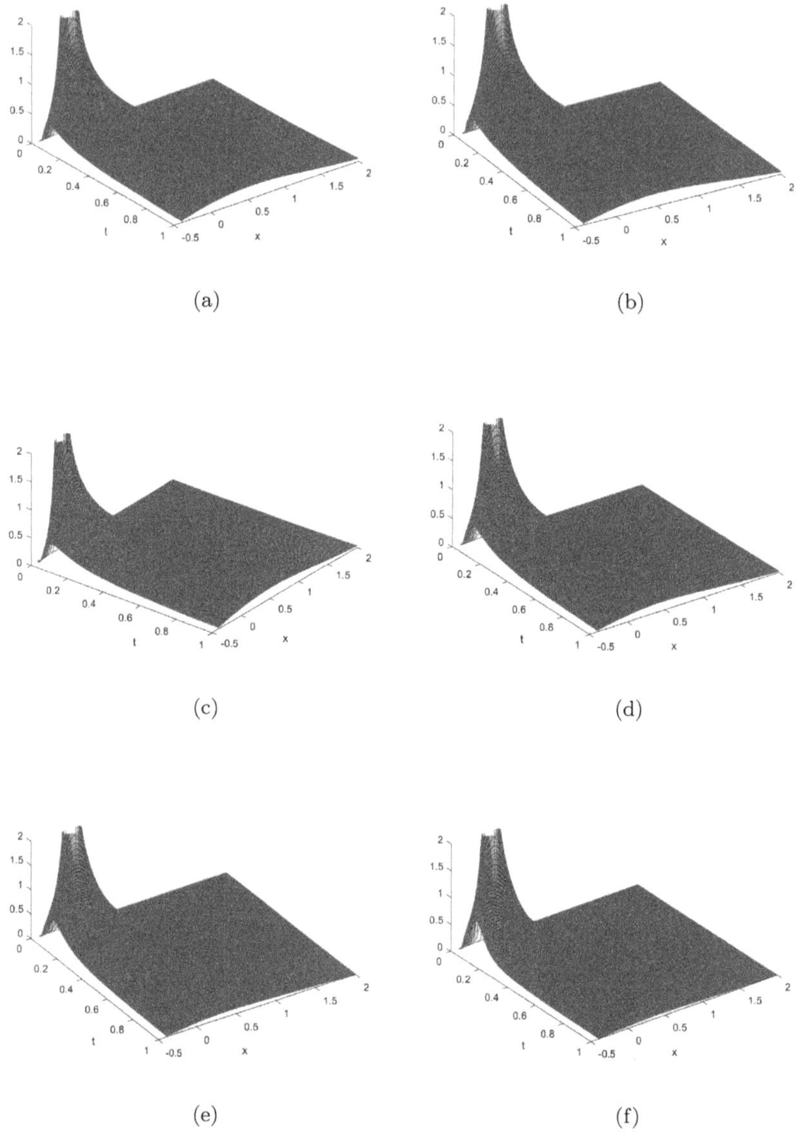

Figure 16.13: The probability density function $p(t, x - z; z)$. We shift the domain down by $-z$ in order to make comparison with the structural default model considered in Chapter 15 more transparent: (a)-(b) analytical and numerical results for $\alpha = 0$; (c)-(f) numerical results for $\alpha = 0.2, 0.4, 0.6, 0.8$, respectively.

Chapter 17

Miscellaneous Problems

In this chapter we present some physics problem that can be naturally solved by using the method of heat potentials and the GIT method. These problems in essence are close to that previously considered in this Part, therefore, we added them to this book to demonstrate the advantage of such an approach not just for finance, but also for physics. In other words, in Sections 17.1 and 17.3, we deal with relatively rare instances when results obtained for financial mathematics can be successfully used in a broader context of applied mathematics, rather than the other way around.

17.1 The supercooled Stefan problem

The Stefan problem is of great theoretical and practical interest, see, e.g., Kamenomostskaja (1961), Rubinstein (1971), Delarue *et al.* (2017) and references therein. The classical Stefan problem studies the evolving boundary between the two phases of the same medium, such as ice and water. Thus, this problem boils down to solving the heat equation with a free boundary, which is determined by a matching condition. The main equations for the supercooled Stefan problem, are very similar to the mean-field banking equations:

$$\begin{gathered}\frac{\partial}{\partial t}p\left(t,x\right)=\frac{1}{2}\frac{\partial^{2}}{\partial x^{2}}p\left(t,x\right), \quad b\left(t\right)\leq x<\infty, \qquad (17.1)\\ p\left(0,x\right)=\delta_{z}\left(x\right), \quad p\left(t,b\left(t\right)\right)=0, \quad p\left(t,x\rightarrow\infty\right)\rightarrow 0,\end{gathered}$$

where p is the negative temperature profile, and b is the liquid-solid boundary. The location of the boundary is determined by the matching condition

$$\frac{d}{dt}b\left(t\right)=\frac{\alpha}{2}\frac{\partial}{\partial x}p\left(t,x\right)\bigg|_{x=b(t)}. \qquad (17.2)$$

As mentioned in Kartashov and Krotov (2009), broadly speaking, the Stefan problem is a collection of mathematical models describing thermal, diffusive, and even thermodiffusive processes accompanied by phase transformations of the media, and by an absorption or release of latent heat. Such processes are observed, for instance, during artificial monocrystal growth, in some modern metallurgical technologies, during the creation and evolution of the internal Earth structure, as well as in some other areas of science and technology, see Kartashov and Krotov (2009) and references therein.

17.1.1 *Solution by the HP method*

As usual, we represent p as $p = H + r$, where r solves the following initial-boundary value problem (IBVP)

$$\frac{\partial}{\partial t} r(t,x) = \frac{1}{2}\frac{\partial^2}{\partial x^2} r(t,x), \quad b(t) \le x < \infty, \tag{17.3}$$
$$r(0,x) = 0, \quad r(t,b(t)) = -H(t, b(t) - z), \quad r(t, X \to \infty) \to 0.$$

By using Eq. (8.63) we get the following system of coupled Volterra equations

$$\nu(t') + \int_0^t \frac{\Theta(t,t')\,\Xi(t,t')\,\nu(t')}{\sqrt{2\pi(t-t')}} dt' + H(t, b(t) - z) = 0, \tag{17.4}$$
$$b(t) + \frac{\alpha}{2}\int_0^t \frac{\Xi(t,t')\,\nu(t')}{\sqrt{2\pi(t-t')}} dt' = 0.$$

Here

$$\Theta(t,t') = \frac{b(t) - b(t')}{(t-t')}, \quad \Xi(t,t') = e^{-\frac{(t-t')\Theta^2(t,t')}{2}}, \tag{17.5}$$
$$\Theta(t,t) = \frac{db(t)}{dt}, \quad \Xi(t,t) = 1.$$

The system of integral equations (17.4) is very similar to Eq. (15.14) and can be solved by the same token.

In Fig. 17.1 we show $b(t)$ for several representative values of α.

17.1.2 *Solution by the GIT method*

The Stefan problem can also be solved by using the GIT method. This was done in Kartashov and Krotov (2009) (which is a translation of the original paper published in Russian in 2008). The authors constructed exact analytical solutions of the Stefan problem with different boundary conditions on

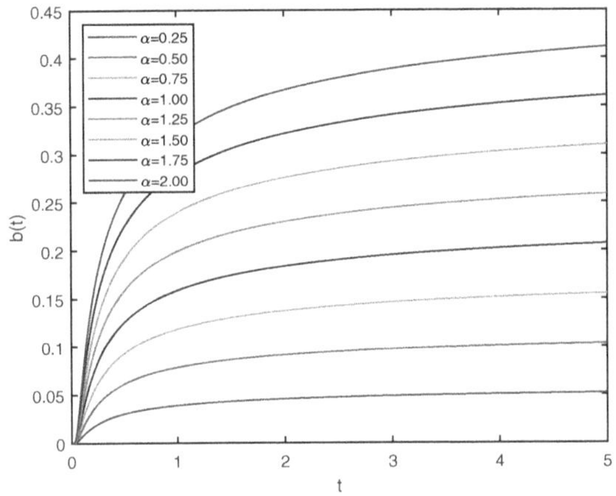

Figure 17.1: The solid-liquid boundaries $b(t)$ for several representative values of α.

the surface $x = 0$. The equivalence of the obtained solutions in the degenerate domain $y(0) = 0$ is also demonstrated. These two methods in use are the method of Fourier-type series and the GIT method. Some particular cases of the obtained general solutions in terms of dimensionless variables were considered together with their diagrams. The proportionality coefficient in the law of the moving boundary was computed with precision of up to ten decimal places.

Below we present this result obtained by using the GIT method by closely following that paper.

Let us consider a general problem with a free boundary generalizing the well-known Stefan problem

$$\frac{\partial T}{\partial t} = a\frac{\partial^2 T}{\partial x^2}, \quad 0 < x < y(t), \quad t > 0, \tag{17.6}$$

$$T(x,t)|_{t=0} = \Phi_0(x), \quad 0 < x < y_0, \quad T(x,t)|_{x=0} = \varphi_1(t), \quad t > 0,$$

$$T(x,t)|_{x=y(t)} = \varphi_2(t), \quad t > 0, \quad \left.\frac{\partial T(x,t)}{\partial x}\right|_{x=y(t)} = \Psi(t), \quad t > 0.$$

Here $T(x,t)$ is the temperature in some medium, x, t are the spatial and temporal coordinates, a is the heat-conductivity coefficient, $y(t)$ is a free boundary between two phases of the medium.

A particular case of the problem Eq. (17.6) is the Stefan problem with $\Phi_0(x) = \text{const} \geq 0, \varphi_2(t) = 0, \Psi(t) = A\frac{dy}{dt}$, $Aa = const$. In the settings of Eq. (17.6) we have to find functions $T(x,t), y(t)$, such that $T(x,t) \in \mathcal{C}^2(\Omega_t)$ and $\text{grad}_x T(x,t) \in \mathcal{C}^0(\bar{\Omega}_t)$. The function $y(t)$ must be differentiable for $t \geq 0, \frac{dy}{dt} \geq \frac{\varphi_2'(t)}{\Psi(t)}$, and $y(0) = y_0 > 0$. Note that this problem does not always have a solution. For instance, in the case when $\Phi(x) = \varphi_1(t) = \varphi_2(t) = \text{const}, \Psi(t) \neq 0$, there is no solution.

To find a solution of Eq. (17.6) by using the GIT method, consider the integral transformation

$$\bar{T}(p,t) = \int_0^{y(t)} T(x,t) \sinh(x\sqrt{p})dx, \tag{17.7}$$

where $p = \sigma + i\omega$ is a complex number with $\text{Re}\, p \geq \beta > 0, |\arg\sqrt{p}| < \frac{\pi}{4}$. Using Eq. (17.7) we obtain a general formula for $T(x,t)$, satisfying the differential equation

$$\frac{\partial T}{\partial t} = a\frac{\partial^2 T}{\partial x^2} + f(x,t), \quad 0 < x < y(t), \quad y(0) = y_0 > 0, \quad t > 0, \tag{17.8}$$

which is valid for a non-degenerate domain.

In the image-space of the transformation Eq. (17.7) the solution of Eq. (17.8) has the following form

$$\begin{aligned}\bar{T}(p,t) = e^{apr} \int_0^{y_0} T(x,0)\sinh(x\sqrt{p})dx + \int_0^1 e^{\text{lap}(t-\text{t})} \\ \times \left\{ a\sinh(y(\tau)\sqrt{p}) \left(\frac{\partial T}{\partial x}\right)_{s=y(\tau)} + \left[\frac{dy(\tau)}{d\tau}\sinh(y(\tau)\sqrt{p})\right.\right. \\ \left.\left. - a\sqrt{p}\cosh(y(\tau)\sqrt{p})\right] T(y(\tau),\tau) + a\sqrt{p}T(0,\tau) + \bar{f}(p,\tau) \right\} d\tau,\end{aligned} \tag{17.9}$$

where

$$\bar{f}(p,t) = \int_0^{y(t)} f(x,t)\sinh(x\sqrt{p})dx. \tag{17.10}$$

The Eq. (17.9) contains the boundary values of the function $T(x,t)$ and of its derivative. The inversion formula for the transformation in Eq. (17.8) may be constructed taking into account the type of the boundary conditions. In the case of the boundary conditions of the first kind, we are looking for the inversion formula in a form of a Fourier-type series, corresponding to the classical solution of the first boundary problem for Eq. (17.8)

$$T(x,t) = \sum_{n=1}^{\infty} a_n(t) e^{-\left(\frac{\sqrt{a}\pi n}{y(n)}\right)^2 t} \sin\frac{n\pi x}{y(t)}. \tag{17.11}$$

Application of Eq. (17.8) to both sides of Eq. (17.11)yields

$$\sum_{n=1}^{\infty} \frac{(-1)^{n+1} n a_n(t) e^{-\left(\frac{\sqrt{a} n\pi}{y(t)}\right)^2 t}}{p + \frac{n^2\pi^2}{y^2(t)}} = \frac{\bar{T}(p,t) y(t)}{\pi \sinh[y(t)\sqrt{p}]}, \tag{17.12}$$

where $\bar{T}(p,t)$ is given by Eq. (17.9).

Using Eq. (17.12) we can calculate the unknown coefficient $a_n(t)$ as follows. First, we integrate both sides of Eq. (17.12) over the contours $\gamma_1, \ldots, \gamma_m, \ldots$. The generic contour γ_n consists of the vertical line $\sigma > 0$, two horizontal lines $\omega_n = \pm\left(2n^2 + 2n + 1\right) \frac{\pi^2}{2y^2(t)}$, and the semicircle with a radius $R_n = \left(2n^2 + 2n + 1\right) \frac{\pi^2}{2y^2(t)}$, centered at the origin of the coordinate system. In more detail, see Chapter 9 and Fig, 9.1.

Integration yields

$$a_1(t) = \frac{y(t)}{\pi} e^{-\left(\frac{\sqrt{a} n\pi}{y(t)}\right)^2 t} \frac{1}{2\pi i} \int_{\gamma_1} \frac{\bar{T}(p,t)dp}{\sinh(y(t)\sqrt{p})} \tag{17.13}$$

$$a_n(t) = \frac{(-1)^{n+1} y(t)}{n\pi} e^{-\left(\frac{\sqrt{a} n\pi}{y(t)}\right)^2 t} \left\{ \frac{1}{2\pi i} \int \frac{\bar{T}(p,t)dp}{\gamma_n} \frac{1}{\sinh(y(t)\sqrt{p})} \right.$$

$$\left. - \frac{1}{2\pi i} \int_{\gamma_{n-1}} \frac{\bar{T}(p,t)dp}{\sinh(y(t)\sqrt{p})} \right\}, \quad n \geq 2.$$

Calculating the contour integrals in Eq. (17.13), we find the coefficients $a_n(t)$ in the form

$$a_n(t) = \frac{2}{y(t)} \left\{ \int_0^{y_0} T(x,0) \sin \frac{n\pi x}{y(t)} dx \right. \tag{17.14}$$

$$+ \int_0^t e^{-\left(\frac{\sqrt{a} n x}{y(t)}\right)^2 \tau} \left(\frac{\partial T}{\partial x}\right)_{x=y(\tau)} \sin\left(\frac{n\pi y(\tau)}{y(t)}\right) d\tau$$

$$+ \frac{n\pi a}{y(t)} \int_0^t T(0,\tau) e^{-\left(\frac{\sqrt{a} n\pi}{y(t)}\right)^2 \tau} d\tau$$

$$+ \int_0^t \left[\frac{dy(\tau)}{d\tau} \sin\left(\frac{n\pi y(\tau)}{y(t)}\right)\right.$$

$$\left. - \frac{n\pi a}{y(t)} \cos\left(\frac{n\pi y(\tau)}{y(t)}\right)\right] e^{-\left(\frac{\sqrt{a} n\pi}{y(t)}\right)^2 \tau} T(y(\tau),\tau) d\tau$$

$$\left. + \int_0^{t'} e^{-\left(\frac{\sqrt{a} n\pi}{y(t)}\right)^2 \tau} d\tau \int_0^{y(\tau)} f(x,\tau) \sin\left(\frac{n\pi x}{y(t)}\right) dx \right\}, \quad n \geq 1.$$

Thus, the solution of Eq. (17.6) is obtained in the domain $x \in [0, y(t)]$, $t > 0$ with the moving boundary which generalizes the classical case of the fixed boundary. In turn, the unknown function $y(t)$ solves the nonlinear integral equation

$$\int_0^\infty e^{-ap\tau}\varphi_2(\tau)\cosh(y(\tau)\sqrt{p})d\tau - \int_0^\infty e^{-ap\tau}\Psi(\tau)\sinh(y(\tau)\sqrt{p})d\tau$$

$$= \sqrt{p}\int_0^\infty e^{-ap\tau}\varphi_1(\tau)d\tau + \frac{1}{a}\int_0^{y_0}\Phi_0(x)\sinh(y(\tau)\sqrt{p})dx, \tag{17.15}$$

where $\Phi_0(t), \Psi(t), \phi_1(t), \phi_2(t)$ are the corresponding initial and boundary functions introduced in Eq. (17.6). As shown in Chapter 9, such a problem with a free boundary is equivalent to pricing of the American option in finance.

17.2 Time-dependent coefficient of heat exchange

In this section we consider the following problem

$$\begin{aligned} \frac{\partial u}{\partial \tau} &= \frac{\partial^2 u}{\partial x^2}, \quad (x,\tau) \in \mathbb{R}_+ \times \mathbb{R}_+, \\ u(0,x) &= u_0, \quad |u(\tau,x)| < \infty, \quad x \ge 0,\ \tau \ge 0, \\ \left.\frac{\partial u}{\partial x}\right|_{x=0} &= h(\tau)u(\tau,x)\Big|_{x=0}, \quad \tau > 0. \end{aligned} \tag{17.16}$$

Here $h(t)$ is the heat-transfer (or heat exchange) coefficient which is time dependent. This dependence is observed in the formation of a thermal boundary layer under the conditions of non-stationary flow of cooling water about solid surfaces; in the heating of bodies by a pulsating flow; in the motion of a ballistic body in a medium with variable density and temperature; in the heat exchange of a rolled metal with rolls and the ambient medium; in studying the phenomena of turbulence in contact measuring of the gas temperature; in non-stationary cooling of thermoelectric devices; in diffusion processes at a variable temperature, etc. There are a number of reasons why the heat-transfer coefficient changes with time: a change in the physical characteristics of a heat-exchange agent (the velocity of motion, the emissivity factor, the density, etc.) or a change in the state of the surface of a heated body with time (oxidation, clogging with dust, cracking, etc.). Again, in more detail, see Kartashov (1999).

The Eq. (17.16) is an initial-boundary value problem for the heat equation with the third (Robin's) boundary condition. This kind of problems were intensively studied in physics (Kartashov, 2001), so below we describe some known solutions by closely following Kartashov (1999).

Let us represent the solution $u(\tau, x)$ as convolution of the heat kernel and an unknown weight $F(\xi)$

$$u(\tau, x) = \frac{1}{2\sqrt{\pi\tau}} \int_{-\infty}^{\infty} F(\xi) e^{-\frac{(x-\xi)^2}{4\tau}} d\xi.$$

Also, let us analytically continuate the function $F(x)$ to the negative semi-axis x such that the Robin's condition is satisfied. It gives rise to the following Fredholm equation for the function $f(x) = F(-x) - u_0$

$$-\frac{1}{2}\int_0^{\infty} f(2\sqrt{x\tau})e^{-x}dx = u_0\gamma(\tau) + \gamma(\tau)\frac{\sqrt{\pi}}{2}\int_0^{\infty} f(2\sqrt{x\tau})x^{1/2}e^{-x}dx,$$

with $\gamma(\tau) = h(\tau)\sqrt{\pi\tau}$. Let us seek for the function $f(x)$ in terms of the following series

$$f(x) = \sum_{n=0}^{\infty} a_n x^n.$$

Let us assume that the function $\gamma(\tau)$ can also be expanded into a power series of $\tau^{1/2}$

$$\gamma(\tau) = \sum_{n=0}^{\infty} \gamma_n \tau^{n/2}.$$

Then the coefficients a_n can be explicitly found and reads

$$a_n = \frac{u_0\gamma_n + \frac{1}{\sqrt{\pi}}\sum_{m=0}^{n-1} 2^{m-1}\Gamma\left(\frac{m+1}{2}\right) a_m \gamma_{n-m}}{2^{n-1}\Gamma\left(1+\frac{n}{2}\right)}.$$

This provides the solution $u(\tau, x)$ to be in the form

$$u(\tau, x) = u_0 + \frac{1}{2\sqrt{\pi\tau}}\sum_{n=1}^{\infty} a_n \int_0^{\infty} \xi^n e^{-\frac{(x+\xi)^2}{4\tau}} d\xi.$$

With this in mind, by using the method of successive approximations (Kartashov, 2001), it can be checked that the function $u(\tau, x)$ defined as

$$u(\tau, x) = \frac{x}{2\sqrt{\pi}}\int_0^{\tau} \frac{A(s)}{(\tau - s)^{3/2}} e^{-\frac{x^2}{4(\tau-s)}} ds, \tag{17.17}$$

$$A(\tau) = \frac{1}{\sqrt{\pi}}\sum_{n=0}^{\infty}\left(\frac{-1}{\sqrt{\pi}}\right)^n \int_0^{\tau}\frac{h(s)ds}{\sqrt{\tau - s}}\int_0^{s}\frac{h(s_1)}{\sqrt{s - s_1}}\cdots\int_0^{s_{n-1}}\frac{h(s_n)\varphi(s_n)ds_n}{\sqrt{s_{n-1} - s_n}},$$

solves the problem in Eq. (17.16) with the initial condition and boundary conditions

$$u(0, x) = 0, \quad \left.\frac{\partial u}{\partial x}\right|_{x=0} = h(\tau)\left(u(\tau, 0) - \varphi(\tau)\right).$$

Under the assumption $|h(t)| < \infty$, $\forall t \in [0, \tau]$, the series in Eq. (17.17) converges absolutely and uniformly $\forall x > 0$ and $\tau > 0$ in any finite interval of their variation. For some parameterizations of $h(\tau)$ and $\varphi(\tau)$ the expression Eq. (17.17) can be sufficiently simplified. For example, let us assume that $h(\tau)$ and $\varphi(\tau)$ have a power structure

$$h(\tau) = h_0 \tau^m, \qquad \varphi(\tau) = T_{med} \tau^r,$$

Here m and r are real numbers such that $r + 1/2 + k(m + 1/2) > 0$, $\forall k = 1 \dots n$, and h_0, T_{med} are some positive constants. With these definitions Eq. (17.17) transforms to

$$u(\tau, x) = \frac{T_{med} x}{2\sqrt{\pi}} \sum_{n=1}^{\infty} (-1)^{n+1} \left(\frac{h_0}{\sqrt{\pi}}\right)^n \prod_{k=1}^{n} B\left(r + \frac{1}{2} + k\left(m + \frac{1}{2}\right), \frac{1}{2}\right)$$
$$\times \int_0^{\tau} \frac{s^{r+n(m+1/2)}}{(\tau - s)^{3/2}} e^{-\frac{x^2}{4(\tau - s)}},$$

where $B(\cdot, \cdot)$ denotes the Beta function (Olver *et al.*, 2020). In the particular case $m = -1/2$, $h(\tau) = h_0 \tau^{-1/2}$ and $r = 0$, this expression simplifies to

$$u(\tau, x) = T_{med} \Phi^* \left(\frac{x}{2\sqrt{t}}\right) \sum_{n=1}^{\infty} (-1)^{n+1} \left(h_0 \sqrt{\pi}\right)^n = \frac{T_{med} h_0 \sqrt{\pi}}{1 + h_0 \sqrt{\pi}} \Phi^* \left(\frac{x}{2\sqrt{t}}\right),$$

with $\Phi^*(\cdot)$ being the complimentary error function

$$\Phi^*(z) = 1 - \Phi(z), \qquad \Phi(z) = \frac{2}{\sqrt{\pi}} \int_0^z e^{-y^2} dy.$$

The problem in Eq. (17.16) can also be treated via integral transforms. Since the domain of definition of Eq. (17.16) has the straight boundary $x = 0$, for doing so one can use the Laplace transform. Further, for the sake of brevity, let us consider a problem with the vanishing initial condition Eq. (17.16), i.e., $u_0 = 0$. We solve this problem by taking the Laplace transform with respect to the temporal variable τ

$$\bar{u}(x, \lambda) = \mathfrak{L} u(\tau, x), \quad \bar{f}(\lambda) = \mathfrak{L}\left(u(\tau, 0) h(\tau)\right).$$

Following the method we used to solve the problem in Eq. (6.24), the function $\bar{u}(x, \lambda)$ can be represent as follows

$$\bar{u}(x, \lambda) = C_1 e^{-\sqrt{\lambda} x} + C_2 e^{\sqrt{\lambda} x}.$$

Taking into account the boundary conditions

$$\left.\frac{d\bar{u}}{dx}\right|_{x=0} = \bar{f}(\lambda), \qquad \lim_{x \to \infty} \bar{u}(x, \lambda) = 0$$

yields the explicit expression for $\bar{u}(x,\lambda)$

$$\bar{u}(x,\lambda) = -\bar{f}(\lambda)\frac{e^{-\sqrt{\lambda}x}}{\sqrt{\lambda}}. \tag{17.18}$$

To inverse Eq. (17.18), we use Eq. (6.19) and the convolution property Eq. (6.23) to obtain the following representation for $u(\tau,x)$

$$u(\tau,x) = -\int_0^\tau h(s)\Psi(s)\frac{e^{-\frac{x^2}{4(\tau-s)}}}{2\pi\sqrt{\tau-s}}ds, \tag{17.19}$$

where $\Psi(\tau) = 2u(\tau,0)$. Note, that $u(\tau,x)$ depends on yet unknown function $\Psi(\tau)$. However, this function can be found by solving the corresponding Volterra equation of the second kind. Indeed, letting $x = 0$ in Eq. (17.19) we obtain the integral equation

$$\frac{\Psi(\tau)}{2} = -\int_0^\tau \frac{h(s)\Psi(s)}{2\pi\sqrt{\tau-s}}ds. \tag{17.20}$$

If $h(\tau) = h_0$, Eq. (17.20) becomes the Abel equation which can be solved explicitly.

One can be curious about applicability of the GIT method to the problem in Eq. (17.16) when the latter has moving boundaries. To design the GIT for this particular case, one should slightly modify the boundary conditions in Eq. (7.8) and construct a Sturm-Liouville expansion associated with the Robin's problem for the corresponding second order ODE. For further details we refer the reader to Elnaqeeb *et al.* (2020).

17.3 The integrate-and-fire neuron excitation model

In this section we briefly describe the famous integrate-and-fire model in neuroscience, see, e.g., Lewis and Rinzel (2003), Ostojic *et al.* (2009) and Carrillo *et al.* (2013). The integrate-and-fire model is a mathematical description of the properties of specific cells (spiking neurons) in the nervous system generating sharp electrical potentials across their cell membrane. These spikes last roughly one millisecond. Spiking neurons are a significant signaling unit of the nervous system as a whole, so that understanding their operation is of great theoretical and practical importance.

The neuron excitation problem has the form

$$\frac{\partial}{\partial t}p(t,x) = \frac{\partial}{\partial x}\left(\left(x-\mu(t)\right)p(t,x)\right) + \frac{1}{2}\frac{\partial^2}{\partial x^2}p(t,x) + \lambda(t)\,\delta_{X_0}(x), \tag{17.21}$$

$$-\infty < x \leq 0, \quad p(0,x) = p_0(x), \quad p(t,-\infty) = 0, \quad p(t,0) = 0,$$
$$X_0 < 0, \quad \lambda(t) = -\frac{1}{2}\frac{\partial}{\partial x}p(t,0), \quad \mu(t) = m_0 + m_1\lambda(t),$$

where $p(t,x) > 0$ is the probability density of finding neurons at a voltage x. Without loss of generality, we choose

$$p_0(x) = \delta_\xi(x), \quad \xi < 0. \tag{17.22}$$

The Eq. (17.21) preserve probability in the sense that

$$\frac{d}{dt}\int_{-\infty}^{0} p(t,x)\,dx = 0. \tag{17.23}$$

Indeed, integration of the main equation yields

$$\frac{d}{dt}\int_{-\infty}^{0} p(t,x)\,dx = \int_{-\infty}^{0} \frac{\partial}{\partial t}p(t,x)\,dx = \int_{-\infty}^{0} \left[\frac{\partial}{\partial x}((x-\mu(t))\,p(t,x))\right.$$
$$\left. + \frac{1}{2}\frac{\partial^2}{\partial x^2}p(t,x) + \lambda(t)\,\delta_{X_0}(x)\right]dx = \frac{1}{2}\frac{\partial}{\partial x}p(t,0) + \lambda(t) = 0.$$

17.3.1 *The stationary problem*

Because the integrate-and-fire equations are probability-preserving, there exists a stationary solution, which solves the time-independent Fokker–Planck problem:

$$\frac{\partial}{\partial x}((x-\mu)\,p(x)) + \frac{1}{2}\frac{\partial^2}{\partial x^2}p(x) + \lambda\delta_{X_0}(x) = 0, \tag{17.24}$$
$$-\infty < x \leq 0, \qquad p(-\infty) = 0, \quad p(0) = 0,$$
$$\lambda = -\frac{1}{2}\frac{\partial}{\partial x}p(0), \quad \mu = m_0 + m_1\lambda.$$

We represent $p(x)$ in the form

$$p(x) = p^{<}(x)\,(1-\Theta(x-X_0)) + p^{>}(x)\,\Theta(x-X_0), \tag{17.25}$$

where $\Theta(.)$ is the Heaviside function, and notice that

$$p^{<}(X_0) = p^{>}(X_0) \equiv \nu, \tag{17.26}$$
$$\frac{1}{2}\left[\frac{\partial}{\partial x}p^{>}(X_0) - \frac{\partial}{\partial x}p^{<}(X_0)\right] = -\lambda,$$

where ν, λ are unknown constants, which have to be determined as part of the solution. In view of the boundary conditions, it is clear that

$$(x-\mu)\,p^{>}(x) + \frac{1}{2}\frac{\partial}{\partial x}p^{>}(x) = c^{>} \equiv -\lambda, \tag{17.27}$$

$$(x-\mu)\,p^{<}(x)+\frac{1}{2}\frac{\partial}{\partial x}p^{<}(x)=c^{<}\equiv 0.$$

Moreover, since p is continuous at $x = X_0$ the second matching condition (17.26) is satisfied automatically.

The method of separation of variables yields

$$p^{<}(x)=\nu e^{(X_0-\mu)^2-(x-\mu)^2}, \tag{17.28}$$

while the method of variation of constants provides

$$p^{>}(x)=2\lambda\left(e^{\mu^2-(x-\mu)^2}D(-\mu)-D(x-\mu)\right), \tag{17.29}$$

where $D(.)$ is Dawson's integral,

$$D(x)=e^{-x^2}\int_0^x e^{y^2}\,dy. \tag{17.30}$$

Thus, we obtain

$$\nu=2\lambda\left(e^{\mu^2-(X_0-\mu)^2}D(-\mu)-D(X_0-\mu)\right), \tag{17.31}$$

and

$$p^{<}(x)=2\lambda\left(e^{\mu^2}D(-\mu)-e^{(X_0-\mu)^2}D(X_0-\mu)\right)e^{-(x-\mu)^2}. \tag{17.32}$$

At the same time, in the stationary case, the probability density $p(x)$ has to integrate to unity

$$\int_{-\infty}^{0}p(x)\,dx=\int_{-\infty}^{X_0}p^{<}(x)\,dx+\int_{X_0}^{0}p^{>}(x)\,dx=1, \tag{17.33}$$

which is a nonlinear equation for λ, because both μ and ν are known functions of λ. Once this equation is solved numerically, the entire profile is determined. It is worth noting that the integral $\int_{-\infty}^{X_0}p^{<}(x)\,dx$ can be computed analytically

$$\begin{aligned}&\int_{-\infty}^{X_0}p^{<}(x)\,dx \\ &\quad=\nu\int_{-\infty}^{X_0}e^{(X_0-\mu)^2-(x-\mu)^2}\,dx=\sqrt{\pi}\nu e^{(X_0-\mu)^2}N\left(\sqrt{2}(X_0-\mu)\right).\end{aligned} \tag{17.34}$$

The second integral $\int_{X_0}^{0}p^{>}(x)\,dx$ can be split into two parts and simplified as follows

$$\begin{aligned}&\int_{X_0}^{0}2\lambda\left[e^{-x(x-2\mu)}D(-\mu)-D(x-\mu)\right]dx \\ &\quad=2\lambda\sqrt{\pi}e^{\mu^2}\left[N\left(-\sqrt{2}\mu\right)-N\left(\sqrt{2}(X_0-\mu)\right)\right]D(-\mu)\end{aligned} \tag{17.35}$$

$$- 2\lambda \int_{X_0-\mu}^{-\mu} D(x)\, dx.$$

As can be seen, the first integral in the LHS of Eq. (17.35) is computed analytically, while the second one may be computed numerically.

Thus, the corresponding nonlinear equation for λ can be written as

$$\sqrt{\pi}\Big[e^{\mu^2} N\left(-\sqrt{2}\mu\right) D(-\mu) - e^{(X_0-\mu)^2} N\left(\sqrt{2}\,(X_0-\mu)\right) D(X_0-\mu)\Big] \\ - \int_{X_0-\mu}^{-\mu} D(x)\, dx - \frac{1}{2\lambda} = 0. \tag{17.36}$$

We show the stationary profile $p(x)$ and its derivative $dp(x)/dx$ in Fig. 17.2. As expected, $dp(x)/dx$ jumps down at $x = X_0$.

17.3.2 *The unsteady problem*

First, we use the following transformation of variables

$$y = x - M(t), \quad M(0) = 0, \tag{17.37}$$

$$\frac{\partial}{\partial t} = \frac{\partial}{\partial t} - M'(t)\frac{\partial}{\partial y}, \quad \frac{\partial}{\partial x} = \frac{\partial}{\partial y},$$

and get the following IBVP

$$\frac{\partial}{\partial t} p(t,y) = \frac{\partial}{\partial y}\Big[(y + M'(t) + M(t) - \mu(t))\, p(t,y)\Big] + \frac{1}{2}\frac{\partial^2}{\partial y^2} p(t,y) \\ + \lambda(t)\,\delta_{X_0-M(t)}(x), \quad -\infty < y \le -M(t), \tag{17.38}$$

$$p(0,y) = \delta_\xi(y), \quad p(t,-\infty) = 0, \quad p(t,-M(t)) = 0,$$

$$\lambda(t) = -\frac{1}{2}\frac{\partial}{\partial y} p(t,-M(t)), \quad \mu(t) = m_0 + m_1\lambda(t).$$

Thus, by choosing M in such a way that

$$M'(t) + M(t) - \mu(t) = 0, \quad M(0) = 0, \tag{17.39}$$

or, explicitly

$$M(t) = \int_0^t e^{-(t-t')}\mu(t')\, dt', \tag{17.40}$$

we get the IBVP for the standard Ornstein-Uhlenbeck process

$$\frac{\partial}{\partial t} p(t,y) = \frac{\partial}{\partial y}(y p(t,y)) + \frac{1}{2}\frac{\partial^2}{\partial y^2} p(t,y) + \lambda(t)\,\delta_{X_0-M(t)}(x), \tag{17.41}$$

$$-\infty < y \le -M(t),$$

$$p(0,y) = \delta_\xi(y), \quad p(t,-\infty) = 0, \quad p(t,-M(t)) = 0,$$

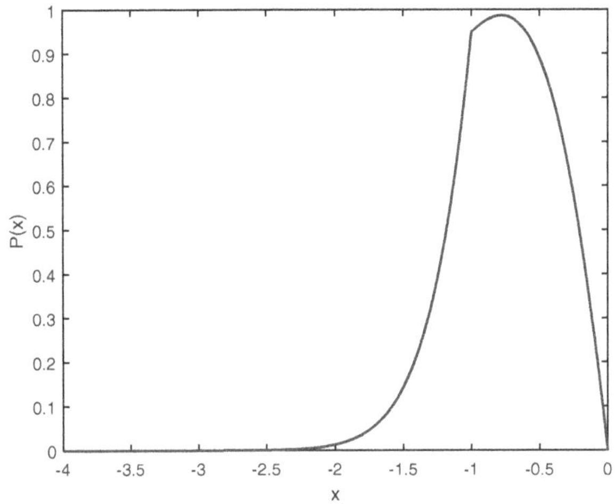

(a)

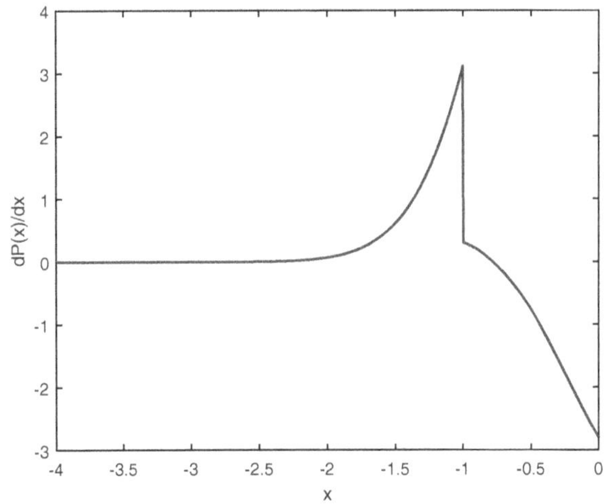

(b)

Figure 17.2: (a) The stationary distribution $p(x)$; (b) $dp(x)/dx$ - for $X_0 = -1, m_0 = 0.5, m_1 = 0.1$. The corresponding value of λ computed as part of the solution is 1.4002.

$$\lambda(t) = -\frac{1}{2}\frac{\partial}{\partial y}p(t, -M(t)), \quad \mu(t) = m_0 + m_1\lambda(t).$$

As usual, we split $p(t,x)$ as follows

$$p(t,x) = e^t H\left(\eta(t), e^t y - \xi\right) + r(t,x), \tag{17.42}$$

where the first term solves the governing equation and satisfies the initial, but not the boundary conditions, while $r(t,x)$ solves the following IBVP

$$\begin{aligned} \frac{\partial}{\partial t}r(t,y) &= \frac{\partial}{\partial y}(yr(t,y)) + \frac{1}{2}\frac{\partial^2}{\partial y^2}r(t,y) + \lambda(t)\,\delta_{X_0-M(t)}(x), \qquad (17.43)\\ r(0,y) &= 0, \quad r(t,-\infty) = 0, \quad r(t,-M(t)) = \chi_0(t),\\ \lambda(t) &= -\frac{1}{2}\frac{\partial}{\partial y}r(t,-M(t)) + \chi_1(t), \quad \mu(t) = m_0 + m_1\lambda(t). \end{aligned}$$

Here

$$\begin{aligned} \chi_0(t) &= -e^t H\left(\eta(t), e^t M(t) + \xi\right), \qquad (17.44)\\ \chi_1(t) &= -\frac{e^{2t}\left(e^t M(t) + \xi\right)}{2\eta(t)} H\left(\eta(t), e^t M(t) + \xi\right). \end{aligned}$$

We apply the familiar change of variables Eq. (10.24) and get the following IBVP for $q(\tau,\theta) = e^{-t}r(t,y)$

$$\frac{\partial}{\partial \tau}q(\tau,\theta) = \frac{1}{2}\frac{\partial^2}{\partial \theta^2}q(t,\theta) + \varkappa(\tau)\,\delta_{X_0-M(t)}(x), \quad -\infty < \theta \le \Gamma(\tau), \tag{17.45}$$

$$\begin{aligned} q(0,\theta) &= 0, \quad q(\tau,-\infty) = 0, \quad q(\tau,\Gamma(\tau)) = \varrho_0(\tau),\\ \varkappa(\tau) &= -\frac{(1+2\tau)}{2}\frac{\partial}{\partial \theta}q(\tau,\Gamma(\tau)) + \varrho_1(\tau), \end{aligned}$$

where

$$\begin{aligned} \Gamma_0(\tau) &= \sqrt{1+2\tau}\left(X_0 - M\left(\ln\left(\sqrt{1+2\tau}\right)\right)\right), \qquad (17.46)\\ \Gamma(\tau) &= -\sqrt{1+2\tau}M\left(\ln\left(\sqrt{1+2\tau}\right)\right),\\ \varrho_0(\tau) &= -H(\tau,\Gamma(\tau) - \xi),\\ \varrho_1(\tau) &= \frac{(\Gamma(\tau) - \xi)}{2\tau}H(\eta(t),\Gamma(\tau) - \xi). \end{aligned}$$

Denoting $q(\tau,\Gamma_0(\tau))$ by $\nu(\tau)$, we can split the IBVP Eq. (17.45) into two IBVP

$$\begin{aligned} \frac{\partial}{\partial \tau}q^{>}(\tau,\theta) &= \frac{1}{2}\frac{\partial^2}{\partial \theta^2}q^{>}(t,\theta), \quad \Gamma_0(\tau) \le \theta \le \Gamma(\tau), \qquad (17.47)\\ q^{>}(0,\theta) &= 0, \quad q(\tau,\Gamma_0(\tau)) = \nu(\tau), \quad q(\tau,\Gamma(\tau)) = \varrho_0(\tau), \end{aligned}$$

and

$$\frac{\partial}{\partial \tau} q^{<}(\tau, \theta) = \frac{1}{2} \frac{\partial^2}{\partial \theta^2} q^{<}(t, \theta), \quad \infty < \theta \leq \Gamma_0(\tau), \tag{17.48}$$
$$q^{<}(0, \theta) = 0, \quad q(\tau, \theta \uparrow -\infty) \to 0, \quad q(\tau, \Gamma_0(\tau)) = \nu(\tau).$$

These two solutions should also obey a matching condition

$$\frac{\partial}{\partial \theta} q^{<}(t, \Gamma_0(\tau)) - \frac{\partial}{\partial \theta} q^{>}(t, \Gamma_0(\tau)) = 2\left(-\frac{(1+2\tau)}{2} \frac{\partial}{\partial \theta} q^{>}(\tau, \Gamma(\tau)) + \varrho_1(\tau)\right). \tag{17.49}$$

We can now use results from Chapter 8 to reduce these equations to a very efficient (but highly nonlinear) system of Volterra integral equations. An analysis of the corresponding system will be presented elsewhere.

PART 7
Multilayer Problems

Chapter 18

Double Barrier Options

In this chapter we consider double barrier options written on the underlying which temporal dynamics is driven by a simple one-factor diffusion process but with time-dependent coefficients. Also, both barriers are assumed to be time dependent. Finally, when the underlying process hits any of the barriers, the holder of, say Call option, gets a rebate-at-hit (different for the upper and lower barriers), and they are also time-dependent. It is important that here we consider only the underlying dynamics whose option pricing problem by using the Feynman-Kac theorem and also some transformations could be reduced to the heat equation. Nevertheless, to the best of our knowledge, even with this simplification a closed-form solution of this problem was unknown until it was published in Itkin and Muravey (2020a). Construction of such models is shortly discussed in Section 18.4.

However, we have to mention Mijatovic (2010), where a similar problem was solved by using a probabilistic approach to obtain a decomposition of the barrier option price into the corresponding European option price minus the barrier premium for a wide class of payoff functions, barrier functions and linear diffusions (i.e., the drift is constant and the local volatility is a function of the underlying only). For this setting it is shown in Mijatovic (2010) that the barrier premium can be expressed as a sum of integrals along the barriers of the option's delta at the barriers, and that those deltas solve a system of Volterra integral equations of the second kind. This is similar to the idea of the generalized integral transform (GIT) method that we use throughout this book, while our setting is more general. Indeed, we allow any diffusion model with time-dependent coefficients and time-dependent barriers and rebates at hit subject to the condition that the pricing partial differential equation (PDE) can be reduced to the heat equation (or, as shown in Carr *et al.* (2020) to the Bessel equation). It can also be checked that the pricing PDE in Mijatovic (2010) by a simple change of the spatial variable can be transformed to the heat equation. Our approach advocated in this book further extends the technique we elaborated in a series of papers which dealt with a similar problem for single barrier options. To remind, in Carr and Itkin (2021) we developed semi-analytic solutions for the barrier (perhaps, time-dependent) and American options where the underlying stock is driven by a time-dependent OU process with a lognormal drift. This model is equivalent to the familiar Hull-White model in Fixed Income that was separately considered in Itkin and Muravey (2020b). In all cases the solution was obtained by using the method of heat potentials (HP) and the GIT method. The GIT method is also known in physics, see Kartashov (1999, 2001) where some problems related to so-called Stefan problem were solved. But it was unknown in finance until the first use in Carr and Itkin (2021). It is important, that the solutions were constructed only for problems where the underlying is defined at the domain $S \in [0, y(t)]$ with S being the stock price, and $y(t)$ being the time-dependent barrier, however, for other domains the solution was yet unknown, even in physics. Then in Itkin and Muravey (2020b) the GIT solution for the first time was constructed for the domain $S \in [y(t), \infty)$.

Latter this technique was elaborated also for the CIR and CEV models, Carr *et al.* (2020), and the Black-Karasinski model, Itkin *et al.* (2020). In particular, in Carr *et al.* (2020), the HP method was further generalized to be capable to solving not just the heat but also the Bessel equations, and was called the Bessel potential (BP) method. In Itkin *et al.* (2020) the PDE is also of a special kind. It is a flavor of the time-dependent Schrödinger equation with the unsteady Morse potential (this can be obtained by the change of variables $x \to -x$ and $\tau \to -\mathrm{i}\tau$, $\mathrm{i} = \sqrt{-1}$).

By presenting our approach we skip the explicit formulation of the underlying model. Instead we define a wide class of models where pricing double barrier options can be translated to solving the heat equation with time-dependent boundaries (barriers) and time-dependent boundary conditions (rebates-at-hit). Note, that the problems considered in the above cited paper - pricing barrier and American options in the time-dependent OU process, pricing barrier options in the Hull-White model, etc., also fit to this class as this is shown in the corresponding papers. Then we construct the solution by using both the GIT and the HP methods. The latter was already shortly presented in Itkin and Muravey (2020b), but for the homogeneous boundary conditions. Also, here we present full derivation of the explicit value of the solution spatial gradient u_x at the lower $x = y(\tau)$ and upper $x = z(\tau)$ boundaries. This derivation differs from that in Lipton and Kaushansky (2020a) (and is closer in sense to Tikhonov and Samarskii (1963)), but provides a similar result. Also, all the results obtained in Itkin and Muravey (2020a) and presented in this chapter are new.

Second, various financial problems, where efficient pricing of double barrier options with rebates at hit is subject of investigation, can be reduced to this setting. As we have mentioned it already in above, they include time-dependent Hull-White and OU models, the time-dependent Black-Scholes model, etc. Carr and Itkin (2021), Itkin and Muravey (2020b) and Itkin *et al.* (2020). Other models can be constructed by using Lie symmetries. Also, for the CIR and CEV models, where the pricing problem is reduced to solving the Bessel PDE with time-dependent boundaries, the latter can also be solved in a similar manner, Carr *et al.* (2020). Also, local volatility models with $\sigma = \sigma(x)$ can be also treated under this setting.

Third, consider a general one-factor model

$$dS_t = \mu(t, S)dt + \sigma(t, S)dW_t, \qquad S_t(t = 0) = S_0. \tag{18.1}$$

where $t > 0$ is the time, S_t is the spot price, $\mu(t, S)$ is the drift, $\sigma(t, S)$ is the volatility of the process, W_t is the standard Brownian motion under the risk-neutral measure. This model can be solved as follows. Let us split the domain of the definition of S into N intervals, and at every interval $i = 1, \dots, N$ approximate the drift by a linear function of S, i.e. $\mu_i(t, S) = a_i(t) + b_i(t)S$, and the volatility — by a quadratic function $\sigma_i(t, S) = c_i(t) + d_i(t)S + e_i(t)S^2$. Then it can be shown that at every interval the corresponding pricing PDE can be transformed to the heat equation with time-dependent boundaries and the boundary conditions. Using continuity of the solution and its gradient at every sub-boundary, this problem can be solved semi-analytically in a similar fashion. In physics this approach is called the method of multilayer heat equation, see, e.g., a nice survey in Dias (2015). In more detail the development of this method as applied to finance will be published elsewhere. Thus, solving (semi-analytically) the heat equation with time-dependent moving boundaries and the boundary conditions is a key element of such a method. Having this method in hands, pricing double barrier options for any financial model of the type Eq. (18.1) can be done semi-analytically.

18.1 Statement of the problem

Let us consider a one-factor diffusion model in Eq. (18.1) By using a standard argument, to price options written on S_t as an underlying, one can apply the Feynman-Kac theorem to obtain the following partial differential equation (PDE) for, e.g., the European Call option price

$$\frac{\partial C}{\partial t} + \frac{1}{2}\sigma^2(t,S)\frac{\partial^2 C}{\partial S^2} + \mu(t,S)S\frac{\partial C}{\partial S} = r(t)C, \qquad t \in [0,T]. \tag{18.2}$$

Here in case of Equities we treat S_t as the stock price, then $r(t)$ is the deterministic interest rate. If S_t is the stochastic interest rate, then $r(t)$ in the RHS of Eq. (18.2) should be replaced with S.

The Eq. (18.2) should be solved subject to the terminal condition at the option maturity $t = T$

$$C(T,S) = (S-K)^+, \tag{18.3}$$

where K is the option strike, and some boundary conditions. Below we are concentrated only on double barrier options with moving barriers: the lower barrier at $S = L(t)$ and the upper barrier at $S = H(t) > L(t)$, so $S \in [L(t), H(t)]$. We assume, that all time-dependent parameters of Eq. (18.2) as well as the barriers are at least one time differentiable functions of t.

Our main assumption is that the PDE in Eq. (18.2) by a series of transformations of the dependent variable $C(S,t) \mapsto U(x,\tau)$ and independent variables $S \mapsto x(t,S)$, $t \mapsto \tau(t)$ can be reduced to the heat equation

$$\frac{\partial U}{\partial \tau} = \frac{\partial^2 U}{\partial x^2}, \tag{18.4}$$

which should be solved at the new domain $x \in [y(\tau), z(\tau)]$, $\tau \in [0, \tau(0)]$, subject to the initial condition

$$U(0,x) = U_0(x), \tag{18.5}$$

and the boundary conditions

$$U(\tau, y(\tau)) = f^-(\tau), \qquad U(\tau, z(\tau)) = f^+(\tau). \tag{18.6}$$

Here $f^{\pm}(\tau), y(\tau), z(\tau)$ are some continuous functions of time τ. From the financial point of view the problem in Eqs. (18.4)–(18.6) (the $\mathcal{B}$ problem) could be viewed as a pricing problem for double barrier options with the moving lower $y(\tau)$ and upper $z(\tau)$ barriers and the rebates $f^{\pm}(\tau)$ paid at hit, i.e., when the underlying process hits either the lower or the upper barrier.

Note, that many well-known financial models fit this framework. For instance, the time dependent OU process used in Carr and Itkin (2021) to model barrier and American options is such an example. Also, the time-dependent Hull-White model considered in Itkin and Muravey (2020b) for pricing barrier options is another example. In general, other models can be constructed from the heat equation by using Lie symmetries and the Feynman-Kac theorem. Also, the number of models that fit this framework could be significantly expanded if one transforms the original PDE in Eq. (18.2) to its multilayer version. This approach is discussed it detail in Itkin *et al.* (2021).

Below we present solution of the $\mathcal{B}$ problem by using two analytic methods — the GIT and HP methods. As mentioned in Introduction, the methods are complementary in a sense that despite both solutions are equal, their convergence properties are different. In particular, the GIT method is more accurate at high strikes and maturities while the HP method — at low strikes and maturities.

It is worth mentioning that the $\mathcal{B}$ problem includes inhomogeneous boundary conditions, hence from the very beginning it is useful to transform it to a similar problem but with homogeneous boundary conditions. This could be done by the change of variables

$$
\begin{aligned}
&u(\tau, x) = U(\tau, x) - A(\tau) - B(\tau)x, \qquad (18.7)\\
&A(\tau) = -\frac{f^+(\tau)y(\tau) - f^-(\tau)z(\tau)}{z(\tau) - y(\tau)}, \qquad B(\tau) = \frac{f^+(\tau) - f^-(\tau)}{z(\tau) - y(\tau)},
\end{aligned}
$$

which transforms the PDE in Eq. (18.2) to the inhomogeneous PDE but with the homogeneous boundary conditions

$$
\begin{aligned}
&\frac{\partial u}{\partial \tau} = \frac{\partial^2 u}{\partial x^2} + g(\tau, x), \qquad (18.8)\\
&g(\tau, x) \equiv -A'(\tau) - B'(\tau)x, \quad (\tau, x) \in \mathbb{R}_+ \times [y(\tau), z(\tau)],\\
&u(0, x) = U_0(x) - A(0) - B(0)x \equiv u_0(x), \qquad u(\tau, y(\tau)) = u(\tau, z(\tau)) = 0.
\end{aligned}
$$

18.2 Solution by the GIT method

In this section we solve the problem in Eq. (18.8) by using the GIT method, see Kartashov (1999), Carr and Itkin (2021), Itkin and Muravey (2020b), Itkin *et al.* (2020) and references therein. However, as mentioned in Kartashov (2001), an analytic solution for the domain with two moving boundaries is yet unknown. Therefore, our solution presented in this section is new, and it extends the results of Carr and Itkin (2021) obtained for the domain $[0, y(\tau)]$.

In Carr and Itkin (2021) the authors used the GIT proposed in Kartashov (1999) which is a map $u(\tau, x) \mapsto \bar{u}(\tau, p)$ of the form

$$\bar{u}(\tau, p) = \int_0^{y(\tau)} u(\tau, x) \sinh(x\sqrt{p}) dx, \tag{18.9}$$

where $p = a + i\omega$ is a complex number with $\Re(p) \geq \beta > 0$, and $-\frac{\pi}{4} < \arg\left(\sqrt{p}\right) < \frac{\pi}{4}$. Here we proceed with a similar idea by introducing the transform

$$\bar{u}(\tau, p) = \int_{y(\tau)}^{z(\tau)} u(\tau, x) \sinh\left(p[x - y(\tau)]\right) dx. \tag{18.10}$$

With a special choice of the lower boundary $y(\tau) = 0$ this transform replicates that one in Eq. (18.9) subject to the point that here we use the spectral parameter p instead of $\sqrt{p}$ as in Carr and Itkin (2021).

Since the kernel of Eq. (18.10) is time-dependent it doesn't make much sense to apply this transform directly to the inhomogeneous heat equation in Eq. (18.8). Therefore, we represent the image $\bar{u}$ as a difference of two other images

$$\bar{u} = \frac{1}{2}(\bar{u}_+ - \bar{u}_-), \qquad \bar{u}_\pm(\tau, p) = \int_{y(\tau)}^{z(\tau)} u(\tau, x) e^{\pm p[x - y(\tau)]} dx. \tag{18.11}$$

To determine $\bar{u}(\tau, p)$ let us multiply both parts of the first line in Eq. (18.8) by $e^{\pm p[x-y(\tau)]}$ and integrate on x. This yields

$$\begin{aligned}
\int_{y(\tau)}^{z(\tau)} \frac{\partial u(\tau, x)}{\partial \tau} e^{\pm p[x-y(\tau)]} dx &= \frac{\partial \bar{u}_\pm(\tau, p)}{\partial \tau} \\
&- u(\tau, z(\tau)) e^{\pm p z(\tau)} z'(\tau) + u(\tau, y(\tau)) e^{\pm p y(\tau)} y'(\tau) \\
&\pm p y'(\tau) \int_{y(\tau)}^{z(\tau)} u(\tau, x) e^{\pm p[x-y(\tau)]} dx = \frac{\partial \bar{u}_\pm}{\partial \tau} \pm p y'(\tau) \bar{u}_\pm, \\
\int_{y(\tau)}^{z(\tau)} \frac{\partial^2 u(\tau, x)}{\partial x^2} e^{\pm p[x-y(\tau)]} dx &= \left[\Phi(\tau) - B(\tau)\right] e^{\pm p[z(\tau)-y(\tau)]} \\
&+ \left[\Psi(\tau) + B(\tau)\right] + p^2 \bar{u}_\pm(\tau, p), \\
\bar{g}_\pm(\tau, p) \equiv \int_{y(\tau)}^{z(\tau)} g(\tau, x) e^{\pm p[x-y(\tau)]} dx &= \frac{B'(\tau)}{p^2}\left(e^{\pm p[z(\tau)-y(\tau)]} - 1\right) \\
&\pm \frac{1}{p}\Big[A'(\tau)\left(1 - e^{\pm p[z(\tau)-y(\tau)]}\right) + B'(\tau)\left(y(\tau) - z(\tau) e^{\pm p[z(\tau)-y(\tau)]}\right)\Big],
\end{aligned} \tag{18.12}$$

where terms proportional to $u(\tau, y(\tau))$ and $u(\tau, z(\tau))$ vanish due to the boundary conditions in Eq. (18.8), and by definition

$$\Psi(\tau) = -\left.\frac{\partial U(\tau,x)}{\partial x}\right|_{x=y(\tau)} \qquad \Phi(\tau) = \left.\frac{\partial U(\tau,x)}{\partial x}\right|_{x=z(\tau)}. \tag{18.13}$$

Collecting terms in Eq. (18.12) yields two initial value problems, one for the function $\bar{u}_+$ and the other one - for $\bar{u}_-$

$$\begin{aligned}\frac{\partial \bar{u}_\pm(\tau,p)}{\partial \tau} + \bar{u}_\pm\left[\pm p y'(\tau) - p^2\right] &= [\Psi(\tau) + B(\tau)] \\ &\quad + [\Phi(\tau) - B(\tau)]\, e^{\pm p[z(\tau)-y(\tau)]} + \bar{g}_\pm(\tau,p), \\ \bar{u}_\pm(0,p) &= \int_{y(0)}^{z(0)} u(0,x) e^{\pm p[x-y(0)]} dx.\end{aligned} \tag{18.14}$$

Each problem in Eq. (18.14) (for the plus and minus signs) can be solved explicitly

$$\begin{aligned}\bar{u}_\pm(\tau,p) &= e^{p^2\tau}\int_{y(0)}^{z(0)} u(0,x) e^{\pm p[x-y(\tau)]} dx \\ &\quad + \int_0^\tau e^{p^2(\tau-s)}\Big[\,[\Phi(s) - B(s)]\, e^{\pm p[z(s)-y(\tau)]} \\ &\quad + \Big(\Psi(s) + B(s) + \bar{g}_\pm(s,p)\Big) e^{\pm p[y(s)-y(\tau)]}\Big] ds.\end{aligned} \tag{18.15}$$

Note that the last term in the second integral in Eq. (18.15) can be re-written in a more convenient form

$$\begin{aligned}\bar{g}_\pm(s,p) e^{\pm p[y(s)-y(\tau)]} &= \frac{B'(s)}{p^2}\left(e^{\pm p[z(s)-y(s)]} - 1\right) e^{\pm p[y(s)-y(\tau)]} \\ &\quad \pm \frac{1}{p}\Big[A'(s)\left(1 - e^{\pm p[z(s)-y(s)]}\right) e^{\pm p[y(s)-y(\tau)]} \\ &\quad + B'(s)\left(y(s) - z(s) e^{\pm p[z(s)-y(s)]}\right) e^{\pm p[y(s)-y(\tau)]}\Big] \\ &= \frac{B'(s)}{p^2}\left(e^{\pm p[z(s)-y(\tau)]} - e^{\pm p[y(s)-y(\tau)]}\right) \\ &\quad \pm \frac{1}{p}\Big[A'(s)\left(e^{\pm p[y(s)-y(\tau)]} - e^{\pm p[z(s)-y(\tau)]}\right) \\ &\quad + B'(s)\left(y(s) e^{\pm p[y(s)-y(\tau)]} - z(s) e^{\pm p[z(s)-y(\tau)]}\right)\Big].\end{aligned}$$

The explicit representation for $\bar{u}$ then follows from its definition in Eq. (18.11)

$$\bar{u}(\tau,p) = e^{p^2\tau}\int_{y(0)}^{z(0)} u(0,x) \sinh\left(p[x - y(\tau)]\right) dx \tag{18.16}$$

$$
\begin{aligned}
&+ \int_0^\tau e^{p^2(\tau-s)}\Big\{ [\Phi(s) - B(s)] \sinh\left(p[z(s) - y(\tau)]\right) \\
&+ [\Psi(s) + B(s)] \sinh(p[y(s) - y(\tau)]) + h(s,p)\Big\} ds, \\
h(s,p) &= \frac{B'(s)}{p^2}\left[\sinh(p[z(s) - y(\tau)]) - \sinh(p[y(s) - y(\tau)])\right] \\
&+ \frac{1}{p}\Big[(A'(s) + B'(s)y(s)) \cosh(p[y(s) - y(\tau)]) \\
&- (A'(s) + B'(s)z(s)) \cosh(p[z(s) - y(\tau)])\Big].
\end{aligned}
$$

18.2.1 *The inverse transform*

General theory of the heat equation tells us that the solution at the space domain $a < x < b,\ a, b \in \Re - const$, can be represented as Fourier series of the form (Polyanin, 2002)

$$
u(\tau, x) = \sum_{n=1}^{\infty} \alpha_n e^{-\frac{\pi^2 n^2}{(b-a)^2}\tau} \sin\left(\frac{\pi n(x-a)}{b-a}\right)
$$

Therefore, by analogy let us look for the inverse transform of $\bar{u}$ (which actually is the solution $u(\tau, x)$ of Eq. (18.8)) to be a generalized Fourier transform of the form (Carr and Itkin, 2021)

$$
u(\tau, x) = \sum_{n=0}^{\infty} A_n(\tau) \sin\left(\pi n \frac{x - y(\tau)}{z(\tau) - y(\tau)}\right), \tag{18.17}
$$

where $A_n(\tau)$ are some yet unknown Fourier coefficients (weights). Applying the direct transform in Eq. (18.10) to the series in Eq. (18.17) yields

$$
\bar{u}(\tau, x) = \int_{y(\tau)}^{z(\tau)} \sum_{n=1}^{\infty} A_n(\tau) \sin\left(\pi n \frac{x - y(\tau)}{z(\tau) - y(\tau)}\right) \sinh\left(p[x - y(\tau)]\right) dx. \tag{18.18}
$$

Using the identity

$$
\int_y^z \sin\left(\pi n \frac{x-y}{z-y}\right) \sinh\left(p[x-y]\right) dx = (-1)^{n+1} \frac{\pi n (z-y) \sinh\left(p[z-y]\right)}{n^2\pi^2 + p^2(z-y)^2}, \tag{18.19}
$$

we obtain another representation for $\bar{u}$

$$
\bar{u}(\tau, x) = \frac{1}{l(\tau)} \sum_{n=1}^{\infty} \frac{(-1)^{n+1} \pi n A_n(\tau) \sinh\left(p l(\tau)\right)}{[p + in\pi/l(\tau)]\,[p - in\pi/l(\tau)]}, \quad l(\tau) = z(\tau) - y(\tau). \tag{18.20}
$$

Combining Eqs. (18.20) and (18.16) yields the equation for $A_n(\tau)$

$$\frac{1}{l(\tau)}\sum_{n=1}^{\infty}\frac{(-1)^{n+1}\pi n A_n(\tau)}{[p+in\pi/l(\tau)]\,[p-in\pi/l(\tau)]}=\frac{1}{\sinh(p\,l(\tau))} \tag{18.21}$$
$$\times\left\{e^{p^2\tau}\int_{y(0)}^{z(0)}u(0,x)\sinh(p[x-y(\tau)])\,dx\right.$$
$$+\int_0^{\tau}e^{p^2(\tau-s)}\Big[[\Phi(s)-B(s)]\sinh(p[z(s)-y(\tau)])$$
$$\left.+[\Psi(s)+B(s)]\sinh(p[y(s)-y(\tau)])+h(s,p)\Big]ds\right\}.$$

The LHS and RHS of Eq. (18.21) as the functions of p are analytic in the whole complex plane domain except the poles

$$p_k^{\pm}=\pm i\pi k/l(\tau),\quad k=1,2,\ldots, \tag{18.22}$$

because $h(s,p)$ is regular and well-behaved at $p\to 0$. Also, as this is easy to check, these poles are common for the LHS and RHS of Eq. (18.21). For what follows we need the following residues

$$\operatorname*{Res}_{p=p_k^{\pm}}\sum_{n=1}^{\infty}\frac{1}{[p+in\pi/l(\tau)]\,[p-in\pi/l(\tau)]}=\pm\frac{l(\tau)}{2i\pi k}, \tag{18.23}$$
$$\operatorname*{Res}_{p=p_k^{\pm}}\frac{1}{\sinh(p\,l(\tau))}=\frac{(-1)^k}{l(\tau)}.$$

The Fourier coefficients $A_k(\tau)$ can now be found from Eq. (18.21) by applying contour integration on p to both sides. We integrate using the contours L_k^+, $k=1,2,\ldots$, where the integration contours look like it is depicted in Fig. 18.1. Thus, we have

$$\frac{1}{l(\tau)}\int_{L_k^+}\sum_{n=1}^{\infty}\frac{(-1)^{n+1}\pi n A_n(\tau)}{[p+in\pi/l(\tau)]\,[p-in\pi/l(\tau)]}dp=\int_{L_k^+}\frac{1}{\sinh(pl(\tau))}$$
$$\left\{e^{p^2\tau}\int_{y(0)}^{z(0)}u(0,x)\sinh(p[x-y(\tau)])\,dx\right. \tag{18.24}$$
$$+\int_0^{\tau}e^{p^2(\tau-s)}\Big[\Phi(s)\sinh(p[z(s)-y(\tau)])$$
$$\left.+\Psi(s)\sinh(p[y(s)-y(\tau)])+h(s,p)\Big]ds\right\}dp.$$

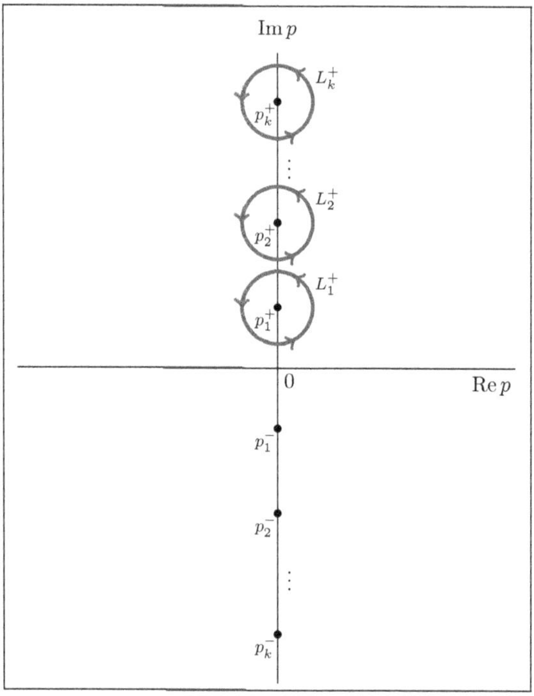

Figure 18.1: Contours of integration of Eq. (18.21) in the complex plane $p \in \mathbb{C}$ with poles at $p_1^{\pm}, p_2^{\pm}, \ldots$.

By the Cauchy's residue theorem each integral in Eq. (18.24) is equal to the sum of the corresponding residues that can be computed with the help of Eq. (18.23). This yields the following formula for $A_k(\tau)$

$$A_k(\tau) = \frac{2}{\mathrm{i}l(\tau)} \bar{u}\left(\tau, \mathrm{i}\frac{\pi k}{l(\tau)}\right). \tag{18.25}$$

With allowance for Eq. (18.16) this can be finally represented as

$$\begin{aligned} A_k(\tau) = \frac{2}{l(\tau)} \Bigg\{ & e^{-\frac{\pi^2 k^2}{l^2(\tau)}\tau} \int_{y(0)}^{z(0)} u(0,x) \sin\left(\frac{\pi k}{l(\tau)}[x - y(\tau)]\right) dx \\ & + \int_0^{\tau} e^{-\frac{\pi^2 k^2}{l^2(\tau)}(\tau - s)} \Bigg[[\Phi(s) - B(s)] \sin\left(\frac{\pi k}{l(\tau)}[z(s) - y(\tau)]\right) \\ & + [\Psi(s) + B(s)] \sin\left(\frac{\pi k}{l(\tau)}[y(s) - y(\tau)]\right) + h_1(k, s, \tau) \Bigg] ds \Bigg\}, \end{aligned} \tag{18.26}$$

with

$$h_1(k,s,\tau) = -\frac{B'(s)l^2(\tau)}{\pi^2k^2}\left[\sin\left(\frac{\pi k}{l(\tau)}[z(s)-y(\tau)]\right)\right.$$
$$\left. - \sin\left(\frac{\pi k}{l(\tau)}[y(s)-y(\tau)]\right)\right]$$
$$- \frac{l(\tau)}{\pi k}\left[(A'(s)+B'(s)y(s))\cos\left(\frac{\pi k}{l(\tau)}[y(s)-y(\tau)]\right)\right.$$
$$\left. - (A'(s)+B'(s)z(s))\cos\left(\frac{\pi k}{l(\tau)}[z(s)-y(\tau)]\right)\right]. \tag{18.27}$$

Keeping in mind that

$$A(\tau)+B(\tau)y(\tau) = f^-(\tau) \qquad A(\tau)+B(\tau)z(\tau) = f^+(\tau)$$

we re-arrange Eq. (18.27) as

$$h_1(k,s,\tau) = \tag{18.28}$$
$$-\frac{B'(s)l^2(\tau)}{\pi^2k^2}\left[\sin\left(\frac{\pi k}{l(\tau)}[z(s)-y(\tau)]\right) - \sin\left(\frac{\pi k}{l(\tau)}[y(s)-y(\tau)]\right)\right]$$
$$-\frac{l(\tau)}{\pi k}\left[((f^-)'(s)-B(s)y'(s))\cos\left(\frac{\pi k}{l(\tau)}[y(s)-y(\tau)]\right)\right.$$
$$\left. - ((f^+)'(s)-B(s)z'(s))\cos\left(\frac{\pi k}{l(\tau)}[z(s)-y(\tau)]\right)\right].$$

Substituting this result into Eq. (18.17), we obtain the solution $u(\tau,x)$ of the problem Eq. (18.8)

$$u(\tau,x) = \frac{2}{l(\tau)}\sum_{n=1}^{\infty}\sin\left(\pi n\frac{x-y(\tau)}{l(\tau)}\right) \tag{18.29}$$
$$\times\left\{e^{-\frac{\pi^2n^2}{l^2(\tau)}\tau}\int_{y(0)}^{z(0)} u(0,\xi)\sin\left(\frac{\pi n}{l(\tau)}[\xi-y(\tau)]\right)d\xi\right.$$
$$+\int_0^{\tau} e^{-\frac{\pi^2n^2}{l^2(\tau)}(\tau-s)}\left[[\Phi(s)-B(s)]\sin\left(\frac{\pi n}{l(\tau)}[z(s)-y(\tau)]\right)\right.$$
$$\left.+[\Psi(s)+B(s)]\sin\left(\frac{\pi n}{l(\tau)}[y(s)-y(\tau)]\right)\right]ds$$
$$\left.+\int_0^{\tau} e^{-\frac{\pi^2n^2}{l^2(\tau)}(\tau-s)}h_1(n,s,\tau)ds\right\}.$$

This expression can be further simplified, see Appendix 18.A. Returning back to the original variable $U(\tau, x)$ yields the final representation

$$U(\tau,x) = \frac{2}{l(\tau)}\sum_{n=1}^{\infty}\sin\left(\pi n\frac{x-y(\tau)}{l(\tau)}\right)\left\{e^{-\frac{\pi^2n^2}{l^2(\tau)}\tau}\int_{y(0)}^{z(0)}U(0,\xi)\right. \tag{18.30}$$
$$\times\sin\left(\frac{\pi n}{l(\tau)}[\xi-y(\tau)]\right)d\xi + \int_0^{\tau}e^{-\frac{\pi^2n^2}{l^2(\tau)}(\tau-s)}\left[\Phi(s)\sin\left(\frac{\pi n}{l(\tau)}[z(s)-y(\tau)]\right)\right.$$
$$\left.\left. + \Psi(s)\sin\left(\frac{\pi n}{l(\tau)}[y(s)-y(\tau)]\right) + \beta(\tau,s,n)\right]ds\right\} + F(\tau,x).$$

where $\beta(\tau, s, n)$ and $F(\tau, x)$ are defined in Eqs. (18.67) and (18.74). Also, as can be checked from the definition in Eq. (18.74) that at $y(\tau) < x < z(\tau)$ the function $F(\tau, x)$ vanishes, and $F(\tau, y(\tau)) = f^-(\tau)$, $F(\tau, z(\tau)) = f^+(\tau)$. Thus, Eq. (18.30) solves the problem in Eq. (18.4) with the initial condition in Eq. (18.5) and the boundary conditions in Eq. (18.6).

It is worth mentioning that the exact same formalism can be developed by using another integral transform

$$\bar{u}(\tau,p) = \int_{y(\tau)}^{z(\tau)}\sinh\left(p[z(\tau)-x]\right)u(\tau,x)dx,$$

with the result being same as in Eq. (18.29).

18.2.2 *Connection to the Jacobi theta function*

As observed in Carr and Itkin (2021), the sums in Eq. (18.29) could be expressed via the Jacobi theta functions of the third kind, Mumford *et al.* (1983).[1] Using their definition

$$\theta_3(z,\omega) = 1 + 2\sum_{n=1}^{\infty}\omega^{n^2}\cos(2nz), \tag{18.31}$$

and the identities

$$\frac{\partial\theta_3(z,\omega)}{\partial z} = \theta_3'(z,\omega) = -4\sum_{n=1}^{\infty}n\omega^{n^2}\sin(2nz). \tag{18.32}$$

we obtain from Eq. (18.29)

$$4\sum_{n=1}^{\infty}e^{-\frac{\pi^2n^2}{l^2(\tau)}\tau}\sin\left(\frac{n\pi(x-y(\tau))}{l(\tau)}\right)\sin\left(\frac{n\pi(\xi-y(\tau))}{l(\tau)}\right) \tag{18.33}$$

[1]Which is not a surprise since it is known that the Jacobi theta functions is the solution of the heat equation with periodic boundary conditions. As applied to the problem considered in this chapter, an example is a double barrier option with zero rebate at hit.

$$
\begin{aligned}
&= \theta_3(\phi_-(x,\xi),\omega_1) - \theta_3(\phi_+(x,\xi),\omega_1),\\
4\sum_{n=1}^{\infty} e^{-\frac{\pi^2 n^2}{l^2(\tau)}(\tau-s)} &\sin\left(\frac{n\pi(x-y(\tau))}{l(\tau)}\right)\sin\left(\frac{n\pi(\xi-y(\tau))}{l(\tau)}\right)\\
&= \theta_3(\phi_-(x,\xi),\omega_2) - \theta_3(\phi_+(x,\xi),\omega_2),\\
8\sum_{n=1}^{\infty} n e^{-\frac{\pi^2 n^2}{l^2(\tau)}(\tau-s)} &\sin\left(\frac{n\pi(x-y(\tau))}{l(\tau)}\right)\cos\left(\frac{n\pi(\xi-y(\tau))}{l(\tau)}\right)\\
&= -\Big(\theta_3'(\phi_-(x,\xi),\omega_2) + \theta_3'(\phi_+(x,\xi),\omega_2)\Big),\\
\omega_1 = e^{-\frac{\pi^2\tau}{l^2(\tau)}}, &\quad \omega_2 = e^{-\frac{\pi^2(\tau-s)}{l^2(\tau)}},\\
\phi_-(x,\xi) = \frac{\pi(x-\xi)}{2l(\tau)}, &\quad \phi_+(x,\xi) = \frac{\pi(x+\xi-2y(\tau))}{2l(\tau)}.
\end{aligned}
$$

With the help of Eq. (18.33) the final formula for $u(\tau,x)$ simplifies

$$
\begin{aligned}
&2l(\tau)\Big[U(\tau,x) - F(\tau,x)\Big] \qquad (18.34)\\
&= \int_{y(0)}^{z(0)} U(0,\xi)\left[\theta_3(\phi_-(x,\xi),\omega_1) - \theta_3(\phi_+(x,\xi),\omega_1)\right] d\xi\\
&+ \int_0^{\tau}\Bigg\{ \left[\Psi(s) - f^-(s)y'(s)\right]\left[\theta_3(\phi_-(x,y(s)),\omega_2) - \theta_3(\phi_+(x,y(s)),\omega_2)\right]\\
&+ \left[\Phi(s) + f^+(s)z'(s)\right]\left[\theta_3(\phi_-(x,z(s)),\omega_2) - \theta_3(\phi_+(x,z(s)),\omega_2)\right]\\
&+ \frac{1}{2}\Big[f^+(s)\left[\theta_3'(\phi_-(x,z(s)),\omega_2) + \theta_3'(\phi_+(x,z(s)),\omega_2)\right]\\
&- f^-(s)\left[\theta_3'(\phi_-(x,y(s)),\omega_2) + \theta_3'(\phi_+(x,y(s)),\omega_2)\right]\Big]\Bigg\} ds.
\end{aligned}
$$

Note, that if rebates at hit are not paid, the boundary conditions become homogeneous, and all terms proportional to $f^-(s) = f^+(s) = 0$ in Eq. (18.34) disappear.

18.2.3 *Determining* $\Psi(\tau)$ *and* $\Phi(\tau)$

Taking the derivative in Eq. (18.34) with respect to x, having in mind that according to Eq. (18.32)

$$
\frac{\partial\theta_3(\phi_\pm(x,\xi),\omega_2)}{\partial x} = \frac{\pi}{l(\tau)}\frac{\partial\theta_3(y,\omega_2)}{\partial y}\bigg|_{y=\phi_\pm(x,\xi)} = \frac{\pi}{l(\tau)}\theta_3'(\phi_\pm(x,\xi),\omega_2), \tag{18.35}
$$

$$\frac{\partial^2 \theta_3(\phi_\pm(x,\xi),\omega_2)}{\partial x^2} = \frac{\pi^2}{l^2(\tau)} \frac{\partial^2 \theta_3(y,\omega_2)}{\partial y^2}\bigg|_{y=\phi_\pm(x,\xi)} = \frac{\pi^2}{l^2(\tau)} \theta_3''(\phi_\pm(x,\xi),\omega_2),$$

and substituting $x = y(\tau)$ and $x = z(\tau)$, we get a system of Volterra integral equations of the second kind to determine $\Psi(\tau)$, $\Phi(\tau)$

$$\begin{aligned}
&-\frac{2l^2(\tau)}{\pi}\Big[\Psi(\tau) + F_x(\tau, y(\tau))\Big] \qquad (18.36)\\
&= \int_{y(0)}^{z(0)} U(0,\xi)\left[\theta_3'(\phi_-(y(\tau),\xi),\omega_1) - \theta_3'(\phi_+(y(\tau),\xi),\omega_1)\right] d\xi\\
&+ \int_0^\tau ds \Bigg\{ \Big[\Psi(s) - f^-(s)y'(s)\Big]\\
&\times \left[\theta_3'(\phi_-(y(\tau),y(s)),\omega_2) - \theta_3'(\phi_+(y(\tau),y(s)),\omega_2)\right]\\
&+ \Big[\Phi(s) + f^+(s)z'(s)\Big]\left[\theta_3'(\phi_-(y(\tau),z(s)),\omega_2) - \theta_3'(\phi_+(y(\tau),z(s)),\omega_2)\right]\\
&+ \frac{2\pi}{l(\tau)}\Big[f^+(s)\left[\theta_3''(\phi_-(y(\tau),z(s)),\omega_2) + \theta_3''(\phi_+(y(\tau),z(s)),\omega_2)\right]\\
&- f^-(s)\left[\theta_3''(\phi_-(y(\tau),y(s)),\omega_2) + \theta_3''(\phi_+(y(\tau),y(s)),\omega_2)\right]\Big]\Bigg\},\\
&\frac{2l^2(\tau)}{\pi}\Big[\Phi(\tau) + F_x(\tau, z(\tau))\Big]\\
&= \int_{y(0)}^{z(0)} U(0,\xi)\left[\theta_3'(\phi_-(z(\tau),\xi),\omega_1) - \theta_3'(\phi_+(z(\tau),\xi),\omega_1)\right] d\xi\\
&+ \int_0^\tau ds \Bigg\{ \Big[\Psi(s) - f^-(s)y'(s)\Big]\\
&\times \left[\theta_3'(\phi_-(z(\tau),y(s)),\omega_2) - \theta_3'(\phi_+(z(\tau),y(s)),\omega_2)\right]\\
&+ \Big[\Phi(s) + f^+(s)z'(s)\Big]\left[\theta_3'(\phi_-(z(\tau),z(s)),\omega_2) - \theta_3'(\phi_+(z(\tau),z(s)),\omega_2)\right]\\
&+ \frac{2\pi}{l(\tau)}\Big[f^+(s)\left[\theta_3''(\phi_-(z(\tau),z(s)),\omega_2) + \theta_3''(\phi_+(z(\tau),z(s)),\omega_2)\right]\\
&- f^-(s)\left[\theta_3''(\phi_-(z(\tau),y(s)),\omega_2) + \theta_3''(\phi_+(z(\tau),y(s)),\omega_2)\right]\Big]\Bigg\}.
\end{aligned}$$

Also, since the theta function $\theta_3(z,\omega)$ solves the heat equation

$$\frac{\partial \theta_3(z, e^q)}{\partial q} = -\frac{1}{4}\frac{\partial^2 \theta_3(z, e^q)}{\partial z^2},$$

the second derivatives with respect to the first argument could be expressed via the first derivatives with respect to the second argument.

However, there exists a problem with the representation in Eq. (18.36). Indeed, using the definition of $F(\tau, x)$ in Eq. (18.74) it can be checked that the derivatives $F_x(\tau, x)$ do not exist at $x = y(\tau)$ and $x = z(\tau)$ as they are proportional to the Dirac Delta $\delta(0)$. Therefore, in the next Section we attack this problem again using an alternative representation of the solution.

18.2.4 *The Poisson summation formula and alternative representations*

It is known that for the fixed spatial domain $x \in [y(\tau), z(\tau)]$, $y(\tau) = 0$, $z(\tau) = const$ there exist two representations of the solution of the heat equation: one - obtained by using the method of images, and the other one - by the Fourier series. Both solutions are equal in a sense of infinite series, but their convergence properties are different, see e.g., Lipton (2002). It turns out that for a curvilinear strip we can also obtain an alternative representation.

The solution $u(\tau, x)$ found in Eq. (18.29) already has the form of the Fourier series. However, applicability of the method of images for the problem Eq. (18.8) is not transparent due to time-dependency of the boundaries. Instead, we can find an alternative representation by using the following property known as the Poisson Summation formula (van der Pol and Bremmer, 1950)

Proposition 18.1 (Poisson Summation formula). *Let $\hat{h}(\nu)$ be the Fourier transform of the appropriate function $h(x)$*

$$\hat{h}(\nu) = \int_{-\infty}^{\infty} h(x) e^{-2\pi i \nu x} dx.$$

The following identity holds

$$\sum_{n=-\infty}^{\infty} h(n) = \sum_{k=-\infty}^{\infty} \hat{h}(k). \tag{18.37}$$

Proof. See van der Pol and Bremmer (1950). □

Applying Eq. (18.37) to the functions

$$h_1(x) = e^{-\frac{\pi^2 x^2}{2}} \cos(\pi x \alpha), \qquad \hat{h}_1(\nu) = \int_{-\infty}^{\infty} e^{-\frac{\pi^2 x^2}{2\beta} - 2\pi i \nu x} \cos(\pi x \alpha)\, dx,$$

$$h_2(x) = x e^{-\frac{\pi^2 x^2}{2\beta}} \sin(\pi x \alpha), \qquad \hat{h}_2(\nu) = \int_{-\infty}^{\infty} x e^{-\frac{\pi^2 x^2}{2\beta} - 2\pi i \nu x} \sin(\pi x \alpha)\, dx,$$

we obtain the following identities

$$\sum_{n=-\infty}^{\infty} e^{-\frac{\pi^2 n^2}{2\beta}} \cos(\pi n\alpha) = \sqrt{\frac{2\beta}{\pi}} e^{-\frac{\alpha^2\beta}{2}} \sum_{n=-\infty}^{\infty} e^{-2n^2\beta} \cosh(2n\alpha\beta) \tag{18.38}$$
$$= \sqrt{\frac{\beta}{2\pi}} \sum_{n=-\infty}^{\infty} \left[e^{-\frac{\beta}{2}(2n-\alpha)^2} + e^{-\frac{\beta}{2}(2n+\alpha)^2} \right]$$
$$= 2\sqrt{\frac{\beta}{2\pi}} \sum_{n=-\infty}^{\infty} e^{-\frac{\beta}{2}(2n+\alpha)^2},$$

$$\sum_{n=-\infty}^{\infty} \pi n e^{-\frac{\pi^2 n^2}{2\beta}} \sin(\pi n\alpha)$$
$$= \frac{\beta^{3/2}}{\sqrt{2\pi}} \sum_{n=-\infty}^{\infty} e^{-\frac{\beta}{2}(2n+\alpha)^2} \left[\alpha + 2n + (\alpha - 2n)e^{4\alpha\beta n}\right]$$
$$= \frac{\beta^{3/2}}{\sqrt{2\pi}} \sum_{n=-\infty}^{\infty} \left[e^{-\frac{\beta}{2}(2n+\alpha)^2}(\alpha + 2n) + e^{-\frac{\beta}{2}(2n-\alpha)^2}(\alpha - 2n) \right]$$
$$= 2\frac{\beta^{3/2}}{\sqrt{2\pi}} \sum_{n=-\infty}^{\infty} e^{-\frac{\beta}{2}(2n+\alpha)^2}(\alpha + 2n).$$

Since each summand in Eq. (18.30) can be represented in the form of the LHS of Eq. (18.38), by using a simple trigonometric formula for the product of sines, we immediately arrive at another form of $U(\tau, x)$, see Appendix 18.B

$$U(\tau, x) = \sum_{n=-\infty}^{\infty} \left\{ \int_{y(0)}^{z(0)} U(0,\xi)\Upsilon_n(x,\tau\,|\,\xi,0)d\xi \right. \tag{18.39}$$
$$+ \int_0^\tau \left[\Phi(s) + f^+(s)z'(s)\right] \Upsilon_n(x,\tau|z(s),s)ds,$$
$$+ \int_0^\tau \left[\Psi(s) - f^-(s)y'(s)\right] \Upsilon_n(x,\tau\,|\,y(s),s)ds$$
$$\left. + \int_0^\tau f^-(s)\Lambda_n(x,\tau\,|\,y(s),s) - f^+(s)\Lambda_n(x,\tau\,|\,z(s),s)ds \right\},$$

$$\Upsilon_n(x,\tau\,|\,\xi,s) = \frac{1}{2\sqrt{\pi(\tau-s)}} \left[e^{-\frac{(2nl(\tau)+x-\xi)^2}{4(\tau-s)}} - e^{-\frac{(2nl(\tau)+x+\xi-2y(\tau))^2}{4(\tau-s)}} \right],$$

$$\Lambda_n(x,\tau\,|\,\xi,s) = \frac{x-\xi+2nl(\tau)}{4\sqrt{\pi(\tau-s)^3}} e^{-\frac{(2nl(\tau)+x-\xi)^2}{4(\tau-s)}}$$
$$+ \frac{x+\xi-2y(\tau)+2nl(\tau)}{4\sqrt{\pi(\tau-s)^3}} e^{-\frac{(2nl(\tau)+x+\xi-2y(\tau))^2}{4(\tau-s)}}.$$

Note that the Fourier series in these expressions usually converge rapidly when n grows. Similarly, taking the derivative of this series on x provides a convenient way of calculating the corresponding derivative $\frac{\partial U(\tau,x)}{\partial x}$, Olver *et al.* (2020).

18.2.5 *A system of Volterra equations for* $\Psi(\tau)$ *and* $\Phi(\tau)$

In Section 18.2.4 we managed to obtain two alternative representations of the solution of the problem, both in a semi-analytical form. These solutions, however, depend on two yet unknown functions gradients $\Psi(\tau), \Phi(\tau)$ that can be found by solving a system of two Volterra equations of the second kind. These equations are obtained by taking the derivative in Eq. (18.30) or Eq. (18.39) with respect to x and substituting $x = y(\tau)$ and $x = z(\tau)$ into thus found expressions. However, at least formally there exist a problem with making the last step, because at these boundaries some integrals in the system of the Volterra equations will contain singularities. Below we describe the resolution of these problems.

Let us again consider Eq. (18.39). It is easy to see that the functions $\frac{\partial \Upsilon_n(x,\tau|\xi,s)}{\partial x}$, $\frac{\partial \Lambda_n(x,\tau|\xi,s)}{\partial x}$ are regular only if $n \neq 0$, $x \in [y(\tau), z(\tau)]$, $\xi \in [y(s), z(s)]$, $s \to \tau$. At $n = 0$ functions $\frac{\partial \Upsilon_0(x,\tau|y(s),s)}{\partial x}$, $\frac{\partial \Lambda_0(x,\tau|y(s),s)}{\partial x}$ have a singularity when $s \to \tau$, $x \to y(\tau)$, and functions $\frac{\partial \Upsilon_0(x,\tau|z(s),s)}{\partial x}$, $\frac{\partial \Lambda_0(x,\tau|z(s),s)}{\partial x}$ - when $s \to \tau$, $x \to z(\tau)$.

Since the functions $\frac{\partial \Upsilon_0(x,\tau|y(s),s}{\partial x}$, $\frac{\partial \Upsilon_0(x,\tau|z(s),s}{\partial x}$ can be represented as a sum of double-layer potentials with a negative sign, the limiting values

$$\lim_{x \to y(\tau)+0} \int_0^\tau \xi(s) \frac{\partial \Upsilon_0(x,\tau|y(s),s)}{\partial x} ds,$$

$$\lim_{x \to z(\tau)-0} \int_0^\tau \xi(s) \frac{\partial \Upsilon_0(x,\tau|z(s),s)}{\partial x} ds$$

can be computed similar to Eq. (18.85).

Applying Eq. (8.30) to the limits corresponding to $\frac{\partial \Lambda_0(x,\tau|y(s),s)}{\partial x}$, $\frac{\partial \Lambda_0(x,\tau|z(s),s)}{\partial x}$ we obtain

$$\begin{aligned}
&\lim_{x \to y(\tau)+0} \int_0^\tau f^-(s) \frac{\partial \Lambda_0(x,\tau|y(s),s)}{\partial x} ds \qquad (18.40)\\
&= -\frac{f^-(\tau)}{\sqrt{\pi\tau}} + \int_0^\tau \frac{f^-(s) e^{-\frac{(y(\tau)-y(s))^2}{4(\tau-s)}} - f^-(\tau)}{2\sqrt{\pi(\tau-s)^3}} ds \\
&- \int_0^\tau f^-(s) \frac{(y(\tau)-y(s))^2 e^{-\frac{(y(\tau)-y(s))^2}{4(\tau-s)}}}{4\sqrt{\pi(\tau-s)^5}} ds,
\end{aligned}$$

$$
\begin{aligned}
\lim_{x\to z(\tau)-0} & \int_0^\tau f^+(s)\frac{\partial\Lambda_0(x,\tau|z(s),s)}{\partial x}ds \\
&= -\frac{f^+(\tau)}{\sqrt{\pi\tau}} + \int_0^\tau \frac{f^+(s)e^{-\frac{(z(\tau)-z(s))^2}{4(\tau-s)}} - f^+(\tau)}{2\sqrt{\pi(\tau-s)^3}}ds \\
&- \int_0^\tau f^+(s)\frac{(z(\tau)-z(s))^2 e^{-\frac{(z(\tau)-z(s))^2}{4(\tau-s)}}}{4\sqrt{\pi(\tau-s)^5}}ds.
\end{aligned}
$$

Finally, taking the derivative of Eq. (18.39) on x, setting $x = y(\tau)$ and $x = z(\tau)$, and using these expressions, we obtain the following system of the Volterra equations of the second kind for the unknown functions $\Psi(\tau), \Phi(\tau)$

$$
\begin{aligned}
-\Psi(\tau) &= -\frac{f^-(\tau)}{\sqrt{\pi\tau}} \qquad (18.41)\\
&+ \int_0^\tau \frac{f^-(s)e^{-\frac{(y(\tau)-y(s))^2}{4(\tau-s)}}C(y,s,\tau) - f^-(\tau)}{2\sqrt{\pi(\tau-s)^3}}ds \\
&- \int_0^\tau \Psi(s)\frac{y(\tau)-y(s)}{2\sqrt{\pi(\tau-s)^3}}e^{-\frac{(y(\tau)-y(s))^2}{4(\tau-s)}}ds + \int_{y(0)}^{z(0)} U(0,\xi)\upsilon^-(\tau\,|\,\xi,0)d\xi \\
&+ \int_0^\tau \Big(\left[\Phi(s) + f^+(s)z'(s)\right]\upsilon^-(\tau\,|\,z(s),s) \\
&\quad + \left[\Psi(s) - f^-(s)y'(s)\right]\upsilon_0^-(\tau\,|\,y(s),s)\Big)ds \\
&+ \int_0^\tau \left(f^-(s)\lambda_0^-(\tau,\,|\,s) - f^+(s)\lambda^-(\tau,\,|\,z(s),s)\right)ds, \\
\Phi(\tau) &= \frac{f^+(\tau)}{\sqrt{\pi\tau}} \\
&- \int_0^\tau \frac{f^+(s)e^{-\frac{(z(\tau)-z(s))^2}{4(\tau-s)}}C(z,s,\tau) - f^+(\tau)}{2\sqrt{\pi(\tau-s)^3}}ds \\
&- \int_0^\tau \Phi(s)\frac{z(\tau)-z(s)}{2\sqrt{\pi(\tau-s)^3}}e^{-\frac{(z(\tau)-z(s))^2}{4(\tau-s)}}ds + \int_{y(0)}^{z(0)} U(0,\xi)\upsilon^+(\tau\,|\,\xi,0)d\xi \\
&+ \int_0^\tau \Big(\left[\Phi(s) + f^+(s)z'(s)\right]\upsilon_0^+(\tau\,|\,z(s),s) \\
&\quad + \left[\Psi(s) - f^-(s)y'(s)\right]\upsilon^+(\tau\,|\,y(s),s)ds\Big) \\
&+ \int_0^\tau \left(f^-(s)\lambda^+(\tau\,|\,y(s),s) - f^+(s)\lambda_0^+(\tau\,|\,s)\right)ds.
\end{aligned}
$$

Here

$$
C(y,s,\tau) = 1 + y'(s)(y(\tau)-y(s))) - \frac{(y(\tau)-y(s))^2}{2(\tau-s)},
$$

$$\upsilon_n(\tau\,|\,\xi,s) = -\frac{y(\tau)-\xi+2nl(\tau)}{2\sqrt{\pi(\tau-s)^3}}e^{-\frac{(y(\tau)-\xi+2nl(\tau))^2}{4(\tau-s)}}, \tag{18.42}$$

$$\lambda_n(\tau\,|\,\xi,s) = \frac{e^{-\frac{(y(\tau)-\xi+2nl(\tau))^2}{4(\tau-s)}}}{2\sqrt{\pi(\tau-s)^3}}\left[1-\frac{(y(\tau)-\xi+2nl(\tau))^2}{2(\tau-s)}\right],$$

$$\upsilon^-(\tau\,|\,\xi,s) = \sum_{n=-\infty}^{\infty}\upsilon_n(\tau\,|\,\xi,s), \qquad \upsilon^+(\tau\,|\,\xi,s) = \sum_{n=-\infty}^{\infty}\upsilon_{n+\frac{1}{2}}(\tau\,|\,\xi,s),$$

$$\upsilon_0^-(\tau\,|\,s) = \sum_{\substack{n=-\infty\\ n\neq 0}}^{\infty}\upsilon_n(\tau\,|\,y(s),s), \qquad \upsilon_0^+(\tau\,|\,s) = \sum_{\substack{n=-\infty\\ n\neq 0}}^{\infty}\upsilon_{n+\frac{1}{2}}(\tau\,|\,z(s),s),$$

$$\lambda^-(\tau\,|\,\xi,s) = \sum_{n=-\infty}^{\infty}\lambda_n(\tau\,|\,\xi,s), \qquad \lambda^+(\tau\,|\,\xi,s) = \sum_{n=-\infty}^{\infty}\lambda_{n+\frac{1}{2}}(\tau\,|\,\xi,s),$$

$$\lambda_0^-(\tau\,|\,s) = \sum_{\substack{n=-\infty\\ n\neq 0}}^{\infty}\lambda_n(\tau\,|\,y(s),s), \qquad \lambda_0^+(\tau\,|\,s) = \sum_{\substack{n=-\infty\\ n\neq 0}}^{\infty}\lambda_{n+\frac{1}{2}}(\tau\,|\,z(s),s).$$

It is worth emphasizing that all summands in Eq. (18.41) are regular. The integrals with respect to the time in the first two lines have weak (integrable) singularities, while other summands are regular.

This system can be further simplified by using Eq. (8.28) and reduction to the Lebesgue-Stieltjes integrals

$$\begin{aligned}
-\Psi(\tau) &= \int_{y(0)}^{z(0)} U(0,\xi)\upsilon^-(\tau\,|\,\xi,0)d\xi \\
&\quad - \frac{f^-(\tau)}{\sqrt{\pi\tau}} + \int_0^{\tau}\frac{f^-(s)-f^-(\tau)}{2\sqrt{\pi(\tau-s)^3}}ds \\
&\quad + \int_0^{\tau}\left[f^-(s)d\left(\eta^-(\tau\,|\,y(s),s)\right) - f^+(s)d\left(\eta^-(\tau\,|\,z(s),s)\right)\right] \\
&\quad - \int_0^{\tau}\Psi(s)\frac{y(\tau)-y(s)}{2\sqrt{\pi(\tau-s)^3}}e^{-\frac{(y(\tau)-y(s))^2}{4(\tau-s)}}ds \\
&\quad + \int_0^{\tau}\left[\Phi(s)\upsilon^-(\tau\,|\,z(s),s) + \Psi(s)\upsilon_0^-(\tau\,|\,s)\right]ds \\
\Phi(\tau) &= \int_{y(0)}^{z(0)} U(0,\xi)\upsilon^+(\tau\,|\,\xi,0)d\xi \\
&\quad + \frac{f^+(\tau)}{\sqrt{\pi\tau}} - \int_0^{\tau}\frac{f^+(s)-f^+(\tau)}{2\sqrt{\pi(\tau-s)^3}}ds \\
&\quad + \int_0^{\tau}\left[f^-(s)d\left(\eta^+(\tau\,|\,y(s),s)\right) - f^+(s)d\left(\eta^+(\tau\,|\,z(s),s)\right)\right]
\end{aligned} \tag{18.43}$$

$$
\begin{aligned}
&- \int_0^\tau \Phi(s) \frac{z(\tau)-z(s)}{2\sqrt{\pi(\tau-s)^3}} e^{-\frac{(z(\tau)-z(s))^2}{4(\tau-s)}} ds \\
&+ \int_0^\tau \left[\Phi(s)\upsilon_0^+(\tau\,|\,s) + \Psi(s)\upsilon^+(\tau\,|\,y(s),s)\right] ds.
\end{aligned}
$$

Here the following notation is used

$$
\eta^-(\tau\,|\,\xi,s) = -\frac{\delta_{\xi,y(s)}}{\sqrt{\pi(\tau-s)}} + \frac{1}{\sqrt{\pi(\tau-s)}} \sum_{n=-\infty}^{\infty} e^{-\frac{(y(\tau)-\xi+2nl(\tau))^2}{4(\tau-s)}}, \quad (18.44)
$$

$$
\eta^+(\tau\,|\,\xi,s) = -\frac{\delta_{\xi,z(s)}}{\sqrt{\pi(\tau-s)}} + \frac{1}{\sqrt{\pi(\tau-s)}} \sum_{n=-\infty}^{\infty} e^{-\frac{(y(\tau)-\xi+(2n+1)l(\tau))^2}{4(\tau-s)}},
$$

$$
\upsilon^-(\tau\,|\,\xi,s) = -\frac{y(\tau)-\xi+2nl(\tau)}{2\sqrt{\pi(\tau-s)^3}} \sum_{n=-\infty}^{\infty} e^{-\frac{(y(\tau)-\xi+2nl(\tau))^2}{4(\tau-s)}},
$$

$$
\upsilon^+(\tau\,|\,\xi,s) = -\frac{y(\tau)-\xi+(2n+1)l(\tau)}{2\sqrt{\pi(\tau-s)^3}} \sum_{n=-\infty}^{\infty} e^{-\frac{(y(\tau)-\xi+(2n+1)l(\tau))^2}{4(\tau-s)}},
$$

where $\delta_{\xi,x}$ is the Kronecker symbol.

The functions υ, η have the following properties

$$
\begin{aligned}
&\lim_{s\to\tau} \upsilon_0^-(\tau\,|\,s) = 0, && \lim_{s\to\tau} \upsilon^-(\tau\,|\,z(s),s) = 0, \\
&\lim_{s\to\tau} \upsilon^+(\tau\,|\,y(s),s) = 0, && \lim_{s\to\tau} \upsilon_0^+(\tau\,|\,s) = 0, \\
&\lim_{s\to\tau} \eta^-(\tau\,|\,y(s),s) = 0, && \lim_{s\to\tau} \eta^-(\tau\,|\,z(s),s) = 0, \\
&\lim_{s\to\tau} \eta^+(\tau\,|\,y(s),s) = 0, && \lim_{s\to\tau} \eta^+(\tau\,|\,z(s),s) = 0.
\end{aligned}
$$

Again, using the Poisson summation formula yields a few alternative representations of the functions $\eta^\pm(\tau\,|\,\xi,s)$ and $\upsilon^\pm(\tau\,|\,\xi,s)$ via the Fourier series

$$
\begin{aligned}
\eta^-(\tau\,|\,\xi,s) &= -\frac{\delta_{\xi,y(s)}}{\sqrt{\pi(\tau-s)}} \qquad (18.45) \\
&+ \frac{1}{l(\tau)}\left[1 + 2\sum_{n=1}^{\infty} e^{-\frac{\pi^2 n^2}{l^2(\tau)}(\tau-s)} \cos\left(\frac{\pi n(\xi-y(\tau))}{l(\tau)}\right)\right], \\
\eta^+(\tau\,|\,\xi,s) &= -\frac{\delta_{\xi,z(s)}}{\sqrt{\pi(\tau-s)}} \\
&+ \frac{1}{l(\tau)}\left[1 + 2\sum_{n=1}^{\infty} e^{-\frac{\pi^2 n^2}{l^2(\tau)}(\tau-s)} (-1)^n \cos\left(\frac{\pi n(\xi-y(\tau))}{l(\tau)}\right)\right], \\
\upsilon^-(\tau\,|\,\xi,s) &= \frac{2\pi}{l^2(\tau)} \sum_{n=1}^{\infty} n e^{-\frac{\pi^2 n^2}{l^2(\tau)}(\tau-s)} \sin\left(\frac{\pi n(\xi-y(\tau))}{l(\tau)}\right),
\end{aligned}
$$

$$\upsilon^+(\tau\,|\,\xi,s) = \frac{2\pi}{l^2(\tau)}\sum_{n=1}^{\infty} n e^{-\frac{\pi^2 n^2}{l^2(\tau)}(\tau-s)}(-1)^n \sin\left(\frac{\pi n(\xi - y(\tau))}{l(\tau)}\right),$$

where $\delta_{\xi,x}$ is the Kronecker symbol.

Finally, using Eq. (18.31) and Eq. (18.33), we obtain another representation of Eq. (18.45) in terms of the Jacobi theta function $\theta_3(z,\omega)$

$$\begin{aligned}
\eta^-(\tau\,|\,\xi,s) &= -\frac{\delta_{\xi,y(s)}}{\sqrt{\pi(\tau-s)}} + \frac{1}{l(\tau)}\theta_3\left(\phi_-(\xi,y(\tau)),\omega_2\right), \qquad (18.46)\\
\eta^+(\tau\,|\,\xi,s) &= -\frac{\delta_{\xi,z(s)}}{\sqrt{\pi(\tau-s)}} + \frac{1}{l(\tau)}\theta_3\left(\phi_-(\xi+l(\tau),y(\tau)),\omega_2\right),\\
\upsilon^-(\tau\,|\,\xi,s) &= -\frac{\pi}{2l^2(\tau)}\theta_3'\left(\phi_-(\xi,y(\tau)),\omega_2\right),\\
\upsilon^+(\tau\,|\,\xi,s) &= -\frac{\pi}{2l^2(\tau)}\theta_3'\left(\phi_-(\xi+l(\tau),y(\tau)),\omega_2\right).
\end{aligned}$$

The formulas Eq. (18.44) and Eq. (18.45) are complementary. Since the exponents in Eq. (18.46) are proportional to the difference $\tau-s$, the Fourier series Eq. (18.46) converge fast if $\tau - s$ is large. Contrary, the exponents in Eq. (18.44) are inversely proportional to $\tau - s$. Therefore, the series Eq. (18.44) converge fast if $\tau - s$ is small.

18.3 Solution by the HP method

Similar to Section 18.2, the HP method (Tikhonov and Samarskii, 1963; Friedman, 1964.; Kartashov, 2001), can be used to price double barrier options by solving the problem in Eq. (18.8). The idea was first proposed and developed in Itkin and Muravey (2020b) and is a generalization of the standard HP method for the case of two moving boundaries. Note, that to the best of authors' knowledge, yet the closed form (or even semi-closed form) solution of this problem was not known in physics, even not mentioning finance. Below we explain our approach paying attention to all intermediate details as the behavior of the solution at the boundaries is not trivial.

Following the main idea of the HP method, let us search for the solution of the $\mathcal{B}$ problem in Eqs. (18.4) and (18.6), Eq. (18.5) in the form

$$U(\tau,x) = q(\tau,x) + \frac{1}{2\sqrt{\pi\tau}}\int_{y(0)}^{z(0)} U(0,x')e^{-\frac{(x-x')^2}{4\tau}}dx', \qquad (18.47)$$

so function $q(\tau,x)$ solves a problem with the homogeneous initial condition

$$\frac{\partial q(\tau,x)}{\partial\tau} = \frac{\partial^2 q(\tau,x)}{\partial x^2}, \qquad (18.48)$$

$$
\begin{aligned}
q(0,x) &= 0, \qquad y(0) < x < z(0), \\
q(\tau, y(\tau)) &= \phi_1(\tau) \equiv f^-(\tau) - \frac{1}{2\sqrt{\pi\tau}} \int_{y(0)}^{z(0)} u(0,x') e^{-\frac{(y(\tau)-x')^2}{4\tau}} dx', \\
q(\tau, z(\tau)) &= \psi_1(\tau) \equiv f^+(\tau) - \frac{1}{2\sqrt{\pi\tau}} \int_{y(0)}^{z(0)} u(0,x') e^{-\frac{(z(\tau)-x')^2}{4\tau}} dx'.
\end{aligned}
$$

In Itkin and Muravey (2020b) it is proposed to search for the solution of Eq. (18.48) in the form of a generalized heat potential

$$
\begin{aligned}
q(x,\tau) = \frac{1}{4\sqrt{\pi}} \int_0^\tau \frac{1}{\sqrt{(\tau-k)^3}} \Big[(x - y(k))\Omega(k) e^{-\frac{(x-y(k))^2}{4(\tau-k)}} \\
+ (x - z(k))\Theta(k) e^{-\frac{(x-z(k))^2}{4(\tau-k)}} \Big] dk,
\end{aligned} \tag{18.49}
$$

where $\Omega(k), \Theta(k)$ are the heat potential densities. In other words, the solution is represented as a sum of two heat potentials: one corresponds to the lower barrier, and the other one — to the upper barrier. It is easy to check, that each such a potential solves the heat equation in Eq. (18.48), see Tikhonov and Samarskii (1963) as the derivative with respect to τ of the RHS of Eq. (18.49) can be pulled into the integral since the value of both integrands at $k = \tau$ vanishes.

To find the unknown functions $\Omega(k), \Theta(k)$ one can substitute into Eq. (18.49) the values $x = y(\tau)$ and $x = z(\tau)$, and get a system of two integral equations that the functions $\Omega(k), \Theta(k)$ solve. However, it is well-known (Tikhonov and Samarskii, 1963), that these integrals at $x \to y(\tau)$ and $x \to z(\tau)$ have a discontinuity, but with the finite value at $x = y(\tau) \pm 0$ and $x = z(\tau) \pm 0$. To investigate this discontinuity in more detail and derive the value of heat potential integral at the boundary $x \to y(\tau) \pm 0$, we consider a problem similar to Eq. (18.48)

$$
\begin{aligned}
\mathcal{L} q(\tau, x) &= 0, \qquad (x,\tau) \in \Omega : [y(\tau), \infty) \times \mathbb{R}_+, \\
q(0,x) &= 0, \quad y(0) < x < \infty, \\
q(\tau, y(\tau)) &= \chi(\tau), \qquad q(\tau, x)\Big|_{x\to\infty} = 0.
\end{aligned} \tag{18.50}
$$

with the operator $\mathcal{L}$ defined as

$$
\mathcal{L} = -\frac{\partial}{\partial \tau} + \sigma^2 \frac{\partial^2}{\partial x},\tag{18.51}
$$

where $\sigma = const$. Using the HP method, the solution of this problem can be expressed as

$$
q(\tau, x) = \int_0^\tau \Omega(k) \frac{x - y(k)}{4\sigma^3 \sqrt{\pi(\tau-k)^3}} e^{-\frac{(x-y(k))^2}{4\sigma^2(\tau-k)}} dk, \tag{18.52}
$$

where $\Omega(\tau)$ is the heat potential density, and $y(\tau)$ is a smooth curve (the moving boundary).

To proceed we need the value of this integral at $x \to y(\tau) \pm 0$, and the gradient $\partial q(\tau, x)/\partial x$ in the same limit, namely

$$\varphi(\tau) = \lim_{x \to y(\tau) \pm 0} q(\tau, x), \qquad \psi(\tau) = \lim_{x \to y(\tau) \pm 0} \frac{\partial q(\tau, x)}{\partial x}. \tag{18.53}$$

However, these expressions have been already derived in Itkin and Muravey (2020a) and also are presented in Chapter 8. Thus, our problem is solved.

18.3.1 *A system of Volterra equations*

With allowance for the representation obtained in Eq. (8.19), by substituting the limiting values $x \to y(\tau)$ and $x \to z(\tau)$ into Eq. (18.49), we obtain a system of two integral equation for functions $\Omega(\tau), \Theta(\tau)$

$$\begin{aligned} 2\phi_1(\tau) &= \Omega(\tau) + \frac{1}{2\sqrt{\pi}} \int_0^\tau \left[\Omega(k) \frac{y(\tau) - y(k)}{(\tau - k)^{3/2}} e^{-\frac{(y(\tau)-y(k))^2}{4(\tau-k)}} \right. \\ &\quad \left. + \Theta(k) \frac{y(\tau) - z(k)}{(\tau - k)^{3/2}} e^{-\frac{(y(\tau)-z(k))^2}{4(\tau-k)}} \right] dk, \\ 2\psi_1(\tau) &= -\Theta(\tau) + \frac{1}{2\sqrt{\pi}} \int_0^\tau \left[\Omega(k) \frac{z(\tau) - y(k)}{(\tau - k)^{3/2}} e^{-\frac{(z(\tau)-y(k))^2}{4(\tau-k)}} \right. \\ &\quad \left. + \Theta(k) \frac{z(\tau) - z(k)}{(\tau - k)^{3/2}} e^{-\frac{(z(\tau)-z(k))^2}{4(\tau-k)}} \right] dk. \end{aligned} \tag{18.54}$$

Each equation in this system is a Volterra equation of the second kind. The system can be solved, e.g., by the Variational Iteration Method (VIM), see Wazwaz (2011) with a linear complexity by using the Fast Gaussian Transform. Once this is done, the solution of our double barrier problem is found.

It is interesting that the representation of the solution gradient in Eq. (8.30) provides connection between the GIT and HP methods. Indeed, by definition in Eq. (18.13) and also using Eqs. (18.7) and (18.47)

$$\begin{aligned} \Psi(\tau) &= -\frac{\partial U(\tau, x)}{\partial x}\bigg|_{x=y(\tau)} \\ &= -\frac{\partial q(\tau, x)}{\partial x}\bigg|_{x=y(\tau)+0} + \frac{1}{4\sqrt{\pi\tau^3}} \int_{y(0)}^{z(0)} U(0, x')(y(\tau) - x') e^{-\frac{(y(\tau)-x')^2}{4\tau}} dx', \end{aligned} \tag{18.55}$$

$$\Phi(\tau) = \left.\frac{\partial U(\tau, x)}{\partial \tau}\right|_{x=z(\tau)}$$
$$= \left.\frac{\partial q(\tau, x)}{\partial x}\right|_{x=z(\tau)-0} + \frac{1}{4\sqrt{\pi\tau^3}} \int_{y(0)}^{z(0)} U(0, x')(z(\tau) - x')e^{-\frac{(z(\tau)-x')^2}{4\tau}} dx'.$$

Therefore, once the pair $\Omega(\tau), \Theta(\tau)$ is known, the other pair $\Psi(\tau), \Phi(\tau)$ can be obtained explicitly from Eq. (18.55). The opposite is also true, i.e., once the pair $\Psi(\tau), \Phi(\tau)$ is known, the heat potential densities $\Omega(\tau), \Theta(\tau)$ can be found by solving this system of Volterra equations of the second kind. Thus, both the GIT and HP methods are interchangeable. But as we have already mentioned several times, despite both solutions are equal, their convergence properties are different.

18.4 Numerical example

As this has been mentioned already several times, construction of one factor models that can be reduced to the solution of the heat equation can be done by using Lie symmetries (see e.g., Craddock and Lennox, 2007; Craddock, 2009 and references therein). To make it possible the model PDE should have 6 Lie symmetry groups, same as the heat equation. In a simplistic way this transformation can be done as follows. Let us use the heat equation in Eq. (18.4) and apply the transformation

$$\tau \mapsto \tau(t), \qquad S \mapsto f(\tau, x), \qquad U(\tau, x) \mapsto V(t, S)g(t, S), \tag{18.56}$$

where (t, S) are new independent variables, $V(t, S)$ is a new dependent variable, and $f(\tau, z)$, $g(t, S)$ are some functions such that their first derivative on the first argument and the second derivative of the second argument exist. If we need to construct a model, e.g., for pricing Equity option, we also need the killing term to be equal to $-r(t)V(t, S)$ with $r(t)$ be the deterministic interest rate.

As an example, we use

$$S \mapsto f(\tau, x) = b(\tau)x^2 - h(\tau), \qquad g(t, S) = \exp(\alpha(t) + \beta(t)S), \tag{18.57}$$
$$h(t) = \frac{1}{4}e^{-\int_t^T \mu(k)dk} \int_0^t e^{\int_s^T \mu(k)dk} \sigma^2(s)ds \geq 0,$$

where $b(t), \alpha(t), \beta(t), \mu(t) \in \mathcal{C}^1$ are some functions of time.

Let us consider a problem of pricing double barrier options written on the stock S_t. Then it is possible to show that the model obtained by these transformations from the heat equation reads

$$dS_t = \mu(t)S_t dt + \sigma(t)\sqrt{S_t + h(t)}dW_t, \tag{18.58}$$

where $\mu(t)S$ is some drift. Hence, this model is the displaced CIR model with zero mean-reversion level (about the CIR model, see Cox *et al.*, 1985; Andersen and Piterbarg, 2010). Also, this model is affine.

Using the Feynman-Kac theorem one can find that the Call option price $C(t,S)$ solves the PDE

$$\frac{\partial C}{\partial t} + \mu(t)S\frac{\partial C}{\partial S} + \frac{1}{2}\sigma^2(t)(S+h(t))\frac{\partial^2 C}{\partial S^2} - r(t)C = 0, \tag{18.59}$$

subject to the terminal and boundary conditions

$$\begin{aligned} C(T,S) &= (S-K)^+, \\ C(t,L(t)) &= C(t,H(t)) = 0, \end{aligned} \tag{18.60}$$

where $L(t), H(t)$ are the lower and upper barriers which we assume to be time-dependent, K is the option strike, T is the maturity, $\sigma(t)$ is the time-dependent volatility. The term $\mu(t)$ under risk-neutral measure reads $\mu(t) = r(t) - q(t)$ with $q(t)$ being the continuous dividends.

To validate, doing the inverse transformation

$$\begin{aligned} W(\tau,z) &= C(t,S)e^{\int_t^T r(k)dk}, \quad z = \sqrt{S+h(t)}e^{\frac{1}{2}\int_t^T \mu(k)dk}, \\ \tau &= \frac{1}{8}\int_t^T \sigma^2(k)e^{\int_k^T \mu(s)ds}, \end{aligned}$$

we reduce the original PDE to the heat equation

$$\frac{\partial W}{\partial \tau} = \frac{\partial^2 W}{\partial z^2}, \qquad \tau \in [0,\tau(0)],\ z \in [y(\tau), z(\tau)], \tag{18.61}$$

which should be solved subject to the transformed initial and boundary conditions

$$\begin{aligned} W(0,z) &= (z^2 - h(T) - K)^+, \\ W(\tau,y(\tau)) &= W(\tau,z(\tau)) = 0. \end{aligned} \tag{18.62}$$

Here $y(\tau)$ and $z(\tau)$ are defined as follows

$$\begin{aligned} y(\tau) &= \sqrt{L(t)+h(t)}e^{\frac{1}{2}\int_t^T \mu(s)ds}, \\ z(\tau) &= \sqrt{H(t)+h(t)}e^{\frac{1}{2}\int_t^T \mu(s)ds}, \qquad t = t(\tau). \end{aligned} \tag{18.63}$$

In our numerical example without any loss of generality we use some artificial dependencies of the model parameters on time

$$\begin{aligned} \sigma(t) = \sigma_0 e^{-\sigma_K t}, \quad r(t) = r, \quad q(t) = 0, \quad z &= \sqrt{S+h(t)}e^{\frac{1}{2}r(T-t)}, \\ \tau &= \frac{\sigma_0^2 e^{rT}\left(e^{-(r+2\sigma_K)t} - e^{-(r+2\sigma_K)T}\right)}{8(2\sigma_K + r)}, \end{aligned} \tag{18.64}$$

$$h(t) = \frac{\sigma_0^2}{4(r + 2\sigma_K)} \left(e^{rt} - e^{-2\sigma_K t} \right),$$
$$y(\tau) = \sqrt{L(t(\tau)) + h(t(\tau))} e^{\frac{1}{2} r(T - t(\tau))},$$
$$z(\tau) = \sqrt{H(t(\tau)) + h(t(\tau)} e^{\frac{1}{2} r(T - t(\tau))}.$$

The dependence $t(\tau)$ can be obtained from Eq. (18.64) by inverting the second line. Let us also model the time-dependent barriers by linear dependencies

$$L(t) = a_L + b_L t, \qquad H(t) = a_H + b_H t. \tag{18.65}$$

Parameters of the model S_0, $\sigma_0, \sigma_K, a_L, b_L, a_H, b_H, r$ used in the numerical experiment are presented in Table 18.1.

S_0	σ_0	σ_K	r	a_L	b_L	a_H	b_H,
50	0.3	0.7	0.01	20	5	71	-2

Table 18.1: Parameters of the test.

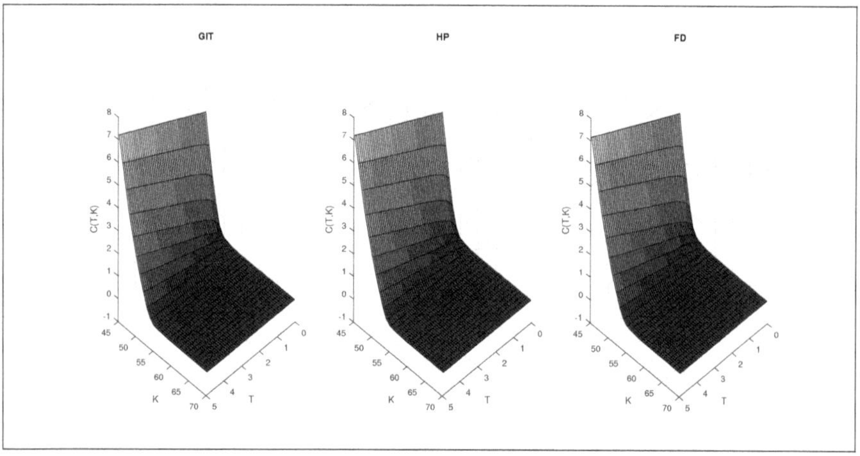

Figure 18.2: Prices of Call double barrier options obtained in the test by using three methods.

We solve this test by using the HP, GIT and FD methods. We use a FD forward scheme to price the options on various strikes simultaneously.

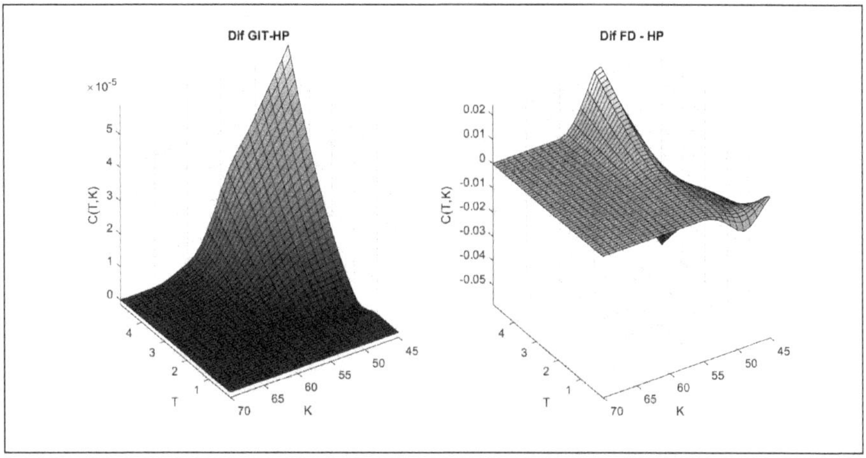

Figure 18.3: Absolute difference in prices of Call double barrier options obtained in the test by using three methods.

We use a set of maturities $T = [14./364, 1./12, 1/3, 0.5, 1, 2., 5.]$ and strikes $T = [45, 50, 55, 60, 65, 70]$. The results of our experiments are presented in Figs. 18.2 and 18.3.

For the HP and GIT method we use $M = 50$ points in time, while for the FD method we use a nonuniform grid of the size 100×100 (using lesser number of nodes provides unsatisfactory results). Details of the FD scheme can be found in Itkin (2017). For approximation of integrals in the GIT and HP solutions we apply a trapezoidal rule. Also note, that if in Eq. (18.36) the second argument ω of $\theta_3'(z, \omega)$ is close to one, this theta function is close to zero. Therefore, the matrix of the system in Eq. (18.36) becomes diagonal. Finally, the elapsed time for these tests are: 6.8 ms for GIT, 4.5 for HP and 112 ms for FD, respectively.

Thus, the speed of our methods is better than that of a FD scheme with the second order of approximation in space and time (e.g., the Crank-Nicolson one). However, by using higher order, e.g., Simpson quadratures one can further reduce M to $\sqrt{M}$ still providing same accuracy, while doing same for the FD method is not trivial.

Another point is the accuracy of all these methods. In Table 18.2 we present the results obtained by all methods as well as by Monte Carlo (MC)

T	0.038	0.083	0.333	0.500	1.000	2.000	5.000	0.038	0.083	0.333	0.500	1.000	2.000	5.000
K	MC							GIT						
45	5.0174	5.0379	5.1500	5.2212	5.4473	5.8825	7.1820	5.0173	5.0375	5.1498	5.2244	5.4478	5.8912	7.1946
50	0.1732	0.2583	0.5216	0.6370	0.8941	1.2815	2.4866	0.1735	0.2587	0.5243	0.6407	0.8984	1.2932	2.5056
55	0.0000	0.0000	0.0000	0.0000	0.0012	0.0075	0.0766	0.0000	0.0000	0.0000	0.0000	0.0012	0.0083	0.0852
K	FD							HP						
45	5.0173	5.0374	5.1483	5.2218	5.4416	5.8804	7.1465	5.0173	5.0375	5.1498	5.2244	5.4478	5.8912	7.1946
50	0.1757	0.2567	0.5224	0.6391	0.8972	1.2918	2.4999	0.1735	0.2587	0.5243	0.6407	0.8984	1.2932	2.5056
55	0.0000	0.0000	0.0000	0.0001	0.0017	0.0103	0.1221	0.0000	0.0000	0.0000	0.0000	0.0012	0.0083	0.0852

Table 18.2: Comparison of double barriers Call option prices obtained by various methods.

where we use $M = 200$ steps in time and $N = 500{,}000$ paths. It can be seen that at maturities $T \le 1$ year the results obtained by all methods are close to each other. However, at long maturities $T > 1$ year the FD results are slightly off. The accuracy of the FD method can be improved by taking more steps in time at high T which, however, takes time. Therefore, under these circumstances the FD methods becomes much slower than the GIT and HP methods. Perhaps, this is not an issue for Equity options, but could definitely be an issue for Fixed Income derivatives that have long maturities.

18.5 Final notes

As the approach described in this chapter and proposed in Itkin and Muravey (2020a) is new, perhaps, the most important point is how robust and efficient it is. As shown in Carr and Itkin (2021), Itkin and Muravey (2020b), Carr *et al.* (2020) and Itkin *et al.* (2020), from the computational point of view similar solutions for pricing single barrier options under various models with time-dependent barriers are very efficient and, at least theoretically, of the same complexity, or even faster than the forward finite-difference (FD) method. Here we use the same technique but applied to pricing double barrier options, therefore, this assessment remains to be valid.

On the other hand, our approach is capable of providing higher accuracy in computing the options prices, as this is regulated by quadrature rule used to discretize the integral kernel in Eqs. (18.36) and (18.41), or in Eq. (18.54). Therefore, the accuracy of the method in x space can be easily increased by using high order quadratures. For instance, using the Simpson instead of the trapezoid rule doesn't affect the complexity of our method but increases the accuracy, while increasing the accuracy for the FD method is not easy (i.e., it significantly increases the complexity of the method, e.g., see Itkin (2017)).

As applied to pricing double barrier options, the difference is that instead of a single Volterra equation of the second kind we now have to solve a system of two equations, either in Eqs. (18.36) and (18.41)), or in Eq. (18.54). This can be done in the same way as for the single barrier problem. The Volterra equation is solved by discretizing the kernel of the integral in time using some quadrature rule which yields a system of linear equations with respect to the discrete values of $\Psi(\tau), \Phi(\tau)$. It can be checked that the matrix of this system is of the form

$$\mathbf{M} = \begin{pmatrix} A & B \\ C & D \end{pmatrix},$$

where A, D are lower triangular matrices with ones on the main diagonal, and B, C are lower triangular matrices with zeros on the main diagonal. Therefore, this system can be solved by a simple Gauss elimination method (by a set of algebraic multiplications and additions) with complexity $O(2N)$ where N is the number of the grid nodes in τ for $\Psi(\tau), \Phi(\tau)$. Alternatively, when using Eq. (18.54) or Eq. (18.41), since the kernel is proportional to Gaussians, the discrete sum approximating the integral can be computed with linear complexity $O(2N)$ using the Fast Gauss Transform, see e.g., [Spivak *et al.* (2010)].

Once the vectors $\Psi(\tau), \Phi(\tau)$ (for the GIT method), or $\Omega(\tau), \Theta(\tau)$ (for the HP method) are found, they can be substituted into Eq. (18.34) or Eq. (18.39) for the GIT method), or into Eq. (18.49) (for the HP method). Then the final solution is obtained by computing the integral(s) numerically. Various numerical examples illustrating this technique for a single barrier pricing problem can be found in Carr and Itkin (2021), Itkin and Muravey (2020b), Carr *et al.* (2020) and Itkin *et al.* (2020). Also, those examples demonstrate that computationally our method is more efficient than both the backward and even forward FD methods (if one uses them to solve this kind of problems), while providing better accuracy and stability.

Somebody could be a bit confused of this terminology, since despite the solution is found explicitly as an integral, the latter depends of the unknown function of time $\Psi(\tau)$. In support of this terminology, we can mention that the solution is definitely of a closed form on variable x. On variable τ the integrand explicitly depends on yet unknown function $\Psi(\tau)$ which solves the Volterra integral equation of the second kind. However, this equation can be solved with no iterations. Indeed, after the function $\Psi(\tau)$ is discretized on some grid in τ (so now it is represented by a finite vector ψ), the integral equation reduces to the linear equation for ψ, with the matrix being low triangular. Thus, the solution can be immediately obtained by a simple Gauss elimination with no iterations. Therefore, this is explicit and as such, the solution is given by a series of algebraic operations (substitutions). The finer is the grid, the closer is the solution to the exact one.

Again, as mentioned in Carr *et al.* (2020), another advantage of the approach advocated in this Chapter is computation of option Greeks. Indeed, in both the HP and GIT methods the option prices are represented in an explicit analytic form on x (via the integrals on τ and the auxiliary variable ξ). This means that an explicit dependence of the option prices on the model parameters is available and transparent. Thus, explicit representations of the option Greeks can be obtained by a simple differentiation under the integrals. This means that the Greek values can be computed simultaneously with the option prices with almost no additional increase in the elapsed time. This is possible because differentiation under the integrals slightly changes the integrands, while these changes could be represented as changes in weights of the quadrature scheme used to compute the integrals.

Also, the integrands in the integral representation of the solution could be treated as a product of some density function and weights. The major computational time is spent for computing the densities as they contain special functions. However, once computed the results can be saved during the calculation of prices, and then reused when computing the Greeks. Therefore, computing Greeks can be done very fast. This is also true e.g., for Vega and other Greeks that cannot be computed by the FD method together with prices and require a separate run of the FD machinery. Here we don't have such a problem as differentiation of the integral representation with respect to the model parameters is done analytically.

Appendices

18.A Simplification of Eq. (18.29)

To simplify Eq. (18.29) we proceed by integrating by parts the last integral in Eq. (18.29)

$$\int_0^\tau e^{-\frac{\pi^2 n^2}{l^2(\tau)}(\tau-s)} h_1(n,s,\tau)ds = -\frac{B(\tau)l^2(\tau)}{\pi^2 n^2}\Big[\sin\left(\frac{\pi n}{l(\tau)}[z(\tau)-y(\tau)]\right) \\ - \sin\left(\frac{\pi n}{l(\tau)}[y(\tau)-y(\tau)]\right)\Big]$$

$$+\frac{B(0)l^2(\tau)}{\pi^2 k^2}e^{-\frac{\pi^2 n^2}{l^2(\tau)}\tau}\left[\sin\left(\frac{\pi n}{l(\tau)}[z(0)-y(\tau)]\right)-\sin\left(\frac{\pi n}{l(\tau)}[y(0)-y(\tau)]\right)\right]$$
$$-\frac{l(\tau)}{\pi n}\left[f^-(\tau)\cos\left(\frac{\pi n}{l(\tau)}[y(\tau)-y(\tau)]\right)-f^+(\tau)\cos\left(\frac{\pi n}{l(\tau)}[z(\tau)-y(\tau)]\right)\right]$$
$$+\frac{l(\tau)}{\pi n}e^{-\frac{\pi^2 n^2}{l^2(\tau)}\tau}\left[f^-(0)\cos\left(\frac{\pi n}{l(\tau)}[y(0)-y(\tau)]\right)\right.$$
$$\left.-f^+(0)\cos\left(\frac{\pi n}{l(\tau)}[z(0)-y(\tau)]\right)\right]$$
$$+\frac{l^2(\tau)}{\pi^2 n^2}\int_0^\tau B(s)e^{-\frac{\pi^2 n^2}{l^2(\tau)}(\tau-s)}\left(\frac{\pi^2 n^2}{l^2(\tau)}\left[\sin\left(\frac{\pi n}{l(\tau)}[z(s)-y(\tau)]\right)\right.\right.$$
$$\left.-\sin\left(\frac{\pi n}{l(\tau)}[y(s)-y(\tau)]\right)\right]$$
$$\left.+\frac{\pi n}{l(\tau)}\left[z'(s)\cos\left(\frac{\pi n}{l(\tau)}[z(s)-y(\tau)]\right)-y'(s)\cos\left(\frac{\pi n}{l(\tau)}[y(s)-y(\tau)]\right)\right]\right)ds$$
$$+\frac{l(\tau)}{\pi n}\int_0^\tau f^-(s)e^{-\frac{\pi^2 n^2}{l^2(\tau)}(\tau-s)}$$
$$\times\left(\frac{\pi^2 n^2}{l^2(\tau)}\cos\left(\frac{\pi n}{l(\tau)}[y(s)-y(\tau)]\right)-\frac{\pi n}{l(\tau)}y'(s)\sin\left(\frac{\pi n}{l(\tau)}[y(s)-y(\tau)]\right)\right)ds$$
$$-\frac{l(\tau)}{\pi n}\int_0^\tau f^+(s)e^{-\frac{\pi^2 n^2}{l^2(\tau)}(\tau-s)}$$
$$\times\left(\frac{\pi^2 n^2}{l^2(\tau)}\cos\left(\frac{\pi n}{l(\tau)}[y(s)-y(\tau)]\right)-\frac{\pi n}{l(\tau)}z'(s)\sin\left(\frac{\pi n}{l(\tau)}[y(s)-y(\tau)]\right)\right)ds$$
$$+\frac{l(\tau)}{\pi n}\int_0^\tau B(s)e^{-\frac{\pi^2 n^2}{l^2(\tau)}(\tau-s)}\left[y'(s)\cos\left(\frac{\pi n}{l(\tau)}[y(s)-y(\tau)]\right)\right.$$
$$\left.-z'(s)\cos\left(\frac{\pi n}{l(\tau)}[z(s)-y(\tau)]\right)\right]ds,$$

or

$$\int_0^\tau e^{-\frac{\pi^2 n^2}{l^2(\tau)}(\tau-s)}h_1(n,s,\tau)ds=\frac{l(\tau)}{\pi n}\left[(-1)^n f^+(\tau)-f^-(\tau)\right] \tag{18.66}$$
$$+\alpha(\tau,n)e^{-\frac{\pi^2 n^2}{l^2(\tau)}\tau}+\int_0^\tau e^{-\frac{\pi^2 n^2}{l^2(\tau)}(\tau-s)}\beta(\tau,s,n)ds$$
$$+\int_0^\tau e^{-\frac{\pi^2 n^2}{l^2(\tau)}(\tau-s)}B(s)\left[\sin\left(\frac{\pi n}{l(\tau)}[z(s)-y(\tau)]\right)\right.$$

$$- \sin\left(\frac{\pi n}{l(\tau)}[y(s) - y(\tau)]\right)\Big] ds,$$

where

$$\alpha(\tau, n) = \frac{B(0)l^2(\tau)}{\pi^2 n^2}\Big[\sin\left(\frac{\pi n}{l(\tau)}[z(0) - y(\tau)]\right) \tag{18.67}$$

$$- \sin\left(\frac{\pi n}{l(\tau)}[y(0) - y(\tau)]\right)\Big] + \frac{l(\tau)}{\pi n} e^{-\frac{\pi^2 n^2}{l^2(\tau)}\tau}\Big[f^-(0)\cos\left(\frac{\pi n}{l(\tau)}[y(0) - y(\tau)]\right)$$

$$- f^+(0)\cos\left(\frac{\pi n}{l(\tau)}[z(0) - y(\tau)]\right)\Big],$$

$$\beta(\tau, s, n) =$$

$$f^-(s)\left(\frac{\pi n}{l(\tau)}\cos\left(\frac{\pi n}{l(\tau)}[y(s) - y(\tau)]\right) - y'(s)\sin\left(\frac{\pi n}{l(\tau)}[y(s) - y(\tau)]\right)\right)$$

$$- f^+(s)\left(\frac{\pi n}{l(\tau)}\cos\left(\frac{\pi n}{l(\tau)}[z(s) - y(\tau)]\right) - z'(s)\sin\left(\frac{\pi n}{l(\tau)}[z(s) - y(\tau)]\right)\right).$$

Now we can transform the whole term

$$\frac{2}{l(\tau)}\sum_{n=1}^{\infty}\sin\left(\pi n\frac{x - y(\tau)}{l(\tau)}\right)\int_0^{\tau} e^{-\frac{\pi^2 n^2}{l^2(\tau)}(\tau - s)} h_1(n, s, \tau) ds, \tag{18.68}$$

which appears in Eq. (18.29). For doing that, first let us consider the integral

$$\int_{y(0)}^{z(0)} u(0, \xi)\sin\left(\frac{\pi n}{l(\tau)}[\xi - y(\tau)]\right) d\xi, \tag{18.69}$$

which is also a part of the RHS in Eq. (18.29). Recalling that by definition in Eq. (18.8) $u(0, x) = U(0, x) - A(0) - B(0)x$, and applying another identity

$$\int_{y(0)}^{z(0)} [A(0) + B(0)\xi]\sin\left(\frac{\pi n}{l(\tau)}[\xi - y(\tau)]\right) d\xi$$

$$= \frac{l(\tau)}{\pi^2 n^2}\Bigg\{\pi n(A(0) + B(0)y(0))\cos\left(\frac{\pi n(y(0) - y(\tau)}{l(\tau)}\right)$$

$$- \pi n\,[A(0) + B(0)z(0)]\cos\left(\frac{\pi n(z(0) - y(\tau)}{l(\tau)}\right)$$

$$+ B(0)l(\tau)\left[\sin\left(\frac{\pi n(z(0) - y(\tau)}{l(\tau)}\right) - \sin\left(\frac{\pi n(y(0) - y(\tau)}{l(\tau)}\right]\right)\Bigg\},$$

we obtain

$$\int_{y(0)}^{z(0)} u(0, \xi)\sin\left(\frac{\pi n}{l(\tau)}[\xi - y(\tau)]\right) d\xi \tag{18.70}$$

$$= \int_{y(0)}^{z(0)} U(0,\xi) \sin\left(\frac{\pi n}{l(\tau)}[\xi - y(\tau)]\right) d\xi - \alpha(\tau, n).$$

Therefore, the terms proportional to $\alpha(\tau, n)$ in Eq. (18.29) are canceling out. Also, substituting Eq. (18.73) into Eq. (18.29) and moving the RHS of Eq. (18.73) into the LHS of Eq. (18.29) results in the change of $u(\tau, x)$ to $U(\tau, x)$ in the LHS, and canceling out the terms proportional to $B(s)$. Finally, introducing the new function $F(\tau, x)$

$$F(\tau,x) = A(\tau)+B(\tau)x-\frac{2}{\pi}\sum_{n=1}^{\infty}\frac{(-1)^{n-1}f^{+}(\tau)+f^{-}(\tau)}{n}\sin\left(\frac{\pi n}{l(\tau)}[x-y(\tau)]\right) \tag{18.71}$$

we obtain the representation of $U(\tau, x)$

$$\begin{aligned} U(\tau,x) &= \frac{2}{l(\tau)}\sum_{n=1}^{\infty}\sin\left(\pi n\frac{x-y(\tau)}{l(\tau)}\right)\left\{e^{-\frac{\pi^2 n^2}{l^2(\tau)}\tau}\right. \\ &\quad\times\int_{y(0)}^{z(0)} U(0,\xi)\sin\left(\frac{\pi n}{l(\tau)}[\xi-y(\tau)]\right)d\xi + \int_0^{\tau} e^{-\frac{\pi^2 n^2}{l^2(\tau)}(\tau-s)} \\ &\quad\times\left[\Phi(s)\sin\left(\frac{\pi n}{l(\tau)}[z(s)-y(\tau)]\right) + \Psi(s)\sin\left(\frac{\pi n}{l(\tau)}[y(s)-y(\tau)]\right)\right. \\ &\quad\left.\left.+\,\beta(\tau,s,n)\right]ds\right\} + F(\tau,x). \end{aligned} \tag{18.72}$$

Further, application of the well-known identities (Gradshtein and Ryzhik, 2007)

$$\sum_{k=1}^{\infty}\frac{\sin kx}{k} = \frac{\pi - x}{2},\ 0 < x < \pi, \quad \sum_{k=1}^{\infty}(-1)^{k-1}\frac{\sin kx}{k} = \frac{x}{2},\ 0 < x < \pi,$$

yields the following relationship

$$\begin{aligned} &\sum_{n=1}^{\infty}\frac{2}{\pi n}\left[(-1)^{n-1}f^{+}(\tau)+f^{-}(\tau)\right]\sin\left(\pi n\frac{x-y(\tau)}{l(\tau)}\right) \\ &\quad= \frac{2}{\pi}\left\{\frac{\pi f^{+}(\tau)}{2}\frac{x-y(\tau)}{l(\tau)} + \frac{f^{-}(\tau)}{2}\left[\pi - \pi\frac{x-y(\tau)}{l(\tau)}\right]\right\} \\ &\quad= \frac{f^{+}(\tau)-f^{-}(\tau)}{l(\tau)}x + \frac{f^{+}(\tau)y(\tau)-f^{-}(\tau)z(\tau)}{l(\tau)} = -\left[A(\tau)+B(\tau)x\right], \\ &\quad x \in (y(\tau), z(\tau)). \end{aligned} \tag{18.73}$$

With the help of Eq. (18.73) we arrive at another formula for $F(\tau, x)$:

$$F(\tau, x) = \begin{cases} f^-(\tau), & x = y(\tau), \\ 0, & x \in (y(\tau), z(\tau)), \\ f^+(\tau), & x = z(\tau). \end{cases} \tag{18.74}$$

Combining Eqs. (18.72) and (18.74) together, and taking into account that the Fourier series in Eq. (18.72) is equal to zero if $x = y(\tau)$ or $x = z(\tau)$, yields

$$U(\tau, x) = \begin{cases} f^-(\tau), & x = y(\tau), \\ \tilde{U}(\tau, x), & x \in (y(\tau), z(\tau)), \\ f^+(\tau), & x = z(\tau). \end{cases} \tag{18.75}$$

Here the function $\tilde{U}(\tau, x) : (y(\tau), z(\tau)) \times \mathbb{R}_+ \to \mathbb{R}$ is defined as follows

$$\tilde{U}(\tau, x) = \frac{2}{l(\tau)} \sum_{n=1}^{\infty} \sin\left(\pi n \frac{x - y(\tau)}{l(\tau)}\right) \left\{ e^{-\frac{\pi^2 n^2}{l^2(\tau)}\tau} \int_{y(0)}^{z(0)} U(0, \xi) \right. \tag{18.76}$$
$$\times \sin\left(\frac{\pi n}{l(\tau)}[\xi - y(\tau)]\right) d\xi + \int_0^\tau e^{-\frac{\pi^2 n^2}{l^2(\tau)}(\tau - s)} \left[\Phi(s) \sin\left(\frac{\pi n}{l(\tau)}[z(s) - y(\tau)]\right)\right.$$
$$\left.\left. + \Psi(s) \sin\left(\frac{\pi n}{l(\tau)}[y(s) - y(\tau)]\right) + \beta(\tau, s, n)\right] ds \right\}.$$

Note, that for the derivative $\frac{\partial F(\tau,x)}{\partial x}$ we have

$$\frac{\partial F(\tau, x)}{\partial x} = B(\tau) - \frac{2}{l(\tau)} \left\{ f^+(\tau) \sum_{n=1}^{\infty} (-1)^{n-1} \cos\left(\frac{\pi n}{l(\tau)}[x - y(\tau)]\right) \right.$$
$$\left. + f^-(\tau) \sum_{n=1}^{\infty} \cos\left(\frac{\pi n}{l(\tau)}[x - y(\tau)]\right) \right\}$$
$$= \frac{f^+(\tau) - f^-(\tau)}{l(\tau)} - \frac{2}{l(\tau)} \left\{ f^+(\tau) \sum_{n=1}^{\infty} (-1)^{n-1} \cos\left(\frac{\pi n}{l(\tau)}[x - y(\tau)]\right) \right.$$
$$\left. + f^-(\tau) \sum_{n=1}^{\infty} \cos\left(\frac{\pi n}{l(\tau)}[x - y(\tau)]\right) \right\}$$
$$= \frac{2}{l(\tau)} \left\{ f^+(\tau) \left[\frac{1}{2} + \sum_{n=1}^{\infty} (-1)^n \cos\left(\frac{\pi n}{l(\tau)}[x - y(\tau)]\right)\right] \right.$$
$$\left. - f^-(\tau) \left[\frac{1}{2} + \sum_{n=1}^{\infty} \cos\left(\frac{\pi n}{l(\tau)}[x - y(\tau)]\right)\right] \right\}.$$

Applying well known representations for the Dirac delta function $\delta(x)$

$$\delta(z(\tau)-x)=\frac{2}{l(\tau)}\left[\frac{1}{2}+\sum_{n=1}^{\infty}(-1)^n\cos\left(\frac{\pi n}{l(\tau)}[x-y(\tau)]\right)\right],$$

$$\delta(x-y(\tau))=\frac{2}{l(\tau)}\left[\frac{1}{2}+\sum_{n=1}^{\infty}\cos\left(\frac{\pi n}{l(\tau)}[x-y(\tau)]\right)\right]$$

yields the following formula for the derivative of $F(\tau,x)$

$$\frac{\partial F(\tau,x)}{\partial x}=f^{+}(\tau)\delta(x-z(\tau))-f^{-}(\tau)\delta(x-y(\tau)). \tag{18.77}$$

Thus, this derivative is defined only in the sense of distributions.

18.B Transformation of Eqs. (18.30) to (18.39)

Applying a product-to-sum trigonometric identities to Eq. (18.76) we obtain

$$\begin{aligned}
\tilde{U}(\tau,x) &= \frac{1}{l(\tau)}\sum_{n=1}^{\infty}\left\{e^{-\frac{\pi^2n^2}{l^2(\tau)}\tau}\int_{y(0)}^{z(0)}d\xi\, U(0,\xi)\right. \qquad (18.78)\\
&\times\left[\cos\left(\frac{\pi n}{l(\tau)}[x-\xi]\right)-\cos\left(\frac{\pi n}{l(\tau)}[x+\xi-2y(\tau)]\right)\right]\\
&+\int_0^{\tau}e^{-\frac{\pi^2n^2}{l^2(\tau)}(\tau-s)}\left[\Phi(s)+f^{+}(s)z'(s)\right]\left[\cos\left(\frac{\pi n}{l(\tau)}[x-z(s)]\right)\right.\\
&\qquad\left.-\cos\left(\frac{\pi n}{l(\tau)}[x+z(s)-2y(\tau)]\right)\right]ds\\
&+\int_0^{\tau}e^{-\frac{\pi^2n^2}{l^2(\tau)}(\tau-s)}\left[\Psi(s)-f^{-}(s)y'(s)\right]\left[\cos\left(\frac{\pi n}{l(\tau)}[x-y(s)]\right)\right.\\
&\qquad\left.-\cos\left(\frac{\pi n}{l(\tau)}[x+y(s)-2y(\tau)]\right)\right]ds\\
&+\frac{\pi n}{l(\tau)}\int_0^{\tau}e^{-\frac{\pi^2n^2}{l^2(\tau)}(\tau-s)}f^{-}(s)\\
&\times\left[\sin\left(\frac{\pi n}{l(\tau)}[x-y(s)]\right)+\sin\left(\frac{\pi n}{l(\tau)}[x+y(s)-2y(\tau)]\right)\right]ds\\
&-\frac{\pi n}{l(\tau)}\int_0^{\tau}e^{-\frac{\pi^2n^2}{l^2(\tau)}(\tau-s)}f^{+}(s)\\
&\times\left.\left[\sin\left(\frac{\pi n}{l(\tau)}[x-z(s)]\right)+\sin\left(\frac{\pi n}{l(\tau)}[x+z(s)-2y(\tau)]\right)\right]ds\right\},
\end{aligned}$$

Since the functions

$$h_1(n)=e^{-\beta n^2}\cos(\alpha n),\qquad h_2(n)=ne^{-\beta n^2}\sin(\alpha n)$$

are even, $h_2(0) = 0$, and in the first three lines of Eq. (18.78) we have a difference of cosines, so at $n = 0$ the difference vanishes, the series in Eq. (18.78) can be slightly modified by replacing

$$\sum_{n=1}^{\infty} h_i(n) = \frac{1}{2} \sum_{n=-\infty}^{\infty} h_i(n), \quad i = 1, 2.$$

Now applying formulae in Eqs. (18.38) to (18.78) and using

$$\alpha = \frac{x-\xi}{l(\tau)}, \qquad \frac{1}{2\beta} = \frac{\tau}{l^2(\tau)}, \quad 2\sqrt{\frac{\beta}{2\pi}} = \frac{l(\tau)}{\sqrt{\pi\tau}},$$

$$\frac{\beta}{2}(2n+\alpha)^2 = \frac{l^2(\tau)}{4\tau}\left(2n + \frac{x-\xi}{l(\tau)}\right)^2 = \frac{(x-\xi+2nl(\tau))^2}{4\tau},$$

we obtain the following identities

$$\begin{aligned} &\frac{1}{2l(\tau)} \sum_{n=-\infty}^{\infty} e^{-\frac{\pi^2 n^2}{l^2(\tau)}(\tau-s)} \cos\left(\frac{\pi n}{l(\tau)}[x-\xi]\right) \\ &\quad = \frac{1}{2\sqrt{\pi(\tau-s)}} \sum_{n=-\infty}^{\infty} e^{-\frac{(x-\xi+2nl(\tau))^2}{4(\tau-s)}}, \\ &\frac{1}{2l(\tau)} \sum_{n=-\infty}^{\infty} e^{-\frac{\pi^2 n^2}{l^2(\tau)}(\tau-s)} \frac{\pi n}{l(\tau)} \sin\left(\frac{\pi n}{l(\tau)}[x-\xi]\right) \\ &\quad = \frac{1}{4\sqrt{\pi(\tau-s)^3}} \sum_{n=-\infty}^{\infty} (x-\xi+2nl(\tau)) e^{-\frac{(x-\xi+2nl(\tau))^2}{4(\tau-s)}}. \end{aligned} \tag{18.79}$$

Observe that each term in Eq. (18.78) can be represented as one of the series in Eq. (18.79). Therefore, assuming $x \in (y(\tau), z(\tau))$, we immediately arrive at the alternative representation for $\tilde{U}(\tau, x)$

$$\begin{aligned} \tilde{U}(\tau,x) &= \sum_{n=-\infty}^{\infty} \left\{ \int_{y(0)}^{z(0)} U(0,\xi)\Upsilon_n(x,\tau\,|\,\xi,0)d\xi \right. \\ &+ \int_0^\tau \left[\Phi(s) + f^+(s)z'(s)\right] \Upsilon_n(x,\tau|z(s),s)ds, \\ &+ \int_0^\tau \left[\Psi(s) - f^-(s)y'(s)\right] \Upsilon_n(x,\tau\,|\,y(s),s)ds \\ &+ \left. \int_0^\tau f^-(s)\Lambda_n(x,\tau\,|\,y(s),s) - f^+(s)\Lambda_n(x,\tau\,|\,z(s),s)ds \right\} + F_1(\tau,x), \\ \Upsilon_n(x,\tau\,|\,\xi,s) &= \frac{1}{2\sqrt{\pi(\tau-s)}}\left[e^{-\frac{(2nl(\tau)+x-\xi)^2}{4(\tau-s)}} - e^{-\frac{(2nl(\tau)+x+\xi-2y(\tau))^2}{4(\tau-s)}}\right], \end{aligned} \tag{18.80}$$

$$\Lambda_n(x,\tau\,|\,\xi,s) = \frac{x-\xi+2nl(\tau)}{4\sqrt{\pi(\tau-s)^3}}e^{-\frac{(2nl(\tau)+x-\xi)^2}{4(\tau-s)}}$$
$$+\frac{x+\xi-2y(\tau)+2nl(\tau)}{4\sqrt{\pi(\tau-s)^3}}e^{-\frac{(2nl(\tau)+x+\xi-2y(\tau))^2}{4(\tau-s)}}.$$

18.B.1 *The limiting values $x \to y(\tau)$ and $x \to z(\tau)$ in Eq. (18.80)*

The Eq. (18.80) provides an alternative representation of the solution $\tilde{U}(\tau,x)$ of the heat equation in Eq. (18.4) with the initial condition in Eq. (18.5) and the boundary conditions in Eq. (18.6) at the time-dependent domain $x \in (y(\tau), z(\tau))$ in terms of the Fourier series. In this section we show that the function $\tilde{U}$ can be analytically continued to the boundary points $y(\tau)$ and $z(\tau)$, and

$$\lim_{x\to y(\tau)+0}\tilde{U}(\tau,x) = f^-(\tau), \qquad \lim_{x\to z(\tau)-0}\tilde{U}(\tau,x) = f^+(\tau). \tag{18.81}$$

It is easy to check that the functions $\Upsilon_n(x,\tau|\xi,s)$ and $\Lambda_n(x,\tau|\xi,s)$ are regular only if $n \neq 0$, $x \in [y(\tau), z(\tau)]$, $\xi \in [y(s), z(s)]$, $s \to \tau$. In this case the following identities hold

$$\lim_{s\to\tau}\Upsilon_n(x,\tau|\xi,s) = 0, \qquad \lim_{s\to\tau}\Lambda_n(x,\tau|\xi,s) = 0, \qquad n \neq 0. \tag{18.82}$$

At $n = 0$ functions $\Upsilon_0(x,\tau|y(s),s)$ and $\Lambda_0(x,\tau|y(s),s)$ have a singularity when $s \to \tau$, $x \to y(\tau)$, and functions $\Upsilon_0(x,\tau|z(s),s)$ and $\Lambda_0(x,\tau|z(s),s)$ - when $s \to \tau$, $x \to z(\tau)$. Note, that the singularity of Υ_0 is integrable and so weak. Therefore, when calculating a corresponding limit of both parts in Eq. (18.80), for the regular terms we can switch the order of the integration and limit operators, and then use the following properties

$$\lim_{x\to y(\tau)+0}\sum_{n=-\infty}^{\infty}\Upsilon_n(x,\tau\,|\,\xi,s) = 0, \quad \lim_{x\to z(\tau)-0}\sum_{n=-\infty}^{\infty}\Upsilon_n(x,\tau\,|\,\xi,s) = 0,$$
$$\lim_{x\to y(\tau)+0}\sum_{\substack{n=-\infty\\ n\neq 0}}\Lambda_n(x,\tau\,|\,\xi,s) = 0, \quad \lim_{x\to z(\tau)-0}\sum_{\substack{n=-\infty\\ n\neq 0}}\Lambda_n(x,\tau\,|\,\xi,s) = 0, \tag{18.83}$$
$$\lim_{x\to y(\tau)+0}\Lambda_0(x,\tau\,|\,z(s),s) = 0, \quad \lim_{x\to z(\tau)-0}\Lambda_0(x,\tau\,|\,y(s),s) \quad = 0,$$

to obtain

$$\lim_{x\to y(\tau)+0}\tilde{U}(\tau,x) = \lim_{x\to y(\tau)+0}\int_0^\tau f^-(s)\Lambda_0(x,\tau\,|\,y(s),s) \tag{18.84}$$

$$\lim_{x\to z(\tau)-0} \tilde{U}(\tau,x) = -\lim_{x\to z(\tau)-0}\int_0^\tau f^+(s)\Lambda_0(x,\tau\,|\,z(s),s).$$

To proceed we need the notion of heat potentials and the results obtained in Section 18.3 (see also Tikhonov and Samarskii, 1963). It can be shown that the functions $\Lambda_0(x,\tau|y(s),s)$ and $\Lambda_0(x,\tau|z(s),s)$ can be represented as a sum of double layer heat potentials. Therefore, we can evaluate the limits Eq. (18.84) with the help of Eqs. (8.10) and (8.19).

In more detail, according to Eq. (18.84) in the explicit form the limits of $\tilde{U}(\tau,x)$ read

$$\begin{aligned}\tilde{U}(\tau,y(\tau)) &= \lim_{x\to y(\tau)+0}\int_0^\tau f^-(s)\left[\frac{x-y(s)}{4\sqrt{\pi(\tau-s)^3}}e^{-\frac{(x-y(s))^2}{4(\tau-s)}}\right.\\ &\quad\left.+\frac{x-2y(\tau)+y(s)}{4\sqrt{\pi(\tau-s)^3}}e^{-\frac{(x-2y(\tau)+y(s))^2}{4(\tau-s)}}\right]ds,\\ \tilde{U}(\tau,z(\tau)) &= -\lim_{x\to z(\tau)-0}\int_0^\tau f^+(s)\left[\frac{x-z(s)}{4\sqrt{\pi(\tau-s)^3}}e^{-\frac{(x-z(s))^2}{4(\tau-s)}}\right.\\ &\quad\left.+\frac{x-2z(\tau)+z(s)}{4\sqrt{\pi(\tau-s)^3}}e^{-\frac{(x-2z(\tau)+z(s))^2}{4(\tau-s)}}\right]ds.\end{aligned}$$

This can also be re-written in the form

$$\begin{aligned}\tilde{U}(\tau,y(\tau)) &= \lim_{x\to y(\tau)+0}\int_0^\tau f^-(s)\frac{x-y(s)}{4\sqrt{\pi(\tau-s)^3}}e^{-\frac{(x-y(s))^2}{4(\tau-s)}}ds \qquad (18.85)\\ &\quad-\lim_{2y(\tau)-x\to y(\tau)-0}\int_0^\tau f^-(s)\frac{2y(\tau)-x-y(s)}{4\sqrt{\pi(\tau-s)^3}}e^{-\frac{(x-2y(\tau)+y(s))^2}{4(\tau-s)}}ds,\\ \tilde{U}(\tau,z(\tau)) &= -\lim_{x\to z(\tau)-0}\int_0^\tau f^+(s)\frac{x-z(s)}{4\sqrt{\pi(\tau-s)^3}}e^{-\frac{(x-z(s))^2}{4(\tau-s)}}ds\\ &\quad+\lim_{2z(\tau)-x\to z(\tau)+0}\int_0^\tau f^+(s)\frac{2z(\tau)-x-z(s)}{4\sqrt{\pi(\tau-s)^3}}e^{-\frac{(x-2z(\tau)+z(s))^2}{4(\tau-s)}}ds.\end{aligned}$$

Using Eq. (8.19), these expressions can be transformed to

$$\tilde{U}(\tau,y(\tau)) = \frac{f^-(\tau)}{2}+\frac{f^-(\tau)}{2} = f^-(\tau),\; \tilde{U}(\tau,z(\tau)) = \frac{f^+(\tau)}{2}+\frac{f^+(\tau)}{2} = f^+(\tau).$$

Since $\tilde{U}(\tau,x)$ has the same limits as the boundary values of $U(\tau,x)$, and at $x\in(y(\tau),z(\tau))$ we have $\tilde{U}(\tau,x)=U(\tau,x)$, Eq. (18.80) allows an alternative form

$$U(\tau,x) = \sum_{n=-\infty}^{\infty}\left\{\int_{y(0)}^{z(0)} U(0,\xi)\Upsilon_n(x,\tau\,|\,\xi,0)d\xi\right.$$

$$
\begin{aligned}
&+ \int_0^\tau \left[\Phi(s) + f^+(s)z'(s)\right] \Upsilon_n(x, \tau|z(s), s)ds, \\
&+ \int_0^\tau \left[\Psi(s) - f^-(s)y'(s)\right] \Upsilon_n(x, \tau \,|\, y(s), s)ds \\
&+ \int_0^\tau f^-(s)\Lambda_n(x, \tau \,|\, y(s), s) - f^+(s)\Lambda_n(x, \tau \,|\, z(s), s)ds \Bigg\}.
\end{aligned}
$$

Chapter 19

Multilayer Heat Equations: Application to Finance

In this chapter, we describe a multilayer (ML) method for solving one-factor parabolic equations, that was developed in Itkin *et al.* (2021). Our approach provides a powerful alternative to the well-known finite difference and Monte Carlo methods. We discuss various advantages of this approach, which judiciously combines semi-analytical and numerical techniques and provides a fast and accurate way of finding solutions to the corresponding equations. To introduce the core of the method, we consider multilayer heat equations, known in physics for a relatively long time but never used when solving financial problems. Thus, we expand the analytic machinery of quantitative finance by augmenting it with the ML method. We demonstrate how one can solve various problems of mathematical finance by using our approach. Specifically, we develop efficient algorithms for pricing barrier options for time-dependent one-factor short-rate models, such as Black-Karasinski and Verhulst. Besides, we show how to solve the well-known Dupire equation quickly and accurately. Numerical examples confirm that our approach is considerably more efficient for solving the corresponding partial differential equations than the conventional finite difference method by being much faster and more accurate than the known alternatives.

Introduction

The problem of solving partial differential equations (PDEs) with moving boundaries appears naturally in various areas of science and technology. As mentioned in Kartashov (2001), such problems have been known in physics for a long time. They arise in several fields, such as (a) nuclear power engineering and safety of nuclear reactors; (b) combustion in solid-propellant rocket engines; (c) laser action on solids; (d) the theory of phase transitions (the Stefan problem and the Verigin problem); (e) the processes of sublimation in freezing and melting; (f) in the kinetic theory of crystal growth; etc. (see Kartashov, 1999 and references therein). Analytical solutions to these problems often require rather sophisticated methods., which were actively developed by the Russian mathematical school in the 20th century starting from A.V. Luikov, and then by B.Ya. Lyubov, E.M. Kartashov, and many others.

As applied to mathematical finance, one of these methods — the method of heat potentials (HP) — was actively utilized by A. Lipton and his co-authors to solve various mathematical finance problems (see Lipton, 2001; Lipton and Lopez de Prado, 2020a,b and references therein). A complementary method of a generalized integral transform (GIT) is developed in Carr and Itkin (2021), Itkin and Muravey (2020b) and Carr *et al.* (2020) to price barrier and American options in the semi-closed form. These authors studied the time-dependent Ornstein-Uhlenbeck (OU), Hull-White, CIR, and CEV models. An extension of the method of heat potentials for the Bessel process called the method of Bessel potentials is developed by Carr *et al.* (2020), who also describe a general scheme of how to construct the potential method for any linear differential operator with time-independent coefficients. Finally, they also extended the method of generalized integral transform to the Bessel process. In all cases, a semi-analytical (or semi-closed form) solution means that first, one needs to solve a linear Volterra equation of the second kind. Then the option price is represented as a one-dimensional integral.

Carr and Itkin (2021), Itkin and Muravey (2020b) and Carr *et al.* (2020) show that the new method is computationally more efficient than the existing ones, such as the backward and forward finite difference methods while providing better accuracy and stability. Also, the heat potential and GIT methods do not duplicate but rather complement each other. The former provides very accurate results for short maturities, and the latter for long maturities.

Even though many new problems have been solved in the above-cited papers, some of the financial models are hard to solve by using these methods directly. For instance, this is the case for the Black-Karasinski model, popular among practitioners. Another problem is the calibration of the local (or implied) volatility surface in various one-factor models. Almost all popular analytic and semi-analytical methods approach the solution of this problem by doing it term-by-term, which, doubtless, produces computational errors. For more details, see Itkin (2020) and references therein.

Here we attack this class of problems (some of them unsolved in the semi-analytical form) by using another method, which we call the method of multilayer (ML) heat equation. An alternative approach is given in Dias (2015), where an innovative technique of recursive images is presented to obtain solutions to the transient diffusion equation in a N-layered material based on the superposition of Green functions for a semi-infinite material. The solution is initially built for a single layer over a substrate by constructing a sequential sum of reflected image functions. These functions are chosen to satisfy in sequence the boundary conditions, first at the front interface, then at the back interface, then again at the front interface, and so on until the added functions' magnitude becomes negligible.

Based on this so-called "1-layer" algorithm, the author also constructs a "2-layer" algorithm by sequential application of the "1-layer" algorithm first to layer 1, then to layer 2, then again to layer 1, and so on. The sequential application of the $N - 1$ algorithm naturally leads to the N-layer algorithm. This scheme works for the first and second kind boundary conditions but does not apply to the case where there is a contact resistance between layers or the convective heat transfer at the end interfaces.

Note that this algorithm as applied to the local volatility calibration problem is similar to the approach used in Lipton and Sepp (2011b), Itkin and Lipton (2018) and Carr and Itkin (2020, 2019).

Since the ML method splits the whole (possibly infinite) domain in the space variable into a sequence of sub-domains, one could extend it naturally to solving parabolic equations with coefficients being functions of time t and location x. At every sub-interval, the corresponding parabolic operator could be either approximated by the operator with the space-homogeneous coefficients or, possibly, reduced to the heat equation by a series of transformations. After either approximation or reduction, the ML method can be applied.

Moreover, the method could be extended further to deal with non-linear volatility, drift, and killing term. Again, piecewise approximations of these terms lead to the parabolic equations at every sub-interval that could be transformed to the heat equation. Then, the application of the ML method solves the problem.

The main idea is to combine the ML method with the method of heat potentials[1] and the GIT method. Since both provide a semi-analytical solution

---

[1]More general potential methods, e.g., the method of Bessel potentials, can also be used in such a scheme.

for sub-interval problems, a combination of these solutions within the ML heat equation method results in the problem's full solution, expressed explicitly via one-dimensional integrals. At each layer, these integrals depend either on the yet unknown potential density (in the HP method) or on the solution gradient at the layer's boundaries (the GIT method). These unknown functions solve the interconnected systems of the integral Volterra equations of the second kind derived in the paper. Once this solution is found (either numerically or, sometimes, analytically), the whole problem is solved. Note that one can transform the system of integral equations to linear equations on a time-space grid, which is lower banded (in our case, block lower triangular). Therefore, the corresponding system can be solved with complexity $O(M^2N)$ where N is the number of layers, and M is the number of time steps (see Itkin and Muravey, 2020b in more detail).

We also propose a particular construction of the layers' internal boundaries, which allows the representation of every integral in the Volterra equation as convolution. Applying the Laplace transform, we obtain a system of linear equations with a block-tridiagonal matrix (it contains four blocks). This system can be efficiently solved numerically (with complexity $O(N)$). In some cases, it can be solved analytically. After this system's solution is found, we use the Gaver-Stehfest method to compute the inverse Laplace transform, also with linear complexity in the number of layers N. This algorithm solves the system of the Volterra equations and thus the whole problem.

We illustrate these novel ideas by representing several significant financial problems in the form suitable for solving them by the ML method. These problems include pricing barrier options in the time-dependent Black-Karasinski and modified Black-Karasinski (Verhulst) models (see Itkin *et al.*, 2020), as well as the solution of the Dupire equation. We also provide several numerical examples to demonstrate our method's high speed and accuracy compared with standard finite-difference (FD) methods.

To the best of our knowledge, all the paper results are new and contribute to the existing financial and physics literature. It is interesting to note that our method is capable of solving similar problems that appear in medicine and biology in addition to finance. For instance, our technique is well-suited for studying (a) the growth of diffusive brain tumors, which considers the brain tissue's heterogeneity (Asvestas *et al.*, 2014); (b) the transdermal drug release from an iontophoretic system (Pontrelli *et al.*, 2016); (c) and many other similar problems. It is imperative to emphasize that our method allows solving the ML problems with time-dependent boundaries and time- and space-dependent diffusion coefficients. In contrast, the method of Carr and March (2018) and all other known approaches operate only with constant boundaries (possibly with time-dependent boundary conditions) and spatially piecewise constant diffusion coefficients. Moreover, their setting corresponds to one of our numerical examples in Section 19.5. Since in Carr and March (2018) the solution is obtained by using spectral (eigenvector) series, while we apply the Laplace transform method, our approach is about 1000 times faster.

The rest of the paper is organized as follows. In Section 19.1, we construct the solution of the ML heat equation by using the method of heat potentials. In Section 19.2, we solve this equation by using the GIT method. In Section 19.3.1.1, we describe the pricing of barrier options in the time-dependent Black-Karasinski (BK) model and also in our modification of this model, which was introduced in Itkin *et al.* (2020) and is called the Verhulst model. In particular, we demonstrate how to reduce the pricing PDEs for both models to the ML heat equation. Also, in Section 19.3.1.1 we provide a generalization of this approach for some other models. In Section 19.3.2, we apply the results of Section 19.2.3 to investigate the case of space-dependent volatility $\sigma(x)$ in conjunction with solving the Dupire equation. Section 19.4 is dedicated to the solution of the Volterra equations. In particular, we describe a construction of the internal boundaries, which allows a transformation of the Volterra equations of the second kind to Abel equations. We solve the latter equations via the Laplace transform. Section 19.5 describes some numerical experiments with the ML method. The final section concludes.

19.1 Solving the ML heat equation via the HP method

Let us consider the following initial-boundary problem

$$\begin{aligned}
\mathcal{L}u(\tau, x) &= 0, \qquad (x,\tau) \in \Omega : \left[y^-(\tau), y^+(\tau)\right] \times \mathbb{R}_+, \\
u(0, x) &= f(x), \quad y^-(0) < x < y^+(0), \\
u(\tau, y^-(\tau)) &= \chi^-(\tau), \qquad u(\tau, y^+(\tau)) = \chi^+(\tau).
\end{aligned} \tag{19.1}$$

Here the operator $\mathcal{L}$ is a partial differential operator of the parabolic type

$$\mathcal{L} = -\frac{\partial}{\partial \tau} + \frac{\partial}{\partial x}\left(\sigma^2(\tau, x)\frac{\partial}{\partial x}\right) + \mu(\tau, x)\frac{\partial}{\partial x} + \nu(\tau, x), \tag{19.2}$$

$\sigma(\tau, x), \mu(\tau, x), \nu(\tau, x)$ are some known functions, Ω is the spatial-temporal domain with curvilinear temporal boundaries, and $\chi^-(\tau), \chi^+(\tau)$ are known functions of time (the boundary conditions).

Similar to Itkin and Muravey (2020b), we represent the solution in the form

$$u(x,\tau) = q(x,\tau) + \int_{y^-(0)}^{y^+(0)} f(\xi)\mathcal{G}(x,\xi,\tau)d\xi, \tag{19.3}$$

where $\mathcal{G}(x,\xi,\tau)$ is Green's function of the problem. Then the function $q(x,\tau)$ solves a problem similar to Eq. (19.1) but with the homogeneous initial condition

$$\begin{aligned}
\mathcal{L}q(\tau,x) &= 0, \qquad (x,\tau) \in \Omega : \left[y^-(\tau), y^+(\tau)\right] \times \mathbb{R}_+, \\
q(0,x) &= 0, \quad y^-(0) < x < y^+(0), \\
q(\tau, y^-(\tau)) &= \chi^-(\tau) - \int_{y^-(0)}^{y^+(0)} f(\xi)\mathcal{G}(y^-(\tau),\xi,\tau)d\xi = \phi^-(\tau), \\
q(\tau, y^+(\tau)) &= \chi^+(\tau) - \int_{y^-(0)}^{y^+(0)} f(\xi)\mathcal{G}(y^+(\tau),\xi,\tau)d\xi = \phi^+(\tau).
\end{aligned} \tag{19.4}$$

If the Green function $\mathcal{G}(x,\xi,\tau)$ is known, the problem in Eq. (19.4) can be solved via the HP method (Itkin and Muravey, 2020a). Otherwise, one can apply the ML method as this is described below.

To use the ML method, suppose the domain Ω could be split into N layers: $\Omega = \bigcup_{i=1}^N \Omega_i$, where each layer is a curvilinear strip

$$\begin{aligned}
&\Omega_i = [y_i(\tau), y_{i+1}(\tau)] \times \mathbb{R}_+, \quad y_i(\tau) < y_{i+1}(\tau), \quad \forall\tau > 0, \quad \forall i = 1,\dots,N, \\
&y_1(\tau) = y^-(\tau), \quad y_{N+1}(\tau) = y^+(\tau).
\end{aligned} \tag{19.5}$$

Let us seek for the solution of the problem Eq. (19.4) in the form

$$u(\tau,x) = \sum_{i=1}^N u_i(\tau,x)\mathbf{1}_{x-y_i(\tau)}\mathbf{1}_{y_{i+1}(\tau)-x}, \qquad \mathbf{1}_x = \begin{cases} 1, & x \ge 0, \\ 0, & x < 0, \end{cases} \tag{19.6}$$

and request that both $u(\tau,x)$ and its flux are continuous functions of x.[2] Using these conditions at every boundary $y_i(\tau)$, $i = 2,\dots,N$ together with the boundary conditions yields the following system of equations

$$\begin{aligned}
u_i(\tau, y_{i+1}(\tau)) &= u_{i+1}(\tau, y_{i+1}(\tau)), \\
\sigma_i^2(\tau, y_{i+1}(\tau))\frac{\partial u_i}{\partial x}\bigg|_{x=y_{i+1}(\tau)} &= \sigma_{i+1}^2(\tau, y_{i+1}(\tau))\frac{\partial u_{i+1}}{\partial x}\bigg|_{x=y_{i+1}(\tau)},
\end{aligned} \tag{19.7}$$

[2]These conditions are natural in physics if by $u(t,x)$ we assume, e.g., the temperature and interpret $\sigma^2\partial_x u(t,x)$ as the heat flux. Therefore, it is standard to require continuity of the heat flux rather than the first derivative $\partial_x u(t,x)$ (Lienhard IV and Lienhard V, 2019)

$$
\begin{aligned}
&i = 1, \ldots, N-1, \\
&u_1(\tau, y^-(\tau)) = \chi^-(\tau), \qquad u_N(\tau, y^+(\tau)) = \chi^+(\tau).
\end{aligned}
$$

The first condition means a continuity of the function u at every boundary y_i, $i = 1, \ldots, N-1$. The second condition is a continuity of the heat flux at the same boundary. The last line follows from the boundary conditions in Eq. (19.4).

Also, let us define the operator $\mathcal{L}$ for the whole domain Ω as follows

$$
\mathcal{L} = \sum_{i=1}^{N} \mathcal{L}_i \mathbf{1}_{x-y_i(\tau)} \mathbf{1}_{y_{i+1}(\tau)-x}, \tag{19.8}
$$

where

$$
\mathcal{L}_i = -\frac{\partial}{\partial \tau} + \frac{\partial}{\partial x}\left(\sigma_i^2(\tau, x)\frac{\partial}{\partial x}\right) + \mu_i(\tau, x)\frac{\partial}{\partial x} + \nu_i(\tau, x), \tag{19.9}
$$

and $\mathcal{L}_i u_i = 0$.

The idea of the ML method is to assume that Green's function $\mathcal{G}_i(x, \tau|\xi, s)$ associated with the operator $\mathcal{L}_i$ can be obtained in the closed form. For an arbitrary dependencies $\sigma_i(\tau, x), \mu_i(\tau, x), \nu_i(\tau, x)$ this is not the case, but for various specific forms of these functions this can be done. For instance, when $\mu(\tau, x) = \nu(\tau, x) = 0$ and $\sigma(\tau, x) = \sigma(\tau)$ or $\sigma(\tau, x) = \sigma(x)$, etc. (Polyanin, 2002). Otherwise, the functions $\sigma_i(\tau, x), \mu_i(\tau, x), \nu_i(\tau, x)$ can be approximated at every layer, e.g., by piecewise constant or linear function in x and an arbitrary function of τ, or by piecewise constant functions in τ and piecewise linear functions in x, etc. These approximations make the ML method somewhat similar to the FD method, however, with some critical distinctions, see Section 19.6.

It is important to mention, that the operator $\mathcal{L}_i$ in Eq. (19.9), while natural for physics where a divergent form of the parabolic equation (e.g., the heat equation) is commonly accepted, is just rarely used in mathematical finance. Instead, in finance it is natural to consider a non-divergent (non-conservative) form, which for the heat equation reads

$$
\mathcal{L}_i = -\frac{\partial}{\partial \tau} + \sigma_i^2(\tau, x)\frac{\partial^2}{\partial x^2}. \tag{19.10}
$$

Obviously, when $\sigma_i = \sigma_i(\tau), \forall i$, i.e. $\sigma_i(\tau, x)$ is a straight line at given τ, both operators in Eq. (19.9) and Eq. (19.10) coincide. However, if one solves Eq. (19.10) by the ML method, it can be unclear what continuity condition should be used. As shown in Lejay (2006), for the divergent heat/diffusion equation with drift this condition remains the same, i.e., this is a continuity of flux over the boundary. Obviously, a non-divergent heat equation can be represented in this form, i.e., the divergent diffusion part plus drift. Therefore, the continuity condition is still represented by the equality of fluxes over the boundary, but the equation now includes an extra drift term.

As applied to the ML method, this can be seen as follows. Suppose we apply the ML method to some parabolic equation, and approximate all coefficients in the drift and killing terms by piecewise constant function at every interval. Then, at the i-th interval this equation reads

$$\frac{\partial u_i(\tau,x)}{\partial \tau} = \sigma^2(\tau,x)\frac{\partial^2 u_i(\tau,x)}{\partial x^2} + \alpha_i \frac{\partial u_i(\tau,x)}{\partial x} + \beta_i u_i(\tau,x), \tag{19.11}$$

where $\alpha_i = const,\ \beta_i = const,\ i = 1, N$. By transforming it to a divergent form we obtain

$$\begin{aligned}\frac{\partial u_i(\tau,x)}{\partial t} &= \frac{\partial}{\partial x}\left(\sigma^2(\tau,x)\frac{\partial u_i(\tau,x)}{\partial x}\right) \\ &\quad + \left[\alpha_i - 2\sigma(\tau,x)\sigma_x(\tau,x)\right]\frac{\partial u_i(\tau,x)}{\partial x} + \beta_i u_i(\tau,x).\end{aligned} \tag{19.12}$$

Hence, again, the continuity condition for this equation is given in Eq. (19.7). Further, Eq. (19.12) by a series of transformations can be reduced to a non-divergent heat equation in Eq. (19.10). Accordingly, these transformations should be applied to Eq. (19.7) as well to obtain the correct continuity conditions.

When the external boundaries are constant, i.e., $y^-(t) = \chi^-(t) = const,\ y^+(t) = \chi^+(t) = const$ one may use an alternative where a non-divergent heat equation can be reduced to a divergent one by a change of variables $x \mapsto y = g(x)$, where $g(x)$ is some function which depends on $\sigma^2(x)$. In more detail this is shown in Appendix 19.A. Accordingly, the operator Eq. (19.10) transforms to

$$\mathcal{L}_i = -\frac{\partial}{\partial \tau} + \frac{\partial}{\partial y}\left(\Xi_i^2(\tau,y)\frac{\partial}{\partial y}\right), \tag{19.13}$$

where $\Xi^2(\tau,y)$ is a new diffusion coefficient, which can be expressed via $\sigma(\tau,x)$, again see Appendix 19.A.

In what follows, we provide our analysis for Eq. (19.9); Eq. (19.10) can be analyzed similarly, as explained above. For simplicity and without loss of generality, we give an exposition of the HP method assuming $\mu_i(\tau,x) = \nu_i(\tau,x) = 0$, and either $\sigma_i(\tau,x) = \sigma_i(\tau)$, or $\sigma_i(\tau,x)$ is a piecewise constant function of x for every layer. In this case each equation $\mathcal{L}_i u_i = 0$ by some change of variables $\tau \mapsto \bar{\tau}, x \mapsto \bar{x}$ can be transformed to the heat equation

Eq. (19.60) with $\sigma_i^2(\bar{\tau}, \bar{x}) = 1$[3] (Polyanin, 2002) and the corresponding Green function $G(\bar{x}, \xi, \bar{\tau})$ reads

$$G(\bar{x}, \xi, \bar{\tau}) = \frac{1}{2\sqrt{\pi\bar{\tau}}} e^{-\frac{(\bar{x}-\xi)^2}{4\bar{\tau}}}. \tag{19.14}$$

Also, these transformations modify the boundary $y(\tau) \mapsto \bar{y}(\bar{\tau})$. Some examples of such transformations are presented in Section 19.3. In Appendix 19.B we also provide some recipes on how to proceed if one needs to generalize this approach by considering a general case $\sigma = \sigma(\tau, x)$.

To the end of this section, for easiness of reading let us drop the bar over new variables. Now, following the general idea of the method of heat potentials for pricing double barrier options (Itkin and Muravey, 2020b; Carr *et al.*, 2020) we represent each function $q_i(\tau, x)$ as

$$\begin{aligned} q_i(\tau_i, x) = \int_0^{\tau_i} \Bigg\{ & \Psi_i(k) \frac{\partial G(x, \xi, \tau_i - k)}{\partial \xi}\Bigg|_{\xi = y_i(k)} \\ & + \Phi_i(k) \frac{\partial G(x, \xi, \tau_i - k)}{\partial \xi}\Bigg|_{\xi = y_{i+1}(k)} \Bigg\} dk. \end{aligned} \tag{19.15}$$

In Eq. (19.15) the second integral is a sum of two single layer potentials with the potential densities $\Psi_i(\tau)$ and $\Phi_i(\tau)$. By writing Eq. (19.15) we take into account that according to Eqs. (19.42) and (19.69), the transformed time τ might differ for each interval, therefore, the notation τ_i is used. However, e.g., for the problem described in Section 19.3.1.1 all new times are equal, i.e. $\tau_i = \tau$, $i = 1, \dots, N+1$.

Since the domain Ω consists of N layers, there are $2N$ unknown density functions $\Psi_i(\tau_i), \Phi_i(\tau_i)$, $i = 1, \dots, N$. To determine them one need to plug the representation of $q_i(\tau_i, x)$ in Eq. (19.15) into Eq. (19.7), and then solve thus obtained system of the integral Volterra equations of the second kind.

However, it is known (Tikhonov and Samarskii, 1963) that the integral in Eq. (19.15) for $x = y_i(\tau_i)$ and $x = y_{i+1}(\tau_i)$ is discontinuous, but with the finite value at $x = y_i(\tau_i) + \varepsilon$, $\forall i = 1, \dots, N+1$ when $\varepsilon \to 0$. Then, as shown in Itkin and Muravey (2020a), Eq. (19.15) should be represented in the form

$$q_i(\tau, y_i(\tau)) = \frac{\Psi_i(\tau)}{2\sigma_i^2(y_i(\tau))} + \int_0^{\tau} \Bigg\{ \Psi_i(k) \frac{\partial G(y_i(\tau), \xi, \tau - k)}{\partial \xi}\Bigg|_{\xi = y_i(k)}$$

[3]Of course, there exist other possible representations of σ_i which give rise to the heat equation, or e.g., to the Bessel equation (Carr *et al.*, 2020).

$$
+ \Phi_i(k) \frac{\partial G(y_i(\tau), \xi, \tau - k)}{\partial \xi}\bigg|_{\xi = y_{i+1}(k)} \Bigg\} dk, \quad \tau = \tau_i, \tag{19.16}
$$

$$
q_i(\tau, y_{i+1}(\tau)) = -\frac{\Phi_i(\tau)}{2\sigma_i^2(y_{i+1}(\tau))} + \int_0^\tau \Bigg\{ \Psi_i(k) \frac{\partial G(y_{i+1}(\tau), \xi, \tau - k)}{\partial \xi}\bigg|_{\xi = y_i(k)}
$$

$$
+ \Phi_i(k) \frac{\partial G(y_{i+1}(\tau), \xi, \tau - k)}{\partial \xi}\bigg|_{\xi = y_{i+1}(k)} \Bigg\} dk. \quad \tau = \tau_{i+1}.
$$

The gradients of $q_i(\tau, x)$ for the heat equation with $\sigma(x) = \sigma = const$ have been derived first in Lipton and Kaushansky (2020a,b), and later in Itkin and Muravey (2020a) by using a different method.[4] The result read

$$
\begin{aligned}
\frac{\partial q_i(\tau, x)}{\partial x}\bigg|_{x = y_i(\tau)} &= -\frac{\Psi_i(\tau)}{2\sigma^3} \left(\frac{1}{\sqrt{\pi \tau}} + \frac{y_i'(\tau)}{\sigma} \right) \\
&+ \int_0^\tau \frac{\Psi_i(k) e^{-\frac{(y_i(\tau) - y_i(k))^2}{4\sigma^2(\tau - k)}} - \Psi_i(\tau)}{4\sigma^3 \sqrt{\pi(\tau - k)^3}} dk \\
&- \int_0^\tau \Psi_i(k) \frac{(y_i(\tau) - y_i(k))^2 e^{-\frac{(y_i(\tau) - y_i(k))^2}{4\sigma^2(\tau - k)}}}{8\sigma^5 \sqrt{\pi(\tau - k)^5}} dk \\
&- \int_0^\tau \Phi_i(k) \frac{\partial^2 G(x, \xi, \sigma^2(\tau - k))}{\partial \xi \partial x}\bigg|_{\substack{\xi = y_{i+1}(k) \\ x = y_i(\tau)}} dk, \quad \tau = \tau_i,
\end{aligned} \tag{19.17}
$$

$$
\begin{aligned}
\frac{\partial q_i(\tau, x)}{\partial x}\bigg|_{x = y_{i+1}(\tau)} &= -\frac{\Phi_i(\tau)}{2\sigma^3} \left(\frac{1}{\sqrt{\pi \tau}} - \frac{y_{i+1}'(\tau)}{\sigma} \right) \\
&- \int_0^\tau \Psi_i(k) \frac{\partial^2 G(x, \xi, \sigma^2(\tau - k))}{\partial \xi \partial x}\bigg|_{\substack{\xi = y_i(k) \\ x = y_{i+1}(\tau)}} dk \\
&+ \int_0^\tau \frac{\Phi_i(k) e^{-\frac{(y_{i+1}(\tau) - y_{i+1}(k))^2}{4\sigma^2(\tau - k)}} - \Phi_i(\tau)}{4\sigma^3 \sqrt{\pi(\tau - k)^3}} dk \\
&- \int_0^\tau \Phi_i(k) \frac{(y_{i+1}(\tau) - y_{i+1}(k))^2 e^{-\frac{(y_{i+1}(\tau) - y_{i+1}(k))^2}{4\sigma^2(\tau - k)}}}{8\sigma^5 \sqrt{\pi(\tau - k)^5}} dk, \quad \tau = \tau_{i+1},
\end{aligned}
$$

where $G(x, \xi, \tau)$ is given in Eq. (19.14).

[4] These results can be naturally generalized for the case $\sigma = \sigma(x)$.

19.2 Solving the ML heat equation via the GIT method

19.2.1 *Background*

An alternative method to construct the ML problem solution is generalized integral transform (GIT). The GIT method is used in physics (Kartashov, 1999, 2001) but was unknown in finance until its first use in Carr and Itkin (2021). The previously known solution to the heat equation, using the GIT method, was obtained only for the domain $S \in [0, y(t)]$. For other domains, the solution was unknown even in physics. Itkin and Muravey (2020b) were the first to construct the GIT solution for the domain $S \in [y(t), \infty)$. The latter technique was extended further for the CIR and CEV models (Carr *et al.*, 2020), the Black-Karasinski model (Itkin *et al.*, 2020), and finally for double barrier options in Itkin and Muravey (2020a). The latter problem deals with the spatial domain determined by two moving in time boundaries, and boundary conditions, which are arbitrary functions of time.

The GIT and HP methods are similar but have an essential difference. In the HP method, the solution is represented in the form of heat potential with the unknown potential density function $\Psi(\tau)$ which solves the corresponding Volterra equation of the second kind, see Section 19.1. In the GIT method, similar to, e.g., the Fourier method, we start with applying some integral transform to the PDE under consideration. The transform has to be such that the transformed equation with $x \mapsto p$ is solvable analytically in time. The second step is to construct an inverse transform, which could be computed analytically using the complex analysis. If this is possible, then the solution can be represented as an explicit integral of some kernel multiplied by the unknown function $\Omega(\tau)$. Hence, this looks pretty similar to the HP method. However, the function $\Omega(\tau)$ has now a transparent meaning — this is the gradient of the solution at the moving boundary. It turns out that this gradient also solves a Volterra equation of the second kind. As mentioned, explicit construction of such forward and inverse transforms is performed in Carr and Itkin (2021), Itkin and Muravey (2020b), Carr *et al.* (2020), Itkin *et al.* (2020) and Itkin and Muravey (2020a) for various models and spatial domains. Also, the authors show that the performance of both methods is the same. Both HP and GIT methods are faster than the finite-difference approach and provide higher accuracy.

As mentioned in Itkin and Muravey (2020a), it is not unreasonable to ask why we need two methods — the HP and GIT, which are used to solve the same problem and demonstrate the same performance. The answer is interesting. As shown in Carr *et al.* (2020), the GIT method produces very accurate results at high strikes and maturities (i.e., when the option price is relatively small), in contrast to the HP method, which struggles under these circumstances. One can verify this fact by looking at the exponents under the GIT solution integral *proportional* to the time τ. Contrary, when the price is higher (short maturities, low strikes), the GIT method is slightly less accurate than the HP method since the exponents in the HP solution integral are *inversely proportional* to τ.

Thus, the GIT and HP methods complement each other for the CIR and CEV models. For other models reducible to the heat equation, the same conclusion holds (see Itkin and Muravey, 2020b). This statement is true because GIT integrals contain the difference of two exponents, which becomes small at large τ. On the contrary, the HP exponent tends to one at large τ. Therefore, the convergence properties of the two methods are different at large τ, so they complement each other.

This situation is well known for the heat equation with constant coefficients (Lipton, 2001). There exist two representations of the solution: one — obtained by using the method of images, and the other one — by the Fourier series. Although both solutions are equal when considered as infinite series, their convergence properties are different.

19.2.2 *Solution of the heat equation*

To apply the GIT method to the solution of Eq. (19.1), we can use the results obtained in Itkin and Muravey (2020b). There it is assumed that $\mathcal{L}_i$ is the heat equation operator

$$\mathcal{L}_i = -\frac{\partial}{\partial \tau} + \frac{\partial^2}{\partial x^2}, \tag{19.18}$$

and then the solution of Eq. (19.1) can be represented in the form

$$\begin{aligned} u_i(\tau, x) = \sum_{n=-\infty}^{\infty} \Bigg\{ & \int_{y_i(0)}^{y_{i+1}(0)} u_i(0,\xi) \Upsilon_{n,i}(x, \tau \,|\, \xi, 0) d\xi \\ & + \int_0^\tau \left[\Omega_i(s) + \chi_i^+(s) y_{i+1}'(s)\right] \Upsilon_{n,i}(x, \tau | y_{i+1}(s), s) ds \\ & + \int_0^\tau \left[\Theta_i(s) - \chi_i^-(s) y_i'(s)\right] \Upsilon_{n,i}(x, \tau \,|\, y_i(s), s) ds \end{aligned} \tag{19.19}$$

$$
+\int_0^\tau \chi_i^-(s)\Lambda_{n,i}(x,\tau\,|\,y_i(s),s) - \chi_i^+(s)\Lambda_{n,i}(x,\tau\,|\,y_{i+1}(s),s)ds\Bigg\},
$$

$$
\Upsilon_{n,i}(x,\tau\,|\,\xi,s) = \frac{1}{2\sqrt{\pi(\tau-s)}}\left[e^{-\frac{(2nl_i(\tau)+x-\xi)^2}{4(\tau-s)}} - e^{-\frac{(2nl_i(\tau)+x+\xi-2y_i(\tau))^2}{4(\tau-s)}}\right],
$$

$$
\begin{aligned}
\Lambda_{n,i}(x,\tau\,|\,\xi,s) &= \frac{x-\xi+2nl_i(\tau)}{4\sqrt{\pi(\tau-s)^3}}e^{-\frac{(2nl_i(\tau)+x-\xi)^2}{4(\tau-s)}} \\
&+ \frac{x+\xi-2y_i(\tau)+2nl_i(\tau)}{4\sqrt{\pi(\tau-s)^3}}e^{-\frac{(2nl_i(\tau)+x+\xi-2y_i(\tau))^2}{4(\tau-s)}}.
\end{aligned}
$$

Here $\chi_i^-(\tau), \chi_i^+(\tau)$ are the boundary conditions at the left and right boundaries of the i-th interval, and

$$
\begin{aligned}
& l_i(\tau) = y_{i+1}(\tau) - y_i(\tau), \quad \tau = \tau_i, \qquad (19.20)\\
& \Omega_i(\tau) = -\left.\frac{\partial u_i(\tau,x)}{\partial x}\right|_{x=y_i(\tau)}, \qquad \Theta_i(\tau) = \left.\frac{\partial u_i(\tau,x)}{\partial x}\right|_{x=y_{i+1}(\tau)}.
\end{aligned}
$$

The functions $\Omega(\tau), \Theta(\tau)$ for the heat equation in Eq. (19.18) can be found by solving a system of the Volterra equations of the second kind. As applied to our problem for the i-th interval with $i \in [1, N]$, it reads (Itkin and Muravey, 2020a)

$$
\begin{aligned}
-\Omega_i(\tau) &= \int_{y_i(0)}^{y_{i+1}(0)} u(0,\xi)\upsilon_i^-(\tau\,|\,\xi,0)d\xi - \frac{\chi_i^-(\tau)}{\sqrt{\pi\tau}} \qquad (19.21)\\
&+ \int_0^\tau \frac{\chi_i^-(s)-\chi_i^-(\tau)}{2\sqrt{\pi(\tau-s)^3}}ds + \int_0^\tau \Big[\chi_i^-(s)d\left(\eta_i^-(\tau\,|\,y_i(s),s)\right) \\
&- \chi_i^+(s)d\left(\eta_i^-(\tau\,|\,y_{i+1}(s),s)\right)\Big] - \int_0^\tau \Omega_i(s)\frac{y_i(\tau)-y_i(s)}{2\sqrt{\pi(\tau-s)^3}}e^{-\frac{(y_i(\tau)-y_i(s))^2}{4(\tau-s)}}ds \\
&+ \int_0^\tau \left[\Theta_i(s)\upsilon_i^-(\tau\,|\,y_{i+1}(s),s) + \Omega_i(s)\upsilon_{0,i}^-(\tau\,|\,s)\right]ds \\
\Theta_i(\tau) &= \int_{y_i(0)}^{y_{i+1}(0)} u(0,\xi)\upsilon_i^+(\tau\,|\,\xi,0)d\xi + \frac{\chi_i^+(\tau)}{\sqrt{\pi\tau}} \\
&- \int_0^\tau \frac{\chi_i^+(s)-\chi_i^+(\tau)}{2\sqrt{\pi(\tau-s)^3}}ds + \int_0^\tau \Big[\chi_i^-(s)d\left(\eta_i^+(\tau\,|\,y_i(s),s)\right) \\
&- \chi_i^+(s)d\left(\eta_i^+(\tau\,|\,y_{i+1}(s),s)\right)\Big] \\
&- \int_0^\tau \Theta_i(s)\frac{y_{i+1}(\tau)-y_{i+1}(s)}{2\sqrt{\pi(\tau-s)^3}}e^{-\frac{(y_{i+1}(\tau)-y_{i+1}(s))^2}{4(\tau-s)}}ds \\
&+ \int_0^\tau \left[\Theta_i(s)\upsilon_{0,i}^+(\tau\,|\,s) + \Omega_i(s)\upsilon_i^+(\tau\,|\,y_i(s),s)\right]ds.
\end{aligned}
$$

Here the following notation is used

$$
\begin{aligned}
\eta_i^-(\tau \,|\, \xi, s) &= -\frac{\delta_{\xi, y_i(s)}}{\sqrt{\pi(\tau - s)}} + \frac{1}{\sqrt{\pi(\tau - s)}} \sum_{n=-\infty}^{\infty} e^{-\frac{(y_i(\tau) - \xi + 2nl_i(\tau))^2}{4(\tau - s)}}, \qquad (19.22)\\
\eta_i^+(\tau \,|\, \xi, s) &= -\frac{\delta_{\xi, y_{i+1}(s)}}{\sqrt{\pi(\tau - s)}} + \frac{1}{\sqrt{\pi(\tau - s)}} \sum_{n=-\infty}^{\infty} e^{-\frac{(y_i(\tau) - \xi + (2n+1)l_i(\tau))^2}{4(\tau - s)}},\\
\upsilon_i^-(\tau \,|\, \xi, s) &= -\sum_{n=-\infty}^{\infty} \frac{y_i(\tau) - \xi + 2nl_i(\tau)}{2\sqrt{\pi(\tau - s)^3}} e^{-\frac{(y_i(\tau) - \xi + 2nl_i(\tau))^2}{4(\tau - s)}},\\
\upsilon_i^+(\tau \,|\, \xi, s) &= -\sum_{n=-\infty}^{\infty} \frac{y_i(\tau) - \xi + (2n+1)l_i(\tau)}{2\sqrt{\pi(\tau - s)^3}} e^{-\frac{(y_i(\tau) - \xi + (2n+1)l_i(\tau))^2}{4(\tau - s)}},\\
\upsilon_{0,i}^-(\tau \,|\, s) &= \upsilon_i^-(\tau | y_i(s), s) + \frac{y_i(\tau) - y_i(s)}{2\sqrt{\pi(\tau - s)^3}} e^{-\frac{(y_i(\tau) - y_i(s))^2}{4(\tau - s)}},\\
\upsilon_{0,i}^+(\tau \,|\, s) &= \upsilon_i^+(\tau | y_{i+1}(s), s) + \frac{y_{i+1}(\tau) - y_{i+1}(s)}{2\sqrt{\pi(\tau - s)^3}} e^{-\frac{(y_{i+1}(\tau) - y_{i+1}(s))^2}{4(\tau - s)}},
\end{aligned}
$$

where $\delta_{\xi,x}$ is the Kronecker symbol. It is worth emphasizing that all summands in Eq. (19.21) are regular. The integrands in the first two lines have weak (integrable) singularities, while other summands are regular.

At the boundaries of the domain where the solution of our problem is defined, we have

$$
\chi_1^- = \chi^-, \qquad \chi_N^- = \chi^+, \qquad (19.23)
$$

where N is the number of intervals.

In Itkin and Muravey (2020a), an alternative system of the Volterra equations of the second kind is also obtained, which has the form of Eq. (19.21), but with a different definition of the coefficients. We have

$$
\begin{aligned}
\eta_i^-(\tau \,|\, \xi, s) &= -\frac{\delta_{\xi, y_i(s)}}{\sqrt{\pi(\tau - s)}} + \frac{1}{l_i(\tau)} \theta_3\left(\phi_i(\xi, y_i(\tau)), \omega_i\right), \qquad (19.24)\\
\eta_i^+(\tau \,|\, \xi, s) &= -\frac{\delta_{\xi, y_{i+1}(s)}}{\sqrt{\pi(\tau - s)}} + \frac{1}{l(\tau)} \theta_3\left(\phi_i(\xi + l_i(\tau), y_i(\tau)), \omega_i\right),\\
\upsilon_i^-(\tau \,|\, \xi, s) &= -\frac{\pi}{2l_i^2(\tau)} \theta_3'\left(\phi_i(\xi, y_i(\tau)), \omega_i\right),\\
\upsilon_i^+(\tau \,|\, \xi, s) &= -\frac{\pi}{2l_i^2(\tau)} \theta_3'\left(\phi_i(\xi + l_i(\tau), y_i(\tau)), \omega_i\right).
\end{aligned}
$$

Here $\theta_3(z,\omega)$ is the Jacobi theta function of the third kind (Mumford *et al.*, 1983), which is defined as follows:

$$\theta_3(z,\omega) = 1 + 2\sum_{n=1}^{\infty} \omega^{n^2} \cos(2nz), \tag{19.25}$$

Also, in Eq. (19.24) the following notation is used

$$\omega_i = e^{-\frac{\pi^2(\tau-s)}{l_i^2(\tau)}}, \qquad \phi_i(x,\xi) = \frac{\pi(x-\xi)}{2l_i(\tau)}, \tag{19.26}$$

$$\frac{\partial\theta_3(z,\omega)}{\partial z} = \theta_3'(z,\omega) = -4\sum_{n=1}^{\infty} n\omega^{n^2} \sin(2nz).$$

Formulas Eqs. (19.24) and (19.22) are complementary. Since the exponents in the definition of the theta functions in Eq. (19.24) are proportional to the difference $\tau - s$, the Fourier series Eq. (19.24) converge fast if $\tau - s$ is large. Contrary, the exponents in Eq. (19.22) are inversely proportional to $\tau - s$. Therefore, the series Eq. (19.22) converge fast if $\tau - s$ is small.

19.2.3 *Solution of Eq. (19.10) when σ is piecewise constant*

Here we assume that $\sigma_i(x) = \sigma_i$, $i = 1,\dots,N$, i.e., the volatility is a piecewise constant function of x. For instance, this is true for the problem described in Section 19.3.2. As shown there, the pricing PDE can be transformed to Eq. (19.9) instead of Eq. (19.68). According to the transformation in Eq. (19.69) the clock will run differently at each ML interval, which is inconvenient. Therefore, instead of a change of temporal variable, below we use a transformation of the spatial variable x. This transformation allows using the same time at each ML interval. To achieve this, we change the definition of $\tau(t)$ in Eq. (19.69) to

$$\tau = \frac{1}{2}\int_0^T e^{-2\int_0^s [r(s)-q(s)]\,dk}\,ds, \tag{19.27}$$

where $r(t), q(t)$ for the problem considered in Section 19.3.2 are the deterministic interest rate and continuous dividends. This converts the problem in Section 19.3.2 and the PDE Eq. (19.67) to

$$\begin{aligned}
&\frac{\partial U(\tau,x)}{\partial \tau} = \sigma^2(x)\frac{\partial^2 U(\tau,x)}{\partial x^2}, \\
&U(0,x) = U_0(x) = (x-S)^+, \\
&U(\tau,0) = 0, \qquad U(\tau,x)\Big|_{x\to\infty} = x - e^{-\int_0^T (r(s)-q(s))ds} S.
\end{aligned} \tag{19.28}$$

And, according to Eq. (19.66)

$$\sigma^2(T,K) = v_i(T), \quad K \in [K_i, K_{i+1}]. \tag{19.29}$$

The boundary and initial conditions in Eq. (19.28) are the direct translation of those conditions in Eqs. (19.63) and (19.64).

Again, as shown in Appendix 19.A, the problem in Eq. (19.28) can be transformed to

$$\begin{aligned}
&\frac{\partial U(\tau,\hat{x})}{\partial \tau} = \frac{\partial}{\partial \hat{x}}\left(\Xi^2(\hat{x})\frac{\partial U(\tau,\hat{x})}{\partial \hat{x}}\right), \qquad \hat{x} = \hat{x}(x), \\
&U(0,\hat{x}) = U_0(\hat{x}) = (x(\hat{x}) - S)^+, \\
&U(\tau,\hat{x}(0)) = 0, \qquad U(\tau, x(\hat{x}))\Big|_{x(\hat{x})\to\infty} = x(\hat{x}) - e^{-\int_0^T (r(s)-q(s))ds} S.
\end{aligned} \tag{19.30}$$

Also, since based on Eq. (19.69)

$$U(\tau, x(\hat{x})) = P(T,K)e^{-\int_0^T q(s)\,ds},$$

the continuity of the solution and its flux at all internal boundaries can be expressed as

$$\begin{aligned}
\chi_i^+(\tau) &= \chi_{i+1}^-(\tau), \\
\Xi_{i+1}^2 \Omega_{i+1}(\tau) &= -\Xi_i^2 \Theta_i(\tau).
\end{aligned} \tag{19.31}$$

To use the results of the previous Section, we proceed by applying the following transformation to Eq. (19.19)

$$\bar{x} = \Xi_i \hat{x}, \qquad \bar{y}(\tau) = \Xi_i y(\tau), \qquad \bar{\xi} = \Xi_i \xi, \qquad \bar{l}(\tau) \equiv \Xi_i l(\tau). \tag{19.32}$$

Note, that the last equality in Eq. (19.32) is actually the definition of $\bar{l}(\tau)$. Another complication which comes due to this transformation is that in new variables $\bar{x}$ the layers stop to be continuous. In other words, the upper boundary of the i-th layer $\bar{y}_i^+(\tau)$ and the lower boundary of the $(i+1)$-th layer $\bar{y}_{i+1}^-(\tau)$ are now not equal. Therefore, in what follows to avoid any confusion we will explicitly use this notation, i.e. the left and right boundaries of the i-th layer are denoted as $y_i^-(\tau)$ and $y_{i+1}^-(\tau)$.

Also, per these transformations, we have

$$\begin{aligned}
&u_i(0,\bar{\xi}) = u_i(0,\Xi_i,\xi), \qquad u_i(0,\Xi_i,\xi)d\xi = \frac{1}{\Xi_i} u_i(0,\bar{\xi})d\bar{\xi}, \\
&\int_{y_i(0)}^{y_{i+1}(0)} u(0,\xi)\upsilon_i^-(\tau\,|\,\xi,0)d\xi = \frac{1}{\Xi_i}\int_{\bar{y}_i(0)}^{\bar{y}_{i+1}(0)} u(0,\bar{\xi})\bar{\upsilon}_i^-(\tau\,|\,\bar{\xi},0)d\bar{\xi}, \\
&\bar{\upsilon}_i^-(\tau\,|\,\bar{\xi},0) = -\sum_{n=-\infty}^{\infty} \frac{\bar{y}_i^-(\tau) - \bar{\xi} + 2n\bar{l}_i(\tau)}{2\Xi_i\sqrt{\pi(\tau-s)^3}} e^{-\frac{(\bar{y}_i^-(\tau)-\bar{\xi}+2n\bar{l}_i(\tau))^2}{4\Xi_i^2(\tau-s)}}.
\end{aligned} \tag{19.33}$$

It is easy to check that this transformation leaves $\phi_i(\hat{x},\xi)$ and $\partial_{\hat{x}}(\phi_i(\hat{x},\xi))$ invariant, but with the new definition

$$\bar{\omega}_i = e^{-\frac{\pi^2 \Xi_i^2(\tau-s)}{l_i^2(\tau)}}.$$

Finally, let us redefine the partial derivatives

$$\bar{\Omega}_i = -\left.\frac{\partial u(\tau,\bar{x})}{\partial \bar{x}}\right|_{\bar{x}=\bar{y}_i^-(\tau)}, \qquad \bar{\Theta}_i = \left.\frac{\partial u(\tau,\bar{x})}{\partial \bar{x}}\right|_{\bar{x}=\bar{y}_i^+(\tau)} \tag{19.34}$$

instead of their definitions in Eq. (19.20), i.e., $\Omega_i = \Xi_i\bar{\Omega}_i$, $\Theta_i = \Xi_i\bar{\Theta}_i$. Then the continuity conditions in Eq. (19.31) change to

$$\begin{aligned} \bar{\chi}_i^+(\tau) &= \bar{\chi}_{i+1}^-(\tau), \\ \Xi_{i+1}^3\bar{\Omega}_{i+1}(\tau) &= -\Xi_i^3\bar{\Theta}_i(\tau). \end{aligned} \tag{19.35}$$

To simplify notation, we omit bars from all new variables assuming this doesn't bring any confusion. Then Eq. (19.19) transforms to

$$\begin{aligned} u_i(\tau,x) = \sum_{n=-\infty}^{\infty} &\left\{ \frac{1}{\Xi_i}\int_{y_i(0)}^{y_{i+1}(0)} u_i(0,\xi)\Upsilon_{n,i}(x,\tau\,|\,\xi,0)d\xi \right. \\ &+ \int_0^\tau \left[\frac{1}{\Xi_i}\Omega_i(s) + \frac{1}{\Xi_{i+1}}\chi_i^+(s)y_i^{+'}(s)\right]\Upsilon_{n,i}(x,\tau|y_i^+(s),s)ds, \\ &+ \frac{1}{\Xi_i}\int_0^\tau \left[\Theta_i(s) - \chi_i^-(s)y_i^{-'}(s)\right]\Upsilon_{n,i}(x,\tau\,|\,y_i^-(s),s)ds \\ &\left. + \int_0^\tau \chi_i^-(s)\Lambda_{n,i}(x,\tau\,|\,y_i^-(s),s) - \chi_i^+(s)\Lambda_{n,i}(x,\tau\,|\,y_i^+(s),s)ds \right\}, \end{aligned} \tag{19.36}$$

$$\Upsilon_{n,i}(x,\tau\,|\,\xi,s) = \frac{1}{2\sqrt{\pi(\tau-s)}}\left[e^{-\frac{(2nl_i(\tau)+x-\xi)^2}{4\Xi_i^2(\tau-s)}} - e^{-\frac{(2nl_i(\tau)+x+\xi-2y_i^-(\tau))^2}{4\Xi_i^2(\tau-s)}}\right],$$

$$\begin{aligned} \Lambda_{n,i}(x,\tau\,|\,\xi,s) &= \frac{x-\xi+2nl_i(\tau)}{4\Xi_i\sqrt{\pi(\tau-s)^3}}e^{-\frac{(2nl_i(\tau)+x-\xi)^2}{4\Xi_i^2(\tau-s)}} \\ &+ \frac{x+\xi-2y_i^-(\tau)+2nl_i(\tau)}{4\Xi_i\sqrt{\pi(\tau-s)^3}}e^{-\frac{(2nl_i(\tau)+x+\xi-2y_i^-(\tau))^2}{4\Xi_i^2(\tau-s)}}. \end{aligned}$$

By analogy, the modified Volterra equations can be obtained from Eq. (19.36).

In Eq. (19.36) the unknown variables are $[\chi_1^-(\tau),\chi_1^+(\tau),\Omega_1(\tau),\Theta_1(\tau),\ldots,\chi_N^-(\tau),\chi_N^+(\tau),\Omega_N(\tau),\Theta_N(\tau)]$, so that there are $4N$ unknowns in total. The boundary conditions in Eq. (19.23) and the continuity conditions in Eq. (19.31) reduce the number of unknown variables to $2N - 2$, because $\chi_i^+(\tau), \Theta_i(\tau)$ can be expressed via $\chi_i^-(\tau), \Omega_i(\tau)$ and substituted into Eq. (19.36). Thus, the GIT method provides a significant simplification of the system of Volterra equations as compared with the HP method.

19.3 Application to finance

In this section, we consider several models that are frequently used in mathematical finance. We provide a short description of each model and demonstrate how to reduce the corresponding pricing problem to the form suitable for solving it by the ML method.

19.3.1 *One-factor short-rate models*

As the first example, we consider one-factor short interest rate (IR) models. Although these models were developed a long time ago, they are still essential and widely used by practitioners. While one can price zero-coupon bonds (ZCB) and European options on the ZCB and swaptions for many of them analytically, this is not true for exotic options. For instance, pricing of barrier options when the barriers are time-dependent and could pay time-dependent rebates has to be done numerically. The same is true for American options.

However, as mentioned in the Introduction, one can find the solution to these problems semi-analytically using the HP and GIT methods for some one-factor models, including the time-dependent OU (Vasicek) model in Carr and Itkin (2021); Lipton and Kaushansky (2020a,b), for the Hull-White model in Itkin and Muravey (2020b), for the CEV and CIR models in Carr *et al.* (2020), and then in a general form for any model that can be reduced to the heat equation — in Itkin and Muravey (2020a). In other words, solving these problems doesn't require the ML method. Therefore, below we consider some other models for which the barrier pricing problems cannot be directly solved by the HP or GIT methods but can be solved by using the ML method.

19.3.1.1 *Pricing zero-coupon bonds and barrier options for the Black-Karasinski and similar models*

The Black-Karasinski (BK) model was introduced in Black and Karasinski (1991), see also Brigo and Mercurio (2006) for a more detailed discussion and also Chapter 2 of this book. The BK model is widely used by practitioners for modeling interest rates and credit and is also known in commodities as the Schwartz one-factor model. The BK model is attractive because it is relatively simple, guarantees non-negativity of the prices (which could be a bad feature in the current environment). It could also be calibrated to the given term-structure of interest rates and the prices or implied volatilities of caps, floors, or European swaptions since the mean-reversion level and volatility are functions of time. However, in the BK model, the price $P(t,T)$ of a ZCB with the maturity T is not known in closed form since this model is not affine. Multiple good approximations have been developed in the literature using asymptotic expansions of various flavors (see, e.g., Antonov and Spector, 2011; Capriotti and Stehlikova, 2014; Horvath *et al.*, 2017), and also survey in Turfus (2020). Also, for exotic options, e.g., liquid barrier options, these prices are not known yet in closed form. Therefore, various numerical methods are used to obtain them.

Here we describe how one can reduce the pricing problem for the ZCB to the ML heat equation. Since this problem is defined at a semi-infinite domain, the corresponding ML heat equation is also defined at a semi-infinite interval. Thus, the number of layers could be infinite. Therefore, truncation of the semi-infinite interval to a finite is needed. Of course, the impact of the remainder should be assessed appropriately.

Along the BK model lines, consider a model where the dynamics of the underlying stochastic variable z_t is the OU process defined in Eq. (2.31). We assume that the interest rate r_t is some deterministic function of z_t

$$r_t = s(t) + f(t, z_t), \qquad z_0 = 0. \tag{19.37}$$

In particular, according to Eq. (2.31) for the BK model we have $f(t, z_t) = Re^{z_t}$, and so $R = r_0 - s_0$.

In terms of z, the corresponding PDE for the ZCB price $F(t,r)$ in Eq. (19.46) and for the option price $C(t,r)$ in Eq. (19.49) reads

$$0 = \frac{\partial V}{\partial t} + \frac{1}{2}\sigma^2(t)\frac{\partial^2 V}{\partial z^2} + \kappa(t)[\theta(t) - z]\frac{\partial V}{\partial z} - [s(t) + f(t,z)]V, \tag{19.38}$$

where $V = V(t,z)$ is either $F(t,z)$ or $C(t,z)$. This equation should be solved subject to the terminal and boundary conditions. For the ZCB price they are given in Eqs. (19.47) and (19.48), and for the Down-and-out barrier Call option price — in Eq. (19.50) and in Eqs. (19.52) and (19.53). Note, that solving Eq. (19.38) for $F(t,z)$ assumes that $z \in \left(f^{-1}(t,-\infty), f^{-1}(t,\infty)\right)$, while for $C(t,z)$ the domain of definition is $z \in [L_z(t), f^{-1}(t,\infty))$, where $L_z(t) = f^{-1}(t, L(t))$, and $f^{-1}(t,r)$ is the inverse function.

To apply the ML method to Eq. (19.38), for instance, when solving the barrier option pricing problem, we truncate the interval $[L(t),\infty)$ from above at $z = z_{\max}$ to make it $[L_z(t), z_{\max}]$. The reason this is possible lies in the fact that when z increases, the ZCB price tends to zero based on the boundary condition. Therefore, the Call option price in this limit vanishes as well. Thus, the contribution of the region $[z_{\max}, f^{-1}(t,\infty))$ to the Call option price becomes negligible.[5]

Now we split the interval $[L_z(t), z_{\max}]$ into $N > 0$ sub-intervals, and at every interval $[z_i, z_{i+1}]$, $i = 1,\dots,N$ assume that $f(t,z) = a_i(t) + b_i(t)z$. Accordingly, at every interval i, $i = 1,\dots,N$ the PDE Eq. (19.38) takes the form

$$0 = \frac{\partial V}{\partial t} + \frac{1}{2}\sigma^2(t)\frac{\partial^2 V}{\partial z^2} + \kappa(t)[\theta(t) - z]\frac{\partial V}{\partial z} - [s(t) + a_i(t) + b_i(t)z]V. \tag{19.39}$$

This PDE can be transformed to the heat equation

$$\frac{\partial U}{\partial \tau} = \frac{\partial^2 U}{\partial x^2}, \tag{19.40}$$

by the change of variables (Polyanin, 2002; Itkin and Muravey, 2020b; Lipton and Kaushansky, 2020b)

$$V(t,z) = \exp[\alpha_i(t)z + \beta_i(t)]U(\tau,x), \quad \tau = \phi(t), \quad x = z\psi(t) + \varrho(t), \tag{19.41}$$

where

$$\begin{aligned}
\psi(t) &= C_1 \exp\left(\int_S^t \kappa(q)dq\right), \\
\phi(t) &= \frac{1}{2}\int_t^S \sigma^2(q)\psi^2(q)dq + C_2, \\
\alpha_i(t) &= \psi(t)\int_S^t \frac{b_i(q)}{\psi(q)}dq + C_3\psi(t), \\
\varrho_i(t) &= -\int_S^t \left[\kappa(q)\theta(q) + \sigma^2(q)\alpha_i(q)\right]\psi(q)dq + C_5, \\
\beta_i(t) &= -\frac{1}{2}\int_S^t \alpha_i(q)\left[2\kappa(q)\theta(q) + \sigma^2(q)\alpha_i(q)\right]dq + \int_S^t [s(q) + a_i(q)]dq + C_4,
\end{aligned} \tag{19.42}$$

where $C_1,\dots,C_5$ are some constants. In our case we can choose $C_1 = 1$, $C_2 = C_3 = C_4 = C_5 = 0$.

[5] For some choices of the functions $f(t,z)$ the value $f^{-1}(t,\infty)$ could be finite which eliminates the need for truncation.

One of the advantages of such an approach is that the new time τ doesn't depend on the specific interval i, i.e., the time τ runs in sync for all intervals $[z_i, z_{i+1}]$, $i = 1, \dots, N$.

Thus, the main trick here is in using the approximation $f(t, z) = a_i(t) + b_i(t)z$. This approximation provides the second order of accuracy in the length of the interval (similar to the finite-difference method of the second order), and allows reduction of the PDE at each interval to the heat equation (while the original PDE doesn't hold this property).

At the end of this Section, note that Eq. (19.39), if used for pricing ZCB, doesn't need the boundary condition at the left boundary $z \to -\infty$, as this is discussed in Itkin and Muravey (2020b) with a reference to Fichera theory (Oleinik and Radkevich, 1973). However, the price of the ZCB at some fixed left boundary $z_{\min}$, i.e., $V(t, z_{\min})$ can be found having in mind that the transformed PDE in Eq. (19.39) is affine, which yields

$$V(z, t, S) = A(t, S)e^{B(t,S)Re^z}. \tag{19.43}$$

With allowance for the terminal condition in Eq. (19.47), the solution reads (Itkin and Muravey, 2020b)

$$\begin{aligned} B(t, S) &= e^{\int_0^t \kappa(m)\, dm} \int_S^t b_i(m)e^{-\int_0^m \kappa(q)\, dq}\, dm, \qquad (19.44)\\ A(t, S) &= e^{\int_S^t \left(a_i(m)+s(m)-\frac{1}{2}B(m,S)\left(2\theta(m)\kappa(m)+B(m,S)\sigma^2(m)\right)\right)dm}. \end{aligned}$$

It can be seen that $B(t, S) < 0$ if $t < S$. Therefore, $F(r, t, S) \to 0$ when $z \to \infty$.

19.3.1.2 *The modified BK (Verhulst) model*

Since the BK model is not fully tractable, in Itkin *et al.* (2020) we introduced a slightly modified version of the model as follows

$$\begin{aligned} dz_t &= k(t)[\bar\theta(t) - e^{z_t}]dt + \sigma(t)dW_t, \qquad (19.45)\\ r_t &= s(t) + Re^{z_t}, \qquad z_0 = 0, \qquad R = r_0 - s(0). \end{aligned}$$

It can be seen, that at small t $|z_t| \ll 1$, and so choosing $\bar\theta(t) = 1 + \theta(t)$ replicates the BK model in the linear approximation on z_t. Similarly, the choice $\bar\theta(t) = e^{\theta(t)}$ replicates the BK model at z_t close the mean-reversion level $\theta(t)$. Thus, this model acquires the properties of the BK model while is a bit more tractable as this will be seen below.

It is worth noting that if by using Itô's lemma we re-write Eq. (19.45) for the stochastic variable r_t, the resulting dynamics can be recognized as the stochastic Verhulst or stochastic logistic model, which are well-known in the population dynamics and epidemiology (see, e.g., Verhulst, 1838; Bacaer, 2011; Giet *et al.*, 2015) and references therein. For more information, see Itkin *et al.* (2020).

By the Itô's lemma and the Feynman–Kac formula any contingent claim written on the r_t as the underlying (for instance, price $F(\bar{r}, t, S)$ of a Zero-coupon bond (ZCB) with maturity S) solves the following partial differential equation

$$0 = \frac{\partial F}{\partial t} + \frac{1}{2}\sigma^2(t)\bar{r}^2\frac{\partial^2 F}{\partial \bar{r}^2} + \kappa(t)\bar{r}[\tilde{\theta}(t) - \bar{r}]\frac{\partial F}{\partial \bar{r}} - (s(t) + R\bar{r})F, \quad (19.46)$$

$$\bar{r}_t = \frac{r_t - s(t)}{r_0 - s(0)} = e^{z_t}, \qquad \tilde{\theta}(t) = \bar{\theta}(t) + \frac{1}{2}\sigma^2(t).$$

This equation should be solved subject to the terminal condition

$$F(\bar{r}, S, S) = 1, \quad (19.47)$$

and the boundary condition

$$F(\bar{r}, t, S)\Big|_{\bar{r}\to\infty} = 0, \quad (19.48)$$

see, e.g., Andersen and Piterbarg (2010).

In the sequel we will also consider a Down-and-Out barrier Call option written on the ZCB. It is known (Andersen and Piterbarg, 2010), that under a risk-neutral measure the option price $C(t, \bar{r})$ solves the same PDE as in Eq. (19.46),

$$0 = \frac{\partial C}{\partial t} + \frac{1}{2}\sigma^2(t)\bar{r}^2\frac{\partial^2 C}{\partial \bar{r}^2} + \kappa(t)\bar{r}[\tilde{\theta}(t) - \bar{r}]\frac{\partial C}{\partial \bar{r}} - (s(t) + R\bar{r})C. \quad (19.49)$$

The terminal condition at the option maturity $T \le S$ for this PDE reads

$$C(T, \bar{r}) = (F(\bar{r}, T, S) - K)^+, \quad (19.50)$$

where K is the option strike.

By a standard contract, the lower barrier $L_F(t)$ (which we assume to be time dependent as well) is set on the ZCB price, and not on the underlying interest rate $\bar{r}$. This means that it can be written in the form

$$C(t, \bar{r}) = 0 \quad \text{if } F(\bar{r}, t, S) = L_F(t). \quad (19.51)$$

This condition can be translated into the $\bar{r}$ domain by solving the equation

$$F(\bar{r}, t, S) = L_F(t),$$

with respect to $\bar{r}$. Denoting the solution of this equation as $L(t)$ we find that Eq. (19.51) in the $\bar{r}$ domain reads

$$C(t, L(t)) = 0. \quad (19.52)$$

The second boundary can be naturally set at $\bar{r} \to \infty$. As at $\bar{r} \to \infty$ the ZCB price tends to zero, the Call option price also vanishes in this limit. This yields

$$C(t, \bar{r})\Big|_{\bar{r}\to\infty} = 0. \tag{19.53}$$

Accordingly, Eq. (19.49) has to be solve at $\bar{r} \in [L(t), \infty)$.

19.3.1.3 *Pricing barrier options in the Verhulst model*

As we have already mentioned, in contrast to other similar one-factor models like the time-dependent Ornstein-Uhlenbeck, Hull-White, CIR and CEV models which have been considered in Carr and Itkin (2021), Itkin and Muravey (2020b) and Carr *et al.* (2020), the solution of the pricing problem for the BK model is not known in closed form.[6] Therefore, we propose an approximation that gives rise to a semi-analytical solution for the barrier Call option price. This approximation is inspired by the ML heat equations which are discussed in Section 19.

Since our problem in Eq. (19.49) is defined at the semi-infinite domain $\bar{r} \in [L(t), \infty)$, using the ML approximation is time-consuming, as we need to split this semi-infinite interval into a fixed number of sub-intervals. Therefore, it is feasible first to make a change of variables

$$C(t, \bar{r}) = V(t, x) e^{\int_0^t s(k) dk}, \quad x = \frac{a(t)}{\bar{r}}, \quad a(t) = e^{\int_0^t \left(\kappa(m)\tilde{\theta}(m) - \sigma^2(m)\right) dm}, \tag{19.54}$$

so the problem to solve in new variables reads

$$0 = \frac{\partial V}{\partial t} + \frac{1}{2}\sigma^2(t) x^2 \frac{\partial^2 V}{\partial x^2} + a(t)\kappa(t)\frac{\partial V}{\partial x} - Ra(t)\frac{V}{x}. \tag{19.55}$$

Thus, now our problem is defined at a fixed domain $x \in [0, 1/L(t)]$, where the upper boundary is time dependent. Accordingly, in the new variables the Down-and-Out barrier Call option becomes the Up-and-Out barrier Call.

Doing the second change of the dependent variable

$$V(t, x) = u(t, x) e^{d(t)/x}, \; d(t) = Re^{-\int_0^t \sigma^2(m) dm} \int_0^t e^{\int_0^y \kappa(m)\tilde{\theta}(m) dm} dy \tag{19.56}$$

yields the equation

$$0 = \frac{\partial u}{\partial t} + \frac{1}{2}\sigma^2(t) x^2 \frac{\partial^2 u}{\partial x^2} + g(t)\frac{\partial u}{\partial x} - f(t)\frac{u}{x^2}, \tag{19.57}$$

[6]Some approaches for doing that are discussed in Itkin *et al.* (2020).

$$f(t) = \frac{1}{2}d(t)\left(2a(t)\kappa(t) - d(t)\sigma^2(t)\right), \qquad g(t) = a(t)\kappa(t) - d(t)\sigma^2(t).$$

Accordingly, in the new variables the initial and boundary conditions read

$$u(T,x) = \exp\left(-\frac{d(T)}{x} - \int_0^T s(k)dk\right)(F(x,T,S) - K)^+, \tag{19.58}$$
$$u(t,0) = 0, \qquad u(t, y(t)) = 0, \quad y(t) = a(t)/L(t).$$

The problem in Eqs. (19.57) and (19.58) cannot be solved in closed form. Therefore, we proceed by borrowing the idea from the ML approach in physics which is described in Sections 19.1 and 19.2. This approach implies that the interval $x \in [0, y(t)]$ we approximate the function $\zeta(x) = x^2$ by using a piecewise constant approximation. In more detail, we split the interval $[0, y(t)]$ into $N > 0$ sub-intervals, and at every interval $[x_i, x_{i+1}]$, $i = 1, \dots, N$ assume that $x^2 \approx \nu_i(t)$.[7] For instance, one can choose the middle value of the function $\zeta(x)$ at each sub-interval, so

$$\nu_i(t) = y^2(t)\frac{(i+1/2)^2}{N^2}.$$

With allowance for this approximation, at every i-th interval Eq. (19.57) takes the form

$$0 = \frac{\partial u}{\partial t} + \frac{1}{2}\sigma^2(t)\nu_i(t)\frac{\partial^2 u}{\partial x^2} + g(t)\frac{\partial u}{\partial x} - \frac{f(t)}{\nu_i(t)}u. \tag{19.59}$$

The Eq. (19.59) can be transformed to the heat equation

$$\frac{\partial U}{\partial \tau} = \frac{\partial^2 U}{\partial \varsigma^2}, \tag{19.60}$$

using the transformation (Polyanin, 2002)

$$u(t,x) = U(\tau,\varsigma)\exp\left(-\int_0^t \frac{f(t)}{\nu_i(t)}\right), \quad l = x - \int_0^t g(k)dk, \tag{19.61}$$
$$\tau = \frac{1}{2}\int_t^T \sigma^2(k)\nu_i(k)dk.$$

Note, that under this approximation the new time τ also becomes a function of the interval i.

[7]Since the upper boundary of the whole interval $y(t)$ is the function of time, we need to put $\nu_i = \nu_i(t)$.

19.3.2 *Local volatility and Dupire's equation*

Calibration of the local volatility model (constructed by using a one-factor Geometric Brownian motion process) to a given set of option prices is a classical problem of mathematical finance. It was considered in multiple papers, and various solutions were proposed (see, e.g., a survey in Itkin, 2020; Itkin and Lipton, 2018) and references therein. In particular, in Itkin and Lipton (2018) an analytical approach to solving the calibration problem is developed. This approach extends the method in Lipton and Sepp (2011b) by replacing a piecewise constant local variance construction with a piecewise linear one and allowing non-zero interest rates and dividend yields. This approach remains analytically tractable as it combines the Laplace transform in time with an analytical solution of the resulting spatial equations in terms of Kummer's degenerate hypergeometric functions.

A similar problem could be formulated not just for the Black-Scholes model but also for other models. For instance, in Carr and Itkin (2020, 2019) two extensions of the Local Variance Gamma model proposed initially in Carr and Nadtochiy (2017) were developed. The first new model (ELVG) considers a Gamma time-changed *arithmetic* Brownian motion with drift and the local variance to be a piecewise linear function of the strike. The second model (GLVG) is a *geometric* version of the ELVG with drift. It also treats various cases by introducing three *piecewise linear* models: the local variance as a function of strike, the local variance as a function of log-strike, and the local volatility as a function of strike (so, the local variance is a *piecewise quadratic* function of the strike). For all these extensions, the authors derive an ordinary differential equation for the option price, which plays the role of Dupire's equation for the standard local volatility model. Moreover, it can be solved in closed form.

In Itkin and Lipton (2018) and Carr and Itkin (2020, 2019) all models were calibrated to the market quotes term-by-term. Therefore, various types of no-arbitrage interpolation were proposed to guarantee no-arbitrage while keeping the model analytically tractable on the other hand; further details are given in Itkin (2020).

Two advantages of the semi-analytical approach, which are essential for calibration the model, should be emphasized. First, the option prices can be found analytically in a semi closed form. Here "semi" means that the analytic solution requires an additional inverse transform to be applied to get the final prices (see Itkin and Lipton, 2018). However, in Carr and Itkin (2020, 2019) since the Dupire-like equation is an ODE and not a

PDE, this step is eliminated. Nevertheless, all these models are calibrated term-by-term.

This idea can be extended by constructing a semi-analytical solution of the ML heat equation, which is analytic in time. Thus, the term-by-term calibration could be eliminated, and the quotes for all strikes and maturities can be used simultaneously. Therefore, this approach allows a further acceleration of the calibration process.

For brevity, let us consider European options, for instance, a Put option on a stock. It is well-known that the price $P(T, K)$ of the option written on the underlying stock price S_t as a function of the option maturity T and strike K solves Dupire's equation (Dupire, 1994)

$$\frac{\partial P}{\partial T} = \frac{1}{2}\sigma^2(T, K)\frac{\partial^2 P}{\partial K^2} - (r - q)K\frac{\partial P}{\partial K} - q(T)P. \tag{19.62}$$

with $\sigma = \sigma(t, S)$ being the local volatility, and $q(t)$ is the dividend yield. This PDE should be solved subject to the initial condition at $T = 0$

$$P(0, K) = (K - S)^+, \tag{19.63}$$

and natural boundary conditions for the put option price that read (Hull, 2011)

$$\begin{aligned} &P(T, K) = 0, && K \to 0, \\ &P(T, K) = D(t, T)(K - S) \approx D(t, T)K, && K \to \infty, \end{aligned} \tag{19.64}$$

where the discount factor $D(t, T)$ is defined as

$$D(t, T) = \exp\left(-\int_t^T r(k)dk\right). \tag{19.65}$$

To proceed further, we use the idea in Lipton and Sepp (2011b), Itkin and Lipton (2018) and approximate the local variance using some piecewise approximation in the strike space. However, in contrast to Lipton and Sepp (2011b) and Itkin and Lipton (2018) we make this approximation a function of time. Further, suppose that for each trading maturity T_j, $j \in [1, M]$ the market quotes are provided at a set of strikes $K_{i,j}$, $i = 1, \ldots, n_j$ where the strikes are assumed to be sorted in the increasing order. Let us construct a finite grid in the strike space $\mathbf{G}(K) : K \in [\min(K_{i,j}), \max(K_{i,j})]$, $j = 1, \ldots, M$, $i = 1, \ldots, n_j$, by splitting the whole interval $K \in [\min(K_{i,j}), \max(K_{i,j})]$ into n sub-intervals. At every interval $[K_i, K_{i+1}]$, $i = 1, \ldots, N$, we approximate the local variance function $\sigma^2(T, K)$ by a piecewise constant function in K as follows:

$$\sigma^2(T, K) = v_i(T), \quad K \in [K_i, K_{i+1}]. \tag{19.66}$$

This approximation is not continuous, so the local variance $\sigma^2(T,K)$ experience a finite jump at every point K_i. However, it is continuous in the maturity T.

Accordingly, at every interval $[K_i, K_{i+1}]$ Eq. (19.62), $V(T,K)$ takes the form

$$\frac{\partial P_i}{\partial T} = \frac{1}{2} v_i(T) \frac{\partial^2 P_i}{\partial K^2} - [r(T) - q(T)] \frac{\partial P_i}{\partial K} - q(T) P_i. \tag{19.67}$$

This equation can be transformed to the heat equation

$$\frac{\partial U_i}{\partial \tau} = \frac{\partial^2 U_i}{\partial x^2}, \tag{19.68}$$

by a change of variables

$$P_i(T,K) = U_i(\tau, x) e^{-\int_0^T (q(s) ds}, \quad x = e^{-\int_0^T (r(s) - q(s)\, ds} K, \tag{19.69}$$

$$\tau = \frac{1}{2} \int_0^T v_i(s) e^{-2\int_0^s (r(s) - q(s))\, dk}\, ds.$$

It is important to note that the new time τ runs differently at every interval $K \in [K_i, K_{i+1}]$ as it depends on the local variance value $v_i(s)$ at this interval.

19.4 Solution of the Volterra equations

An efficient solution of the derived systems of Volterra equations is a problem that requires some attention and extended description. Therefore, it will be published elsewhere. Instead, here we show that a particular choice of the internal boundaries can reduce this problem to a linear system with a tridiagonal matrix allowing the inverse Laplace transform. In this section, we explain this approach in detail.

We start with Eq. (19.21). Using the definitions of $\eta^\pm, \upsilon^\pm, \upsilon_0^\pm$ in Eq. (19.22), we observe that under all integrals in Eq. (19.21) the functions $f_i(s), \Theta_i(s), \Omega_i(s)$ are functions of s, while functions $\eta^\pm, \upsilon^\pm, \upsilon_0^\pm$ are functions of $t - s$ and $y_i(t) - y_i(s)$ and $z_i(t) - z_i(s)$. Recall that functions $y_1(t) = \chi^-$, $y_N(t) = \chi^+$ define the external boundaries of the computational domain, while functions $y_i(t)$, $i = 2, \ldots, N-1$ define the boundaries of the internal layers (the internal boundaries).

Since the internal boundaries are artificial, we can construct them as we wish. For instance, we can use polynomial functions, such as $y_i(s) = a_i s^2 + b_i s + c_i$, where a_i, b_i, c_i are some constants. Then $y_i(t) - y_i(s)$ can also be represented as a certain function $g(t-s)$. Indeed

$$y_i(t) - y_i(s) = -a_i(t-s)^2 + (b_i + 2a_i t)(t-s). \tag{19.70}$$

A similar representation can be obtained for $y_{i+1}(s) - y_i(s)$

$$\begin{aligned} y_{i+1}(s) - y_i(s) &= (a_{i+1} - a_i)s^2 + (b_{i+1} - b_i)s + (c_{i+1} - c_i) \\ &= A(t-s)^2 + B(t-s) + C, \\ A &= a_{i+1} - a_i, \ B = 2At + b_{i+1} - b_i, \\ C &= c_i - c_{i+1} + t[b_i - b_{i+1} + (a_i - a_{i+1})t]. \end{aligned} \tag{19.71}$$

Same can be done for a polynomial of any degree.

All coefficients a_i, b_i, c_i can be precomputed given the external boundaries. An example of this construction is given in Fig. 19.4.1.

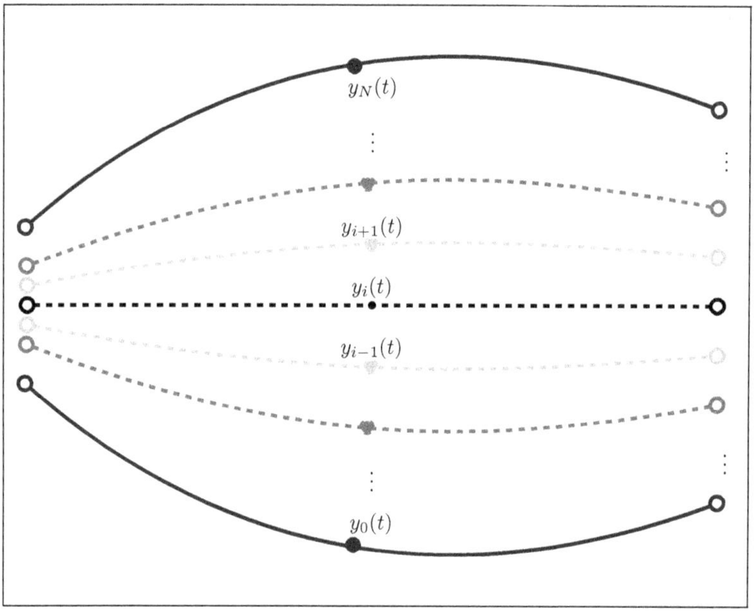

Figure 19.4.1: Internal layers constructed for the given external boundaries $y_0(t)$ and $y_N(t)$, and the number of layers N, by using 3 points for each boundary $y_i(t)$ and polynomial curves.

In more detail, suppose that for the given functions $y_1(t) = \chi^-(t)$, $y_{N+1} = \chi^+(t)$, we want to have N layers. We use N uniform nodes to split the interval $[\chi^-(0), \chi^+(0)]$ into N subintervals. We do the same for the interval $[\chi^-(t), \chi^+(t)]$. If the boundaries $\chi^-(t), \chi^+(t)$ are smooth enough, we can connect points $y_i(0), y_i(t)$, $i = 2, \dots, N$ by straight lines in such a way that all boundaries don't cross each other. Suppose this is not possible because the external boundaries are too convex or concave. In that case, we can find some $s = \tau$ where the distance between the external boundaries is minimal and put N nodes there. Then, we can connect all points $y_i(0), y_i(\tau), y_i(t)$ by parabolas, and again check that all the boundaries don't cross each other. We can continue this process by using polynomials of a higher degree to provide the final representation of the boundaries.

Thus, we can find a polynomial of the necessary degree to guarantee that all boundaries don't cross each other unless the external curves have a very peculiar shape, which does not happen in the context of financial applications. Otherwise, we need to use a general approach to the computation of the integrals in Eq. (19.21), which will be published elsewhere.

Provided that all the boundaries are constructed in such a way, one can observe that all integrals in Eq. (19.21) are convolutions. Therefore, we can apply the Laplace transform $\mathfrak{L}(f|s, \lambda)$

$$\mathfrak{L}(f|s, \lambda) = \int_0^\infty e^{-\lambda s} f(s) ds \tag{19.72}$$

to both parts of each equation in Eq. (19.21). Taking into account that

$$\mathfrak{L}(f * g) = \mathfrak{L}(f)\mathfrak{L}(g), \qquad \mathfrak{L}(f') = \lambda\mathfrak{L}(f) - f(0), \tag{19.73}$$

and obtain from Eq. (19.21) a linear system of equations for $\mathfrak{L}(\chi_i^+), \mathfrak{L}(\chi_i^-), \mathfrak{L}(\Omega_i), \mathfrak{L}(\Theta_i)$. Using the conditions at the internal boundaries (such as in Eq. (19.35)), this system can be reduced to a linear system for only $\mathfrak{L}(\chi_i^-), \mathfrak{L}(\Omega_i)$. One can check that the resulting system is block-diagonal, with all blocks being tridiagonal matrices. Once this system is solved, the functions $\chi_i^-(t), \Omega_i(t)$ is found by using the inverse Laplace transform. Consider the heat equation in a strip with a piecewise constant thermal conductivity coefficient to illustrate our approach.

19.4.1 *The heat equation in a strip*

Consider the following problem for the heat equation with a piecewise constant thermal conductivity coefficient (the ML problem):

$$\begin{aligned} &\frac{\partial}{\partial x}\left(\sigma^2(x)\frac{\partial U}{\partial x}\right) = \frac{\partial U}{\partial t}, \qquad (x,t) \in [y_0, y_N] \times \mathbb{R}_+, \\ &\qquad U(t, y_0) = 0, \qquad U(t, y_N) = 0. \\ &\qquad U(0, x) = \delta(x - x_0). \end{aligned} \tag{19.74}$$

Here the thermal conductivity coefficient $\sigma(x)$ is a piecewise constant function of x, which changes from layer to layer

$$\sigma(x) = \sum_{i=1}^{N} \mathbf{1}(y_{i-1} < x \le y_i)\sigma_i, \tag{19.75}$$

$$y_0 < y_1 < y_2 < \ldots < y_i < \ldots < y_N.$$

and $\delta(x)$ denotes the Dirac delta function. As before y_i are the boundaries of the layers in x space. Without loss of generality, we assume that $x_0 \in [y_{j-1}, y_j),\ 1 < j < N$.

Due to the initial condition in Eq. (19.74) the solution of this problem is Green's function for Eq. (19.74).

We represent the solution $U(t,x)$ in the form

$$U(t,x) = \sum_{i=1}^{N} \mathbf{1}(y_{i-1} < x \le y_i)[U_i(t,x) + H_i(t,x)], \tag{19.76}$$

where the functions $U_i(t,x)$ and $H_i(t,x)$ solve the following problems

$$\begin{aligned}
&\frac{\partial}{\partial x}\left(\sigma_i^2 \frac{\partial U_i}{\partial x}\right) = \frac{\partial U_i}{\partial t}, \qquad (x,t) \in (y_{i-1}, y_i] \times \mathbb{R}_+, \\
&\lim_{x \to y_{i-1}} U_i(t,x) = \chi_i^-(t), \qquad U_i(t, y_i) = \chi_i^+(t), \\
&\qquad U(0,x) = 0,
\end{aligned} \tag{19.77}$$

and

$$\begin{aligned}
&\frac{\partial}{\partial x}\left(\sigma_i^2 \frac{\partial H_i}{\partial x}\right) = \frac{\partial H_i}{\partial t}, \qquad (x,t) \in (y_{i-1}, y_i] \times \mathbb{R}_+, \\
&\lim_{x \to y_{i-1}} H_i(t,x) = 0, \qquad H_i(t, y_i) = 0, \\
&\qquad H(0,x) = \delta(x - x_0).
\end{aligned} \tag{19.78}$$

A well-known physical argument shows that the solution and its flux must be continuous at the layers' boundaries. The first condition yields

$$\begin{aligned}
&U_1(t, y_0) = 0, \\
&\lim_{x \to y_i} U_i(t,x) = U_{i+1}(t, y_i), \quad i = 1 \ldots N-1, \\
&\lim_{x \to y_N} U_N(t,x) = 0.
\end{aligned} \tag{19.79}$$

According to Itkin and Muravey (2020a), the function $H(t,x) \neq 0$ only at that interval which contains the point x_0, i.e., $[y_{j-1}, y_j)$. Therefore, the flux continuity conditions could be written as

$$\lim_{x \to y_i} \sigma_i^2 \frac{\partial U_i}{\partial x}(t, y_i) = \sigma_{i+1}^2 \frac{\partial U_{i+1}}{\partial x}(t, y_i), \qquad i \neq j-1, j,$$

$$\lim_{x \to y_{j-1}} \sigma_{j-1}^2 \frac{\partial U_{j-1}}{\partial x}(t, y_{j-1}) = \sigma_j^2 \frac{\partial U_j}{\partial x}(t, y_{j-1}) + \sigma_j^2 \frac{\partial H_j}{\partial x}(t, y_{j-1}),$$

$$\lim_{x \to y_j} \left[\sigma_j^2 \frac{\partial U_j}{\partial x}(t, y_j) + \sigma_j^2 \frac{\partial H_j}{\partial x}(t, y_j) \right] = \sigma_{j+1}^2 \frac{\partial U_{j+1}}{\partial x}(t, y_j). \tag{19.80}$$

It follows from Eq. (19.79) that

$$\chi_1^-(t) = \chi_N^+(t) = 0, \qquad \chi_i^+(t) = \chi_{i+1}^-(t), \quad i = 1, \dots, N-1. \tag{19.81}$$

To simplify the notation let us introduce new functions $f_i(t)$, such as

$$\chi_i^+(t) = \chi_{i+1}^-(t) = f_i(t), \quad i = 0, \dots, N, \tag{19.82}$$

so obviously $f_0(\tau) = f_{N+1}(\tau) = 0$. Using Eq. (19.21) (or Eq. (3.33) in Itkin and Muravey (2020a)), one can get an explicit representation for the derivatives of $U(t, x)$ at each interval

$$\begin{aligned}
\left.\frac{\partial U_i}{\partial x}\right|_{x=y_i} &= \frac{f_i(t)}{\sigma_i \sqrt{\pi t}} - \int_0^t \frac{f_i(s) - f_i(t)}{2\sigma_i \sqrt{\pi (t-s)^3}} ds \\
&\quad + \int_0^t \left[f_{i-1}(s) \lambda_i^+(t|y_{i-1}, s) - f_i(s) \lambda_{0,i}^+(t|y_i, s) \right] ds, \\
\left.\frac{\partial U_{i+1}}{\partial x}\right|_{x=y_i} &= -\frac{f_i(t)}{\sigma_{i+1} \sqrt{\pi t}} + \int_0^t \frac{f_i(s) - f_i(t)}{2\sigma_{i+1} \sqrt{\pi (t-s)^3}} ds \\
&\quad + \int_0^t \left[f_i(s) \lambda_{0,i+1}^-(t|y_i, s) - f_{i+1}(s) \lambda_{i+1}^-(t|y_{i+1}, s) \right] ds.
\end{aligned} \tag{19.83}$$

Here

$$\begin{aligned}
\lambda_i^-(t \,|\, \xi, s) &= \sum_{n=-\infty}^{\infty} \frac{e^{-\frac{(y_i - \xi + 2nl_i)^2}{4\sigma_i^2 (t-s)}}}{2\sigma_i^3 \sqrt{\pi (t-s)^3}} \left[1 - \frac{(y_i - \xi + 2nl_i)^2}{2\sigma_i^2 (t-s)} \right] \\
&= -\frac{\pi^2}{4 l_i^3} \theta_3'' \left[\frac{\pi (y_i - \xi)}{2 l_i}, q_i(s) \right],
\end{aligned} \tag{19.84}$$

$$\begin{aligned}
\lambda_i^+(\tau \,|\, \xi, s) &= \sum_{n=-\infty}^{\infty} \frac{e^{-\frac{(y_i - \xi + (2n+1)l_i)^2}{4\sigma_i^2 (t-s)}}}{2\sigma_i^3 \sqrt{\pi (t-s)^3}} \left[1 - \frac{(y_i - \xi + (2n+1)l_i)^2}{2\sigma_i^2 (t-s)} \right] \\
&= -\frac{\pi^2}{4 l_i^3} \theta_3'' \left[\frac{\pi (y_i + l_i - \xi)}{2 l_i}, q_i(s) \right],
\end{aligned}$$

$$\lambda_{0,i}^-(t \,|\, \xi, s) = \lambda_i^-(t \,|\, \xi, s) - \frac{e^{-\frac{(y_{i-1} - \xi)^2}{4\sigma_i^2 (t-s)}}}{2\sigma_i^2 \sqrt{\pi (t-s)^3}} \left[1 - \frac{y_{i-1} - \xi}{2\sigma_i^2 (t-s)} \right],$$

$$\lambda_{0,i}^{+}(t\,|\,\xi,s) = \lambda_i^{+}(t\,|\,\xi,s) - \frac{e^{-\frac{(y_i-\xi+l_i)^2}{4\sigma_i^2(t-s)}}}{2\sigma_i^2\sqrt{\pi(t-s)^3}}\left[1-\frac{(y_i-\xi+l_i)^2}{2\sigma_i^2(t-s)}\right],$$

$$q_i(s) = e^{-\pi^2\sigma_i^2(t-s)/l_i^2}.$$

Here $\theta_i(z,p)$, $i=2,3$ are the Jacobi theta functions of the second and third kind (Mumford *et al.*, 1983), and $\theta_3''(z,p)$ is the second derivative of $\theta_3(z,p)$ on the first argument.

Substituting $\xi = y_{i-1}, y_i, y_{i+1}$ into Eq. (19.84) we obtain

$$\lambda_{i+1}^{-}(t\,|\,y_{i+1},s) = -\frac{\pi^2}{4l_{i+1}^3}\theta_3''(0,q_{i+1}(s)), \tag{19.85}$$

$$\lambda_i^{+}(\tau\,|\,y_{i-1},s) = -\frac{\pi^2}{4l_i^3}\theta_3''(\pi,q_i(s)),$$

$$\lambda_{0,i+1}^{-}(t\,|\,y_i,s) = \sum_{\substack{n=-\infty\\ n\neq 0}}^{\infty}\frac{e^{-\frac{(2nl_{i+1})^2}{4\sigma_{i+1}^2(t-s)}}}{2\sigma_{i+1}^3\sqrt{\pi(t-s)^3}}\left[1-\frac{(2nl_{i+1})^2}{2\sigma_{i+1}^2(t-s)}\right],$$

$$\lambda_{0,i}^{+}(t\,|\,y_i,s) = -\frac{\pi^2}{4l_i^3}\theta_3''(\pi,q_i(s)) - \frac{e^{-\frac{l_i^2}{4\sigma_i^2(t-s)}}}{2\sigma_i^2\sqrt{\pi(t-s)^3}}\left[1-\frac{l_i^2}{2\sigma_i^2(t-s)}\right].$$

Since at $s\to t$ we have $q(s)\to 1$, all RHSs in Eq. (19.85) are regular in this limit and vanish. The latter is due to the fact that $\lim_{q\to 1}\theta_3''(0,q) = \theta_3''(\pi,q) = 0$. Therefore, all integral kernels in Eq. (19.83) are regular. We also assume that functions $f_i(s)$ are smooth enough, so that in the limit all the integrals vanish.

Applying integration by parts to Eq. (19.83), we get the following simplified system

$$\left.\frac{\partial U_i}{\partial x}\right|_{x=y_{i-1}} = -\int_0^t\left[\eta_i^{even}(t,s)d\left(f_{i-1}(s)\right) - \eta_i^{odd}(t,s)d\left(f_i(s)\right)\right] \tag{19.86}$$
$$-\frac{f_{i-1}(0)}{\sigma_i\sqrt{\pi t}} - f_{i-1}(0)\eta_i^{even}(t,0) + f_i(0)\eta_i^{odd}(t,0),$$

$$\left.\frac{\partial U_i}{\partial x}\right|_{x=y_i} = -\int_0^t\left[\eta_i^{odd}(t,s)d\left(f_{i-1}(s)\right) - \eta_i^{even}(t,s)d\left(f_i(s)\right)\right]$$
$$+\frac{f_i(0)}{\sigma_i\sqrt{\pi t}} + f_i(0)\eta_i^{even}(t,0) - f_{i-1}(0)\eta_i^{odd}(t,0),$$

where

$$\eta_i^{even}(t-s) = \frac{1}{\sigma_i\sqrt{\pi(t-s)}}\sum_{n=-\infty}^{\infty}e^{-\frac{(2nl_i)^2}{4\sigma_i^2(t-s)}}, \tag{19.87}$$

$$\eta_i^{odd}(t-s) = \frac{1}{\sigma_i\sqrt{\pi(t-s)}} \sum_{n=-\infty}^{\infty} e^{-\frac{((2n+1)l_i)^2}{4\sigma_i^2(t-s)}}.$$

Indeed, since $\eta_i^-(t|y_{i-1},t) = 0$, $\eta_i^-(t|y_i,t) = 0$, $f_{i-1}(0) = 0$, $f_i(0) = 0$ we have

$$\begin{aligned}
&-\frac{f_{i-1}(t)}{\sigma_i\sqrt{\pi t}} + \int_0^t \frac{f_{i-1}(s)-f_{i-1}(t)}{2\sigma_i\sqrt{\pi(t-s)^3}} \qquad (19.88)\\
&+\int_0^t \left[f_{i-1}(s)d\left(\eta_i^-(t|y_{i-1},s)\right) - f_i(s)d\left(\eta_i^-(t|y_i,s)\right)\right]\\
&= -\frac{f_{i-1}(t)}{\sigma_i\sqrt{\pi t}} - \frac{f_{i-1}(t)}{\sigma_i\sqrt{\pi(t-s)}}\Bigg|_{s=0}^{s=t} + \frac{f_{i-1}(s)}{\sigma_i\sqrt{\pi(t-s)}}\Bigg|_{s=0}^{s=t}\\
&-\int_0^t \frac{1}{\sigma_i\sqrt{\pi(t-s)}} d\left(f_{i-1}(s)\right) + f_{i-1}(s)\eta_i^-(t|y_{i-1},s)\Bigg|_{s=0}^{s=t}\\
&-f_i(s)\eta_i^-(t|y_i,s)\Bigg|_{s=0}^{s=t} - \int_0^t \left[\eta_i^-(t|y_{i-1},s)d\left(f_{i-1}(s)\right) - \eta_i^-(t|y_i,s)d\left(f_i(s)\right)\right]\\
&= -\int_0^t \left[\left(\eta_i^-(t|y_{i-1},s) + \frac{1}{\sigma_i\sqrt{\pi(t-s)}}\right) d\left(f_{i-1}(s)\right) - \eta_i^-(t|y_i,s)d\left(f_i(s)\right)\right]\\
&= -\int_0^t \left[\eta_i^{even}(t,s)d\left(f_{i-1}(s)\right) - \eta_i^{odd}(t,s)d\left(f_i(s)\right)\right].
\end{aligned}$$

In turn, as shown in Itkin and Muravey (2020a), Eq. (3.31), the function $H_j(t,x)$ can be represented as follows

$$H_j(t,x) = \frac{1}{2\sigma_j\sqrt{\pi t}} \sum_{n=-\infty}^{\infty} \left[e^{-\frac{(2nl_j+x-x_0)^2}{4\sigma_j^2 t}} - e^{-\frac{(2nl_j+x+x_0-2y_{j-1})^2}{4\sigma_j^2 t}}\right]. \quad (19.89)$$

Hence, the gradients at the boundaries are

$$\begin{aligned}
&\frac{\partial H_j(t,x)}{\partial x}\bigg|_{x=y_{j-1}} \equiv \upsilon^-(t|x_0,0) \qquad (19.90)\\
&= \frac{1}{2l_j}\frac{\partial}{\partial x}\left\{\theta_3\left[\frac{\pi(x-x_0)}{2l_j}, q_j(0)\right] - \theta_3\left[\frac{\pi(x+x_0-2y_{j-1})}{2l_j}, q_j(0)\right]\right\}\Bigg|_{x=y_{j-1}}\\
&= \frac{1}{4l_j^2}\left\{\theta_3'\left[\frac{\pi(y_{j-1}-x_0)}{2l_j}, q_j(0)\right] - \theta_3'\left[\frac{\pi(x_0-y_{j-1})}{2l_j}, q_j(0)\right]\right\},\\
&\frac{\partial H_j(t,x)}{\partial x}\bigg|_{x=y_j} = \frac{1}{4l_j^2}\left\{\theta_3'\left[\frac{\pi(y_j-x_0)}{2l_j}, q_j(0)\right] - \theta_3'\left[\frac{\pi(x_0-y_j)}{2l_j}, q_j(0)\right]\right\}.
\end{aligned}$$

Thus, we obtain the following system of Volterra equations

$$\int_0^t \Bigg[-\sigma_i^2 \eta_i^{odd}(t-s) d\left(f_{i-1}(s)\right) + \Big(\sigma_i^2 \eta_i^{even}(t-s) \tag{19.91}$$
$$+ \sigma_{i+1}^2 \eta_{i+1}^{even}(t-s)\Big) d\left(f_i(s)\right) - \sigma_{i+1}^2 \eta_{i+1}^{odd}(t-s) d\left(f_{i+1}(s)\right) \Bigg] = h_i(t),$$

$$h_i(t) = 0, \ i \neq j-1, j,$$
$$h_{j-1}(t) = \sigma_j^2 \upsilon^-(t|x_0, 0), \quad h_j(t) = -\sigma_j^2 \upsilon^+(t|x_0, 0).$$

Since the kernels depend only on $t-s$ one can rewrite the above equations as a convolution

$$\Big(-\sigma_i^2 \eta_i^{odd}(\cdot) * f'_{i-1}(\cdot) + \left[\sigma_i^2 \eta_i^{even}(\cdot) + \sigma_{i+1}^2 \eta_{i+1}^{even}(\cdot)\right] * f'_i(\cdot) \tag{19.92}$$
$$- \sigma_{i+1}^2 \eta_{i+1}^{odd}(\cdot) * f'_{i+1}(\cdot)\Big)(t) = h_i(t).$$

19.4.2 *The Laplace transform*

Applying the Laplace transform to Eq. (19.92), we get

$$\sqrt{\lambda}\Big[-\sigma_i^2 \mathfrak{L}(\eta_i^{odd})\, \mathfrak{L}(f_{i-1}) + \left(\sigma_i^2 \mathfrak{L}(\eta_i^{even}) + \sigma_{i+1}^2 \mathfrak{L}(\eta_{i+1}^{even})\right) \mathfrak{L}(f_i) \tag{19.93}$$
$$- \sigma_{i+1}^2 \mathfrak{L}(\eta_{i+1}^{odd})\, \mathfrak{L}(f_{i+1})\Big] = \frac{\mathfrak{L}(h_i)}{\sqrt{\lambda}},$$

or, in the matrix form

$$\mathbf{M}\mathbf{g} = \frac{\sigma_j^2}{\sqrt{\lambda}} \left[\mathfrak{L}(\upsilon^-(t|x_0, 0))\mathbf{1}_{j-1} - \mathfrak{L}(\upsilon^+(t|x_0, 0))\mathbf{1}_j\right]. \tag{19.94}$$

Here $\mathbf{1}_j$ denotes the indicator vector, i.e.

$$\mathbf{1}_j = \Big(\underbrace{0, 0, ..0}_{j-1}, 1, 0, ...0\Big)^\top, \tag{19.95}$$

the vector $\mathbf{g}$ is the column vector

$$\mathbf{g} = \left(\mathfrak{L}f_1, \ldots, \mathfrak{L}f_{N-1}\right)^\top, \tag{19.96}$$

and the matrix $\mathbf{M}$ is a symmetric tridiagonal matrix

$$\mathbf{M} = \begin{pmatrix} D_1 & -\beta_1 & & & \\ -\beta_1 & D_2 & -\beta_2 & & \\ & -\beta_2 & \ddots & \ddots & \\ & & \ddots & \ddots & -\beta_{N-2} \\ & & & -\beta_{N-2} & D_{N-1} \end{pmatrix}, \tag{19.97}$$

Coefficients of the matrix $\mathbf{M}$ have the form

$$\beta_i = \sqrt{\lambda}\sigma_{i+1}^2 \mathfrak{L}(\eta_{i+1}^{odd}),\; D_i = \sqrt{\lambda}\left[\sigma_i^2 \mathfrak{L}(\eta_i^{even}) + \sigma_{i+1}^2 \mathfrak{L}(\eta_{i+1}^{even})\right], \tag{19.98}$$

and can be found explicitly, see Appendix 19.C

$$\begin{aligned}
\mathfrak{L}(\eta_i^{even}) &= \frac{1}{\sigma_i\sqrt{\lambda}} \coth\left(\frac{\sqrt{\lambda} l_i}{\sigma_i}\right), \\
\mathfrak{L}(\eta_i^{odd}) &= \frac{1}{\sigma_i\sqrt{\lambda}} \frac{1}{\sinh\left(\frac{\sqrt{\lambda} l_i}{\sigma_i}\right)}, \\
\mathfrak{L}(\upsilon^-(t|x_0,0)) &= \frac{1}{\sigma_j^2} \frac{\sinh\left(\frac{(y_{j+1}-x_0)\sqrt{\lambda}}{\sigma_j}\right)}{\sinh\left(\frac{l_j\sqrt{\lambda}}{\sigma_j}\right)}, \\
\mathfrak{L}(\upsilon^+(t|x_0,0)) &= -\frac{1}{\sigma_j^2} \frac{\sinh\left(\frac{(x_0-y_j)\sqrt{\lambda}}{\sigma_j}\right)}{\sinh\left(\frac{l_j\sqrt{\lambda}}{\sigma_j}\right)}.
\end{aligned} \tag{19.99}$$

Finally, introducing the notation

$$\omega_i = \frac{l_i}{\sigma_i}, \qquad \gamma_1 = \frac{y_{j+1} - x_0}{l_j}, \qquad \gamma_2 = \frac{x_0 - y_j}{l_j}, \tag{19.100}$$

the system Eq. (19.94) can be represented as

$$\begin{aligned}
\mathbf{Mg} &= \frac{1}{\sqrt{\lambda}}\left[\frac{\sinh\left(\gamma_1\omega_j\sqrt{\lambda}\right)}{\sinh\left(\omega_j\sqrt{\lambda}\right)}\mathbf{1}_{j-1} + \frac{\sinh\left(\gamma_2\omega_j\sqrt{\lambda}\right)}{\sinh\left(\omega_j\sqrt{\lambda}\right)}\mathbf{1}_j\right], \\
D_i &= \sigma_i \coth\left(\omega_i\sqrt{\lambda}\right) + \sigma_{i+1}\coth\left(\omega_{i+1}\sqrt{\lambda}\right),\; \beta_i = \frac{\sigma_{i+1}}{\sinh\left(\omega_{i+1}\sqrt{\lambda}\right)}.
\end{aligned} \tag{19.101}$$

19.5 Numerical experiments

In this section, we solve the problem in Eq. (19.74) by using the ML method. Note that such problems appear both in physics and in finance. A simple example of a financial problem is finding Green's function for pricing double barrier options written on the underlying S_t with local volatility

$$dS_t = \sigma(S_t)dW_t, \tag{19.102}$$

where $\sigma(S)$ is a piecewise constant function. Below we describe two numerical experiments.

19.5.1 *Constant volatility* σ_i

To start with, we assume that $\sigma_i = const$, $i = 1, \dots, N$, and hence, σ_i^2 in Eq. (19.74) can be pulled out of the derivative in x. This problem has an analytic solution (see Lipton, 2001 and references therein), which can be represented as the Fourier series. Re-writing it by using the definition of Jacobi theta functions yields

$$
\begin{aligned}
U(T, y) &= \frac{1}{2l}\left[\theta_3\left(\frac{\pi(y - x_0)}{2l}, q\right) - \theta_3\left(\frac{\pi(y + x_0 - 2y_0)}{2l}, q\right)\right], \qquad (19.103) \\
l &= y_N - y_0, \quad q = e^{-\frac{\pi^2\sigma^2}{l^2}T}.
\end{aligned}
$$

To solve this problem, we need first to solve the linear system in Eq. (19.101) numerically, and then use the corresponding Laplace images to find the function $f_i = f(y_i)$, $i = 1, \dots, N$ by applying an inverse Laplace transform. For the latter step, we use the Gaver-Stehfest method

$$
f(T, y) = \Lambda \sum_{s=1}^{[m]} St_s^{(m)} \mathfrak{L}(f_i(\Lambda)), \quad \Lambda = \frac{\log 2}{T}. \qquad (19.104)
$$

This algorithm was widely studied (see, e.g., Kuznetsov, 2013 and references therein), and, provided that the resulting function is non-oscillatory, converges very quickly. For instance, choosing $m = 12$ terms in the series representing the solution is usually sufficient. The coefficients St_s can be found explicitly in advance.

Table 19.5.1: Parameters of the test.

y_0	y_N	σ	T	N	m
-1.0	1.0	0.5	1.0	20	16

The model parameters for this test are given in Table 19.5.1, and the results are depicted in Fig. 19.5.1a. Here, the left vertical axis shows the values of $y_i(T)$, and the right vertical axis shows the relative error (in percent) of the solution compared with the analytic one.

The ML solution coincides with the analytic one with high accuracy. The elapsed time of the experiment is 8.8 ms (we run our code, written in Matlab, on a PC with two Quadcore CPU Intel i7-4790 3.60 GHz). The elapsed time doesn't depend on the option maturity T, so the calculation is fast even for long maturities. Note that the ML solution's computation

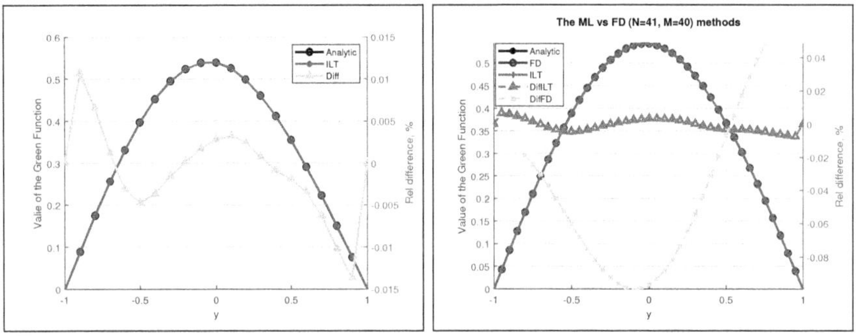

(a) (b)

Figure 19.5.1: Comparison of the Analytic and ML solutions (a), and Analytic, ML and FD solutions (grid with 41×40 nodes) (b) for $\sigma_i = 0.5, T = 1$. Here *Analytic* denotes the analytic solution of the problem, *ILT* — the ML solution, *FD* — the FD solution, *DiffILT* — the relative error of the ML solution with respect to the analytic one, *DifFD* — same for the FD method.

takes only 2.3 ms, while the remaining time is used for computing the Gaver-Stehfest coefficients (but those can be precomputed if so desired). Since the matrix $\mathbf{M}$ is tridiagonal, the ML method's complexity is $O(mN)$. By comparison, the complexity of the finite difference (FD) method is $O(MN)$, M is the number of time steps. Obviously, for long maturities $M \gg m$, so the FD method is slower.

To validate this, we also implemented an FD method to solve the same problem. The FD solver runs on a uniform grid and is a Crank-Nicolson scheme after four steps, while for the first four steps it uses an implicit Euler scheme. In other words, we start with four Ranacher steps (see Itkin, 2017 and references therein).

To have the same spatial approximation in y, we need to run both the ML and FD methods with the same number of nodes. While the ML method provides an accurate result even at $N = 10$, the FD method fails and needs at least 40 nodes to converge to the solution. Therefore, we choose $N = 41$. The same is true for the time step, so the minimal number of FD time steps in our experiment is $M = 40$, while the ML method provides the solution at $t = T$ just at once. The results of the comparison of both methods with the analytic solution are given in Fig. 19.5.1b. Again, the left vertical axis shows the values of $y_i(T)$, and the right vertical axis is the relative error (in percent) of the solution compared with the analytic one.

It is clear that the accuracy of the FD method is worse than that of the ML method. By increasing N and M, one can improve the FD method's accuracy, but it takes time. The elapsed time for the FD method with $N = 41, M = 40$ is 41 ms, so it is about 18 times slower than the ML method. Obviously, for longer maturities, more time-steps are necessary, so the FD method becomes even slower.

It is also known that for small T and volatilities, the FD method's error increases. To illustrate this fact, we run the same experiment, but now with $T = 0.5$, $\sigma = 0.3$. The results are presented in Fig. 19.5.2.

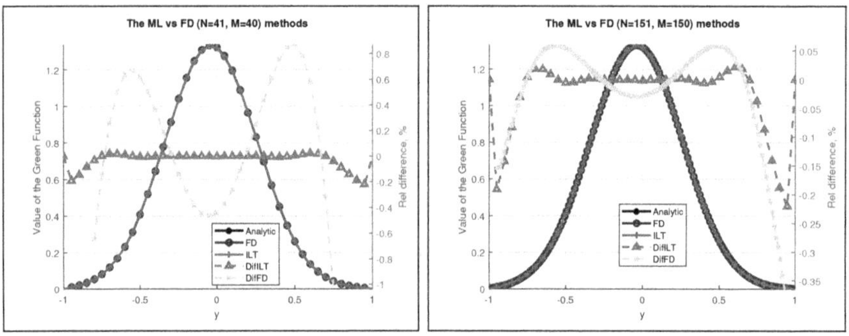

(a) (b)

Figure 19.5.2: Comparison of the Analytic, ML and FD solutions for $\sigma_i = 0.3, T = 0.5$. Here *Analytic* denotes the analytic solution of the problem, *ILT* — the ML solution, *FD* — the FD solution, *DiffILT* — the relative error of the ML solution with respect to the analytic one, *DifFD* — same for the FD method.

Fig. 19.5.2a shows that the accuracy of the ML method is still good with the same number of internal layers $N = 40$, while the error of the FD method with $N = 41, M = 40$ is quite significant. The error can be reduced by running the FD method with $N = 151, M = 150$ (see Fig. 19.5.2b); however, the corresponding elapsed time is 86 ms. Thus, for the same accuracy, the ML method is 37 times faster.

19.5.2 *Piecewise constant volatility* σ_i

Here $\sigma(x)$ is a piecewise constant function defined in Eq. (19.75). In this case, there is no analytic solution of the problem[8], hence as a benchmark, we use an FD method, namely the same FD solver as in the previous experiment. However, since we solve the problem in Eq. (19.74), an FD scheme has to be implemented for the conservative heat equation. While such an implementation is possible, we prefer to rewrite Eq. (19.77) and Eq. (19.78) in a non-divergent form. For instance,

$$\frac{\partial U_i}{\partial t} = \Xi_i^2(x)\frac{\partial^2 U_i}{\partial x^2} + \frac{\partial \Xi^2(x)}{\partial x}\frac{\partial U_i}{\partial x}, \tag{19.105}$$
$$\Xi_i(x) = \sigma_i[\Theta_H(x_{i+1} - x) - \Theta_H(x_i - x)],$$

where $\Theta_H(x)$ is the Heaviside theta function (Abramowitz and Stegun, 1964) with $\Theta_H(0) = 1$. Accordingly, on the interval $x \in (y_i, y_{i+1}]$ we have

$$\frac{\partial \Xi^2(x)}{\partial x} = \sigma_{i+1}^2[\delta(x_i - x) - \delta(x_{i+1} - x)]. \tag{19.106}$$

At the point $x = x_{i+1}$ this gives $\partial\Xi^2(x)/\partial x = -\sigma_{i+1}^2\delta(0)$. In turn, $\delta(0)$ can be numerically approximated as

$$\delta(0) = \frac{2}{y_N - y_0}, \tag{19.107}$$

which provides the correct normalization of the Dirac delta function. Indeed, the integral over the interval $[y_0, y_n]$ of the test function equal to 1 at $x = y_{i+1}$ and 0 otherwise computed by using a trapezoidal rule is equal to $(y_N - y_0)/2$. Therefore, we need to use Eq. (19.107) to provide the correct numerical normalization.

y_0	y_N	T	N	m	M
-1.0	4.0	2.0	50	16	100

Table 19.5.2: Parameters of the second experiment.

In this experiment, we use parameters of the model given in Table 19.5.2, and the piecewise constant volatility σ_i, which is defined as follows

$$\sigma_i(s) = e^{-i/N}, \quad s \in (y_i, y_{i+1}],\ i = 1, \ldots, N. \tag{19.108}$$

[8]Based on Eq. (19.101) it is possible to derive an explicit series representation of the solution. It will be published elsewhere.

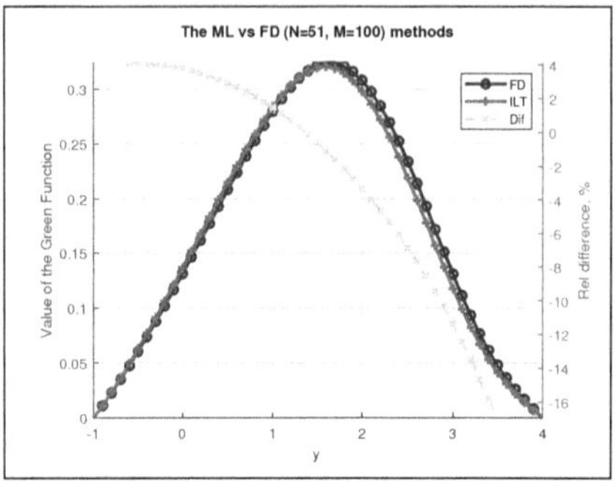

(a)

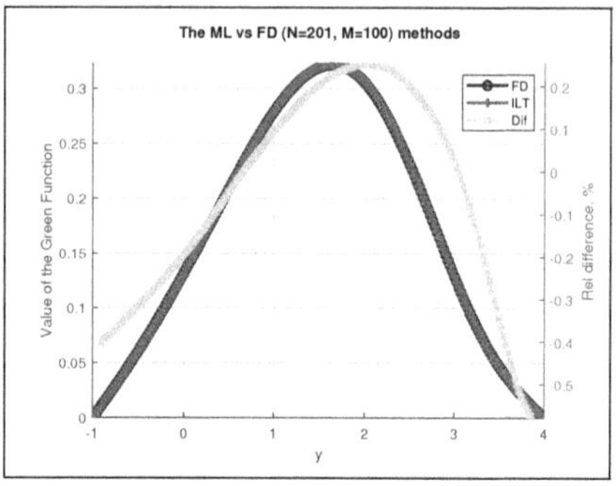

(b)

Figure 19.5.3: Comparison of the ML and FD solutions for a piecewise constant $\sigma(x)$. Here *ILT* denotes the ML solution, *FD* — the FD solution, *Dif* — the relative error of the FD solution with respect to the ML one.

The results of the test are presented in Fig. 19.5.3a. Again, the number of nodes for the ML and FD methods is the same. The difference between the two solutions reaches 16% at the right external boundary, 4% at the left external boundary, and changes in this range in between. The elapsed time is 2.2 ms for the ML and 50 ms for the FD methods.

To check the convergence of the solution, we rerun the calculation with $N = 200$. The results are presented in Fig. 19.5.3b. The relative error ε drops down to be in $\varepsilon \in [-0.6, 0.2]$ percent. The elapsed time is 5 ms for the ML and 88 ms for the FD methods. Note that in this case, to reduce the error, we also need to increase the number of layers for the ML method. This effect is explained in Section 19.6; it is due to the non-smoothness of σ in this experiment.

The physics meaning of the obtained results is as follows. Suppose we consider diffusion rather than heat conduction. According to Eq. (19.108) the diffusion coefficient $\sigma^2(x)$ is a decreasing function of x when moving from y_0 to y_N. Since we request continuity of the flux at the internal boundaries, the gradient of the solution increases with x when moving from left to right. The maximum of the solution, which is located at $x = x_0$ when $t = 0$, travels to the right when t increases. Recall that the solution is the Green function of our problem. This behavior was also observed in Lançon *et al.* (2001), where the authors studied particles trapped between two nearly parallel walls making their confinement position dependent. They not only measured a diffusion coefficient which depended on the particles' position but also reported and explained a new effect: a drift of the particles individual positions (so change in concentration) in the direction of the diffusion coefficient gradient, in the absence of any external force or concentration gradient.

19.6 Discussion

In the previous section, we have demonstrated that the ML method's complexity is linear in N. The same is true in the general case because, as mentioned at the end of Section 19.4, from Eq. (19.21) we obtain a linear system of equations for $\mathfrak{L}(\chi_i^-), \mathfrak{L}(\Omega_i)$ which has a block diagonal matrix with all blocks being tridiagonal matrices. Therefore, the ML method's complexity remains linear and approximately is $O(4mN)$, so it doesn't depend on T. Hence, if $4m$ is of the order of M, the ML and the FD methods have the same complexity. For typical values $m = 12$ we have $4m = 48$. Therefore, for short maturities $T < 1$ year, both methods' complexity is roughly the same. However, our method has an obvious advantage for the long maturities occurring in the Fixed Income context.

The ML method has some other advantages as well. First, the FD construction provides only the values of the unknown function at the grid nodes in space, and at intermediate points they can be found only by interpolation. In contrast, using the ML method, we obtain an analytic representation of the solution at any x (once the values at the layers' boundaries are found). Second, the Greeks, i.e., derivatives of the solution, can be expressed semi-analytically by differentiating the solution with respect to x or some parameter of the model and performing numerical integration, provided that the values at the internal boundaries are found. For the FD method, the Greeks can be found only numerically. Moreover, to compute the Vega, a new run of the FD method is required, while for the ML method, all Greeks can be calculated in one go, as described.

As far as an approximation with respect to x is concerned, the following observation holds. Using the ML method, we obtain an analytical solution at every interval i, $i = 1, \ldots, N$. However, to do this, we need to approximate the corresponding coefficient, e.g., $\sigma(x)$ over layers by piecewise constant or linear functions. For the linear approximation, the solution's accuracy is $O((\Delta x)^2)$, i.e., same as for the FD method of the second order. Therefore, it seems that the spatial accuracies of both ML and FD methods are the same.

On the other hand, the error of both methods is also proportional to the second derivative. For the FD method, this is the second derivative of the solution; for the ML method — the second derivative of the coefficient, e.g., $\sigma_{xx}(x)$. If the latter is smaller than the second derivative of the solution (say, the option Gamma), then the number of layers N can be decreased while providing the same accuracy. This reduction provides an additional speedup of our method as compared with the FD method. This fact is illustrated by our first experiment where function $\sigma(x)$ is smooth, so even a small number of layers is sufficient to obtain a very accurate solution. In the second experiment, $\sigma(x)$ jumps at the layer's boundaries, and, therefore, one needs to increase the number of layers to provide the same accuracy.

Note that for the FD method, the difficulties caused by sharp gradients can be alleviated by using nonuniform grids where the nodes are condensed in the area where gradients are high. The same approach could be applied to the construction of internal layers in the ML method.

Overall, we can conclude that the new ML method proposed in Itkin *et al.* (2021) and presented in this Chapter is significantly faster than the FD method, provides better accuracy, and represents the solution in a semi-analytical form. The method's speed is close to that for the Radial Basis Functions (RBF) approach (Hon and Mao, 1999; Fasshauer *et al.*, 2004; Pettersson *et al.*, 2008), while other properties listed above are superior to the RBF.

Appendices

19.A Transformation of a non-divergent heat equation to a divergent form

Consider the PDE in Eq. (19.30) which is a divergent form of the heat equation

$$\frac{\partial U(t,x)}{\partial t} = \frac{\partial}{\partial x}\left(\Xi^2(x)\frac{\partial U(t,x)}{\partial x}\right). \tag{19.109}$$

In this Section we show how to transform it to a non-divergent form as in Eq. (19.28) when the external boundaries are constant, i.e., $y_0(t) = \chi^-(t) = const$, $y_N(t) = \chi^+(t) = const$. We start with making a change of variables $x \mapsto z = f(x)$ with

$$f(x) = c_1 + c_2 \int_0^x \frac{1}{\Xi^2(k)} dk, \tag{19.110}$$

where c_1, c_2 are some constants. This transformation reduces Eq. (19.30) to

$$\begin{aligned}\frac{\partial U(t,z)}{\partial t} &= \sigma^2(z)\frac{\partial^2 U(t,z)}{\partial z^2}, \\ \sigma(z) &= \frac{\Xi(x(z))}{x'(z)} = \frac{c_1}{\Xi(x(z))}.\end{aligned} \tag{19.111}$$

The Eq. (19.111) is a non-divergent form of the heat equation. The only thing which remains to be done is finding the dependence $x(z)$. Obviously, it solves the equation

$$z = f(x) = c_2 + c_1 \int_0^x \frac{1}{\Xi^2(k)} dk. \tag{19.112}$$

Given $\Xi(x)$, it can be solved either numerically (so this dependence can be precomputed), or in some cases analytically. As an example, assume that $\Xi(x) = e^{-ax}$, $a = const \neq 0$, and also let $c_2 = 0$. Then,

$$\begin{aligned}x &= \frac{1}{a}\log\left(1 + \frac{a}{c_1}z\right), \\ \sigma^2(z) &= c_1(c_1 + az).\end{aligned} \tag{19.113}$$

Reverting these steps, we obtain the inverse transformation from a non-divergent heat equation to a divergent one.

Also, the second continuity condition for Eq. (19.109) (an equality of fluxes over the boundary) is given by Eq. (19.7) which, by using our notation in this section, can be re-written as

$$\Xi_i^2(y_{i+1})\frac{\partial U_i}{\partial x}\bigg|_{x=y_{i+1}} = \Xi_{i+1}^2(y_{i+1})\frac{\partial U_{i+1}}{\partial x}\bigg|_{x=y_{i+1}}, \quad i = 1,\dots,N-1. \tag{19.114}$$

Using Eq. (19.111) this can be transformed to

$$\begin{aligned}\frac{\partial U_i}{\partial z}\bigg|_{z=z(y_{i+1})} &= \frac{\partial U_{i+1}}{\partial z}\bigg|_{z=z(y_{i+1})}, \quad i = 1,\dots,N-1, \\ z(y_{i+1}) &= c_2 + c_1\int_0^{y_{i+1}} \frac{1}{\Xi^2(k)}dk = c_2 + \frac{1}{c_1}\int_0^{y_{i+1}} \sigma^2(k)dk.\end{aligned} \tag{19.115}$$

This is the continuity condition for Eq. (19.111).

19.B Multilayer method for time-inhomogeneous coefficients and the domain

In this section we generalize the ML method to the case $\sigma = \sigma(\tau, x)$. We again consider the initial-boundary problem Eq. (19.4) for the differential operator $\mathcal{L}_i$ of the form Eq. (19.8) where now each operator $\mathcal{L}_i$ reads

$$\mathcal{L}_i = -\frac{\partial}{\partial \tau} + \sigma_i^2(\tau, x)\frac{\partial^2}{\partial x^2}. \tag{19.116}$$

As before, we look for the solution of the problem Eq. (19.4) in the form Eq. (19.6) such that the conditions Eq. (19.7) still hold. Our goal is to show that under certain assumptions the problem Eq. (19.4) with time and space dependent volatility $\sigma(\tau, x)$ can be reduced to the corresponding time-homogeneous problem.

Suppose that for each sub-domain Ω_i we can construct a map $\mathcal{M}_i$ transforming Eq. (19.116) into PDE of the form Eq. (19.9). Then the solution of the transformed PDE can be represented in the form of the heat potential Eq. (19.15), and then transformed back by inverting the map $\mathcal{M}_i$. More precisely, consider a collection of maps $\{\mathcal{M}_i\}_{i=1}^N$ acting on triplets $(\tau, x, u_i(\tau, x))$

$$(\tau, x, u_i(\tau, x)) \xmapsto{\mathcal{M}_i} (\mathcal{T}_i(\tau), \mathcal{X}_i(\tau, x), \mathcal{U}_i(\mathcal{T}_i, \mathcal{X}_i)), \quad \mathcal{T}_i(0) = 0, \tag{19.117}$$

such that the function $\mathcal{U}_i(\mathcal{T}_i, \mathcal{X}_i)$ solves the following PDE with time-independent coefficients

$$-\frac{\partial \mathcal{U}_i}{\partial \mathcal{T}_i} + A^2(\mathcal{X}_i)\frac{\partial^2 \mathcal{U}_i}{\partial \mathcal{X}_i^2} = 0. \tag{19.118}$$

Also, let us denote the inverse map as $\Upsilon_i(\tau, x)$, such that the following representation holds

$$u_i(\tau, x) = \Upsilon_i(\tau, x)\mathcal{U}_i(\mathcal{X}_i(\tau, x), \mathcal{T}_i(\tau)). \tag{19.119}$$

The map $\mathcal{M}_i$ transforms the sub-domain Ω_i to the sub-domain Ξ_i

$$\Xi_i : \left[\mathcal{Y}_i^-(\mathcal{T}_i), \mathcal{Y}_i^+(\mathcal{T}_i)\right] \times \mathbb{R}_+$$

bounded by the curves $\mathcal{Y}_i^-(\mathcal{T}_i)$ and $\mathcal{Y}_i^+(\mathcal{T}_i)$ which are defined as

$$\mathcal{Y}_i^-(\mathcal{T}_i) = \mathcal{X}_i(\lambda_i(\mathcal{T}_i), y_i(\lambda_i(\mathcal{T}_i))), \quad \mathcal{Y}_i^+(\mathcal{T}_i) = \mathcal{X}_i(\lambda_i(\mathcal{T}_i), y_{i+1}(\lambda_i(\mathcal{T}_i))).$$

Here λ_i is the inverse map $\mathcal{T}_i^{-1}$, i.e. $\lambda_i(\mathcal{T}_i) = \tau(\mathcal{T}_i) = \mathcal{T}_i^{-1}$. Since the new time variables $\mathcal{T}_i$ are different for each layer Ξ_i the transformed boundaries are different as well, i.e., $\mathcal{Y}_i^+(\mathcal{T}_i) \neq \mathcal{Y}_{i+1}^-(\mathcal{T}_{i+1})$. Also, the initial value function $f(x)$ is transformed to the function $\mathcal{F}_i$

$$f(x) \xrightarrow{\mathcal{M}_i} \mathcal{F}_i(\mathcal{X}_i), \qquad \mathcal{F}_i(\mathcal{X}_i) = f(\eta_i(\mathcal{X}_i))/\Upsilon_i(\eta_i(\mathcal{X}_i)),$$

where $\eta_i(x)$ solves the equation

$$\eta_i(x) : \mathcal{X}_i(0, \eta_i(x)) = x.$$

Since the equations Eq. (19.118) are time-homogeneous, we can represent their solutions in the form of Eq. (19.15)

$$\mathcal{U}_i(\mathcal{T}_i, \mathcal{X}_i) = \int_0^{\mathcal{T}_i} \left\{ \Phi_i(k) \frac{\partial G(\mathcal{X}_i, \xi, \mathcal{T}_i - k)}{\partial \xi}\bigg|_{\xi=\mathcal{Y}_i^-(k)} + \Psi_i(k) \frac{\partial G(\mathcal{X}_i, \xi, \mathcal{T}_i - k)}{\partial \xi}\bigg|_{\xi=\mathcal{Y}_i^+(k)} \right\} dk. \tag{19.120}$$

Then making the inversion in Eq. (19.119), applying the chain rule

$$\frac{\partial u_i}{\partial x} = \mathcal{U}_i(\mathcal{T}_i(\tau), \mathcal{X}_i(\tau, x)) \frac{\partial \Upsilon(\tau, x)}{\partial x} + \Upsilon(\tau, x) \frac{\partial \mathcal{X}_i(\tau, x)}{\partial x} \frac{\partial \mathcal{U}_i(\mathcal{T}_i(\tau), \mathcal{X})}{\partial \mathcal{X}}\bigg|_{\mathcal{X}=\mathcal{X}_i(\tau,x)},$$

and taking into account the discontinuity of the layer potentials on the boundaries, we arrive at the system of Volterra equations in Eq. (19.7).

The map Eq. (19.117) can be explicitly found via two different approaches. The first is by application of Lie symmetry analysis. It is well known, that if Eq. (19.118) has six or four independent groups of symmetries, it can be reduced to the heat or Bessel PDE (see Craddock, 2009).

Another method is based on the theory of diffusion processes. Since any PDE of the form Eq. (19.116) and Eq. (19.118) can be associated with some diffusion process, say $X = \{X_t, t \geq 0\}$ for Eq. (19.116) and $Y = \{Y_t, t \geq 0\}$ for Eq. (19.118), the map in Eq. (19.117) can be found via reduction methods (see Lipton and Kaushansky, 2020a and references therein). The terms $\mathcal{T}(\tau), \mathcal{X}_i(\tau, x)$ and $\Upsilon_i(\tau, x)$ are interpreted as a scale, time and measure changes.

19.C Coefficients of Eq. (19.94)

By using the definitions of coefficients of Eq. (19.94) given in Eq. (19.87) and Eq. (19.24) and tables of Laplace transforms we find

$$\begin{aligned}\mathfrak{L}(\eta_i^{even}) &= \mathfrak{L}\left\{\sum_{n=-\infty}^{\infty}\frac{e^{-\frac{(2nl_i)^2}{4\sigma_i^2 t}}}{\sigma_i\sqrt{\pi t}}\right\} = \frac{1}{\sigma_i\sqrt{\lambda}}\sum_{n=-\infty}^{\infty}e^{-\frac{\sqrt{\lambda}|2nl_i|}{\sigma_i}} \\ &= \frac{1}{\sigma_i\sqrt{\lambda}}\left(1+2\sum_{n=1}^{\infty}e^{-\frac{2\sqrt{\lambda}nl_i}{\sigma_i}}\right) \\ &= \frac{1}{\sigma_i\sqrt{\lambda}}\left(1+\frac{2e^{-\frac{2\sqrt{\lambda}l_i}{\sigma_i}}}{1-e^{-\frac{2\sqrt{\lambda}l_i}{\sigma_i}}}\right) = \frac{1}{\sigma_i\sqrt{\lambda}}\coth\left(\frac{\sqrt{\lambda}l_i}{\sigma_i}\right),\end{aligned} \tag{19.121}$$

$$\begin{aligned}\mathfrak{L}(\eta_i^{odd}) &= \mathfrak{L}\left\{\sum_{n=-\infty}^{\infty}\frac{e^{-\frac{((2n+1)l_i)^2}{4\sigma_i^2 t}}}{\sigma_i\sqrt{\pi t}}\right\} = \frac{1}{\sigma_i\sqrt{\lambda}}\sum_{n=-\infty}^{\infty}e^{-\frac{\sqrt{\lambda}|(2n+1)l_i|}{\sigma_i}} \\ &= \frac{2e^{-\frac{\sqrt{\lambda}nl_i}{\sigma_i}}}{\sigma_i\sqrt{\lambda}}\sum_{n=0}^{\infty}e^{-\frac{2\sqrt{\lambda}nl_i}{\sigma_i}} \\ &= \frac{1}{\sigma_i\sqrt{\lambda}}\frac{2e^{-\frac{\sqrt{\lambda}l_i}{\sigma_i}}}{1-e^{-\frac{2\sqrt{\lambda}l_i}{\sigma_i}}} = \frac{1}{\sigma_i\sqrt{\lambda}}\frac{1}{\sinh\left(\frac{\sqrt{\lambda}l_i}{\sigma_i}\right)},\end{aligned}$$

$$\begin{aligned}\mathfrak{L}(\upsilon^-(t|x_0,0)) &= -\frac{1}{\sigma_j^2}\sum_{n=-\infty}^{\infty}(y_j-x_0+2nl_i)\,e^{-\frac{\sqrt{\lambda}}{\sigma_j}|y_j-x_0+2nl_j|} \\ &= -\frac{1}{\sigma_j^2}\sum_{n=1}^{\infty}e^{-\frac{\sqrt{\lambda}}{\sigma_j}(y_j-x_0)-\frac{\sqrt{\lambda}}{\sigma_j}2nl_j}+\frac{1}{\sigma_j^2}\sum_{n=0}^{\infty}e^{\frac{\sqrt{\lambda}}{\sigma_j}(y_j-x_0)-\frac{\sqrt{\lambda}}{\sigma_j}2nl_j} \\ &= \frac{1}{\sigma_j^2}\left[e^{\frac{\sqrt{\lambda}}{\sigma_j}(y_j-x_0)}-e^{-\frac{\sqrt{\lambda}}{\sigma_j}(y_j-x_0)-2\frac{\sqrt{\lambda}}{\sigma_j}l_j}\right]\sum_{n=0}^{\infty}e^{-\frac{\sqrt{\lambda}}{\sigma_j}2nl_j} \\ &= \frac{1}{\sigma_j^2}\frac{e^{\frac{\sqrt{\lambda}}{\sigma_j}(y_j-x_0)}-e^{-\frac{\sqrt{\lambda}}{\sigma_j}(y_j-x_0)-2\frac{\sqrt{\lambda}}{\sigma_j}l_j}}{1-e^{-\frac{2\sqrt{\lambda}}{\sigma_j}l_j}} \\ &= \frac{1}{\sigma_j^2}\frac{\sinh\left(\frac{(y_j-x_0+l)\sqrt{\lambda}}{\sigma_j}\right)}{\sinh\left(\frac{l_j\sqrt{\lambda}}{\sigma_j}\right)} = \frac{1}{\sigma_j^2}\frac{\sinh\left(\frac{(y_{j+1}-x_0)\sqrt{\lambda}}{\sigma_j}\right)}{\sinh\left(\frac{l_j\sqrt{\lambda}}{\sigma_j}\right)},\end{aligned}$$

$$\mathfrak{L}(\upsilon^+(t|x_0,0)) = -\frac{1}{\sigma_j^2}\sum_{n=-\infty}^{\infty}(y_{j+1}-x_0+2nl_i)\,e^{-\frac{\sqrt{\lambda}}{\sigma_j}|y_{j+1}-x_0+2nl_j|}$$

$$
\begin{aligned}
&= -\frac{1}{\sigma_j^2}\sum_{n=0}^{\infty} e^{-\frac{\sqrt{\lambda}}{\sigma_j}(y_{j+1}-x_0)-\frac{\sqrt{\lambda}}{\sigma_j}2nl_j} + \frac{1}{\sigma_j^2}\sum_{n=1}^{\infty} e^{\frac{\sqrt{\lambda}}{\sigma_j}(y_{j+1}-x_0)-\frac{\sqrt{\lambda}}{\sigma_j}2nl_j} \\
&= \frac{1}{\sigma_j^2}\left[e^{\frac{\sqrt{\lambda}}{\sigma_j}(y_{j+1}-x_0)-2\frac{\sqrt{\lambda}}{\sigma_j}l_j} - e^{-\frac{\sqrt{\lambda}}{\sigma_j}(y_{j+1}-x_0)}\right]\sum_{n=0}^{\infty} e^{-\frac{\sqrt{\lambda}}{\sigma_j}2nl_j} \\
&= \frac{1}{\sigma_j^2}\frac{e^{\frac{\sqrt{\lambda}}{\sigma_j}(y_{j+1}-x_0)} - e^{-\frac{\sqrt{\lambda}}{\sigma_j}(y_{j+1}-x_0)-2\frac{\sqrt{\lambda}}{\sigma_j}l_j}}{1-e^{-\frac{2\sqrt{\lambda}}{\sigma_j}l_j}} \\
&= \frac{1}{\sigma_j^2}\frac{\sinh\left(\frac{(y_{j+1}-x_0-l)\sqrt{\lambda}}{\sigma_j}\right)}{\sinh\left(\frac{l_j\sqrt{\lambda}}{\sigma_j}\right)} = -\frac{1}{\sigma_j^2}\frac{\sinh\left(\frac{x_0-y_j)\sqrt{\lambda}}{\sigma_j}\right)}{\sinh\left(\frac{l_j\sqrt{\lambda}}{\sigma_j}\right)}.
\end{aligned}
$$

Bibliography

Abate, J., Choudhury, G. L., and Whitt, W. (2000). An introduction to numerical transform inversion and its application to probability models, in *Computational probability* (Springer), pp. 257–323.

Abazari, R. and Kilicman, A. (2013). Numerical Study of Two-Dimensional Volterra Integral Equations by RDTM and Comparison with DTM, *Abstract and Applied Analysis* , 929478.

Abramowitz, M. and Stegun, I. (1964). *Handbook of Mathematical Functions* (Dover Publications, Inc.).

Addison, J., Howison, S. D., and King, J. (2006). Ray methods for free boundary problems, *Quarterly of Applied Mathematics* **64**, 1, pp. 41–59.

Adomian, G. (1994). *Solving Frontier Problems of Physics: The Decomposition Method*, Vol. 60 (Luwer Academic, Dordrecht, The Netherlands).

Alili, L. and Patie, P. (2010). Boundary-crossing identities for diffusions having the time-inversion property. *Journal of Theoretical Probability* **23**, 1, pp. 65–85.

Alili, L., Patie, P., and Pedersen, J. L. (2005). Representations of the first hitting time density of an Ornstein-Uhlenbeck process, *Stochastic Models* **21**, 4, pp. 967–980.

Andersen, L., Lake, M., and Offengenden, D. (2016). High-performance American option pricing, *Journal of Computational Finance* **20**, 1, pp. 39–87.

Andersen, L. and Piterbarg, V. (2010). *Interest Rate Modeling*, no. v. 2 in Interest Rate Modeling (Atlantic Financial Press), ISBN 9780984422111.

Antonelli, F., Kohatsu-Higa, A., *et al.* (2002). Rate of convergence of a particle method to the solution of the McKean–Vlasov equation, *The Annals of Applied Probability* **12**, 2, pp. 423–476.

Antonov, A. and Spector, M. (2011). General short-rate analytics, *Risk*, pp. 66–71.

Asvestas, M., Sifalakis, A., Papadopoulou, E., and Saridakis, Y. (2014). Fokas method for a multi-domain linear reaction-diffusion equation with discontinuous diffusivity, *Journal of Physics: Conference Series* **490**, 012143.

Avellaneda, M. and Lee, J. (2010). Statistical arbitrage in the US equities market, *Quantitative Finance* **10**, 7, pp. 761–782.

Avellaneda, M. and Zhu, J. (2001). Distance to default, *Risk* **14**, 12, pp. 125–129.

Bacaer, N. (2011). *A short history of mathematical population dynamics*, chap. 6 (Springer-Verlag, London), ISBN 978-0-85729-114-1, pp. 35–39.

Bai, Y. and Wu, L. (2018). Analytic value function for optimal regime-switching pairs trading rules, *Quantitative Finance* **18**, 4, pp. 637–654.

Bailey, D. and Prado, M. L. D. (2013). The sharpe ratio efficient frontier, *Journal of Risk* **15**, 2.

Bailey, D. and Prado, M. L. D. (2014). The deflated sharpe ratio: Correcting for selection bias, backtest overfitting and non-normality, *Journal of Portfolio Management* **40**, 5, pp. 94–107.

Baratella, P. and Orsi, A. (2004). A new approach to the numerical solution of weakly singular Volterra integral equations, *Journal of Computational and Applied Mathematics* **163**, 2, pp. 401–418.

Bateman, H. and Erdélyi, A. (1953). *Higher Transcendental Functions, Bateman Manuscript Project California Institute of Technology*, Vol. 1 (McGraw-Hill).

Baviera, R. and Baldi, T. (2019). Stop-loss and leverage in optimal statistical arbitrage with an application to energy market, *Energy Economics* **79**, pp. 130–143.

Bergomi, L. (2016). *Stochastic Volatility Modeling*, CRC Financial Mathematics Series (Chapman and Hall).

Bertram, W. (2009). Optimal trading strategies forito diffusion processes, *Physica A: Statistical Mechanics and its Applications* **388**, 14, pp. 2865–2873.

Bertram, W. (2010). Analytic solutions for optimalstatistical arbitrage trading, *Physica A: Statistical Mechanics and itsApplications* **389**, 11, pp. 2234–2243.

Bielecki, T. R. and Rutkowski, M. (2013). *Credit Risk: Modeling, Valuation and Hedging* (Springer Science & Business Media).

Black, F. and Cox, J. (1976a). Valuing corporate securities: Some effects of bond indenture provisions, *Journal of Finance* **31**, 2, pp. 351–367.

Black, F. and Cox, J. C. (1976b). Valuing corporate securities: Some effects of bond indenture provisions, *The Journal of Finance* **31**, 2, pp. 351–367.

Black, F. and Karasinski, P. (1991). Bond and option pricing when short rates are lognormal, *Financial Analysts Journal* , pp. 52–59.

Black, F. and Scholes, M. (1973). The pricing of options and corporate liabilities, *The Journal of Political Economy* **81**, pp. 637—659.

Bluman, G. (1980). On the transformation of diffusion, *SIAM Journal on Applied Mathematics*, **39**, 2, pp. 238–247.

Bluman, G. and Cole, J. (1974). *Similarity Methods for Differential Equations*, *Appl. Math. Sci.*, Vol. 13 (Springer-Verlag, New York, Heidelberg, Berlin).

Borkar, V. and Suresh Kumar, K. (2010). McKean–Vlasov limit in portfolio optimization, *Stochastic Analysis and Applications* **28**, 5, pp. 884–906.

Borodin, A. N. and Salminen, P. (2012). *Handbook of Brownian motion-facts and formulae* (Birkhäuser).

Borovkov, K. and Novikov, A. (2005). Explicit bounds for approximation rates of boundary crossing probabilities for the Wiener process, *Journal of Applied Probability* **42**, 1, pp. 82–92.

Bossu, S. (2014). *Advanced Equity Derivatives: Volatility and Correlation*, Wiley Finance (Wiley), ISBN 9781118774717.

Bossy, M. and Talay, D. (1997). A stochastic particle method for the McKean-Vlasov and the Burgers equation, *Mathematics of Computation* **66**, 217, pp. 157–192.

Bouchouev, I. (2020). Negative oil prices put spotlight on investors, *Risk.net* .

Breiman, L. (1967). First exit times from a square root boundary, in *Fifth Berkeley Symposium*, Vol. 2, pp. 9–16.

Brigo, D. and Mercurio, F. (2006). *Interest Rate Models – Theory and Practice with Smile, Inflation and Credit*, 2nd edn. (Springer Verlag).

Brogan, R. (2020). Options traders adapt to electronic markets in pandemic, https://flextrade.com/options-traders-adapt-to-electronic-markets-in-pandemic/.

Brunner, H. (1985). The numerical solution of weakly singular Volterra integral equations by collocation on graded meshes, *Mathematics of Computation* **45**, 172, pp. 417–437.

Brunner, H. (2004). *Collocation Methods for Volterra Integral and Related Functional Differential Equations*, *Cambridge Monographs on Applied and Computational Mathematics*, Vol. 15 (Cambridge University Press, Cambridge).

Brunner, H. (2017). *Volterra Integral Equations. An Introduction to Theory and Applications* (Cambridge University Press, UK).

Brunner, H. and van der Houwen, P. (1986). *The Numerical Solution of Volterra Equations*, no. XVI in CWI Monographs (Centrum voor Wiskunde en Informatica, Amsterdam).

Bujok, K. and Reisinger, C. (2012). Numerical valuation of basket credit derivatives in structural jump-diffusion models, *Journal of Computational Finance* **15**, 4, p. 115.

Butuzov, V. and Butuzova, M. (2016). *Integrals depending on parameters* (Moscow State University, Moscow), in Russian.

Cannon, J. and Browder, F. (1984). *The One-Dimensional Heat Equation*, Encyclopedia of Mathematics and its Applications (Cambridge University Press), ISBN 9780521302432.

Capriotti, L. and Stehlikova, B. (2014). An Effective Approximation for Zero-Coupon Bonds and Arrow-Debreu Prices in the Black-Karasinski Model, *International Journal of Theoretical and Applied Finance* **17**, 6, p. 1650017.

Carr, E. and March, N. (2018). Semi-analytical solution of multilayer diffusion problems with time-varying boundary conditions and general interface conditions, *Applied Mathematics and Computation* **333**, 15, pp. 286–303.

Carr, P. and Itkin, A. (2019). Geometric local variance gamma model, *The Journal of Derivatives* **27**, 2, pp. 7–30.

Carr, P. and Itkin, A. (2020). An expanded local variance gamma model, *Compoutational Economics* doi:10.1007/s10614-020-10000-w.

Carr, P. and Itkin, A. (2021). Semi-closed form solutions for barrier and American options written on a time-dependent Ornstein Uhlenbeck process, *Journal of Derivatives* **Fall**.

Carr, P., Itkin, A., and Muravey, D. (2020). Semi-closed form prices of barrier options in the time-dependent CEV and CIR models, *Journal of Derivatives* **28**, 1, pp. 26–50.

Carr, P. and Linetsky, V. (2006). A jump to default extended CEV model: an application of Bessel processes, *Finance and Stochastics* **10**, pp. 303–330.

Carr, P., Lipton, A., and Madan, D. (2018). The reduction method for valuing derivative securities, in A. Lipton (ed.), *Financial Engineering, Selected Works of Alexander Lipton*, chap. 9 (WSPC), pp. 135–184.

Carr, P. and Nadtochiy, S. (2017). Local Variance Gamma and explicit calibration to option prices, *Mathematical Finance* **27**, 1, pp. 151–193.

Carrillo, J., González, M., Gualdani, M., and Schonbek, M. (2013). Classical Solutions for a Nonlinear Fokker-Planck , *Comm. Partial Differential Equations* **38**, 3, pp. 385–409.

Chen, S. (2010). *Decisions modeling the dynamics of commodity prices for investement decisions under uncertainty*, Ph.D. thesis, University of Waterloo, Ontario, Canada.

Cheridito, P. and Xu, Z. (2015). Pricing and hedging CoCos, *Available at SSRN: https://ssrn.com/abstract=2201364*.

Cherkasov, I. (1957). On the transformation of the diffusion process to a Wiener process, *Theor. Probab. Appl.* **2**, 3, pp. 373–377.

CME Clearing (2020). Switch to bachelier options pricing model, Tech. rep., CME Group, available at https://www.cmegroup.com/content/dam/cmegroup/notices/clearing/2020/04/Chadv20-171.pdf.

Coculescu, D., Geman, H., and Jeanblanc, M. (2008). Valuation of default-sensitive claims under imperfect information, *Finance and Stochastics* **12**, 2, pp. 195–218.

Coddington, E. and Levinson, N. (1955). *Theory of ordinary differential equations* (McGraw-Hill, New York).

Cohen Jr., A. (1940). The numerical computation of the product of conjugate imaginary gamma functions, *Ann. Math. Statist.* **11**, 2, pp. 213–218.

Collin-Dufresne, P. and Goldstein, R. S. (2001). Do credit spreads reflect stationary leverage ratios? *The Journal of Finance* **56**, 5, pp. 1929–1957.

Costabel, M. (1990). Boundary integral operators for the heat equation, *Integral Equations and Operator Theory* **13**, 4, pp. 498–552.

Courant, R. and Hilbert, D. (1962). *Methods of Mathematical Physics*, Vol. —— (Interscience (Wiley), New York, USA).

Cox, J. (1975). Notes on option pricing i. constant elasticity of variance diffusions, Tech. rep., Stanford University working paper.

Cox, J., Ingersoll, J., and Ross, S. (1985). A theory of the term structure of interest rates, *Econometrica* **53**, 2, pp. 385–408.

Craddock, M. (2009). Fundamental solutions, transition densities and the integration of Lie symmetries, *Journal of Differential Equations* **246**, pp. 2538–2560.

Craddock, M. and Lennox, K. (2007). Lie group symmetries as integral transforms of fundamental solutions, *J. Differential Equations* **232**, pp. 652–674.

Cummins, M. and Bucca, A. (2012). Quantitative spread trading on crude oil and refined products markets, *Quantitative Finance* **12**, 12, pp. 1857–1875.

Daniels, H. E. (1996). Approximating the first crossing-time density for a curved boundary, *Bernoulli* , pp. 133–143.

David, A. and Lehar, A. (2014). Why are banks highly interconnected? Available at SSRN 1108870.

Davis, H. (1962). *Introduction to nonlinear differential and integral equations* (Dover publications, New York).

Davydov, D. and Linetsky, V. (2001). Pricing and hedging path-dependent options under the CEV process, *Management Science* **47**, 7, pp. 949–965.

Debnath, L. and Bhatta, D. (2014). *Integral Transforms and Their Applications*, 3rd edn. (Taylor & Francis).

Delarue, F., Inglis, J., Rubenthaler, S., and Tanré, E. (2015). Particle systems with a singular mean-field self-excitation. Application to neuronal networks, *Stochastic Processes and their Applications* **125**, 6, pp. 2451–2492.

Delarue, F., Nadtochiy, S., and Shkolnikov, M. (2017). Global solutions to the supercooled Stefan problem with blow-ups: regularity uniqueness, ArXiv:1902.05174.

Derman, E., Miller, M., and Park, D. (2016). *The Volatility Smile*, Wiley Finance (Wiley), ISBN 9781118959169.

Detemple, J. (2006). *American-Style Derivatives: Valuation and Computation*, Financial Mathematics Series (Chapman & Hall/CRC, Boca Raton, London, New York).

Dias, C. (2015). A method of recursive images to solve transient heat diffusionin multilayer materials, *International Journal of Heat and Mass Transfer* **85**, pp. 1075–1083.

Ding, M. and Rangarajan, G. (2004). First Passage Time Problem: A Fokker-Planck Approach, in L. T. Wille (ed.), *New Directions in Statistical Physics: Econophysics, Bioinformatics, and Pattern Recognition*, chap. 3 (Springer-Verlag, Berlin Heidelberg), pp. 31–46.

Do, B. and Faff, R. (2012). Are pairs trading profits robust to trading costs? *Journal of Financial Research* **35**, 2, pp. 261–287.

Doff, R. (2020). Valuing scenarios with real option pricing, *Risk.net* .

Duffy, D. J. (2013). *Finite Difference methods in financial engineering: A Partial Differential Equation approach* (John Wiley & Sons).

Dupire, B. (1994). Pricing with a smile, *Risk* **7**, pp. 18–20.

Einstein, A. (1905). Über die von der molekularkinetischen theorie der wärme geforderte bewegung von in ruhenden flüssigkeiten suspendierten teilchen, *Annalen der physik* **322**, 8, pp. 549–560.

Eisenberg, L. and Noe, T. (2001). Systemic risk in financial systems, *Management Science* **47**, 2, pp. 236–249.

Ekstrom, E., Lindberg, C., and Tysk, J. (2011). Optimal liquidation of a pairs trade, in G. D. Nunno and B. E. B. Oksendal (eds.), *Advanced mathematical methods for finance* (Springer, Berlin, Heidelberg), pp. 247–255.

Ekstrom, E., Lindberg, C., Tysk, J., and Wanntorp, H. (2010). Optimal liquidation of a call spread, *Journal of Applied Probability* **47**, pp. 586–593.

Ekström, E. and Tysk, J. (2011). Boundary conditions for the single-factor term structure equation, *The Annals of Applied Probability* **21**, 1, pp. 332–350.

Elliott, R., Hoek, J. V. D., and Malcolm, W. (2005). Pairs trading, *Quantitative Finance* **5**, 3, pp. 271–276.

Elnaqeeb, T., Shah, N., and Vieru, D. (2020). Weber-Type Integral Transform Connected with Robin-Type Boundary Conditions, *Mathematics* **8**, 8, p. 1335.

Emanuel, D. and Macbeth, J. (1982). Further Results on the Constant Elasticity of Variance Call Option Pricing Model, *Journal of Financial and Quantitative Analysis* **17**, pp. 533–554.

Endres, S. and Stübinger, J. (2019). Optimal trading strategies for Lévy-driven Ornstein-Uhlenbeck processes, *Applied Economics* **51**, 29, pp. 3153–3169.

Ernzerhof, M. (1994). Taylor-series expansion of density functionals, *Phys. Rev. A* **50**, p. 4593.

Farrington, S. and Cesa, M. (2020). Podcast: Kaminski and Ronn on negative oil and options pricing, *Risk.net* .

Fasano, A., Primicerio, M., and Hadeler, K. P. (1983). A critical case for the solvability of Stefan-like problems, *Mathematical Methods in the Applied Sciences* **5**, 1, pp. 84–96.

Fasano, A., Primicerio, M., Howison, S. D., and Ockendon, J. R. (1990). Some remarks on the regularization of supercooled one-phase Stefan problems in one dimension, *Quarterly of Applied Mathematics* **48**, 1, pp. 153–168.

Fasshauer, G. E., Khaliq, A. Q. M., and Voss, D. A. (2004). Using meshfree approximation for multi-asset American option problems, *J. Chinese Inst. Engrs.* **27**, 4, pp. 563–571.

Finkelstein, V. and Lardy, J. (2001). Assessing default probabilities from equity markets: simple closed-form solution, in *ICBI Global Derivatives Conference* (Juan les Pins).

Fomin, S. V. and Kolmogorov, A. N. (1999). *Elements of the Theory of Functions and Functional Analysis* (Dover Books on Mathematics, USA).

Fortet, R. (1943). Les fonctions aléatoires du type Markoff associées à certaines équations linéaires aux dérivées partielles du type parabolique, *J. Math. Pures Appl.* **22**, pp. 177–243.

Friedman, A. (1964.). *Partial Differential Equations of Parabolic Type* (Prentice-Hall, New Jersey,).

Fuhry, M. (2001). *A new Tikhonov regularization method*, Master's thesis, Kent State University Honors College.

Gatev, E., Goetzmann, W., and Rouwenhorst, K. (2006). Pairs trading: Performance of a relative-value arbitrage rule, *Review of Financial Studies* **19**, 3, pp. 797–827.

Gatheral, J. (2002). Stochastic volatility and local volatility, http://web.math.ku.dk/~rolf/teaching/ctff03/Gatheral.1.pdf.

Gatheral, J. (2006). *The volatility surface* (Wiley finance).

Geman, H. (2009). *Commodities and Commodity Derivatives: Modeling and Pricing for Agriculturals, Metals and Energy*, The Wiley Finance Series (Wiley).

Giet, J., Vallois, P., and Wantz-Mezieres, S. (2015). The logistic sde, *Theory of Stochastic Processes* **20**, 36, pp. 28–62.

Gobet, E. and Menozzi, S. (2010). Stopped diffusion processes: boundary corrections and overshoot, *Stochastic Processes and Their Applications* **120**, 2, pp. 130–162.

Gobet, E. and Pagliarani, S. (2018). Analytical approximations of non-linear SDEs of McKean–Vlasov type, *Journal of Mathematical Analysis and Applications* **466**, 1, pp. 71–106.

Göing-Jaeschke, A. and Yor, M. (2003). A clarification note about hitting times densities for Ornstein–Uhlenbeck processes, *Finance Stoch.* , 3, pp. 413–415.

Goldenberg, D. (1991). A unified method for pricing options on diffusion-processes, *Journal of Financial Economics* **29**, pp. 3–34.

Goncu, A. and Akyldirim, E. (2016). Statistical arbitrage with pairs trading, *International Review of Finance* **16**, 2, pp. 307–319.

Govender, K. (2011). Statistical arbitrage in south african financial markets, Tech. rep., University of Cape Town.

Gradshtein, I. and Ryzhik, I. (2007). *Table of Integrals, Series, and Products* (Elsevier).

Granas, A. and Dugundji, J. (2003). *Fixed point theory* (Springer Verlag, New York).

Gregoty, I., Ewald, C., and Knox, P. (2011). Analytical pairs trading under different assumptions on the spread and ratio dynamics, University of Sydney.

Groetsch, C. (1984). *The Theory of Tikhonov Regularization for Fredholm Equations of the First Kind.* (Pitman, Boston).

Guinter, N. (1967). *Potential Theory and Its Applications to Basic Problems of MathematicalPhysics* (Frederick Ungar, New York).

Hairer, E., Lubich, C., and Schlichte, M. (1985). Fast numerical solution of nonlinear Volterra convolution equations, *SIAM Journal on Scientific and Statistical Computing* **6**, 3, pp. 532–541.

Haji-Ali, A.-L. and Tempone, R. (2018). Multilevel and Multi-index Monte Carlo methods for the McKean–Vlasov equation, *Statistics and Computing* **28**, 4, pp. 923–935.

Halperin, I. and Feldshteyn, I. (2018). Market self-learning of signals, impact and optimal trading:invisible hand inference with free energy, SSRN:3174498.

Hambly, B., Ledger, S., and Sojmark, A. (2018). A McKean–Vlasov equation with positive feedback and blow-ups, ArXiv:1801.07703.

Hansen, P. (1998). *Rank-defficient and discrete ill-posed problems* (SIAM, Philadelphia).

Hilberink, B. and Rogers, L. (2002). Optimal capital structure and endogenous default, *Finance and Stochastics* **6**, pp. 237–263.

Hon, Y. C. and Mao, X. Z. (1999). A radial basis function method for solving options pricing model, *Financial Engineering* **8**, 1, pp. 31–49.

Horowitz, J. (1985). Measure-valued random processes, *Zeitschrift für Wahrscheinlichkeitstheorie und Verwandte Gebiete* **70**, 2, pp. 213–236.

Horvath, B., Jacquier, A., and Turfus, C. (2017). Analytic option prices for the Black-Karasinski short rate model, https://papers.ssrn.com/sol3/papers.cfm?abstract_id=3253833, SSRN: 3253833.

Howison, S. (1995). Barrier options, Available at https://people.maths.ox.ac.uk/howison/barriers.pdf.

Huang, M., Malhamé, R. P., Caines, P. E., *et al.* (2006). Large population stochastic dynamic games: closed-loop McKean–Vlasov systems and the Nash certainty equivalence principle, *Communications in Information & Systems* **6**, 3, pp. 221–252.

Huck, N. and Afawubo, K. (2015). Pairs trading and selection methods: Is cointegration superior? *Applied Economics* **47**, 6, pp. 599–613.

Hull, J. (2011). *Options, Futures, and Other Derivatives*, 8th edn. (Prentice Hall).

Hull, J. and White, A. (1990). Pricing interest-rate-derivative securities, *Review Of Financial Studies* **3**, pp. 573–592.

Hull, J. and White, A. (2001). Valuing credit default swaps, ii, *Journal of Derivatives* **8**, 3, pp. 12–22.

Hyer, T., Lipton, A., Pugachevsky, D., and Qui, S. (1998). A hidden-variable model for risky bonds, Bankers Trust working paper.

Hyer, T., Lipton-Lifschitz, A., and Pugachevsky, D. (1997). Passport to successx, *Risk Magazine* **10**, 9, pp. 127–131.

Ichiba, T., Ludkovski, M., and Sarantsev, A. (2018). Dynamic contagion in a banking system with births and defaults, ArXiv:1807.09897.

Itkin, A. (2017). *Pricing derivatives under Lévy models*, 1st edn., no. 12 in Pseudo-Differential Operators (Birkhauser, Basel).

Itkin, A. (2020). *Fitting Local Volatility: Analytic and Numerical Approaches in Black-Scholes and Local Variance Gamma Models*, 11623 (World Scientific Publishing Co. Pte. Ltd.).

Itkin, A. and Lipton, A. (2015). Efficient solution of structural default models with correlated jumps and mutual obligations, *International Journal of Computer Mathematics* **92**, 12, pp. 2380–2405.

Itkin, A. and Lipton, A. (2017). Structural default model with mutual obligations, *Review of Derivatives Research* **20**, 1, pp. 15–46.

Itkin, A. and Lipton, A. (2018). Filling the gaps smoothly, *Journal of Computational Sciences* **24**, pp. 195–208.

Itkin, A., Lipton, A., and Muravey, D. (2020). From the Black-Karasinski to the Verhulst model to accommodate the unconventional Fed's policy, https://arxiv.org/abs/2006.11976.

Itkin, A., Lipton, A., and Muravey, D. (2021). Multilayer heat equations: application to finance, ArXiv:2102.08338.

Itkin, A. and Muravey, D. (2020a). Semi-analytic pricing of double barrier options with time-dependent barriers and rebates at hit, https://arxiv.org/abs/2009.09342.

Itkin, A. and Muravey, D. (2020b). Semi-closed form prices of barrier options in the Hull-White model, *Risk*.

Johansson, F. (2007). Mpmath library, http://mpmath.org/.

Kac, M. (1956). Foundations of kinetic theory, in *Proceedings of the Third Berkeley Symposium on Mathematical Statistics and Probability*, Vol. 3 (University of California Press Berkeley and Los Angeles, California), pp. 171–197.

Kamenomostskaja, S. (1961). On Stefan's problem, *Matematicheskii Sbornik (Novaya Seriya)* **53**, 95, pp. 489–514.

Kanwal, R. P. (1997). *Linear integral equations: Theory and technique*, 2nd edn. (Birkhauser, Boston).

Karatzas, I. and Shreve, S. (1991). *Brownian Motion and Stochastic Calculus*, Graduate Texts in Mathematics (Springer, New York).

Kartashov, E. (2001). *Analytical Methods in the Theory of Heat Conduction in Solids* (Vysshaya Shkola, Moscow).

Kartashov, E. and Krotov, G. (2009). Analytical Solution of the Single-Phase Stefan Problem, *Mathematical Models and Computer Simulations* **1**, 2, pp. 180–188.

Kartashov, E. M. (1999). Analytical methods for solution of non-stationary heat conductance boundary problems in domains with moving boundaries, *Izvestiya RAS, Energetika* , 5, pp. 133–185.

Kaushansky, V., Lipton, A., and Reisinger, C. (2018). Numerical analysis of an extended structural default model with mutual liabilities and jump risk, *Journal of Computational Science* **24**, 218–231.

Kaushansky, V. and Reisinger, C. (2019). Simulation of particle systems interacting through hitting times, *Discrete & Continuous Dynamical Systems - B* **24**, 10, pp. 5481–5502.

Khvedelidze, B. (2001). Fredholm theorem for integral equations, in M. Hazewinkel (ed.), *Encyclopaedia of Mathematics* (Springer, Netherlands).

Klebaner, F. (2005). *Introduction to stochastic calculus with applications* (Imperial College Press, London, UK.).

Kolk, M. and Pedas, A. (2009). Numerical solution of Volterra integral equations with weakly singular kernels which may have a boundary singularity, *Mathematical Modelling and Analysis* **14**, 1, pp. 79–89.

Kolk, M. and Pedas, A. (2013). Numerical solution of Volterra integral equations with singularities, *Frontiers of Mathematics in China* **8**, 2, pp. 239–259.

Kolk, M., Pedas, A., and Vainikko, G. (2009). High-order methods for Volterra integral equations with general weak singularities, *Numerical Functional Analysis and Optimization* **30**, 9-10, pp. 1002–1024.

Kontorovich, M. and Lebedev, N. (1939). On a method of solution of some problems in diffraction theory, *J. Phys. Moscow* **1**, pp. 229–241.

Krauss, C. (2015). Statistical arbitrage pairs trading strategies: Review and outlook, IWQW Discussion Papers 09/2015, Friedrich-Alexander-Universität Erlangen-Nürnberg.

Kristensson, G. (2009). Jump conditions for single and double layer potentials, https://www.eit.lth.se/fileadmin/eit/courses/eit070f/Documents/JumpConditions.pdf, Lund University.

Kuan, G. and Webber, N. (2003). Pricing barrier bond options with one-factor interest rate models, *Journal of Derivatives* **4**, 10, pp. 33–50.

Kuznetsov, A. (2013). On the convergence of the Gaver-Stehfest algorithm, *SIAM J. Numerical Analysis* **51**, 6, pp. 2984–2998.

Kythe, P. K. (2011). *Green's Functions and Linear Differential Equations: Theory, Applications and Computation* (CRC Press).

Lançon, P., Batrouni, G., Lobry, L., and Ostrowsky, N. (2001). Drift without flux: Brownian walker with a space-dependent diffusion coefficient, *Europhysics Letters (EPL)* **54**, 1, pp. 28–34.

Landis, E. (1997). *Second Order Equations of Elliptic and Parabolic Type* (American Mathematical Society, US).

Larsson, S., Lindberg, C., and Warfheimer, M. (2013). Optimal closing of a pair trade with a model containing jumps. *Applications of Mathematics* **58**, 3, pp. 249–268.

Lataillade, J. D., Deremble, C., Potters, M., and Bouchaud, J. (2012). Optimal trading with linear costs, *Journal of Investment Strategies* **1**, 13, pp. 91–115.

Lawler, G. F. (2018). Notes on the bessel process, http://www.math.uchicago.edu/~lawler/bessel18new.pdf, corpus ID: 52200396.

Leblanc, B., Renault, O., and Scaillet, P. (2000). A correction note on the first passage time of an Ornstein-Uhlenbeck process to a boundary, *Finance Stoch.* **4**, 1, pp. 109–111.

Leblanc, B. and Scaillet, P. (1998). Path dependent options on yields in the affine term structure model, *Finance Stoch.* **2**, 4, pp. 349–367.

Lejay, A. (2006). On the constructions of the skew Brownian motion, *Probability Surveys* **3**, pp. 413–466.

Leung, T. and Li, X. (2015a). *Optimal mean reversion trading: Mathematical analysis and practical applications* (World Scientific, New Jersey, USA).

Leung, T. and Li, X. (2015b). Optimal mean reversion trading with transaction costs and stop-loss exit, *International Journal of Theoretical and Applied Finance* **18**, 3, p. 1550020.

Lewis, T. and Rinzel, J. (2003). Dynamics of spiking neurons connected by both inhibitory and electrical coupling, *J. Comput. Neurosci.* **14**, 3, pp. 283–309.

Li, X. (2015). *Optimal multiple stopping approach to mean reversion trading*, Ph.D. thesis, The Graduate School of Arts and Sciences, Columbia University.

Lienhard IV, J. and Lienhard V, J. (2019). *A Heat Transfer Textbook*, 5th edn. (Phlogiston Press, Cambridge, MA).

Lindberg, C. (2014). Pairs trading with opportunity cost, *Journal of Applied Probability* **51**, pp. 282–286.

Linetsky, V. (2004). Computing Hitting Time Densities for CIR and OU diffusions: Applications to mean-reverting models, *J. Comput. Finance* **7**, pp. 1–22.

Linetsky, V. and Mendoza, R. (2010). Encyclopedia of quantitative finance, in *Constant Elasticity of Variance (CEV) Diffusion Model* (John Wiley & Sons), ISBN 9780470061602.

Linz, P. (1985). *Analytical and Numerical Methods for Volterra Equations* (SIAM).

Lipton, A. (2001). *Mathematical Methods For Foreign Exchange: A Financial Engineer's Approach* (World Scientific).

Lipton, A. (2002). The vol smile problem, *Risk* **15**, 2, pp. 61–65.

Lipton, A. (2016). Modern monetary circuit theory, stability of interconnected banking network, and balance sheet optimization individual banks, *International Journal of Theoretical and Applied Finance* **19**, 6, p. 1650034.

Lipton, A. (2018). *Financial Engineering: Selected Works of Alexander Lipton* (World Scientific, Singapore).

Lipton, A. and Kaushansky, V. (2018). On the First Hitting Time Density of an Ornstein-Uhlenbeck Process, https://arxiv.org/pdf/1810.02390.pdf.

Lipton, A. and Kaushansky, V. (2020a). On the first hitting time density for a reducible diffusion process, *Quantitative Finance,* **5**, DOI: 10.1080/14697688.1713394.

Lipton, A. and Kaushansky, V. (2020b). On three important problems in mathematical finance, *The Journal of Derivatives. Special Issue* **28**, 1, pp. 123 – 142.

Lipton, A., Kaushansky, V., and Reisinger, C. (2019). Semi-analytical solution of a McKean-Vlasov equation with feedback through hitting boundary, *Euro. Jnl of Applied Mathematics* , pp. 1–34.

Lipton, A. and Lopez de Prado, M. (2020a). A closed-form solution for optimal Ornstein-Uhlenbeck driven trading strategies, *International Journal of Theoretical and Applied Finance* **23**, 8, DOI: 10.1142/S0219024920500569.

Lipton, A. and Lopez de Prado, M. (2020b). A closed-form solution for optimal mean-reverting trading strategies, *Risk* **33**, 6.

Lipton, A. and Savescu, I. (2014). Pricing credit default swaps with bilateral value adjustments, *Quantitative Finance* **14**, 1, pp. 171–188.

Lipton, A. and Sepp, A. (2009). Credit value adjustment for credit default swaps via the structural default model, *The Journal of Credit Risk* **5**, 2, pp. 123–146.

Lipton, A. and Sepp, A. (2011a). Credit value adjustment in the extended structural default model, in *The Oxford Handbook of Credit Derivatives* (Oxford University Oxford), pp. 406–463.

Lipton, A. and Sepp, A. (2011b). Filling the gaps, *Risk Magazine* **24**, 10, pp. 86–91.

Liu, B., Chang, L., and Geman, H. (2017). Intraday pairs trading strategies on high frequency data: The case of oil companies, *Quantitative Finance* **17**, 1, pp. 87–100.

Londono, J. and Sandoval, J. (2015). A new logistic-type model for pricing european options, *SpringerPlus* , 4, p. 762.

Lopez De Prado, M. (2018). *Advances in Financial Machine learning* (John Wiley & Sons, Hoboken, New Jersey, USA).

Lyapunov, A. (1949). *Works on the theory of potential* (Technical and Theoretical State Publishing House, Moscow - Leningrad), in Russian.

Lyubov, B. Y. (1978). *Kinetic Theory of Phase Transformations* (Amerind Publishing), ASIN: B0000EGG4O.

Maleknejad, K., Hashemizadeh, E., and Ezzati, R. (2011). A new approach to the numerical solution of Volterra integral equations by using Bernstein's approximation, *Communications in Nonlinear Science and Numerical Simulation* **16**, 2, pp. 647–655.

Maleknejad, K., Rostami, Y., and Kalalagh, H. (2016). Numerical Solution for First Kind Fredholm Integral Equations by Using Sinc Collocation Method, *International journal of applied physcis and mathematics* **6**, 3, pp. 120–128.

Maleknejad, K., Sohrabi, S., and Rostami, Y. (2007). Numerical solution of nonlinear Volterra integral equations of the second kind by using Chebyshev polynomials, *Applied Mathematics and Computation* **188**, 1, pp. 123–128.

Martin, R., Craster, R., and Kearney, M. (2015). Infinite product expansion of the Fokker–Planck equation with steady-state solution, *Proc. R. Soc. A* **471**, 2179, p. 20150084.

Martin, R., Kearney, M., and Craster., R. (2019). Long-and short-time asymptotics of the first-passage time of the Ornstein–Uhlenbeck and other mean-reverting processes, *Journal of Physics A: Mathematical and Theoretical* **52**, p. 134001.

Martin, R. and Schöneborn, T. (2011). Mean reversion pays, but costs, *Risk Magazine* **24**, 2, pp. 96–101.

McKean, H. P. (1966). A class of Markov processes associated with nonlinear parabolic equations, *Proceedings of the National Academy of Sciences* **56**, 6, pp. 1907–1911.

Mercer, A. (1962). On integral transform pairs arising from second order differential equations, *Proc. Edin. Math. Soc.* **13**, 1, pp. 63–68.

Merton, R. (1974). On the pricing of corporate debt: The risk structure of interest rates, *Journal of Finance* **29**, pp. 449—470.

Mijatovic, A. (2010). Local time and the pricing of time-dependent barrier options, *Finance and Stochastics* **14**, 1, pp. 13–48.

Mitrinovic, D. and Keckic, J. (1984). *The Cauchy method of residues: Theory and Applications*, Mathematics and its Applications (Springer, Netherlands), ISBN 978-90-277-1623-1.

Mumford, D., Nori, C. M. M., Previato, E., and Stillman, M. (1983). *Tata Lectures on Theta*, Progress in Mathematics (Birkhäuser Boston), ISBN 9780817631093.

Nadtochiy, S. and Shkolnikov, M. (2017). Particle systems with singular interaction through hitting times: application in systemic risk modeling, ArXiv:1705.00691.

Nadtochiy, S. and Shkolnikov, M. (2018). Mean field systems on networks, with singular interaction through hitting times, ArXiv:1807.02015.

Nardo, E. D., Nobile, A., Pirozzi, E., and Ricciardi, L. (2001). A computational approach to first-passage-time problems for Gauss–Markov processes, *Advances in Applied Probability* **33**, 2, pp. 453–482.

Nedaiasl, K., Bastani, A., and Rafiee, A. (2019). A product integration method for the approximation of the early exercise boundary in the American option pricing problem, *Mathematical methods in the applied science* doi:10.1002/mma.5553.

Noble, B. (1969). Instability when solving Volterra integral equations of the second kind by multistep methods, in *Conference on the numerical solution of differential equations* (Springer), pp. 23–39.

Norberg, R. (2004). Vasicek beyond the normal, *Mathematical Finance* **14**, 4, pp. 585–604.

Novikov, A., Frishling, V., and Kordzakhia, N. (1999). Approximations of boundary crossing probabilities for a Brownian motion, *Journal of Applied Probability* **36**, 4, pp. 1019–1030.

Novikov, A. A. (1971). On stopping times for a Wiener process, *Theory of Probability & Its Applications* **16**, 3, pp. 449–456.

Ohyama, Y. (1995). Differential relations of theta functions, *Osaka Journal of Mathematics* **32**, 2, pp. 431–450.

Oldham, K. and Spanier, J. (2006). *The Fractional Calculus: Theory and Applications of Differentiation and Integration to Arbitrary Order*, Dover books on mathematics (Dover Publications, Chicago).

Oldham, K. B. and Spanier, J. (1974). *The Fractional Calculus; Theory and Applications of Differentiation and Integration to Arbitrary Order*, Mathematics in Science and Engineering, V (Academic Press).

Oleinik, O. A. and Radkevich, E. V. (1973). *Second order equations with nonnegative characteristic form* (Kluwer Academic Publishers).

Olver, F. W. J., Olde Daalhuis, A. B., Lozier, D. W., Schneider, B. I., Boisvert, R. F., Clark, C. W., Miller, B. R., Saunders, B. V., Cohl, H. S., and McClain, M. A. (2020). NIST Digital Library of Mathematical Functions, http://dlmf.nist.gov/, release 1.0.28 of 2020-09-15.

Ostojic, S., Brunel, N., and Hakim, V. (2009). Synchronization properties of networks of electrically coupled neurons in the presence of noise and heterogeneities, *J. Comput. Neurosci.* **26**, 3, pp. 369–392.

Peskir, G. (2002). On integral equations arising in the first-passage problem for Brownian motion, *The Journal of Integral Equations and Applications* , pp. 397–423.

Pettersson, U., Larsson, E., Marcusson, G., and Persson, J. (2008). Improved radial basis function methods for multi-dimensional option pricing, *J. Comput. Appl. Math.* **222**, 1, pp. 82–93, doi:10.1016/j.cam.2007.10.038.

Polyanin, A. (2002). *Handbook of linear partial differential equations for engineers and scientists* (Chapman & Hall/CRC).

Polyanin, P. and Manzhirov, A. (2008). *Handbook of Integral Equations: Second Edition*, Handbooks of mathematical equations (CRC Press), ISBN 9780203881057.

Pontrelli, G., Lauricella, M., Ferreira, J., and Pena, G. (2016). Iontophoretic transdermal drug delivery: A multi-layered approach, *Mathematical Medicine and Biology* **00**, pp. 1–18.

Pötzelberger, K. and Wang, L. (2001). Boundary crossing probability for Brownian motion, *Journal of Applied Probability* **38**, 01, pp. 152–164.

Prado, M. L. D. (2019). Tactical investment algorithms, https://ssrn.com/abstract=3459866.

Privault, N. (2012). *An Elementary Introduction to Stochastic Interest Rate Modeling*, Advanced series on statistical science & applied probability (World Scientific), ISBN 9789814390859.

Quaife, B. (2011). *Fast Integral Equation Methods for the Modified Helmholtz Equation*, Ph.D. thesis, University of Calgary.

Redner, S. (2001). *A guide to first-passage processes* (Cambridge University Press, Cambridge).

Reis, G. d., Smith, G., and Tankov, P. (2018). Importance sampling for McKean-Vlasov SDEs, ArXiv preprint arXiv:1803.09320.

Revuz, D. and Yor, M. (1999). *Continuous Martingales and Brownian Motion*, 3rd edn. (Springer, Berlin, Germany).

Ricciardi, L. (1976). On the Transformation of diffusion processes into the Wiener process, *J. Math. Anal. Appl.* **54**, 1, pp. 185–199.

Ricciardi, L. and Sato, S. (1988). First-passage-time density and moments of the Ornstein–Uhlenbeck process, *Journal of Applied Probability* **25**, 1, pp. 43–57.

Ricketson, L. (2015). A multilevel Monte Carlo method for a class of McKean–Vlasov processes, *arXiv preprint arXiv:1508.02299*.

Risken, H. and Haken., H. (1989). *The Fokker-Planck Equation: Methods of Solution and Applications Second Edition* (Springer).

Ross, S. (2010). *Introduction to Probability Models*, 10th edn. (Academic Press, San Diego,CA, USA).

Rubinstein, L. (1971). *The Stefan Problem, Translations of Mathematical Monographs*, Vol. 27 (American Mathematical Society, Providence, RI).

Rudebusch, G. (2018). A review of the fed's unconventional monetary policy, *FRBSF Economic Letter* , 27.

Samko, S., Kilbas, A., and Marichev, O. (1993). *Fractional Integrals and Derivatives. Theory and Applications* (Gordon and Breach Science Publishers, Amsterdam).

Schwartz, E. (1997). The stochastic behavior of commodity prices: Implications for valuation and hedging, *Journal of Finance,* **52**, 3, pp. 923–973.

Shepp, L. A. (1967). A first passage problem for the Wiener process, *The Annals of Mathematical Statistics* **38**, 6, pp. 1912–1914.

Smith, C. (1991). A Laguerre series approximation for the probability density of the first passage time of the Ornstein–Uhlenbeck process, in *Noise in Physical Systems and 1/f Fluctuations* (Tokyo: Ohmsha).

Spivak, M., Veerapaneni, S., and Greengard, L. (2010). The fast generalized gauss transform, *SIAM Journal on Scientific Computing* **32**, 5, pp. 3092–3107.

Strang, G. (2006). *Linear algebra and its applications*, 4th edn. (Thomson, Brooks/Cole, Belmont, CA).

Suzuki, K. (2018). Optimal pair-trading strategy over long/short/square positions – empirical study, *Quantitative Finance* **18**, 1, pp. 97–119.

Szpruch, L., Tan, S., and Tse, A. (2017). Iterative particle approximation for McKean–Vlasov SDEs with application to Multilevel Monte Carlo estimation, arXiv preprint arXiv:1706.00907.

Tavella, D. and Randall, C. (2000). *Pricing Financial Instruments. The Finite-Difference method.*, Wiley series in financial engineering (John Wiley & Sons, Inc., New York,).

Temme, N. (1978). Uniform asymptotic expansions of confluenthypergeometric functions, *J. Inst. Maths Applies* **22**, pp. 215–223.

Thomson, I. (2016). Option pricing model: Comparing louis bachelier with black-scholes merton, SSRN 2782719, https://papers.ssrn.com/sol3/papers.cfm?abstract_id=2782719.

Tikhonov, A. and Samarskii, A. (1963). *Equations of mathematical physics* (Pergamon Press, Oxford).

Titchmarsh, E. (1962). *Eigenfunction Expansions Associated with Second-Order Differential Equations* (Oxford University Press, Oxford, UK).

Torabi, S., Tari, A., and Shahmorad, S. (2019). Two-step collocation methods for two-dimensional Volterra integral equations of the second kind, *Journal of Applied Analysis* **25**, 1, pp. 1–11.

Trefethen, L. (2013). *Approximation Theory and Approximation Practice* (Society for Industrial andApplied Mathematics, Philadelphia).

Turfus, C. (2020). Analytic swaption pricing in the black-karasinski model, https://papers.ssrn.com/sol3/papers.cfm?abstract_id=3253866, SSRN: 3253866.

van der Pol, B. and Bremmer, H. (1950). *Operational calculus based on the two-sided Laplace integral* (Cambridge University Press, Cambridge, UK).

Van Kampen, N. (2007). *Stochastic processes in physics and chemistry* (North Holland).

Verges-Liahi, J. (2005). *Color Constancy and Image Segmentation Techniques for Applications to Mobile Robotics*, Ph.D. thesis, Universitat Politecnica de Catalunya.

Verhulst, P. (1838). Notice sur la loi que la population suit dans son accroisseement, *Correspondance mathematique et physique* **10**, pp. 113–121.

Vidyamurthy, G. (2004). *Pairs trading: Quantitative methods and analysis* (John Wiley & Sons, Hoboken, NY, USA.).

Von Smoluchowski, M. (1906). Zur kinetischen theorie der brownschen molekularbewegung und der suspensionen, *Annalen der physik* **326**, 14, pp. 756–780.

Watson, N. (2012). *Introduction to Heat Potential Theory*, no. 182 in Mathematical Surveys and Monographs (Pergamon Press, Providence, RI).

Wazwaz, A.-M. (2011). *Linear and Nonlinear Integral Equations* (Higher Education Press, Beijing and Springer-Verlag GmbH Berlin Heidelberg).

Wilmott, P., Lewis, A., and Duffy, D. (2014). Modeling volatility and valuing derivatives under anchoring, *Wilmott* , 73, pp. 48–57.

Wimp, J. (1964). A class of integral transforms, *Proc. Edinburgh Math. Soc. (2)* **14**, 1, pp. 33–40.

Yakubovich, S. (1996). *Index Transforms* (World Scientific, Singapore).

Yi, C. (2010). On the first passage time distribution of an Ornstein–Uhlenbeck process, *Quantitative Finance* **10**, 9, pp. 957–960.

Zeng, Z. and Lee, C. (2014). Pairs trading: Optimal thresholds and profitability, *Quantitative Finance* **14**, 11, pp. 1881–1893.

Zenyuk, D. (2014). Numerical simulation of random walks on regular fractal sets, *Matem. Mod.* **26**, 11, pp. 101–104.

Zhang, K. and Yang, X. (2017). Pricing European options on zero-coupon bonds with a fitted finite volume method, *International Journal of Numerical Analysis and Modeling* **14**, 3, pp. 405–418.

Zhou, C. (2001). An analysis of default correlations and multiple defaults, *Review of Financial Studies* **14**, 2, pp. 555–576.

Index

www.ingramcontent.com/pod-product-compliance
Lightning Source LLC
LaVergne TN
LVHW010326220926
848484LV00004BB/848
* 9 7 8 9 8 1 1 2 3 1 7 3 5 *